An Introduction to the
World's
Oceans

Fourth Edition

An Introduction to the *World's* *Oceans*

Alyn C. Duxbury
University of Washington

Alison B. Duxbury
Seattle Community College

WCB **Wm. C. Brown Publishers**
Dubuque, Iowa • Melbourne, Australia • Oxford, England

Book Team

Executive Editor *Jeffrey L. Hahn*
Developmental Editor *Robert Fenchel*
Production Editor *Cathy Ford Smith*
Designer *Mark Elliot Christianson / Eric Engelby*
Art Editor *Carla Goldhammer*
Photo Editor *Lori Gockel*
Permissions Coordinator *Mavis M. Oeth*
Art Processor *Joyce Watters*

Wm. C. Brown Publishers
A Division of Wm. C. Brown Communications, Inc.

Vice President and General Manager *Beverly Kolz*
Vice President, Publisher *Earl McPeek*
Vice President, Director of Sales and Marketing *Virginia S. Moffat*
Marketing Manager *Christopher T. Johnson*
Advertising Manager *Janelle Keeffer*
Director of Production *Colleen A. Yonda*
Publishing Services Manager *Karen J. Slaght*
Permissions/Records Manager *Connie Allendorf*

Wm. C. Brown Communications, Inc.

President and Chief Executive Officer *G. Franklin Lewis*
Corporate Senior Vice President, President of WCB Manufacturing *Roger Meyer*
Corporate Senior Vice President and Chief Financial Officer *Robert Chesterman*

Cover photo © *David Muench*
Chapter opener background photo: © *Steve M. Alden*
Unless otherwise credited, all photographs © *Alyn and Alison Duxbury*

Copyedited by *Barbara Walsh*

The credits section for this book begins on page 464 and
is considered an extension of the copyright page.

To Andrew S. Duxbury,
Alison Jean Duxbury,
and Alec R. Duxbury.

Contents

Preface x
Overture by Richard H. Fleming xii

Prologue
The History of Oceanography 2

The Early Times 4
The Middle Ages 5
Voyages of Discovery 6
The Beginnings of Ocean Science 8
The Importance of Charts and
 Navigational Information 9
Ocean Science Begins 11
The *Challenger* Expedition 13
Oceanography As Science 13
U.S. Oceanography Moves into the
 Twentieth Century 18
World War II 19
Ocean Science Expands in the United
 States 19
Oceanography of the Recent Past 20
The Present and the Future 22
Box: Marine Archaeology 23
Study Questions 24
Suggested Readings 24

1
The Water Planet 26

1.1 The Beginnings 28
 Origin of the Solar System 28
 The Early Planet Earth 28
1.2 Age and Time 30

The Age of the Earth 30
Geologic Time 30
Natural Time Periods 31
1.3 The Shape of the Earth 33
1.4 Location Systems 34
 Latitude and Longitude 34
 Chart Projections 35
 Measuring Latitude 37
 Longitude and Time 37
1.5 Earth: The Water Planet 39
 Water on the Earth's Surface 39
 The Hydrologic Cycle 41
 *Reservoirs and Residence
  Time 42*
 *Distribution of Land and
  Water 42*
 The Oceans 42
 The Hypsographic Curve 44
1.6 Practical Considerations: Modern
 Navigational Techniques 45
Box: Satellite Oceanography 46
Summary 48
Key Terms 49
Study Questions 49
Study Problems 49
Suggested Readings 49

2
The Seafloor 50

2.1 Bathymetry of the Seafloor 52
 The Continental Margin 52
 The Ocean Basin Floor 57
 The Ridges and Rises 58
 The Trenches 59
2.2 Measuring the Depths 61
2.3 Sediments 64
 Sediment Sources 64
**Box: Bathymetrics: New Techniques
 for Surveying the Seafloor 66**

Biogenous Oozes 68
Red Clay 69
Patterns of Deposit 70
Particle Size 73
Rates of Deposit 74
Sampling Methods 74
2.4 Practical Considerations: Seabed
 Resources 75
 Sand and Gravel 75
 Phosphorite 77
 Sulfur 77
 Coal 77
 Oil and Gas 77
 Manganese Nodules 78
 Sulfide Mineral Deposits 79
 Laws and Treaties 80
Summary 80
Key Terms 81
Study Questions 81
Study Problems 82
Suggested Readings 82

3
The Not-So-Rigid Earth 84

3.1 The Interior of the Earth 86
 Internal Layers 86
 *Evidence for the Internal
  Structure 86*
3.2 The Lithosphere 88
 The Layers 88
 Isostasy 89
3.3 The Movement of the
 Continents 89
 History of a Theory 89
 Evidence for a New Theory 91
 Evidence for Crustal Motion 91
3.4 Plate Tectonics 96
 Plate Boundaries 96
 Rifting 97

Subduction 99
Continental Margins 99
Driving Forces for Plate
Motion 101
3.5 The Motion of the Plates 102
Rates of Motion 102
Hot Spots 102
Polar Wandering Curves 103
3.6 The History of the Continents 104
The Breakup of Pangaea 104
Before Pangaea 104
Terranes 109
3.7 Current Research 110
Drilling Plans 110
Project FAMOUS 111
Hydrothermal Vents 111
Box: Megaplumes 114
3.8 Practical Considerations: Ocean
Waste Management 115
Summary 116
Key Terms 116
Study Questions 116
Study Problems 117
Suggested Readings 117

4

The Properties of Water 118

4.1 The Water Molecule 120
4.2 Changes of State 120
4.3 Heat Capacity 122
4.4 Cohesion, Surface Tension, and
Viscosity 123
4.5 Compressibility 123
4.6 Density 123
The Effect of Temperature 124
The Effect of Salt 124
The Effect of Pressure 125
4.7 Dissolving Ability 125
4.8 Transmission of Energy 126
Heat 126
Light 126
Sound 128
4.9 Practical Considerations: Ice and
Fog 130
Sea Ice 130
Icebergs 133
Fog 135
Box: The Heard Island
Experiment 136
Summary 137
Key Terms 137
Study Questions 138
Study Problems 138
Suggested Readings 138

5

The Salt Water 140

5.1 The Salts 142
Ocean Salinities 142
Dissolved Salts 142
Sources of Salt 143
Regulating the Salt Balance 145
Residence Time 146
Constant Proportions 147
Determining Salinity 147
5.2 The Gases 148
Distribution with Depth 148
Carbon Dioxide as a Buffer 149
Box: The Messages in Polar Ice 150
The Carbon Dioxide Cycle 151
The Oxygen Balance 152
Measuring the Gases 152
5.3 Other Substances 152
Nutrients 152
Box: The Effect of Sunlight on
Seawater 153
Organics 153
5.4 Practical Considerations: Salt and
Water 153
Chemical Resources 153
Desalination 155
Summary 157
Key Terms 158
Study Questions 158
Study Problems 158
Suggested Readings 158

6

Structure of the Oceans 160

6.1 Heating and Cooling the Earth's
Surface 162
The Heat Budget 162
Annual Cycles of Solar
Radiation 163
Heat Capacity 164
6.2 Evaporation and Precipitation
Patterns 166
6.3 Density Structure and Vertical
Circulation 167
Surface Processes 167
Changes with Depth 167
Density-Driven Circulation 168
Sigma-t 169
6.4 Upwelling and Downwelling 169
Chemical Tracers 170
6.5 The Layered Oceans 171

Structure of the Atlantic
Ocean 171
Structure of the Pacific Ocean 172
Structure of the Indian Ocean 172
Mediterranean Sea and Red Sea
Water 172
Comparing the Oceans 172
6.6 T-S Curves 172
T-S Curves and Water Masses 172
6.7 Measurement Techniques 175
Determining Water Properties 175
Temperature Measurements 177
Box: New Ways to Measure the
Oceans 179
6.8 Practical Considerations: Energy
Sources 180
Osmotic Pressure 180
Ocean Thermal Energy
Conversion 181
Summary 183
Key Terms 184
Study Questions 184
Suggested Readings 184

7

*The Ocean and the
Atmosphere* 186

7.1 The Atmosphere 188
Structure of the Atmosphere 188
Composition of Air 188
Atmospheric Pressure 189
7.2 Atmospheric Gases of Global
Concern 189
Changing Levels of Carbon
Dioxide 189
The Ozone Problem 191
7.3 The Role of Sulfur
Compounds 192
7.4 The Atmosphere in Motion 192
Box: Clouds and Climate 194
Winds on a Nonrotating Earth 194
The Effects of Rotation 195
Wind Bands 197
7.5 Modifying the Wind Bands 198
Seasonal Changes 199
The Monsoon Effect 199
The Topographic Effect 199
The Jet Streams 202
7.6 Hurricanes 203
7.7 El Niño 204
7.8 Practical Considerations: Storm
Tides and Storm Surges 206
Box: A North Atlantic Cool Pool 207
Summary 208
Key Terms 210
Study Questions 210
Suggested Readings 211

8

The Currents 212

8.1   Ocean Surface Currents 214
  The Pacific Ocean Currents 214
  The Atlantic Ocean Currents 215
  The Indian Ocean Currents 216
8.2   Gyres and Current Flow 216
  *The Ekman Spiral and Ekman
    Transport 216*
  Geostrophic Flow 216
  Current Speed 216
  Western Intensification 217
8.3   Eddies 217
8.4   Convergence and Divergence 219
  Permanent Zones 219
  Seasonal Zones 219
8.5   Global Circulation Changes 220
8.6   Measuring the Currents 223
**Box: Looking for Carbon: A
    Worldwide Effort 224**
8.7   Practical Considerations: Energy
    from the Currents 226
Box: The Great Sneaker Spill 227
Summary 229
Key Terms 229
Study Questions 229
Suggested Readings 229

9

The Waves 232

9.1   How a Wave Begins 234
9.2   Anatomy of a Wave 235
9.3   Wave Motion 235
9.4   Wave Speed 236
9.5   Deep-Water Waves 236
  Storm Centers 237
  Swell 237
  Dispersion 237
  Group Speed 237
  Wave Interaction 238
9.6   Wave Height 238
  Episodic Waves 239
  Wave Energy 240
  Wave Steepness 241
9.7   Shallow-Water Waves 241
Box: Universal Sea State Code 242
  Refraction 242
  Reflection 243

  Diffraction 244
  *Navigation from Wave
    Direction 245*
9.8   The Surf Zone 245
  Breakers 245
  Water Transport 246
  Energy Release 247
9.9   Tsunamis 247
9.10 Internal Waves 248
9.11 Standing Waves 249
9.12 Practical Considerations: Energy
    from Waves 250
Summary 252
Key Terms 252
Study Questions 253
Study Problems 253
Suggested Readings 253

Photoessay: Going to Sea 255

10

The Tides 260

10.1 Tide Patterns 262
10.2 Tide Levels 263
10.3 Tidal Currents 264
10.4 Equilibrium Tidal Theory 264
  The Moon Tide 265
  The Tidal Day 266
  The Tide Wave 266
  The Sun Tide 266
  Spring Tides and Neap Tides 267
  Declinational Tides 267
  Elliptical Orbits 268
10.5 Dynamic Tidal Analysis 268
  The Tide Wave 268
  Progressive Tides 269
  Standing Wave Tides 269
  Tide Waves in Narrow Basins 272
10.6 Tidal Bores 273
10.7 Predicting Tides and Tidal
    Currents 273
  Tide Tables 274
  Tidal Current Tables 275
10.8 Practical Considerations: Energy
    from Tides 276
**Box: Undersea Robotic
    Technology 280**
Summary 281
Key Terms 282
Study Questions 282
Study Problems 282
Suggested Readings 283

11

Coasts, Shores, and Beaches 284

11.1 Major Zones 286
11.2 Types of Coasts 287
  Primary Coasts 288
  Secondary Coasts 291
Box: Rising Sea Level 294
11.3 Anatomy of a Beach 295
11.4 Beach Types 298
11.5 Beach Dynamics 299
  Natural Processes 300
  Coastal Circulation 301
  Artificial Processes 303
11.6 Practical Considerations: Case
    Histories of Two Harbors 304
  The Santa Barbara Story 304
  The History of Ediz Hook 305
Summary 306
Key Terms 306
Study Questions 307
Suggested Readings 307

12

Bays and Estuaries 308

12.1 Estuaries 310
  Types of Estuaries 310
  Circulation Patterns 311
  Temperate Zone Estuaries 312
12.2 Embayments with High
    Evaporation Rates 313
12.3 Flushing Time 313
12.4 Degradation of Coastal
    Environments 314
  Water and Sediment Quality 314
  The Plastic Trash Problem 319
  Oil Spills 320
12.5 Marine Wetlands 322
Box: Going, Going, Gone 324
12.6 Practical Considerations: Case
    Histories 325
  *The Development of San Francisco
    Bay 325*
  *The Situation in Chesapeake
    Bay 328*
Summary 330
Key Terms 330
Study Questions 330
Study Problems 331
Suggested Readings 331

13

Oceans: Environment for Life 332

13.1 Buoyancy and Flotation 334
13.2 Osmotic Processes 335
13.3 Temperature 336
13.4 Pressure 336
13.5 Gases 337
13.6 Nutrients 337
13.7 Light and Color 337
   Sunlight 337
   Bioluminescence 338
   Color 338
13.8 Circulation 339
13.9 Barriers and Boundaries 340
13.10 Bottom Types 340
13.11 Environmental Zones 340
13.12 Classification of Organisms 341
13.13 Practical Considerations:
   Modification and
   Mitigation 341
**Box: *Spartina:* Valuable and
   Productive or Invasive and
   Destructive? 343**
Summary 345
Key Terms 345
Study Questions 346
Suggested Readings 346

14

Production and Life 348

14.1 Primary Production 350
   Gross and Net 350
   Standing Crop 350
14.2 Controls on Primary
   Production 350
   Light 350
   Nutrients 351
   Nutrient Cycles 352
14.3 Global Primary Production 353
14.4 Measuring Primary Production 354
14.5 Total Production 356
   Food Chains and Food Webs 356
Box: Satellite Measurements 358
   Trophic Pyramids 359
   Other Systems 361
14.6 Practical Considerations: Human
   Concerns 362
Summary 362
Key Terms 364
Study Questions 364
Study Problems 364
Suggested Readings 364

15

The Plankton: Drifters of the Open Ocean 366

15.1 The Kinds of Plankton 368
   Phytoplankton 368
   Zooplankton 373
Box: A Krill-Based Ecosystem 382
15.2 Bacteria 382
15.3 Classification Summary of the
   Plankton 383
15.4 Sampling the Plankton 383
Box: Viruses in the Ocean 384
15.5 Practical Considerations: Marine
   Toxins 385
   Red Tides 385
   Other Toxic Blooms 386
   Ciguatera Poisoning 387
Summary 387
Key Terms 388
Study Questions 388
Suggested Readings 389

16

The Nekton: Free Swimmers of the Sea 390

16.1 The Mammals 392
   Whales and Whaling 392
**Box: Echolocation and
   Communication 397**
   Dolphins and Porpoises 398
   *Seals, Sea Lions, and
      Walruses 398*
   Sea Otters 401
   Sea Cows 401
   *Marine Mammal Protection
      Act 402*
16.2 The Reptiles 402
   Sea Snakes 402
   Sea Turtles 402
16.3 The Squid 404
16.4 The Fish 404
   Sharks and Rays 406
   *Commercial Species of Bony
      Fish 408*
   *Deep-Sea Species of Bony
      Fish 408*
16.5 Classification Summary of the
   Nekton 410
16.6 Practical Considerations:
   Commercial Fisheries 410
   Anchovies 411

   Tuna 412
   Salmon 412
   Redfish 414
   Shark 414
   Problems and Policies 414
   Fish Farming 415
Summary 417
Key Terms 418
Study Questions 418
Suggested Readings 418

17

The Benthos: Dwellers of the Seafloor 420

17.1 The Plants 422
   *General Characteristics of Benthic
      Algae 422*
   Kinds of Seaweeds 423
   Other Plants 423
17.2 The Animals 425
   Animals of the Rocky Shore 425
   Symbiotic Relationships 430
   Tide Pools 430
   Animals of the Soft Substrates 431
   The Deep Seafloor 432
   *Fouling and Boring
      Organisms 434*
17.3 Classification Summary of the
   Benthos 435
17.4 The Tropical Coral Reefs 436
17.5 High-Energy Environments 439
17.6 Deep-Ocean Chemosynthetic
   Communities 440
17.7 Sampling the Benthos 442
Box: Mussel Watch 443
17.8 Practical Considerations:
   Harvesting the Benthos 444
   The Animals 444
   The Algae 445
**Box: Genetic Manipulation of Fish and
   Shellfish 446**
   Kelp Bioconversion 446
   Biomedical Products 446
Summary 450
Key Terms 450
Study Questions 451
Suggested Readings 451

Appendix 452
Glossary 456
Credits 464
Index 466

Preface

This fourth edition, like its predecessors, is the result of many experiences. Lectures heard, journals read, classes taught, and projects accomplished have all influenced and continue to influence our approach to the study and teaching of oceanography. Our goals in this edition remain the same as with previous editions: to write an introductory oceanography text for the student without a background in mathematics, chemistry, physics, geology, or biology, and to emphasize the role of basic scientific principles over those processes that govern the oceans and the earth.

A new edition requires meeting again the challenge of stimulating student interest and curiosity by blending contemporary information and research with basic principles to form an integrated introduction to the sciences of the oceans. The preparation of this fourth edition provided the impetus to review the ever-increasing stream of new information, ideas, and concepts produced by the ever-accelerating changes in today's ocean sciences. Because of the increasing emphasis on air-sea interaction, a new chapter, Chapter 7—The Ocean and the Atmosphere, has been added. This chapter includes sections on the structure and composition of the atmosphere, changing levels of carbon dioxide and ozone, the role of sulfur compounds, material on hurricanes and jet streams, as well as the previously included sections on winds and the El Niño effect. This edition highlights new techniques and new projects: side-scan acoustical imaging and swath bathymetry for viewing the seafloor, acoustic tomography to describe seawater properties, increasing use of remote sensing and robotic devices, the Heard Island Experiment to investigate global ocean warming, and the Joint Global Ocean Flux Study tracing the carbon cycle in the oceans. New ideas in ocean waste management, new evidence linking the oceans and world climate, and new data on ocean bacteria and viruses are featured. The degradation of coastal and oceanic environments and the problems of marine fisheries are discussed and updated. An additional aim of this fourth edition has been to review each chapter, seeking to make it as clear and readable as possible.

Many of the techniques and concepts being explored in the text are involved and complex. We have not attempted to provide easy explanations but rather seek to encourage a thoughtful discussion of each item, leading to a more balanced understanding of earth and oceanic phenomena. Scientists and students share a global outlook as all people recognize that the presence of our species is modifying earth's governing processes and changing the rates at which these processes proceed. In this edition we emphasize concepts that encourage understanding how, possibly why, and in what directions the earth and its oceans are changing.

Choices made in preparing this edition include order of presentation, methods of explanation, and types of information to be included. Because the breadth of the material is so great and the interdependency between subject areas is so strong, there is a continuous struggle over the order in which to present the material. We have not changed the order in this edition, but we have continued to make each chapter stand as independently as possible. Cross references from one chapter to another indicate more detailed discussion of topics elsewhere in the text. Keeping in mind nature's feedback mechanisms and spiderweb relationships, we have continued to present a text that does not lock the user into a fixed order but allows flexibility and adaptation.

Methods of explanation have been keyed to the students who will read this book. For example, we have chosen to use centrifugal force to explain tidal principles, since most students do not have the background to handle vectors. Also, more emphasis has been placed on the physical and geological aspects of the oceans than on the chemical and geochemical aspects, which require more specific background knowledge. An ecological approach as well as necessary descriptive material is used to integrate the biological chapters with other subjects. The end-of-chapter "practical consideration" sections have been updated and continue to emphasize the relationships among people, ocean resources, and the marine environment. Choices are made to minimize oceanography as a collection of subjects gathered under a marine umbrella and to emphasize oceanography as the study of the oceans as a cohesive and united whole.

As in all introductory courses, there is an emphasis on the vocabulary of the discipline. Students must learn to speak the language if they are to be able to comprehend the explanation, interpret the result, and apply what is learned. Newspapers, magazines, television programs, and even radio bombard students with information concerning their planet at a time of rapid and increasing change. If students are to be able to make intelligent decisions, they must be able to determine the worth of this information. To do so they need an understanding of language as well as process and principle. At the same time recognizing that vocabulary does not excite many students, we have tried not to expand its scope in this edition, preferring to explain new material with the considerable terminology already available. All terms are defined in the text; a student does not need a prior understanding of the terminology to use this text. Terms that are particularly important are printed in boldface at the point when they are first used and defined. A word list of important terms is at the end of each chapter, and a glossary of terms is included at the end of the book.

At a time in which new information of varying quality appears daily at all levels, it is easy to forget that students are approaching this subject for the first time. Under these circumstances students need to learn a set of principles that explain why oceanographic processes work the way they do, but they also want to learn the oceanography that will help them to understand the information presented in today's media. For this reason we have tried to integrate topics of current interest and practical concern with a topic's basic information, instead of grouping all such items in one or two chapters at the end of the book. For example, seabed resources and ocean solid-waste management are discussed in the context of the geology of ocean basins and sediments, water quality is a topic in the chapter on estuaries and bays, and problems of overfishing and an overview of sea farming are presented in the appropriate biological chapters for the organisms discussed. Special interest items, such as marine archaeology, genetic manipulation of shellfish, and the history of the earth's changing atmosphere recorded in ice cores, are placed in boxes throughout the text. In this edition, with the generous cooperation of the researchers, we have been able to obtain illustrations and photographs from many current research projects to complement written information. The computer graphics illustrating primary production and ocean temperatures during an El Niño episode were produced for this edition.

This book is designed for a one-quarter or one-semester course. Since the experience and knowledge of those using this book will differ, it is expected that each instructor will emphasize and elaborate on some topic at the expense of others. The lists of additional readings at the end of each chapter can help extend and enrich subjects. The fourth edition reading lists have been extensively revised and updated to guide students to recent publications on text topics. Among the listed readings are articles on topics of current interest as well as some of oceanography's historical and classical books and papers. We recommend supplementing the text with films and videos from the earth sciences according to the resources available on individual campuses.

Summaries at the end of each chapter provide a quick review of the key concepts for that chapter. Problems are included in most chapters, and questions are included at the end of each chapter. These questions and problems are not intended merely for review but to challenge the student to think further about the lessons of the chapter. The answers to the questions appear in the instructor's manual.

WCB provides a computerized test generator for use with this text. It allows you to quickly create tests based on questions provided by WCB and requires no programming experience to use. The questions are provided on diskette in a test item file. WCB also provides support services, via mail or phone, to assist in the use of test generator software, as well as in the creation and printing of tests.

A computerized grade management is also available for instructors. This allows you to track student performance on exams and assignments. Reports based on this information can be generated for your review.

Software to generate quizzes can also be provided. These quizzes can be used to allow students to prepare for exams on their own.

Another item we hope you will find useful is a set of overhead transparencies. The transparencies reproduce selected illustrations from the text to support and enhance your lectures.

Acknowledgments

We begin this edition with an overture written by Richard H. Fleming. Professor Fleming, with H. U. Sverdrup and M. Johnson, published in 1942 the 1087-page *The Oceans,* oceanography's first comprehensive textbook, which included physical, chemical, biological, and geological aspects of the science. He was associated with marine education for over fifty years, pioneering the development of undergraduate education in oceanography, fostering graduate studies, and encouraging the application of the marine sciences to public planning and policy. He brought this experience and perspective to his introductory essay. We consider ourselves especially privileged to have had a long association with Professor Fleming as our dear friend and mentor. Professor Fleming died in October 1989. This overture is the last writing of his long career.

As a book is the product of many experiences, it is also the product of people other than the authors. We extend many thanks to our friends and colleagues who have graciously answered our questions and provided us with information and access to their photo files. We owe very special thanks to the faculty and staff of the School of

Oceanography, College of Ocean and Fishery Sciences, University of Washington, who helped us research and secure many of the illustrations in this edition.

We would also like to acknowledge the valuable contributions of the reviewers, whose comments and suggestions over the last several years have helped shape the present edition: John E. Mylroie, Mississippi State University; Stan Ulanski, James Madison University; Keith Sverdrup, University of Wisconsin–Milwaukee; David Schwartz, Cabrillo College; Dr. Bernard Oostdam, Millersville University; John W. Winchester, Florida State University; Russ Flynn, Cypress College; Christopher Schmidt, Western Michigan University; James Reese, Orange Coast College; Ed Stroup, University of Hawaii at Manoa; Christina Emerick, University of Washington; James Ogg, Purdue University; Laurie Brown, University of Massachusetts; David Schwartz, Cabrillo College in Aptos, California; Stephen A. Macko, University of Virginia; John T. Merrill, Assoc. Prof., University of Rhode Island, Graduate School Oceanography; T.C. Moore, Jr., Center for Great Lakes and Aquatic Sciences, University of Michigan, and Joan C. Stover, South Seattle Community College.

We thank again Wm. C. Brown Publishers, without whose help, enthusiasm, and coordinated efforts this fourth edition could not have been completed.

Overture

by Richard H. Fleming

The past fifty years have been notable for rapidly growing interests in the ocean. These concerns range in scale from international and national legislatures and organizations to single individuals. That you are reading this book is an indication of such interest. Scientific curiosity about the unique features of our planet is obvious, but the levels of scientific and technological effort devoted to the ocean are driven by political, sociological, economic, and practical motives. It is worthwhile to identify some of these that are pertinent to the increased interests in the ocean.

In the late 1940s the United Nations Organization began a series of worldwide studies of human populations and of their rates of growth. These investigations awakened the world to the realities of the population explosion. The global average rate of population growth is about 2% per year. This is similar to compound interest and, if the rate was continued, would lead to a doubling of the population in thirty-five years. Although this may appear to be relatively small, it means that during the next century the population could increase to eight times the present numbers and in the next 350 years could reach the vast total of one thousand times the present numbers. From this it should be obvious that the biggest problem facing the world is how to curb the population explosion.

Even today there are many land areas that cannot raise sufficient food to feed their human populations. On a global basis we can grow enough food to feed the population of the earth. There are major efforts to increase the amount of land cultivation to raise not only food crops but also nonfood crops needed for many purposes. These projects are often hampered by inadequate fresh water in marginal areas, and droughts are a major threat.

Coupled with the population explosion is the technological explosion. This can easily be identified with the industrial revolution that began about two hundred years ago. This explosion involves rapidly rising demands for energy and a wide variety of minerals and other substances that are the raw materials of industry. The per capita consumption of these raw materials has been rising at a rapid rate. The technological explosion has raised the standard of living in the industrialized countries and is a major objective of the lesser-developed countries of the world. A basic question confronting us is, what can the ocean contribute to these growing demands? The ocean has long been a medium for the transportation of goods, and the merchant fleet plays an essential role in the movement of raw materials and manufactured goods. It is still the most cost-effective means of transport as represented by the supertankers, bulk carriers, and container vessels.

Fishing for human food and for other purposes is a traditional use of the ocean. The improved techniques and increased activity have led to growing catches that provide a significant fraction of the world's supply of protein. The question remains as to whether or not the ocean can ever be farmed in the way that the land is used for agriculture. Attempts have been made to raise certain types of fish and shellfish in shallow coastal areas. Some have been successful but they are limited to high-priced commodities. There are no marine plants that can supplement the cereals (carbohydrate producers) that form the major component of agriculture. The algae are in some cases edible, but they are low in nutritional value.

Petroleum (both liquid and natural gas) is a major source of energy and raw materials for many synthetic compounds. The land reserves of petroleum have been depleted and their discovery beneath the shallow coastal areas stimulated the first great industrial exploitation of the ocean's resources. Drilling techniques have been extended into deeper and deeper water, and the instruments used in the search for oil deposits have greatly contributed to scientific as well as practical needs.

The successes in the search for petroleum led to two important consequences. First, all the countries bordered by the ocean laid territorial claims to the resources of the seabed underlying their continental shelves. This involved the governments in expanded scientific interests. Second, the industrial interests were awakened to the possibilities of other potential resources. These included the design of mechanisms to collect the manganese nodules that carpet certain deep areas of the ocean. Studies were made to investigate the ways in which energy might be obtained from the tides, from surface waves, from the heat content of the water, and from the digestion of algae. Last but not least were studies of the various ways in which fresh water could be produced by desalination of seawater. A wide variety of minerals was studied, including such romantic items as gold and diamonds.

The ocean has always been and will always be an environment hostile to man. Stormy seas are hazardous to vessels and their cargoes. Surf can damage and destroy such shore structures as seawalls, piers, and breakwaters. Floodings can cause havoc and large loss of life in the low-lying coastal plains. These inundations can be caused by certain violent storms such as hurricanes. Tsunamis, caused by submarine volcanic explosions and by earthquakes on the ocean floor, can generate giant waves that wreak havoc on the coastal plain. There is obvious need to study these phenomena and to learn when and where they will occur and to be able to provide adequate warning so the endangered coastal population can be evacuated. Over long periods, sea level is affected by the accumulation and melting of glacial ice on the landmasses. The magnitude of the danger is only a fraction of one inch per year, but over time spans of tens of thousands of years this change can amount to 100 feet or more. Storm surges, tsunamis, and sea-level changes are natural in origin but should enter into urban planning for the siting of developments in low-lying regions.

It was long believed that human activities could have little or no effect upon the ocean environment. The first indication this was not true came from the decreased abundance of certain fish, marine mammals, and birds. The improved fishing techniques and greater fishing effort have reduced certain populations to critical levels and in some cases have led to their extinction. Various groups of conservationists (some governmental and some private) have endeavored to protect species under heavy pressure. In some cases these efforts have been successful, but in many others they have been thwarted in their attempts to stop or reduce the slaughter. The principal difficulty is the reluctance of those harvesting the ocean to give up their traditional freedom to exploit the resources of the oceans. It is exceedingly difficult to police the vast areas of the ocean and to prosecute the offenders.

The freedom of the seas has, in recent decades, given way to national territorial claims to the resources of the seabed beneath the continental shelf and now to the fisheries out to distances of 200 nautical miles from the coast. It is worth noting that sea boundaries do not have the characteristics of territorial land boundaries. They cannot be marked with visible signs, and they cannot be guarded by watchtowers nor claims reinforced by settlers. (No one yet lives permanently on or in the ocean.) Fish and marine mammals freely pass across such boundaries.

Other types of human activities that are having an increased impact on both the ocean and the atmosphere, and hence on the landmasses, are those associated with waste disposal and pollution. Coal, wood, and petroleum combustion products are released into the atmosphere. Most

Richard H. Fleming, 1909–89
Professor Emeritus
Oceanography and Marine Studies, University of Washington

of these products are gases, principally carbon dioxide (CO_2), with small quantities of carbon monoxide and compounds of nitrogen and sulphur. There are also small particles of solids that, when concentrated near the sources, we identify as smoke. All of these gases and particles can remain in the atmosphere for long periods and can be distributed worldwide. There has been a measurable increase in the CO_2 in the atmosphere, and this has raised concern about the so-called "greenhouse effect." The quantity of CO_2 in the atmosphere is believed to affect the temperatures of the atmosphere and the ocean surface. Increased CO_2 content will tend to trap more of the solar radiation and result in a small but significant rise in environmental temperatures. Although this might appear to be beneficial to those who live in the higher latitudes, it could lead to accelerated melting of glacial ice with a consequent rapid rise in sea level. It may also alter the amount and distribution of rainfall. The ocean is involved in this debate because it tends to dissolve the CO_2, but this is apparently a relatively slow process.

A second topic of concern is referred to as "acid rain," apparently caused by certain products of combustion and from some industrial processes that contain oxides of nitrogen and sulphur. These materials dissolve in the water droplets in the clouds and hence fall as acid rain. This rain tends to kill off forest trees and destroy fish and other life in rivers and lakes downwind from industrial areas. Acid rain has much less effect in the oceans.

A third topic of concern is the destruction of the ozone layer. Ozone (O_3) is a compound of oxygen that accumulates in the high atmosphere. It is produced naturally and it protects the life on earth by absorbing harmful ultraviolet radiation received from the sun. It is believed that the ozone is destroyed by certain synthetic organic substances used as solvents and propellants employed in aerosol containers. These are not products of combustion. Their role in the ocean is not yet understood.

These three situations have been introduced because they are examples of the consequences of increased human activities, they are widespread, and they can be expected to increase unless there is international agreement to limit the sources.

Ocean dumping of trash and garbage has been a practice of some coastal cities. Sewage, as well as effluents from factories and industrial plants, has commonly been directly released into the coastal waters. Biological decay rendered much of this material harmless, but increased rates of disposal and the presence of more and more non-biodegradable materials have raised the concern of conservationists, public health officials, and others. Many coastal areas have lost their pristine conditions, and fish have been killed or declared unfit for human consumption. This type of waste disposal can be controlled by state or local agencies. Much still remains to be done to reduce the heedless introduction of human wastes and trash into coastal waters.

Some of the most obvious and troublesome types of pollution are the result of accidents. Oil tanker groundings and collisions can release large volumes of crude oil or petroleum products; other vessels may lose their fuel oil. Failure of pipelines may also be the cause of near-shore pollution. These massive point-source introductions of liquid pollutants may have catastrophic localized effects on marine life and may coat the beaches and shorelines, making them unsightly and unusable by humans. Accidents on shallow-water drilling rigs are another source of petroleum pollution. The oil pollutants are gradually degraded and destroyed by marine bacteria, but their local consequences may persist for many years.

Radioactive wastes, the by-products of the nuclear industry, represent major disposal problems. Many of the materials are extremely hazardous to organisms, including humans, and their half-lives can exceed several thousand years. Proper safe storage of these substances has been debated for forty years and has still not been resolved. One of the early suggestions was that these wastes be dumped on the seabed in the deep ocean. However, strong arguments were made against ocean dumping. International agreements were reached that place strict controls on such practices.

Let us now turn to the private or personal interests in the ocean. These have been growing rapidly since World War II. With greater affluence and more leisure time, there is increasing ocean travel and visits to coastal areas and beaches. Shoreline property or even a building site with a distant view of the water is highly prized. Swimming, sunbathing, and surfing draw crowds to the warm tropical and subtropical areas. Free diving with scuba gear or snorkels makes it possible for amateurs to become underwater explorers and treasure hunters. Fish-watchers have joined the bird-watchers, and whale watching from land or from special excursion boats has attracted many viewers.

So called wet suits have made it possible to dive in otherwise cold water. Yachts, small boats, kayaks, and windsurfers take more and more people out on the water. Recreational fishing and shellfish collecting are, in some areas, competing with commercial fishermen for limited stocks. Marine preserves and parks have been established in some areas where scuba diving and snorkeling are popular. It is inevitable that recreational uses of the coastal waters will come in conflict with commercial activities and we may expect to see more "traffic control."

Added to those who enjoy the ocean firsthand are the vicarious participants who read about the ocean and watch documentary films and TV shows. The traditional aquaria have been expanded to provide marine mammal acts that perform in outdoor pools. There are many paths that lead to the ocean and all of them are drawing more traffic. Understanding of the ocean by those who enjoy it is imperative, if you wish to preserve it for these uses now and in the future.

Those interested in natural history of the ocean may focus their attention on the birds in and on the ocean, particularly those common in coastal waters. Fishes as well as the exotic invertebrates that inhabit the shallow seabed and the intertidal zone of the shoreline attract the interest of many. Most marine plants are microscopic and hence are invisible to the naked eye. Lush growths of marine algae (seaweeds) compare to the jungles on land, but the plants themselves have few of the characteristics and variety of land plants.

The biological features of interest are mostly visible to the unaided eye but, as divers will tell you, underwater vision is not very good. Only in a relatively few localities are the coastal waters clear enough to permit good "seeing" and allow underwater photography to produce good results. In most shallow-water areas the water is turbid due to small solid particles, organic debris, and the yellow pigments released from decaying organic matter.

If you visit different coastal regions you will notice differences in the plants and animals that make up the biological community. There are many other features that can differ markedly from place to place. What are the colors, texture, and composition of the sands forming the beaches? What are the characteristics of the waves that come in from the open sea? How can one describe the features of the tides and tidal currents? There are many interesting features of the ocean that invoke aspects of geology, physics, and chemistry that can be easily seen or measured. The presence of dissolved salts in the oceanic waters modify some of the properties of pure water and create some new ones. However, it should be remembered that naturally occurring water on land always contains at least small amounts of salts, although the relative proportions of the solutes are variable and differ from those in seawater. Ocean water contains about 3.5% of its weight in salts although the *salinity,* a measure of the total dissolved solids, can vary from almost zero where rivers enter the ocean to values of 4.0% or more in regions of excessive evaporation.

Over the entire planet there must be a balance between evaporation and precipitation. There is, however, a net loss of water from the ocean because the evaporation over the ocean exceeds the precipitation falling on it. Over the land the reverse is true, with the precipitation exceeding the evaporation. The surplus of precipitation supplies the fresh water on and in the land, in other words the lakes and rivers, the groundwater, and the glacial ice. The net loss of water from the ocean is balanced by the runoff from the land, chiefly by rivers. This is known as the hydrologic cycle. Obviously the annual evaporation and precipitation vary from place to place. Fresh water is one of the essential resources for human survival and includes that needed for human consumption and household cleanliness, but by far the greatest amounts are employed in agriculture and industrial processes. It is interesting to note that, unlike other resources where the raw material is destroyed or greatly altered in its nature (as in industrial processes), the total volume of water remains the same. The use of water may involve the addition or removal of heat, and the addition of dissolved salts and particles of inorganic and organic matter and pollutants of a wide variety. This may make it unfit for human use. However, given time, the water will lose much of its added particles by their sinking, and biological processes will destroy the organic debris. In the ocean the river water will be mixed with surrounding seawater and lose its identity.

Understanding the ocean can be approached at many levels of sophistication. This volume is intended for students with limited backgrounds in the basic sciences and with careful study will provide an excellent introduction to the science of the ocean and its relation to man. It will help you understand the processes that make the ocean what it is and the role it plays on our earth.

During the last fifty years the scope in the topics studied and the amount of effort devoted to the ocean have both grown rapidly. The reason for this can be traced to attitudes of people, both in national governments and other organizations, and to the concern of individuals. A generation ago the ocean was considered to have little value except for shipping and fishing. Today all the countries of the world, including those that are landlocked, are claiming that they should have access to the marine resources, whatever these may be. What will happen in the next half-century? We can anticipate a growing concern with territorial claims and with what might be called the management of the ocean. The management involves political, economic, legal, and regulatory matters for which there may be no precedents. For a variety of reasons, procedures that have evolved to deal with land management cannot be transferred to the ocean.

We must solve the dilemmas of the population explosion and learn how the oceans can help in the production of food and other human needs. We must learn how to conserve the living resources and to limit pollution. To achieve these goals we must learn to look further ahead in time (in centuries rather than years) and to treat the entire earth as a single system. This is obviously a task for generalists with broad interests and an understanding of the natural systems of the earth and how they interact.

July 6, 1989

Prologue
The History of Oceanography

The Early Times
The Middle Ages
Voyages of Discovery
The Beginnings of Ocean Science
The Importance of Charts and Navigational
    Information
Ocean Science Begins
The *Challenger* Expedition
Oceanography As Science
U.S. Oceanography Moves into the Twentieth
    Century
World War II
Ocean Science Expands in the United States
Oceanography of the Recent Past
The Present and the Future

Box: Marine Archaeology

Study Questions
Suggested Readings

Breaking waves, Oregon coast.

*O*ceanography is a broad field in which many sciences are focused on the common goal of understanding the oceans. Geology, geography, geophysics, physics, chemistry, geochemistry, mathematics, meteorology, botany, and zoology have all played roles in expanding our knowledge of the oceans. The field is so broad that oceanography today is usually broken down into a number of subdisciplines.

Geological oceanography includes the study of the earth at the sea's edge and below its surface and the history of the processes that formed the ocean basins. Physical oceanography investigates how and why the oceans move; marine meteorology, the study of heat transfer, water cycles, and air-sea interactions, is often included in this discipline. Chemical oceanography studies the composition and history of the water, its processes, and its interactions. Biological oceanography concerns the marine organisms and the relationship between these organisms and the environment of the oceans. Ocean engineering is the discipline that designs and plans equipment and installations for use at sea.

The study of the oceans was promoted by intellectual and social forces as well as by our needs for marine resources, trade and commerce, and national security. Oceanography started slowly and informally; it began to develop as a modern science in the mid-1800s and has grown dramatically, even explosively, in the last forty years. Our progress toward the goal of understanding the oceans has been uneven, and it has frequently changed direction. The interests and needs of nations as well as the intellectual curiosity of scientists have controlled the rate at which we study the oceans, the methods we use to study it, and the priority we give to certain areas of study. To gain some perspective on the current state of knowledge about the oceans, we need to know something of the events and incentives that guided people's investigations of the oceans.

The Early Times

People have been gathering information about the oceans for millennia, accumulating bits and pieces of knowledge, passing it on by word of mouth. Curious individuals must have acquired their first ideas of the oceans from wandering the seashore, wading in the shallows, gathering food from the ocean's edges. During the Paleolithic period humans developed the barbed spear or harpoon and the gorge. The gorge was a stick with a string attached, pointed at both ends and enclosed in bait. At the beginning of the Neolithic period the bone fishhook was developed and later the net. By 5000 B.C. copper fishhooks were in use (see traditional Native American implements in Prologue fig. I). As early humans moved slowly away from their inland centers of development, they were prepared to take advantage

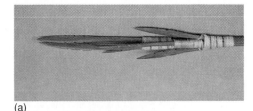

(a)

(b)

(c)

Prologue I

Traditional fishing and hunting implements from coastal Native American cultures of the Pacific Northwest: (a) A duck spear made of cedar. (b) A bone harpoon point and lanyard made of sinew, hemp, and twine. (c) A fishhook made of bone and steam-bent cedar root.

Photos courtesy Thomas Burke Memorial Washington State Museum, Seattle, WA.

of the sea's food sources when they first explored and later settled along the ocean shore. The remains of shells and other refuse, in piles known as kitchen middens, have been found at the sites of ancient shore settlements. These remains show that our early ancestors gathered shellfish, and fish bones found in some middens suggest that they also began to use rafts or some type of boat for offshore fishing. Drawings on ancient temple walls show fishnets; on the tomb of the Egyptian Pharaoh Ti, Fifth Dynasty (five thousand years ago), is a drawing of the poisonous pufferfish with a hieroglyphic description and warning. As long ago as 1200 B.C. or earlier dried fish were traded in the Persian Gulf; in the Mediterranean the ancient Greeks caught, preserved, and traded fish while the Phoenicians founded fishing settlements, such as "the fisher's town" Sidon, which grew into important trading ports.

We know that early information about the oceans was mainly collected by explorers and traders. Around 1500 B.C. the Phoenicians, well known as excellent sailors and navigators who traded across the Mediterranean Sea from North Africa with Italy, Spain, and Greece, sailed out of the

Prologue II

A navigational chart (*rebillib*) of the Marshall Islands. Sticks represent a series of regular wave patterns (swells). Curved sticks show waves bent by the shorelines of individual islands. Islands are represented by shells.
Chart courtesy of Tom Rice of Sea and Shore Museum, Port Gamble, WA.

Mediterranean Sea and north along the Atlantic coast of Europe to trade in the British Isles. At about the same time Queen Hatshepsut of Egypt went by sea to the "Land of Punt," probably Somalia. The Phoenicians may have circumnavigated Africa around 600 B.C., but there is no record of this voyage.

Between 1500 and 500 B.C. the Arab traders explored the Indian Ocean, and the Polynesians made long voyages of discovery in the Pacific Ocean. These voyages left little in the way of recorded information. Using descriptions passed down from one mariner to another, early voyagers piloted their way from one landmark to another, sailing close to shore and often bringing their boats up onto the beach each night. As they began to move away from shore they used birds, waves, cloud formations, and the observation of astronomical bodies in their travels (see Prologue fig. II). The distinctive smells of land such as flowers and burning wood alerted them to possible landfalls.

The Greeks called the Mediterranean "Thalassa" and believed that it was encompassed by land, which in turn was surrounded by the endlessly circling river Oceanus. In 325 B.C. Alexander the Great reached the deserts of the Mekran Coast, now a part of Pakistan. He sent his fleet down the coast in an apparent effort to probe the mystery of Oceanus. He and his troops had expected to find a dark, fearsome sea of whirlpools and water spouts inhabited by monsters and demons; they did find tides that were unknown to them in the Mediterranean Sea. His commander, Nearchus, took the first Greek ships into the ocean, explored the coast, and brought them safely to the port of Hormuz eighty days later. Pytheas (350–300 B.C.), a navigator, geographer, astronomer, and contemporary of Alexander, made one of the earliest recorded voyages from the Mediterranean to England. From there he sailed north to Scotland, Norway, and Germany. He navigated by the sun, stars, and wind, although he may have had some form of sailing directions. He recognized a relationship between the tides and the moon and made early attempts at determining latitude and longitude. These early sailors did not investigate the oceans; for them the oceans were only a dangerous road, a pathway from here to there, a situation that continued for hundreds of years. However, the information that they accumulated began to build into a body of lore to which sailors and voyagers added from year to year.

While the Greeks traded and warred throughout the Mediterranean, they observed and also asked themselves questions about the sea. Aristotle (384–322 B.C.) believed the oceans occupied the deepest parts of the earth's surface; he knew that the sun evaporated water from the sea surface, which condensed and returned as rain. He also began to catalog marine organisms. The brilliant Eratosthenes (c. 265–194 B.C.) of Alexandria, Egypt, mapped his known world and calculated the circumference of the earth to be 40,250 kilometers or 25,000 miles (today's measurement is 40,067 km or 24,881 mi). Posidonius (c. 135–50 B.C. reportedly measured an ocean depth to about 1800 meters (6000 feet) near the island of Sardinia according to the Greek geographer Strabo (c. 63 B.C.–A.D. 21). Pliny the Elder (c. A.D. 23–79) related the phases of the moon to the tides and reported on the currents moving through the Strait of Gibraltar. Ptolemy (c. A.D. 127–51) produced the first world atlas and established world boundaries; to the north the British Isles, Northern Europe, and the unknown lands of Asia; to the south an unknown land, "Terra Australis Incognita," including Ethiopia, Libya, and the Indian Sea; to the east China; and to the west the great Western Ocean reaching around the earth to China on the other side (see Prologue fig. III). His atlas listed more than eight thousand places by latitude and longitude, but his work contained a major flaw. He had accepted a value of 29,000 kilometers (18,000 miles) for the earth's circumference. This shortened earth distances and allowed Columbus, more than one thousand years later, to believe that he had reached the eastern shore of Asia when he landed in the Americas.

The Middle Ages

After Ptolemy, intellectual activity and scientific thought declined in northern Europe for about one thousand years. However, shipbuilding improved during this period; vessels became more seaworthy and sailors extended their voyages. The mainly unrecorded voyages of the Vikings (A.D. 700–1000) colonized Iceland by 900 and settled Greenland where they remained until the fourteenth century. Extending their voyages farther west they reached Vineland, or northeastern North America, in 985. To the south, in the region of the Mediterranean after the fall of the Roman Empire, the Arabs acquired the knowledge of the Greeks and the Romans on which they continued to build. The Arabic writer El-Mas'údé (d. 956) gives the first description of the reversal of the currents due to the seasonal

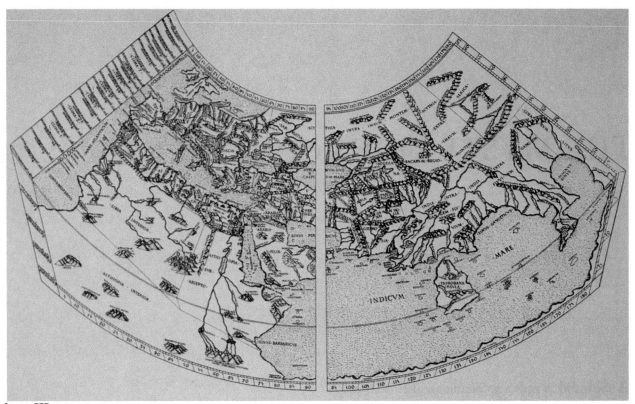

Prologue III

A chart from an Italian fifteenth-century edition of Ptolemy's *Geographia*.
R.V. Tooley, Charles Bricker, and Gerald Crone, *Landmarks of Mapmaking*.
Copyright © 1968 Elsevier Science Publishing Co., Inc., New York, New York.

monsoon winds. Using this knowledge of winds and currents, the Arabs established regular trade routes across the Indian Ocean.

During the Middle Ages, while scholarship about the sea remained static, the knowledge of navigation increased. Harbor-finding charts, or *portolanos,* appeared. These charts carried a mileage scale and noted hazards to navigation, but they did not have latitude or longitude. With the introduction of the magnetic compass to Europe from Asia in the thirteenth century, compass directions were added. A Dutch navigational chart from Johannes van Keulen's *Great New and Improved Sea-Atlas or Water-World* of 1682–84 is shown in Prologue figure IV. The compass course directions follow the pattern used in early fourteenth-century portolanos.

Although tides were not understood, the Venerable Bede (673–735) illustrated his account of the tides with data from the British coast. His calculations were followed in the tidal observations collected by the British Abbot Wallingford of Saint Alban's Monastery in about 1200. His tide table, titled "Flod at London Brigge," documented the times of high water. Sailors made use of Bede's calculations until the seventeenth century.

As scholarship was reestablished in Europe, Arabic translations of early Greek studies were in turn translated into Latin, which made them available to European scholars. The study of tides continued to absorb the medieval scientists, and they were also interested in the saltiness of the sea. By the 1300s Europeans had established successful trade routes, including some partial ocean crossings. An appreciation of the importance of navigational techniques grew as trade routes were extended.

Prince Henry of Portugal founded a school of navigation in 1416 for mariners and craftsmen, where they could develop and improve instruments, charts, and ship designs. Henry himself, however, never went to sea.

Voyages of Discovery

Although people feared the unknown oceans, their desire for the riches from new lands persuaded wealthy individuals, often representing their countries, to underwrite the costs of long voyages to all the oceans of the world. The great age of discovery began when, in 1487, Bartholomeu Dias (1450?–1500) sailed round the Cape of Good Hope into the Indian Ocean, looking for new and faster routes to the spices and silks of the east. Christopher Columbus (1451–1506) made four voyages across the Atlantic Ocean, believing he had found a way to the riches of Cathay or China (Prologue fig. V). Vasco da Gama (c. 1469?–1524) journeyed south and east around the Cape of Good Hope to India, searching for a sea route to the same lands. The Italian navigator Amerigo Vespucci (1454–1512) made several voyages to the New World (1499–1504) for Spain and Portugal, exploring six thousand miles of South American coastline. He accepted South America as a new continent not part of Asia, and in 1507 the German

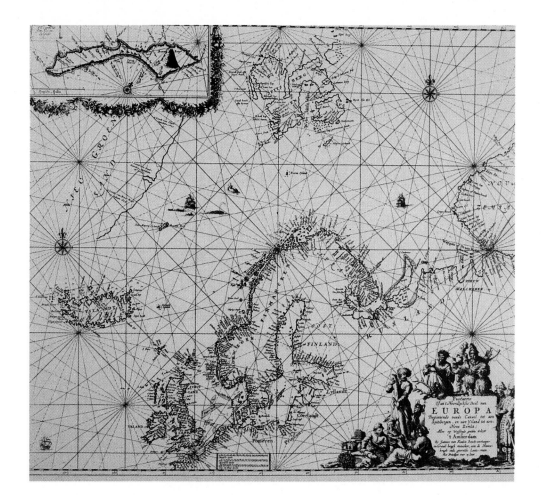

Prologue IV

A navigational chart of northern
Europe from Johannes van
Keulen's *Sea-Atlas* of 1682–84.

R.V. Tooley, Charles Bricker, and
Gerald Crone, *Landmarks of
Mapmaking*. Copyright © 1968
Elsevier Science Publishing Co.,
Inc., New York, New York.

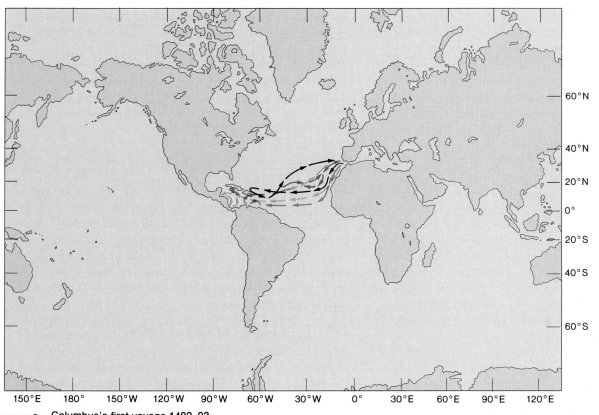

Prologue V

The four voyages of
Christopher
Columbus. Because
of administrative
problems in the new
colony, Columbus
was arrested in
1500 and returned
to Spain as a
prisoner. This
return is not
considered one of
his voyages.

→ Columbus's first voyage 1492–93

→ Columbus's second voyage 1493–96

→ Columbus's third voyage 1498

→ Columbus's fourth voyage 1502–04

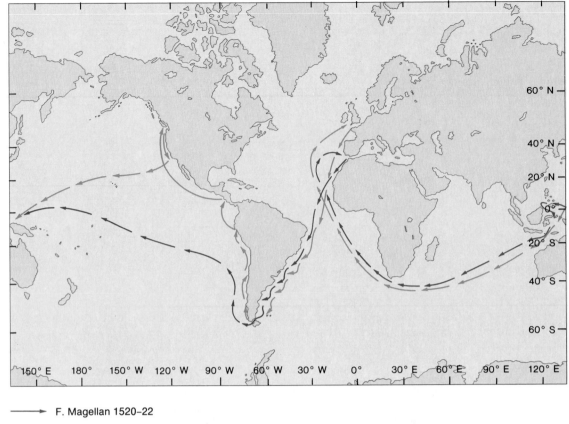

→ F. Magellan 1520–22

→ F. Drake 1577–80

cartographer Martin Waldseemüller applied the name "America" to the continent in Vespucci's honor. Vasco Núñez de Balboa (1475–1519) crossed the isthmus of Panama and found the Pacific Ocean in 1513, and in the same year Juan Ponce de León (1460?–1521) discovered Florida and the Florida current. All claimed the new lands they found for their home countries. Although they had sailed for fame and riches, not knowledge, they more accurately documented the extent and properties of the oceans, and the news of their travels stimulated others to follow.

Ferdinand Magellan (1480?–1521) left Spain in 1519 with five vessels and discovered and passed through the Strait of Magellan rounding South America in 1520. He crossed the Pacific Ocean with three ships in 1520–21. Magellan was killed in the Philippines, and another ship was lost, but the third ship, *Victoria,* made its way across the Indian Ocean and around the Cape of Good Hope, completing the first circumnavigation of the earth (see Prologue fig. VI). Magellan's skill as a navigator makes his voyage probably the most outstanding single contribution to the early charting of the oceans. In addition, during the voyage he established the length of a degree of latitude and measured the circumference of the earth. It is said that Magellan tried to test the midocean depth of the Pacific with a hand line, but this idea seems to come from writings by a nineteenth-century German oceanographer; writings from Magellan's time do not support this story.

By the latter half of the sixteenth-century adventure, curiosity and hopes of finding a trading shortcut to China spurred efforts to find a sea passage around North America. Sir Martin Frobisher (1535?–94) made three voyages in 1576, 1577, and 1578, and Henry Hudson (d. 1611) made four voyages (1607, 1608, 1609, and 1610), dying with his son when set adrift in Hudson Bay by his mutinous crew. The Northwest Passage continued to beckon, and in 1615 and 1616 William Baffin (1584–1622) made two unsuccessful attempts.

While European countries were setting up colonies and claiming new lands, Francis Drake (1540?–96) set out in 1577 to show the English flag around the world (Prologue fig. VI). In 1580 he completed his circumnavigation and returned home in the *Golden Hind* with a cargo of Spanish gold, to be knighted and treated as a national hero. Queen Elizabeth I encouraged her sea captains' exploits as explorers and raiders because, when needed, their ships and their knowledge of the sea brought military victories as well as economic gains.

The Beginnings of Ocean Science

New ideas and new knowledge had stimulated the practical exploration of the oceans during the fifteenth and sixteenth centuries, but most of the thinking about the sea was still rooted in the ideas of Aristotle and Pliny. In the seventeenth century, although the practical needs of commerce,

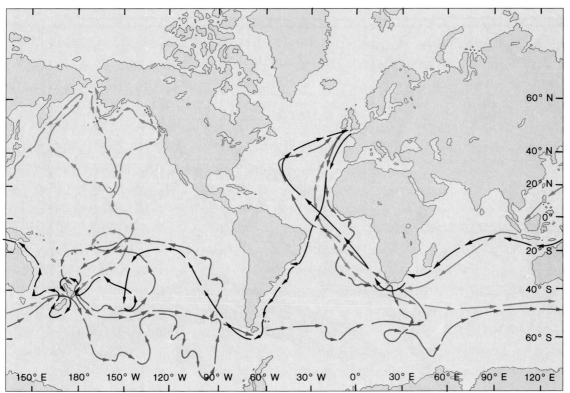

→ Cook's first voyage 1768–71

→ Cook's second voyage 1772–75

→ Cook's third voyage 1776–80

national security, and economic and political expansionism guided events at sea, scientists on land were beginning to show an interest in experimental science and the study of the specific properties of substances. Curiosity about the earth flourished; scientists wrote pamphlets and formed societies in which to discuss their discoveries. The works of Johannes Kepler (1571–1630) on planetary motion and those of Galileo Galilei (1564–1642) on mass, weight, and acceleration would in time be used to understand the oceans. Although both Kepler and Galileo had theories of tidal motion, it was not until Sir Isaac Newton (1642–1727) wrote his *Principia* in 1687, which gave the world the unifying law of gravity, that the processes governing the tides were explained. Edmund Halley (1656–1742), the astronomer of comet fame and a contemporary and friend of Newton, also had an interest in the oceans; he made a voyage to measure longitude and study the variation of the compass in 1698 and suggested that the age of the oceans could be calculated by determining the rate at which rivers carry salt to the sea.

The Importance of Charts and Navigational Information

As colonies were established far away from their home countries, and as trade and travel expanded, there was renewed interest in developing better charts and more accurate navigational techniques. In 1714 Queen Anne of England authorized a public reward for a practical method

of keeping time at sea in order to measure longitude using a method suggested by Gemma Frisius in 1530. The first hydrographic office dedicated to mapping the oceans was established in France in 1720 and was followed in 1795 by the British Admiralty's appointment of a hydrographer.

Captain James Cook (1728–79) made his three great voyages to chart the Pacific Ocean between 1768 and 1779 (Prologue fig. VII). In 1768 he left England in command of the *Endeavour* on an expedition to chart the transit of Venus; he returned in 1771 having circumnavigated the globe and having explored and charted the coasts of New Zealand and eastern Australia. Between 1772 and 1775 he commanded an expedition of two ships, the *Resolution* and the *Adventure,* to the South Pacific, charted many islands, explored the Antarctic Ocean, and by controlling his sailors' diet prevented vitamin C deficiency and scurvy, the disease that had decimated the crews of vessels that spent long periods of time at sea. Cook sailed on his third and last voyage in 1776 in the *Resolution* and *Discovery.* He spent a year in the South Pacific and then sailed north, discovering the Hawaiian Islands in 1778. He continued on to the northwest coast of North America and into the Bering Strait, searching for a passage to the Atlantic. He returned to Hawaii for the winter and was killed by the native islanders at Kealakekua Bay on the island of Hawaii in 1779. Cook takes his place not only as one of history's greatest navigators and seamen but also as a fine scientist. He made soundings to depths of 400 meters (1200 feet) and accurate

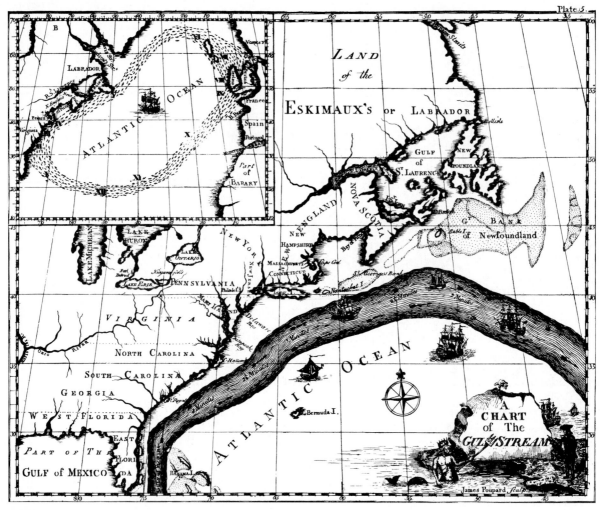

Prologue VIII
The Franklin-Folger map of the Gulf Stream, 1769.

observations of winds, currents, and water temperatures. Cook's careful and accurate observations produced much valuable information and made him one of the founders of oceanography.

In the United States Benjamin Franklin (1706–90) became concerned about the amount of time required for news and cargo to travel between England and America. With Captain Timothy Folger, his cousin and a whaling captain from Nantucket, he constructed the 1769 Franklin-Folger chart of the Gulf Stream current (Prologue fig. VIII), which, when published, encouraged captains to sail within the Gulf Stream en route to Europe and to avoid it on the return passage. In 1802 Nathaniel Bowditch (1773–1838), another American, published the *New American Practical Navigator*. In this book Bowditch made the techniques of celestial navigation available for the first time to every competent sailor and set the stage for U.S. supremacy of the seas during the years of the Yankee Clippers. When Bowditch died his copyright was bought by the U.S. Navy, and his book has been printed by the U.S. government since 1867. It continues in print today, its information updated and expanded with each edition, continuing to serve generations of mariners and navigators.

In 1807 the United States Congress, at the direction of President Thomas Jefferson, formed the Survey of the Coast under the Treasury Department, later named the Coast and Geodetic Survey and now known as the National Ocean Survey. The U.S. Naval Hydrographic Office, now the U.S. Naval Oceanographic Office, was set up in 1830. Both were dedicated to exploring the oceans and producing better coast and ocean charts. In 1842 Lieutenant Matthew F. Maury (1806–73), seen in Prologue figure IX, who had worked with the Coast and Geodetic Survey, was assigned to the Hydrographic Office and founded the Naval Depot of Charts. He began a systematic collection of wind and current data from ships' logs. He produced his first wind and current charts of the North Atlantic in 1847. At the 1853 Brussels Maritime Conference, Maury issued a plea for international cooperation in data collection, and from the ships' logs he received he produced the first atlases of sea conditions and sailing directions. His work was enormously useful, and as a result ships sailed more safely and were able to take days off their sailing times between major ports around the world. The British estimated that Maury's sailing directions took thirty days off the passage from the British Isles to California, twenty days off the voyage to

(a)

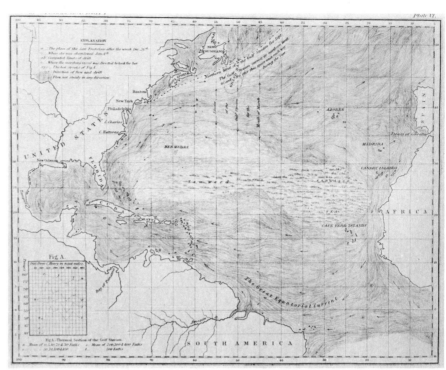

(b)

Prologue IX

(a) Lieutenant Matthew Fontaine Maury of the U.S. Navy, 1806–73. (b) The chart of the Gulf Stream and North Atlantic Ocean surface currents from *Physical Geography of the Sea,* 1855.

Australia, and ten days off the sailing time to Rio de Janeiro. In 1855 he published *The Physical Geography of the Sea.* This work includes chapters on the Gulf Stream, the atmosphere, currents, depths, winds, climates, and storms, and the first bathymetric chart of the North Atlantic with contours at 6000, 12,000, 18,000 and 24,000 feet. (See Prologue fig. IX for the Gulf Stream chart from this book and compare the change in detail and style with the Franklin-Folger chart in Prologue fig. VIII.) Many consider Maury's book the first textbook of what we now call oceanography and consider Maury the first true oceanographer. Again national and commercial interests were the driving forces behind the study of the oceans.

Ocean Science Begins

As charts became more accurate and as information about the oceans increased, the oceans captured the interest of naturalists and biologists. Baron Alexander von Humboldt (1769–1859) made observations on a five-year (1799–1804) cruise to South America; he was particularly fascinated with the vast numbers of animals inhabiting the current flowing northward along the west coast of South America, the current that now bears his name. Charles Darwin (1809–82) joined the survey ship *Beagle* and served as the ship's naturalist from 1831 to 1836 (Prologue fig. X). He described, collected, and classified organisms from the land and sea. His theory of atoll formation is still the accepted explanation. At approximately the same time, another English naturalist, Edward Forbes (1815–54), began a systematic survey of marine life around the British Isles and the Mediterranean and Aegean seas. He collected organisms in deep water and, based on his observations,

proposed a system of ocean depth zones, each characterized by specific animal populations. However, he also mistakenly theorized that there was an azoic or lifeless environment below 550 meters (1800 feet). His announcement is certainly curious, since twenty years earlier the Arctic explorer Sir John Ross (1777–1856), looking again for the Northwest Passage, had taken bottom samples at over 1800 meters (6000 feet) depth in Baffin Bay with a "deep-sea clamm" or bottom grab and had found worms and other animals living in the mud. His nephew, Sir James Clark Ross (1800–62) took even deeper samples from Antarctic waters and noted their similarity to the Arctic species recovered by his uncle. Still, Forbes's systematic attempt to make orderly predictions about the oceans, his enthusiasm, and his influence make him another candidate for a founder of oceanography.

Christian Ehrenberg (1795–1876), the German naturalist, found the skeletons of minute organisms in seafloor sediments and also recognized that the same organisms were alive at the sea surface; he concluded that the sea was filled with microscopic life and that the skeletal remains of these tiny organisms were still being added to the seafloor. The investigation of the minute drifting plants and animals of the ocean was not seriously undertaken until the German scientist Johannes Müller (1801–58) began his work in 1846. He used an improved fine mesh tow net similar to that used by Charles Darwin to collect these organisms, which he examined microscopically. This work was pursued by Victor Hensen (1835–1924), who improved the Müller net, introduced the quantitative study of these minute drifting sea organisms, and gave them the name plankton in 1887. A portion of a plate from an 1899 publication describing these organisms is seen in Prologue figure XI.

Prologue X

The voyage of
H.M.S. *Beagle*,
1831–36.

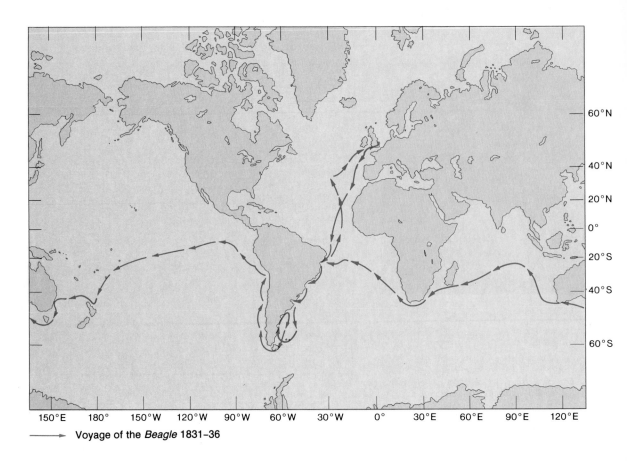

60°N

40°N

20°N

0°

20°S

40°S

60°S

150°E   180°   150°W   120°W   90°W   60°W   30°W   0°   30°E   60°E   90°E   120°E

Voyage of the *Beagle* 1831–36

Prologue XI

This illustration of
microscopic drifting animals
is from volume 25 of *Fauna
and Flora des Golfes von
Neapel (Fauna and Flora of
the Gulf of Naples)*, published
in 1899.

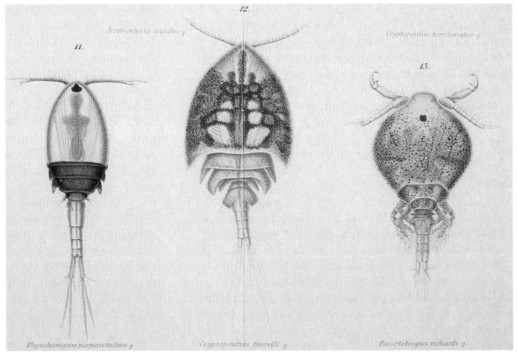

Although science blossomed in the seventeenth and eighteenth centuries, there was little scientific interest in the sea except as we have seen for the practical reasons of navigation, tide prediction, and safety. In the early nineteenth century ocean scientists were still few and usually only temporarily attracted to the sea. Some historians believe that the subject and study of the oceans were so vast, requiring so many people and such large amounts of money, that it required the interest and support of government before the growth of oceanography as a science could occur. This did not happen until the nineteenth century in Great Britain.

In the last part of the nineteenth century the laying of transatlantic telegraph cables made a better knowledge of the deep sea a necessity. Engineers needed to know about seafloor conditions, including bottom topography, currents, and organisms that might dislodge or destroy the cables. The British began a series of deep-sea studies stimulated by the retrieval of a damaged cable from more than 1500 meters (5000 feet) deep, well below Forbes's azoic zone. When the cable was brought to the surface it was found to be covered with organisms, many of which had never been seen before. In 1868 the *Lightning* dredged between Scotland and the Faroe Islands at depths of 915 meters (3000 feet) and found many animal forms. The British Admiralty continued these studies with the *Porcupine* during the summers of 1869 and 1870, dredging up animals from depths of more than 4300 meters (14,000 feet). Wyville Thomson (1830–82), like Forbes a professor of natural history at Edinburgh University, was one of the scientific leaders of these two expeditions, and based on their results he wrote *The Depths of the Sea,* published in 1873, which became very popular and is regarded by some as the first book on oceanography.

The Challenger *Expedition*

With public interest running high, the Circumnavigation Committee of the British Royal Society was able to persuade the British Admiralty to organize the most comprehensive single oceanographic expedition ever undertaken. The Society obtained the use of the naval corvette *Challenger,* a sailing vessel with auxiliary steam power. All but two of the corvette's guns were removed, and the ship was refitted with laboratories, winches, and equipment, including 144 miles of sounding rope. The leadership was offered to Wyville Thomson and his assistant was a young geologist, John Murray (1841–1914). The *Challenger* sailed from Portsmouth, England, on December 21, 1872, for a voyage that was to last nearly three-and-a-half years, during which time the vessel logged 110,840 kilometers (68,890 miles) (see Prologue fig. XII). The first leg of the voyage took her to Bermuda, then to the South Atlantic island of Tristan da Cunha, around the Cape of Good Hope, and east across the southernmost part of the Indian Ocean, the first steamship to cross the Antarctic Circle. She continued on to Australia, New Zealand, the Philippines, Japan, and China. Turning south to the Marianas Islands, she took her deepest sounding, 8180 meters (26,850 feet). Sailing

across the Pacific to Hawaii, Tahiti, and through the Strait of Magellan, she returned to England on May 24, 1876. Queen Victoria conferred a knighthood on Thomson, and the *Challenger* expedition was over. The *Challenger* expedition's purpose was scientific research; during the voyage the crew had taken soundings at 361 ocean stations, collected deep-sea water samples, investigated deep-water motion, and made temperature measurements at all depths. Thousands of biological and sea-bottom samples were collected. She brought back evidence of an ocean teeming with life at all depths and opened the way for the era of descriptive oceanography that followed.

Although the *Challenger* expedition ended in 1876, the work of organizing and compiling information continued for twenty years, until the last of the fifty-volume *Challenger Reports* was issued. John Murray (later Sir John Murray) edited the reports after Thomson's death and wrote many of them himself. He is considered the first geological oceanographer. William Dittmar (1833–92) prepared the information on seawater chemistry for the *Challenger Reports.* He identified the major elements present in the water and confirmed the findings of earlier chemists that in a seawater sample the proportion of the major dissolved elements to each other is constant. Oceanography as a modern science is usually dated from the *Challenger* expedition. The *Challenger Reports* laid the foundation for the science of oceanography.

The *Challenger* expedition stimulated other nations to mount ocean expeditions. Although their avowed purpose was the scientific exploration of the sea, in large measure national prestige was at stake. Norway explored the North Atlantic with the *Voringen* in the summers of 1876 to 1878; Germany studied the Baltic and North seas in the SS *Pomerania* in 1871 and 1872 and in the *Crache* in 1881, 1882, and 1884. The French government financed cruises by the *Travailleur* and the *Talisman* in the 1880s. The Austrian ship *Pola* worked in the Mediterranean and the Red seas in the 1890s. The United States vessel *Enterprise* circumnavigated the earth between 1883 and 1886, as did Italian and Russian ships between 1886 and 1889.

Oceanography As Science

During the nineteenth century and the early twentieth century, intellectual interest in the oceans increased. Oceanography was changing from a descriptive science to a quantitative one. Oceanographic cruises now had the goal of testing hypotheses by gathering data. Theoretical models of ocean circulation and water movement were developed. The Scandinavian oceanographers were particularly active in the study of water movement. One of them, Fridtjof Nansen (1861–1930), Prologue figure XIII, a well-known athlete, explorer, and zoologist, was interested in the current systems of the polar seas. This extraordinary man decided to test his ideas about the direction of ice drift in the Arctic by freezing a vessel into the polar ice pack and drifting with it to reach the North Pole. To do so he had to design a special vessel that would be able to survive the great pressure from the ice; the 39-meter (128-foot) wooden

Prologue XII

The *Challenger* Expedition: December 21, 1872–May 24,1876. Engravings from the *Challenger Reports,* volume 1, 1885. (a) "H.M.S. *Challenger*—Shortening Sail to Sound," decreasing speed to take a deep-sea depth measurement. (b) "H.M.S. *Challenger*." (c) "Dredging and Sounding Arrangement on board *Challenger*." Rigging is hung from the ship's yards to allow the use of the over-the-side sampling equipment. A biological dredge can be seen hanging outboard of the rail. The large cylinders in the rigging are shock absorbers. (d) Sieving bottom samples for organisms. (e) "Deep Sea Deposits." This plate shows the shells of microscopic organisms making up different kinds of muds and clays from the floor of the deep sea. (f) "H.M.S. *Challenger* at St. Paul's Rocks," in the equatorial mid-Atlantic. (g) "Zoological Laboratory on the Main Deck." (h) "Chemical Laboratory." (i) A biological dredge used for sampling bottom organisms. Note the frame and skids that keep the mouth of the net open and allow it to slide over the seafloor. (j) The cruise of the *Challenger,* 1872–76, the first major oceanographic research effort.

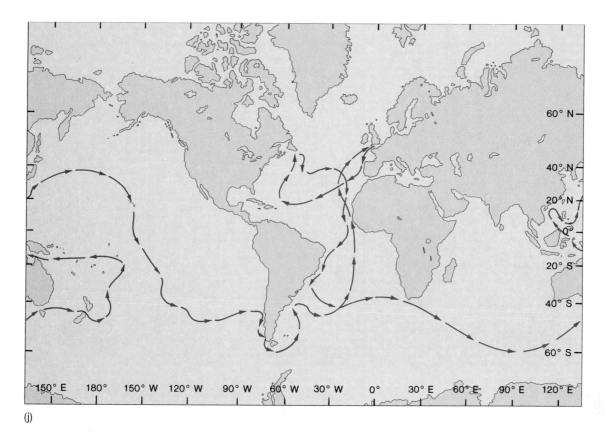

(j)

Fram ("to push forward"), shown in Prologue figure XIII, was built with a smoothly rounded hull and planking over 60 centimeters (2 feet) thick. Nansen with thirteen men departed from Oslo in June of 1893. The ship was frozen into the ice nearly 1100 kilometers (700 miles) from the North Pole and remained there for thirty-five months. During this period measurements were made through holes in the ice that showed that the Arctic Ocean was a deep ocean basin and not the shallow sea that had been expected. Water and air temperatures were recorded, water chemistry was checked, and the great plankton blooms of the area were observed. Nansen became impatient with the slow rate of drift and with F. H. Johansen left the *Fram* locked in the ice some 500 kilometers (300 miles) from the pole. He set off with dogsleds toward the pole, but after four-and-a-half weeks they were still more than 300 kilometers (200 miles) from the pole, with provisions running low and the condition of their dogs deteriorating. The two men turned back and spent the winter of 1895–96 on the ice living on seals and walrus. They were found by another British polar expedition in June of 1896 and returned to Norway in August of that year. The crew of the *Fram* continued to drift with the ship until they freed the vessel from the ice, also in 1896, and returned home. Nansen's expedition had laid the basis for future Arctic work.

After the expedition's findings had been published Nansen continued to be active in oceanography, and his name is familiar today from the Nansen bottle, which he designed to isolate and collect water samples taken from deep water. In 1905 he turned to a career as a statesman working for the peaceful separation of Norway from Sweden. During the years after World War I he worked with the League of Nations to resettle refugees, for which he received the 1922 Nobel Peace Prize. The well-designed *Fram* paid another visit to polar waters, carrying the Norwegian explorer Roald Amundsen (1872–1928) to the Antarctic continent on his successful 1911 expedition to the South Pole. It was also Amundsen who finally made a northwest passage entirely by water in the *Gjoa,* leaving Norway in 1903 and arriving in Nome, Alaska, three years later (Prologue fig. XIV).

Fluctuations in the abundance of commercial fish in the North Atlantic and adjacent seas, and the effect of these changes on national fishing programs, stimulated oceanographic research and international cooperation. As early as 1870, researchers began to realize their need for knowledge of ocean chemistry and physics in order to understand ocean biology. The study of the ocean and its fisheries required the crossing of national boundaries, and in 1902 Germany, Russia, Great Britain, Holland, and the Scandinavian countries formed the International Council for the Exploration of the Sea (ICES) to coordinate and sponsor research in the ocean and in fisheries.

Advances in theoretical oceanography sometimes could not be verified with practical knowledge until new instruments and equipment were developed. Lord Kelvin (1824–1907) invented a tide-predicting machine in 1872 that made it possible to combine tidal theory with astronomic predictions to produce predicted tide tables. Deep-sea circulation could not be systematically explored until approximately 1910, when Nansen's water sampling bottles were combined with thermometers designed for deep-sea

(a)

(b)

Prologue XIII

(a) Fridtjof Nansen, Norwegian scientist, explorer, and statesman (1861–1930), using a sextant to determine his ship's position. (b) The *Fram* frozen in the ice. As the ice pressure increased, it lifted her specially designed and strengthened hull so that she escaped being crushed.

(a)

(b)

Prologue XIV

(a) Roald Amundsen, Norwegian explorer of the Arctic and Antarctic (1872–1928). (b) The *Gjoa* preparing for her journey through the Northwest Passage (1903).

Prologue XV

The U.S. Fish Commission steamer *Albatross* was built in 1882, the first
vessel built especially for marine research by any government and the first
ship fitted throughout with electric lights.

temperature measurements, and an accurate method for determining salinity was devised by the chemist Martin Knudsen (1871–1949). The reliable and accurate measurement of ocean depths had to wait until the development of the echo sounder, which was given its first scientific use on the 1925–27 German cruise of the *Meteor*. Interestingly, although the *Meteor* expedition was supposedly sent out for purely oceanographic reasons, it was also an attempt by the German government to find an affordable way to separate dissolved gold from seawater. While the expedition failed to find a cheap way to produce gold, it did accumulate a great deal of information about the South Atlantic.

U.S. Oceanography Moves into the Twentieth Century

In the United States, government agencies related to the oceans proliferated during the nineteenth century. These agencies were concerned with gathering information to further commerce, fisheries, and the Navy. After the Civil War, the replacement of sail by steam lessened government interest in studying winds and currents and in surveying the ocean floor. Private institutions and wealthy individuals took over the support of oceanography in the United States. Alexander Agassiz (1835–1910), mining engineer, developer of copper

mines, and marine scientist, financed a series of expeditions that greatly expanded knowledge of deep-sea biology. Agassiz served as the scientific director on the first ship built especially for scientific ocean exploration, the U.S. Fish Commission's *Albatross* (Prologue fig. XV), commissioned in 1882. He designed and financed much of the deep-sea sampling equipment that allowed the *Albatross* to bring up more specimens of the deep-sea fishes in one haul than the *Challenger* had collected during her entire three-and-a-half years at sea. Also, in the first twenty years of this century, the Carnegie Institution funded a series of exploratory cruises, including investigations of the earth's magnetic field (Prologue fig. XVI), and maintained a biological laboratory. In 1927 the National Academy of Sciences established its first committee on oceanography, and in 1930 the Rockefeller Foundation allocated funds to stimulate programs and construct laboratories for marine research. Oceanography in the United States began to move onto university campuses. In order to teach oceanography, the subject material had to be consolidated, and in 1942, *The Oceans* by Harald U. Sverdrup, Martin W. Johnson, and Richard H. Fleming was published. It captured between its covers nearly all the world's knowledge of oceanographic processes and was used to train a generation of ocean scientists.

Prologue XVI

The *Carnegie,* built in 1909 by the Carnegie Institute of Washington, was constructed of nonmagnetic materials for use in mapping magnetic forces over the oceans. She was lost in 1929 at Samoa in a gasoline explosion.

World War II

Oceanography mushroomed during World War II, when practical problems of military significance had to be solved quickly. The United States and its allies needed to move men and materials by sea to remote locations, to predict ocean and shore conditions for amphibious landings, to know how explosives behaved in seawater, to chart beaches and harbors from aerial reconnaissance, and to find and destroy submarines. Academic studies ceased as oceanographers pooled their knowledge in the national effort.

Ocean Science Expands in the United States

After the war, oceanographers returned to their classrooms and laboratories with an array of new, sophisticated instruments, including radar, improved sonar, automated wave detectors, and temperature-depth recorders. They also returned with large-scale government funding for research and education. The earth sciences in general and oceanography in particular blossomed during the 1950s. The numbers of scientists, students, educational programs, research institutes, and professional journals all increased. The Office of Naval Research (ONR) funded applied research programs and research vessels; the National Science Foundation (NSF) underwrote basic research; the Atomic Energy Commission (AEC) financed ocean work at the South Pacific atoll sites of atomic tests. During the 1950s the Coast and Geodetic Survey expanded its operations and began its seismic sea wave warning system. In 1965 a major reorganization of governmental agencies occurred; the Environmental Science Services Administration (ESSA) was formed by consolidating the Coast and Geodetic Survey and the Weather Bureau among others. Under ESSA, federal environmental research institutes and laboratories were established, and satellites and their data became a major area of emphasis. In 1970 the U.S. government reorganized its earth science agencies once more. The National Oceanic and Atmospheric Administration (NOAA) was formed under the Department of Commerce. NOAA combines the National Ocean Survey and a fleet of research vessels (Prologue fig. XVII), National Weather Service, National Marine Fisheries Service, Environmental Data Service, National Environmental Satellite Service, and Environmental Research Laboratories. NOAA also administers the National Sea Grant College Program. NOAA has a broad mandate in many fields relating to the overall objectives of the Department of Commerce—"to aid in the

Prologue XVII
The NOAA research vessel *Discoverer*.

Oceanography of the Recent Past

International cooperation brought about the 1957–58 International Geophysical Year (IGY) program, in which sixty-seven nations cooperated to explore the seafloor and made discoveries that completely revolutionized geology and geophysics. In 1963–64 another multinational endeavor, the Indian Ocean Expedition, was in progress. This was followed by the ten-year International Decade of Ocean Exploration (IDOE) in the 1970s, a multinational effort to survey seabed mineral resources, improve environmental forecasting, investigate coastal ecosystems, and modernize and standardize the gathering and use of marine data.

The decade of the 1960s brought giant strides in programs and equipment. As a direct result of the IGY exploration program, special research vessels and submersibles were built to be used by both federal agencies and university research programs. The Deep Sea Drilling Program, a cooperative venture between research institutions and universities, began to sample the earth's crust beneath the sea (see Prologue fig. XVIII). Electronics developed for the space program were applied to ocean research. Computers went on board research vessels, and for the first time data could be sorted, analyzed, and interpreted at sea; experiments could be adjusted while in progress. Government funding allowed large-scale ocean experiments; fleets of oceanographic vessels representing many institutions and nations studied ocean chemistry, water motion, and air-sea interaction (see Prologue fig. XIX).

Oceanography in the 1970s faced a reduction in funding for ships and basic research; nonetheless, the discovery of deep-sea hot water vents and their associated animal life and mineral deposits renewed the excitement over deep-sea biology, chemistry, geology, and ocean exploration in general. Instrumentation continued to become more sophisticated and expensive as deep-sea moorings, deep-diving submersibles, and the remote sensing of the ocean by satellite became possible. Cooperation among institutions increased, leading to the integration of research at sea between subdisciplines and resulting in large-scale, multi-faceted research programs. While collection of oceanographic data expanded at sea, data collected by satellites in the 1970s and 1980s increasingly presented researchers with the ability to observe sea-surface changes in space and time. Currents, eddies, plant production, sea level, waves, thermal properties, and air-sea interactions were all monitored, allowing scientists to compare theoretical models with actual phenomena.

industrial revitalization or economic growth of the nation by increasing exports and the U.S. competitive position in the world market."

Prologue XVIII

The *Glomar Challenger*, the Deep Sea Drilling Program drill ship used from 1968 to 1983.

Prologue XIX

The research vessel *Atlantis II* operated by the Woods Hole Oceanographic Institute.

During these years of expanding programs, earth scientists began to recognize the signs of global degradation and the need for policy and management of living and nonliving resources. Students were attracted to the marine sciences in record numbers, and marine policy programs and ocean management courses were added to curricula. As more and more nations were turning to the sea for food and as technology was increasing our ability to harvest the sea, problems of resource ownership, dwindling fish stocks, and the need for fishery management had to be faced.

The Present and the Future

The scientific targets for investigation in the 1990s include the effect of ocean circulation on the earth's climate balance, the management of living and nonliving resources, the transport of materials from the land to the deep ocean basins, the chemistry of the interaction of seawater with the earth's crust, the dynamics of the continental margins and the ocean seabed, the energy sources of the sea, the exchange of gases between the oceans and the atmosphere, methods of decreasing the cost of ocean transport, and increasing food availability. At the same time, scientists using refined sensors and techniques will continue to seek answers to the basic questions of how and why ocean processes occur and the relationship of these processes to sea resources and to ourselves.

Scientists increasingly recognize the earth as a complex of systems and subsystems acting as a whole. Driving the unified study of global change are (1) the recognition that individual sciences need to cross disciplines in order to advance, (2) the recognition that more-sophisticated satellite sensing has the potential to move the sciences toward an understanding of the total earth, and (3) the worldwide concern over accelerated changes in the earth's environment due to human causes. The success of an integrated approach to earth studies will require that governments, agencies, universities, and national and international programs set priorities and agree on a joint program of global study. In 1986 the General Assembly of the International Council of Scientific Unions (ICSU) endorsed the outline of an international program on global change. At present twenty nations have formed national committees in support of the International Geosphere-Biosphere Program. The ICSU effort begins in the early 1990s and will continue into the next century.

The United States is responding to this new emphasis on global oceanography under the multiagency U.S. Global Changes Research Program. To better understand the role of the oceans in processes of the atmosphere-ocean-land system, a number of oceanographic programs have been developed. These programs are providing data for models that scientists use to predict the evolution of the earth's environment as well as the consequences of human-driven changes.

To better understand global ocean circulation, the World Ocean Circulation Experiment (WOCE) plans a decade-long study of the ocean using computer models and chemical tracers to model the oceans' present state and predict its evolution in relation to long-term changes in the atmosphere. This effort combines sampling by ship, floating independent sensors, and satellites. Much of the work is being done in little-studied areas of the South Pacific and Indian oceans. The U.S. Joint Global Ocean Flux Study (JGOFS) is studying the relationship between ocean plant production and solar radiation. Scientists are monitoring the worldwide abundance of plant life by ship and satellite to understand how carbon and other biologically active elements move between the ocean, atmosphere, and land margin. The Tropical Ocean and Global Atmosphere (TOGA) program studies the energy transferred to the atmosphere by the tropical oceans. From this will come a better understanding of El Niño and its effects as well as improved large-scale climate prediction. Data from these programs will help the Global Ocean Ecosystem Dynamic (GLOBEC) program to understand the response of marine plant and animal populations to changes in ocean circulation and chemistry. Continued exploration of the world's underwater mountain range systems with their associated exchanges of energy, as well as the biological, physical, and chemical consequences, are the missions of the Ridge Interdisciplinary Global Experiment, or RIDGE program.

Although cooperative ventures have achieved much and will achieve much more, it is important to remember that studies driven by the specific research interests of individual scientists are essential to point out new directions for oceanography and the other earth sciences. In the following chapters you will learn more about scientists with such ideas, and you will follow the development of the ideas that have allowed us to build an understanding of the dynamic and complex systems that are the earth's oceans.

Scattered across the seafloor, shipwrecks hold a great storehouse of information about human life in ancient times. Today's archaeologists use techniques developed for oceanographic research to find, explore, recover, and preserve wrecks and other artifacts lying under the sea. Sound beams that sweep the seafloor produce images that are viewed on board ship to locate an object of interest. Once the initial contact has been made, divers can be sent down in shallow water; in deeper water, the wreck can be located very precisely with computer-controlled positioning systems, and a research submersible (submarine) carrying observers can be dispatched to verify the find. Unmanned, towed camera and instrument sleds or remotely operated vehicles (ROVs) equipped with underwater video cameras explore in deep water or in areas that are difficult or unsafe for divers and submersibles. ROVs and submersibles also collect samples to help identify wrecks.

The oldest known shipwreck, from the fourteenth century B.C., a Bronze Age merchant vessel, was discovered in 1983 more than 33 meters (100 feet) down in the Mediterranean Sea off the Turkish coast. Divers have recovered thousands of artifacts from its cargo, including copper and tin ingots, pottery, ivory, and amber. Using these items archaeologists have been able to learn about the life and culture of the period, trace the ship's trade route, and understand more about Bronze Age people's shipbuilding skills.

The waters around northern Europe have claimed thousands of wrecks over more than two thousand years of wars and trading. Two of these vessels, found in shallow water, have provided quantities of information for historians and archaeologists. The Swedish warship *Vasa* sank in 1628 in Stockholm Harbor at the beginning of her maiden voyage. Raised in 1961, she is one of the few complete ships recovered. The English vessel *Mary Rose* sank in the summer of 1545 as she sailed out to engage the French fleet. She was studied in place and then raised in 1982. More than 17,000 objects were salvaged from her, giving

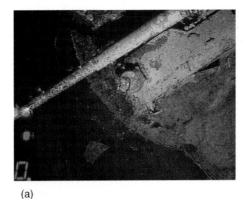

(a)

(b)

Box figure P.I.

The World War II German battleship *Bismarck* was sunk in 1941 and lies 4750 m (15,600 ft) below the surface in the North Atlantic. In 1989, the towed camera sled *Argo* photographed sections of the wreck: (a) a 4.1 inch antiaircraft gun, (b) part of the ship's super structure.

archaeologists and naval historians insights into the personal as well as the working lives of the officers and crew of a naval vessel at this time.

Wrecks that lie in deep water are initially much better preserved than those in shallow water because these deep-water wrecks lie below the depths of the strong currents and waves that break up most shallow-water wrecks. Shallow-water wrecks are also often the prey of treasure hunters who destroy the history of the site while they search for adventure and items of market value. Over long periods the cold temperatures and low oxygen content of deep water favor preservation of wooden vessels by slowing decomposition and excluding the marine organisms that bore into wood in shallow areas.

Also the rate at which muds and sands falling from above cover objects on the deep seafloor is much slower than the same processes in shallow water.

Robert Ballard of the Woods Hole Oceanographic Institute has led two expeditions to find more modern vessels sunk in the very deep sea. One of the world's most famous wrecks, the *Titanic,* which sank in 1912 after striking an iceberg on her maiden voyage, was located 4000 meters (13,000 feet) down in the North Atlantic in 1985. The contact was made by a towed underwater camera sled, computer controlled from the vessel above. Observers descended via submersible for direct observation and further inspection of the vessel using a ROV that could be maneuvered to take pictures both inside and outside the wreck. The *Titanic* has been left as is to serve as a memorial to the 1522 passengers and crew killed in the disaster. A more recent wreck, the World War II German battleship *Bismarck,* was also located using an underwater camera sled. She was sunk in 1941 after one of the most famous sea battles of World War II. In 1989 the *Bismarck* was located, after a search of 120 square miles of ocean, 4750 meters (15,600 feet) below the surface of the North Atlantic. A photographic survey documented both battle damage and destruction due to the sinking (see Box fig. P.I.). These expeditions show that with today's oceanographic instrumentation all shipwrecks are now available to archaeologists.

A number of sunken shore towns have been explored. A major find for marine archaeologists was the remains of a city and harbor built by Herod the Great, King of Judaea from 37 to 4 B.C. This city, Caesarea Maritima, was built between present-day Haifa and Tel Aviv on the Mediterranean Sea. Archaeological studies began in 1978 and have documented the considerable engineering and construction skills required to build the harbor's complex breakwater system and an associated sluice system to flush the harbor of silt.

Study Questions

1. Eratosthenes estimated the circumference of the earth at approximately 25,000 miles. Compare this estimate to the circumference used by Ptolemy. What difference would it have made to later voyages of discovery if Eratosthenes' measurement had been used rather than that of Ptolemy?
2. Who first assigned the name America to the "New World"? For whom was it named?
3. Why was there such great interest in finding and establishing a northwest passage?
4. Who first understood the tides and published an explanation of them?
5. What were Captain James Cook's contributions to our understanding of the oceans?
6. Why did Benjamin Franklin consider it so important to chart the Gulf Stream current?
7. Who was Matthew F. Maury and why is he considered by many to be the "founder of oceanography"?
8. Why do you think that Edward Forbes concluded there was no life in the oceans below 1800 feet?
9. What did the engineers who laid the first transatlantic cable need to know about the oceans?
10. The *Challenger* and its expedition are often called unique. Why is this term used? What were the benefits of this expedition to the science of the oceans?
11. What was Fridtjof Nansen trying to prove by freezing the *Fram* into the polar ice?
12. The amount of ocean data has been expanding at an ever-increasing rate since the first years of ocean exploration. Why?
13. How has each of the following affected twentieth-century oceanography? (a) Economics; (b) Commerce and transportation; (c) Military needs.
14. In what ways have computers altered oceanography?
15. What are the reasons for the increased global interest in resources of the sea? What types of management do you think may be required for these resources in the future?

Suggested Readings

Bailey, H. S. 1953. The Voyage of the Challenger. In *Ocean Science*, readings from *Scientific American* 188 (5): 8–12.

Ballard, R. D. 1986. A Long Last Look at *Titanic*. *National Geographic* 170 (6): 698–727.

Ballard, R. D. 1989. Finding the Bismarck. *National Geographic* 176 (5): 622–37.

Ballard, R. D. 1985. How We Found *Titanic*. *National Geographic* 168 (6): 698–722.

Bass, G. F. 1987. Oldest Known Shipwreck Reveals Splendors of the Bronze Age. *National Geographic* 172 (6): 693–734.

Brosse, J. 1983. *Great Voyages of Discovery, Circumnavigators and Scientists, 1764–1843*. Facts on File Publications, New York. 232 pp.

Charnock, H. 1973. H.M.S. Challenger and the Development of Marine Science. In *Oceanography, Contemporary Readings in Ocean Sciences*, 2d ed. Pirie, R. G., ed. (1977). Oxford Univ., New York. pp. 24–33.

Deacon, M. 1971. *Scientists and the Sea 1650–1900, A Study of Marine Science*. Academic Press, New York. 445 pp.

Deacon, M., ed. 1978. *Oceanography, Concepts and History*. Dowden, Hutchinson & Ross, Stroudsburg, Pa. 394 pp. (Source book of milestone papers in facsimile form.)

Gerard, S. 1992. The Caribbean Treasure Hunt. *Sea Frontiers* 38 (3): 48–53.

Hohlfelder, R. L. 1987. Herod's City on the Sea. *National Geographic* 171 (2): 261–80.

Marx, R. F. 1990. In Search of the Perfect Wreck. *Sea Frontiers* 36 (5): 46–51. New techniques in marine archaeology.

Oceanus. Winter 1985–1986, 28 (4). (Issue on discovery of the *Titanic*.)

Oceanus. Winter 1990–1991, 33 (4). (Issue devoted to naval oceanography.)

Oceanus. Spring 1991, 34 (1). (Issue devoted to ocean engineering and technology.)

Oceanus. Summer 1991, 34 (2). (Issue devoted to Soviet-American cooperation.)

Schlee, S. 1973. *The Edge of an Unfamiliar World, A History of Oceanography*. Dutton, New York. 398 pp.

Sears, M., and D. Merriman, eds. 1980. *Oceanography: The Past*. Springer-Verlag, New York. 812 pp.

Wachsmann, S. 1990. Ships of Tarshish to the Land of Ophir. *Oceanus* 33 (1): 70–82. (Biblical seafaring in the Mediterranean Sea.)

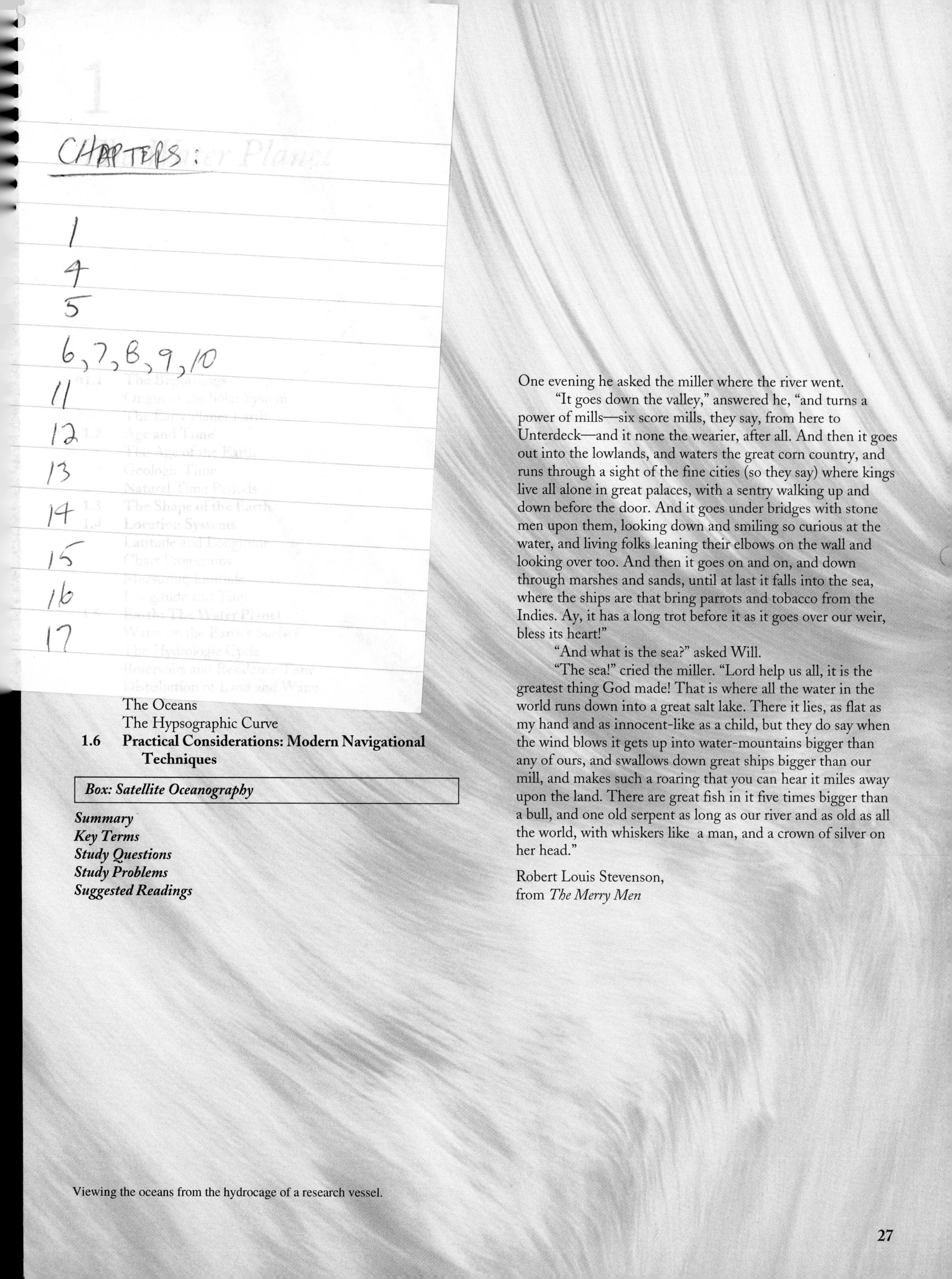

The Oceans
The Hypsographic Curve

**1.6   Practical Considerations: Modern Navigational
Techniques**

Box: Satellite Oceanography

Summary
Key Terms
Study Questions
Study Problems
Suggested Readings

One evening he asked the miller where the river went.

"It goes down the valley," answered he, "and turns a power of mills—six score mills, they say, from here to Unterdeck—and it none the wearier, after all. And then it goes out into the lowlands, and waters the great corn country, and runs through a sight of the fine cities (so they say) where kings live all alone in great palaces, with a sentry walking up and down before the door. And it goes under bridges with stone men upon them, looking down and smiling so curious at the water, and living folks leaning their elbows on the wall and looking over too. And then it goes on and on, and down through marshes and sands, until at last it falls into the sea, where the ships are that bring parrots and tobacco from the Indies. Ay, it has a long trot before it as it goes over our weir, bless its heart!"

"And what is the sea?" asked Will.

"The sea!" cried the miller. "Lord help us all, it is the greatest thing God made! That is where all the water in the world runs down into a great salt lake. There it lies, as flat as my hand and as innocent-like as a child, but they do say when the wind blows it gets up into water-mountains bigger than any of ours, and swallows down great ships bigger than our mill, and makes such a roaring that you can hear it miles away upon the land. There are great fish in it five times bigger than a bull, and one old serpent as long as our river and as old as all the world, with whiskers like a man, and a crown of silver on her head."

Robert Louis Stevenson,
from *The Merry Men*

Viewing the oceans from the hydrocage of a research vessel.

illions of shimmering masses of stars, known as galaxies,
B *move through the space we call the universe. About one-
third of the way toward the center of one whirling mass of some
100 billion stars called the Milky Way galaxy there is a fairly
ordinary star, the sun. Around the sun move nine specks, or
planets, following predictable and nearly constant orbits. The
sun and its nine planets are called the solar system, and the third
planet from the sun is called earth (fig. 1.1). Although we call
our planet earth, when we consider the qualities of this planet
that set it apart from the other planets in the solar system, it is
more accurate to consider it as the water planet. Water did not
exist on the earth in the beginning, but its formation on a planet that was not too far from the sun nor too close, not too hot nor
too cold, changed the earth and allowed the development of life.
In this chapter we begin at the solar system's beginning to
investigate the earth, the water planet, on which the largest
bodies of water are known as the oceans.*

1.1 The Beginnings

Origin of the Solar System

Present theories attribute the beginning of our solar system
to the collapse of a rotating interstellar cloud of gas and
dust about 4.6 billion years ago. As the cloud collapsed, its
speed of rotation increased and, heated by its own gravitational energy, its temperature rose. The gas and dust, spinning faster and faster, contracted parallel to the axis of spin,
forming a disk. At the center of the disk a star, our sun, was
formed. Self-sustaining nuclear reactions kept the sun hot,
but the outer regions began to cool, and in this cooler outer
portion of the rotating disk molecules of gas began to collide and chemically interact. These collisions and interactions produced particles of matter, which grew from collisions with other particles and became large enough to have
sufficient gravity to attract still other particles. The planets
of our solar system had begun to form. After a few million
years the sun was orbited by nine planets (in order from the
sun): Mercury, Venus, Earth, Mars, Jupiter, Saturn, Uranus,
Neptune, and Pluto.

If Mercury, Venus, Earth, and Mars are compared
with Jupiter, Saturn, Uranus, and Neptune, the four planets
closer to the sun are seen to be much smaller in diameter
and mass. (See table 1.1. Note use of metric units; see
Appendix for further information.) These four inner planets
are rich in metals and rocky materials. The four outer planets are cold giants, dominated by ices of water, ammonia,
and methane. Their atmospheres are made up of helium and
hydrogen; the planets located nearer the sun lost these
lighter gases because the higher temperature and intensity
of solar radiation tends to push these gases out and away

from the center of the solar system. If the mass of each
planet in table 1.1 is divided by its volume, the results will
show that the outer planets are composed of lighter or less
dense materials than the inner planets.

Pluto, a little-known small planet at least five hundred
times less massive than the earth, has an elliptical orbit that
takes it inside the orbit of Neptune. It has a reflective surface that may be primarily composed of frozen methane.
Because of its unusual orbit, it has been suggested that Pluto
might at one time have been a satellite of Neptune.

The Early Planet Earth

Before the beginning of the geologic record, during the first
billion years of the earth's existence, the earth is thought to
have been a mixture of silicon compounds, iron and magnesium oxides, and small amounts of other naturally occurring elements. According to this model, the earth formed
originally from cold matter, but events occurred that raised
the earth's temperature and initiated processes that obliter-
ated its earlier history and resulted in its present form. The
early earth was bombarded by particles of all sizes, and a
portion of their energy was converted into heat on impact.
Each new layer of accumulated material buried the material
below it, trapping the heat and raising the temperature of
the earth's interior. At the same time, the growing weight
of the accumulating layers compressed the interior, and the
energy of compression was converted to heat, raising the
earth's internal temperature to approximately 1000°C.
Atoms of radioactive elements, such as uranium and thori-
um, disintegrated by emitting subatomic particles that were
absorbed by the surrounding matter, further raising the
temperature.

Some time during the first few hundred million years
after the earth formed, its interior reached the melting point
of iron and nickel. When the iron and nickel in the planet
melted, they migrated toward the center. Frictional heat
was generated and lighter substances were displaced. In
this way the temperature of the earth was raised to an average of 2000°C. Material from the partially molten interior
moved upward and spread over the surface, cooling and
solidifying. The melting and solidifying probably happened
repeatedly, separating the lighter, less dense compounds
from the heavier, more dense substances in the interior of
the planet. In this way the earth became completely reorganized and differentiated into a layered system that will be
explored in greater detail in chapter 3.

The earth's oceans and atmosphere are probably both
by-products of this heating and differentiation. As the earth
warmed and partially melted, water locked in the minerals
as hydrogen and oxygen was released and carried to the surface as water vapor mixed with other gases. As the earth's
surface cooled, the water condensed to form the oceans.

At first the earth must have been too small and had
too little gravity to have accumulated an atmosphere. It is
generally believed that during the process of differentiation, gases released from the earth's hot, chemically active
interior formed the first atmosphere, which was primarily

Figure 1.1

The water planet earth as seen from space.

Table 1.1

Features of the Planets in the Solar System

Planet	Mean distance from sun (10^6 km)	Diameter (km)	Mass relative to earth mass	Rotation period[1] (hours, days)	Orbit period (years)	Mean temperature of surface (°C)	Principal atmospheric gases[2]
Mercury	57.9	4,878	0.055	58.6 d	0.24	−170 night 430 day	Na
Venus	108.2	12,104	0.815	−243 d	0.62	− 23 clouds 480 surface	CO_2, N
Earth	149.6	12,756	1.000	23.94 h	1.00	16	N, O
Mars	227.9	6,787	0.107	24.62 h	1.88	− 50 (av)	CO_2, N, Ar
Jupiter	778.3	142,800	317.8	9.93 h	11.86	−150	H, He, CH_4, NH_3
Saturn	1429.0	120,000	95.2	10.5 h	29.48	−180	H, He, CH_4, NH_3
Uranus	2875	50,800	14.5	−17.24 h	84.01	−210	H, He, CH_4
Neptune	4504	48,600	17.2	16 h	164.8	−220	H, CH_4, He?
Pluto	5900	2,245	0.002	6.4 d	247.71	−230	CH_4 (temporary)

[1]Negative rotation period indicates rotation opposite to orbit direction about the sun.

[2]H = hydrogen; He = Helium; CO_2 = carbon dioxide; N = nitrogen; O = oxygen; Ar = argon; Na = sodium; CH_4 = methane; NH_3 = ammonia.

made up of water vapor, hydrogen gas, hydrogen chloride, carbon monoxide, carbon dioxide, and nitrogen. Any free oxygen present would have combined with the metals of the crust to form compounds such as iron oxide. Oxygen gas could not accumulate in the atmosphere until its production exceeded its loss by chemical reactions with the earth's crust. This did not occur until life evolved to a level of complexity in which green plants could convert carbon dioxide and water with the energy of sunlight into organic matter and free oxygen. This process and its significance to life is discussed in chapters 5 and 14.

1.2   Age and Time

The Age of the Earth

Over the centuries people have asked the question, "How old is the earth?" In the seventeenth century Archbishop Ussher of Ireland attempted to answer the question by counting the generations listed in the Bible; he arrived at 9:00 A.M., October 26, 4004 B.C. for the beginning of the earth. In the late 1800s the English physicist Lord Kelvin calculated the time necessary for molten rock to cool to present temperatures, and dated the earth as twenty million to forty million years old. In 1899 another physicist, John Joly, used the amount of salt in the oceans to determine the earth's age and calculated the time for the rivers to wash the salt from the land to be 100 million years.

It was not until scientists understood radioactive decay, and a method was developed to apply it to the dating of rock samples, that reliable age data became available. This method, known as **radiometric dating,** uses radioactive **isotopes** of certain elements.

An atom of radioactive isotope has an unstable nucleus. This unstable nucleus changes or decays and emits one or more particles plus energy. For example, the isotope carbon-14 decays or changes to nitrogen-14; uranium-235 decays to lead-207; and potassium-40 decays to argon-40. The time at which any single nucleus will decay is unpredictable, but if large numbers of atoms of the same isotope are present it is possible to predict that a certain fraction of the isotope will decay over a certain period of time. The time over which one-half of the atoms of an isotope decays or changes from one element to another element is known as the isotope's **half-life.** The half-life of each isotope is characteristic and constant. For example, the half-life of carbon-14 is 5730 years, uranium-235 is 704 million years, and potassium-40 is 1.3 billion years. Therefore, if a substance is found that was originally made up only of atoms of uranium-235, in 704 million years, by a series of reactions, the substance is one-half uranium-235 and one-half lead-207. In another 704 million years three-quarters of the substance will be lead-207 and only one-quarter uranium-235. Because each isotopic system behaves differently in nature, data must be carefully tested, compared, and evaluated. The best data are those in which different isotopic systems give the same date.

The earth is an active planet and its original surface rocks no longer exist. The oldest rocks on the surface of the earth have been dated at 3.8 billion years old. Moon samples returned by the Apollo mission are dated at 4.2 billion years old. Meteorites that have survived the journey from space through the earth's atmosphere have been dated between 4.5 billion and 4.6 billion years old. The substances found in many meteorites represent materials that condensed out of hot gases thought to be present at the beginning of the solar system. These ages agree with theoretical calculations made for the age of the sun. This information sets the accepted age of the earth at about 4.6 billion years.

Geologic Time

To refer to events in the history and formation of the earth, scientists use geologic time (see fig. 1.2 and table 1.2). The principal divisions are the four eons: the Hadean (3.9 to 4.6 billion years ago), the Archean (3.9 to 2.5 billion years ago), the Proterozoic (2.5 billion to 570 million years ago), and the Phanerozoic (since 570 million years). Fossils are known from other eons but are common only from the Phanerozoic. This eon is divided into three eras: the Paleozoic era of ancient life, the Mesozoic era of middle life (popularly called the Age of Reptiles), and the Cenozoic era of recent life (the Age of Mammals). Each of these eras is subdivided into periods and epochs; today, for instance, we live in the Holocine epoch of the Quaternary period of the Cenozoic era. The appearance or disappearance of fossil types was used to set the boundaries of the time units before radiometric dating allowed scientists to set these time scale boundaries more accurately.

Very long periods of time are incomprehensible to most of us. We often have difficulty coping with time spans of more than ten years—what were you doing exactly ten years ago today? We have nothing with which to compare the 4.6 billion-year age of the earth or the 500 million years since the first **vertebrates** appeared on this planet. In order to place geologic time in a framework we can understand, let us divide the earth's age by one hundred million. If we do so, then we can consider that Mother Earth is forty-six years old, a middle-aged lady in whose past we have a strong and intimate interest. What has happened to her over that forty-six years?

The first seven years of her life leave no record; as she entered her eighth year a few fragments of her history can be read in some rocks of Africa and Greenland. By the time she was 12 years old the first living cells of bacteria-type organisms had appeared. Approximately ten or eleven years later, between ages 22 and 23, oxygen production by living cells began, but it took eight or more years, until about the earth's thirty-first year, to change the atmosphere sufficiently to support the first complex oxygen-requiring cells. The first hard-shelled fossils were laid down only six years ago, in her fortieth year, and by the time she was 41 the first vertebrates had developed. Eight-and-a-half months later the first land plants appeared, quickly followed by an age in which the fish were the characteristic

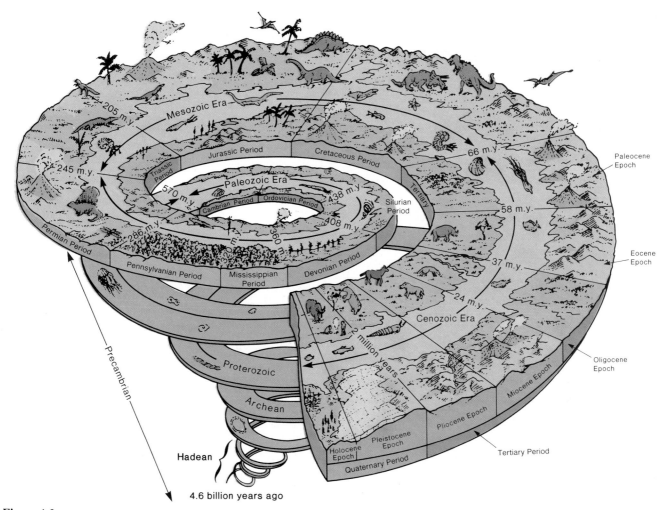

Figure 1.2
The history of the earth showing evolution of life-forms.
After U. S. Geological Survey publication, Geologic Time.

life form. By age 43 the first reptiles had arrived; dinosaurs became abundant at the beginning of her forty-fourth year, disappearing shortly after her forty-fifth birthday. A little over a year ago plants with flowers began to develop; seven months ago the mammals, birds, and insects became the dominant land animals. Our first human ancestors appeared twenty-five days ago, followed two weeks later by the first identifiable member of the genus *Homo*. About half an hour ago modern humans began the long process we know as recorded civilization, and only one minute ago the Industrial Revolution began changing the earth and our relationship to her for all time.

Natural Time Periods

People first defined time by the natural motions of the earth, sun, and moon. Later, people grouped natural time periods for their convenience, and still later, artificial time periods were created for people's special uses. Time is used to determine the starting point of an event, the event's duration, and the rate at which the event proceeds. An accurate measurement of time is required to determine location or position. This is discussed in the Location Systems section of this chapter.

The year is the time required by the earth to complete one orbit about the sun. The time required for this orbit is 365 1/4 days, adapted for convenience to 365 days with an extra day added every four years, except years ending in hundreds and not divisible by 400. As the earth follows its orbit around the sun, those who live in temperate zones and polar zones are very conscious of the seasons and of the differences in the lengths of the periods of daylight and darkness. The reason for these seasonal changes is seen in figure 1.3. The earth moves along its orbit with its axis tilted 23 1/2° from the vertical. Therefore, during the year the earth's North Pole is sometimes tilting toward the sun and sometimes tilting away from it. As it moves along its orbit the Northern Hemisphere receives the maximum hours of sunlight when the North Pole is tilted toward the sun; this is the Northern Hemisphere's summer. During the same period, the South Pole is tilted away from the sun, so that the Southern Hemisphere receives the least sunlight; this period is winter in the Southern Hemisphere. At the other side of the earth's orbit, the North Pole is tilted away from the sun, creating the Northern Hemisphere's winter, and the South Pole is inclined toward the sun, creating the Southern Hemisphere's summer. Note also that during summer in the

Table 1.2
The Geologic Time Scale

Eon	Era	Period	Epoch	Began millions of years ago	Life-forms/Events
Phanerozoic		Quaternary	Holocene	0.01	Modern humans
			Pleistocene	1.6	Stone-age humans First humans
	Cenozoic	Tertiary	Pliocene	5.3	
			Miocene	23.7	
			Oligocene	36.6	Flowering plants
			Eocene	57.8	Mammals, birds, and insects dominant
			Paleocene	66.4	
	Mesozoic	Cretaceous		144	Last of dinosaurs; flowering plants begin
		Jurassic		208	Dinosaurs abundant; first birds
		Triassic		245	First mammals First dinosaurs
	Paleozoic	Permian		286	Age of reptiles
		Carboniferous		360	Age of amphibians; first reptiles
		Devonian		408	First seed plants Age of fishes
		Silurian		438	First land plants
		Ordovician		505	Marine algae; vertebrate fish
		Cambrian		570	Primitive marine algae and invertebrates
Proterozoic		Precambrian		2500	Earliest bacteria and algae
Archean				3900	Oldest surface rocks
Hadean				4600	Oldest meteorites Formation of the earth (assumed)

Northern Hemisphere it is light around the North Pole and dark around the South Pole; the opposite is true during the Northern Hemisphere's winter.

As we are carried along by the spinning earth orbiting the sun, we are not conscious of any movement. What we sense is that the sun rises in the east and sets in the west daily and slowly moves up in the sky from south to north and back during one year. The periods of daylight in the Northern Hemisphere increase as the sun moves north in relation to the earth, to stand 23 1/2° above the equator over the **Tropic of Cancer.** This position occurs at the **summer solstice** on or about June 22, the day with the longest period of daylight and the beginning of summer in the Northern Hemisphere; on this day the sun does not sink below the horizon above 66 1/2° north of the equator, the

Arctic Circle, nor does it rise above 66 1/2° south of the equator, the **Antarctic Circle.** Then the sun appears to move southward until about September 23, the **autumnal equinox,** when it stands directly above the equator. On this day the periods of daylight and darkness are equal all over the world. The sun continues its southward movement until about December 21, when it stands 23 1/2° below the equator over the **Tropic of Capricorn;** this position marks the **winter solstice** and the beginning of winter in the Northern Hemisphere. On this day the daylight period is the shortest in the Northern Hemisphere, and above the Arctic Circle the sun does not rise. In the Southern Hemisphere the reverse is true, and above the Antarctic Circle the sun does not set. The sun then begins to move northward, and about March 21, the **vernal equinox,** it stands again above the

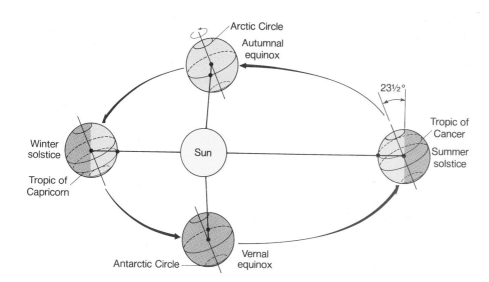

Figure 1.3

The earth's seasons. The fixed orientation of the earth's axis during the earth's orbit of the sun causes different portions of the earth to remain in shadow at different seasons. One year is the time increment between successive vernal equinoxes.

equator; spring begins in the Northern Hemisphere, and the periods of daylight and darkness are once more equal around the world. Follow figure 1.3 around again, checking the position of the South Pole, and note how the seasons of the Southern Hemisphere are reversed from those of the Northern Hemisphere. Note also that the greatest annual variation in the intensity of direct solar illumination occurs in the temperate zones. In the polar regions the seasons are dominated by the long periods of light and dark, but the direct heating of the earth's surface is small as the sun is always low on the horizon. Between the Tropics of Cancer and Capricorn there is little seasonal change in solar radiation levels, as the sun never moves beyond these boundaries.

Weeks and months, as they presently exist, modify natural time periods. It requires 27 1/3 days for the moon to orbit the earth, but a period of 29 1/2 days defines the **lunar month.** In the lunar month the moon passes through four phases: new moon, first quarter, full moon, and last quarter. The four phases approximately match the four weeks of the month. Days are grouped into twelve months of unequal length in order to form one calendar year. The present arrangement is known as the Gregorian calendar after Pope Gregory XIII, who in the sixteenth century made the changes necessary to correct the old Julian calendar, adopted in 46 B.C. and named after Julius Caesar. The Gregorian calendar was adopted in the United States in 1752, by which time the Julian calendar was eleven days in error. In that year, by parliamentary decree, in both Great Britain and the United States, September 2 was followed by September 14, accompanied by riots in which people demanded their eleven days back. Also in 1752, the beginning of the calendar year was changed from the original date of the vernal equinox, March 25, to January 1; 1751 had no months of January and February.

The day is derived from the earth's rotation on its axis as it orbits the sun. The average time for the earth to make one rotation relative to the sun is twenty-four hours; this is the mean **solar day,** our clock day. Another measure of a day is the time required for the earth to make a complete rotation with respect to a far-distant point in space. This is known as the **sidereal day** and is about four minutes shorter than the mean solar day; it gives the true rotational period of the earth. The sidereal day is useful in astronomy and navigation.

Living organisms respond to these natural cycles. In temperate zones flowers bloom and die back; forest trees lose their leaves, enter a period of dormancy, and then produce new leaves and buds as the length of the periods of daylight and darkness change and the temperatures increase or decrease; but tropic forests remain lush year-round. Some animals migrate and alternate periods of activity and hibernation with the seasons. Other animals set their internal clocks to the day-night pattern, hunting in the dark and sleeping in the light; still others do the reverse. Plants and animals of the sea also react to these rhythms, as do the physical processes that move the atmosphere and circulate the water in the oceans. An understanding of these cycles helps us understand processes that occur at the ocean surface and are discussed in later chapters.

1.3 The Shape of the Earth

As the earth cooled and turned in space, gravity and the forces of rotation produced its nearly spherical shape. The earth sphere has a mean, or average, radius of 6371 km (3959 mi). It has a shorter polar radius (6356.9 km; 3950 mi) and a longer equatorial radius (6378.4 km; 3963 mi). This difference of 21.5 km, or about 13 mi, occurs because the earth is not a rigid sphere. As the earth spins it tends to bulge at the equator, much as a ball of potter's clay bulges when spun on a stick (see fig. 1.4). Because the landmasses are presently concentrated in the middle region of the Northern Hemisphere and centered on the South Pole in the Southern Hemisphere, the earth's surface is depressed slightly in these areas, while it is elevated at the North Pole and in the middle region of the Southern Hemisphere. This

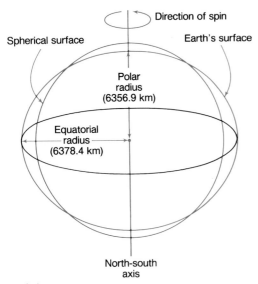

Figure 1.4

The rotation of the earth on its axis stretches the equator. Notice that the equatorial radius is larger than the polar radius.

land distribution causes the earth to have a very slight pear shape, about 15 m (50 ft) between depressions and elevations. The earth is a nearly perfect sphere.

The earth is also quite smooth. The top of the earth's highest mountain, Mount Everest in the Himalayas, is about 8840 km (29,000 ft) above sea level, while the deepest ocean depth, the bottom of the Challenger Deep in the Mariana Trench of the Pacific Ocean, is about 11,000 m (36,000 ft). If these measurements are divided by the mean radius of the earth (6371 km or 20,896,000 ft), the resulting ratios are 0.00139 for the mountain and 0.00173 for the trench. On a scale model of the earth with a radius of 50 centimeters (20 inches), Mount Everest would be about 0.07 cm (.027 in) high and the Challenger Deep would be about 0.086 cm (.034 in) in depth. The earth model's surface would feel rather like the skin of a grapefruit or the surface of a basketball. The earth's high mountains and deep oceans are minor compared to the size of the planet.

1.4 Location Systems

Latitude and Longitude

In order to find our way around on the surface of our planet we need a reference (or location) system. Most of us use such a system daily consisting of the city name, street name or number, and building number. Armed with a city map, we confidently navigate to areas never visited before. Most of the earth's surface, however, is not provided with streets and building numbers, and we must use another system. To determine the location of a position on the earth we use a grid of reference lines that are superimposed on the earth's surface and cross at right angles. These grid lines are called lines of **latitude** and **longitude.** Lines of latitude, also known as **parallels,** begin at the **equator.** The equator is created by passing a plane through the earth halfway

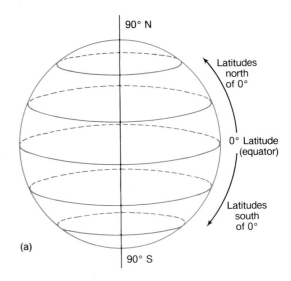

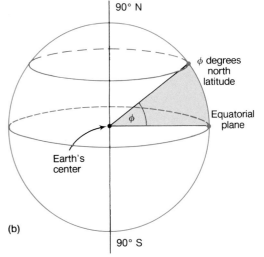

Figure 1.5

(a) Latitude lines are drawn parallel to the equatorial plane. (b) The value of a latitude line is expressed in angular degrees determined by the angle formed between the equatorial plane and the latitude line to the earth's center. This is the angle ϕ (phi). The degree value of ϕ must be noted as north or south of the equator.

between the poles and at right angles to the earth's axis. This process is much like cutting an orange in two pieces halfway between the depressions marking the stem and the navel. The equator is marked at 0° latitude, and other latitude lines are drawn around the earth parallel to the equator, northward to 90°N, or the North Pole, and southward to 90°S, or the South Pole (see fig. 1.5a). Notice that the parallels of latitude describe increasingly smaller circles as the poles are approached. Notice also that all parallels of latitude must be designated as an angle either north or south of the equator. The latitude value is determined by the internal angle (ϕ or phi) between the latitude line, the earth's center, and the equatorial plane (see fig. 1.5b). The previously mentioned Tropics of Cancer and Capricorn correspond to latitudes 23 1/2°N and S, respectively. Latitudes 66 1/2°N and S, respectively, correspond to the Arctic and Antarctic Circles.

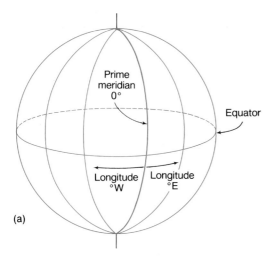

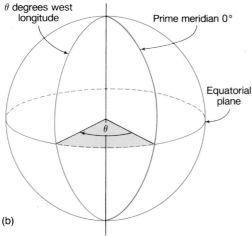

Figure 1.6

(a) Longitude lines are drawn with reference to the prime meridian.
(b) The value of a longitude line is expressed in angular degrees
determined by the angle formed between the prime meridian and the
longitude line to the earth's center. This is the angle θ (theta). The value
of θ is given in degrees east or west of the prime meridian.

Figure 1.7

The Royal Naval Observatory at Greenwich, England. The brass strip set
into the courtyard marks the prime meridian, the division between east and
west longitudes.

Lines of longitude, or **meridians,** are formed at right
angles to the latitude grid (see fig. 1.6a). Longitude begins
at an arbitrarily chosen point: 0° longitude is a line on the
earth's surface extending from the North Pole to the South
Pole and passing directly through the Royal Naval
Observatory in Greenwich, England, just outside London.
The 0° longitude line is shown outside the Greenwich
Observatory in figure 1.7. On the other side of the earth,
180° longitude is directly opposite 0°. The 0° longitude line
is known as the **prime meridian.** The 180° longitude line
approximates the **international date line.** Longitude lines
are identified by their angular displacement (θ or theta) to the
east and west of 0° longitude, as shown in figure 1.6b. Note
in figure 1.6a that meridians are the same size, much like the
lines marking the segments of an orange. The meridians
mark the intersection of the earth's surface with a plane passing through the earth's center at right angles to the parallels
of latitude. Any circle at the earth's surface with its center at
the earth's center is a **great circle.** All longitude lines form

great circles; only the equator is a great circle of latitude. A
great circle connecting any two points on the earth's surface
defines the shortest distance between them.

To identify any location on the earth's surface, we use
the crossing of the latitude and longitude lines; for example,
158°W, 21°N is the location of the Hawaiian Islands, and
20°E, 33°S identifies the Cape of Good Hope at the southern tip of Africa. Since the distance expressed in whole
degrees is large (1 degree of latitude equals 60 nautical
miles), each degree of arc is divided into 60 minutes, each
minute into 60 seconds, and each second into tenths of a
second. One **nautical mile** is equal to one minute of arc
length of latitude or longitude at the equator, or 1.85 km
(1.15 land miles; see Appendix). Using this system, positions on the earth are specified with great accuracy.

Chart Projections

Charts and maps are used to show the earth's three-
dimensional surface on a flat, or two-dimensional, surface. Maps usually portray the earth's land features or

Figure 1.8

The three basic types of map projections: (a) An equatorial cylindrical projection. (b) A simple polar conic projection. (c) A polar tangent plane projection.

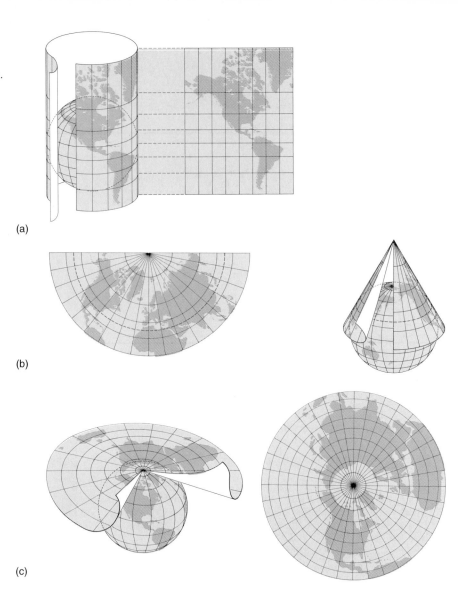

(a)

(b)

(c)

land and sea relationships, while similar displays of the sea and the sky are called charts. Any chart or map produces a distorted image of the curved surface of the earth. The task of the mapmaker, or cartographer, is to produce the most accurate, most convenient, and least distorted picture for the task for which the map or chart is to be used.

Maps are made by projecting the features of the earth, along with the latitude and longitude system, onto a surface; the resulting picture is called a map or chart **projection.** Consider a transparent globe with the continents and latitude and longitude lines painted on its surface. Place a light in the center of the globe and let the light rays shine out through it. The light will project the shadows of the continents and the latitude and longitude lines onto a piece of paper held up to the outside of the globe. Different projections are obtained by varying the position of the light and the type of surface on which the projection is made. Many chart and map projections have been constructed, but most of them are modifications of three basic types: cylindrical, conic, and tangent plane, all shown in figure 1.8. In these projections, the surface that is to be the map is rolled

around the globe as a cylinder (fig. 1.8a), made into a cone (fig. 1.8b), or laid flat (tangent) against the sphere (fig. 1.8c). Although the cylindrical and conic surfaces may be placed around, over, or tangent to the earth at any location, they are usually placed so that the cylinder touches the earth at its equator and the cone is centered on the polar axis.

Compare the three parts of figure 1.8 and notice that distortion on a map or chart increases as the distance from the place of contact to the sphere increases. Consider Greenland. In the tangent plane projection (fig. 1.8c) its size and shape are very close to its true form on the earth's sphere. In the conic projection (fig. 1.8b) the island has grown larger, and in the cylindrical projection (fig. 1.8a) both its size and its shape are greatly distorted. The traditional and familiar world map used in many books and school classrooms is the **Mercator projection,** an adjusted form of the cylindrical type shown in figure 1.8a. Although distortion is great at high latitudes and poles cannot be shown, the Mercator projection, unlike other projections, has the advantage that a straight line as drawn on the map is a line of true direction or constant compass heading, and

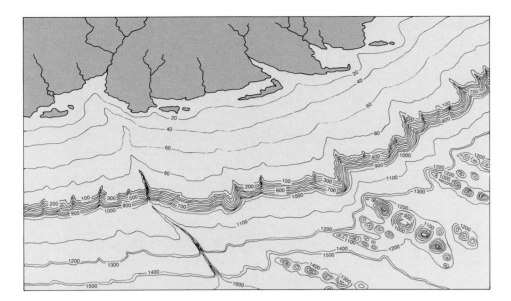

Figure 1.9

A bathymetric, or contour, chart of the seafloor along a section of generalized coast. Changes in the pattern and the spacing between contour lines indicate changes in depth.

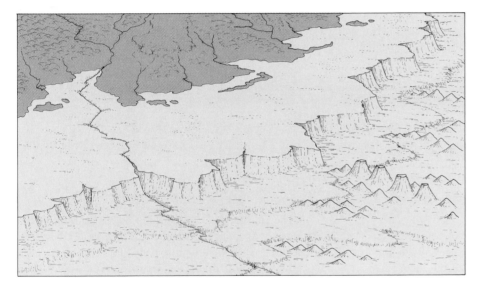

Figure 1.10

A physiographic map of the same area shown in figure 1.9.

therefore the Mercator projection is useful in navigation. Each type of chart or map has its own characteristics. The user must select the projection with the least distortion for his or her purpose.

Maps that show lines connecting points of similar elevation on land (known as **contours** of elevation) are **topographic** maps. Charts of the ocean showing contour lines connecting points of the same depth below the sea surface are **bathymetric** charts (figure 1.9). Color, shading, and perspective drawings may be used to indicate elevation changes, so as to produce a visual representation, or bird's-eye view, of the earth's elevation features. These are **physiographic maps.** See figure 1.10 for an example of a physiographic map, and compare it with the bathymetric chart in figure 1.9.

Measuring Latitude

Early maps show us that the first cartographers and navigators had considerable difficulty in precisely describing and locating the then known earth features. When accurate measurement failed, artistic license appeared to fill the gaps (see fig. 1.11). The major problem was that early navigators were not able to determine their position accurately. As navigational techniques improved, so did the maps. It was known by early navigators that the North Star, **Polaris,** appeared to hang in the sky above the North Pole and did not appreciably move from this spot. In the Northern Hemisphere, therefore, measuring the angle of elevation of Polaris above the horizon gave a good estimate of one's latitude. Once a ship was out of sight of land, it sailed north or south until reaching the desired latitude, then sailed east or west along this line of latitude until again reaching land. Adjustments north and south to reach the desired landfall of the voyage's end were made close to shore and according to visible landmarks.

Longitude and Time

Determining longitude was a much more difficult task. Since the longitude lines rotate with the turning earth, 360° in twenty-four hours, it becomes necessary to know the

Figure 1.11

A map of the Americas from Flemish geographer Abraham Ortelius's 1570 atlas, *Theatre of the World.*

R. V. Tooley, Charles Bricker, and Gerald Crone, *Landmarks of Mapmaking.* Copyright © 1968 Elsevier Science Publishing Co., Inc., New York, New York.

position of the sun or the stars with time relative to one's longitude line. Because early clocks did not work well on rolling ships, precise longitude measurements were not possible until the eighteenth century, although the theory for using time to determine longitude was available in the sixteenth century.

The relationship between time and longitude was proposed by the Flemish astronomer Gemma Frisius in 1530. If a clock is set to exactly noon when the sun is at its **zenith,** or highest elevation above a reference longitude, and if that clock is then carried to a new location and the zenith time of the sun is determined at this location, the clock's time difference between the sun's zenith at the reference longitude and at the new location is used to determine the longitude at the new location. Using this technique, a position that is 15° of longitude west of the reference longitude is directly under the sun one hour later, or at 1 P.M. by the clock, because the earth has turned eastward 15° during that hour. A position 15° east of the reference longitude is directly under the sun at 11 A.M., because it requires an hour to turn the 15° to bring the sun to its zenith over the reference longitude (see fig. 1.12).

The importance of longitude determination by this technique was quickly realized. Commerce, exploration, and accurate chart making all needed this ability. As early as 1598 King Phillip III of Spain offered a reward of 100,000 crowns to any clock maker who could build a clock that would keep accurate time on board ship. In 1714 the British Parliament offered 20,000 pounds sterling for a seagoing clock that could keep time with an error not greater than two minutes on a voyage to the West Indies from England. A Yorkshire clock maker, John Harrison, accepted the challenge. He built his first **chronometer** in 1735, but it was not until 1761 that his fourth model met the test, losing only 51 seconds on the 81-day voyage. In 1772, Captain James Cook took a copy of the fourth version of Harrison's chronometer (fig. 1.13) on his famous voyage of discovery to the south seas and with it was able to produce accurate charts of new areas and to correct previously charted positions. Harrison was awarded only a portion of the prize after his success in 1761, and it was not until 1775, at the age of eighty-three, that he received the remainder from the reluctant British government. Accurate navigation based on time and celestial motions relative to the earth had become commonplace.

Figure 1.12

The time on a meridian relative to the sun changes by one hour for each 15° change in longitude.

Sun

10 11 12 1 2
A.M. A.M. P.M. P.M.

Noon

Figure 1.13

John Harrison's fourth chronometer met the conditions required for time accuracy on board ship.

The reference longitude in use today is the prime meridian, or 0° longitude. The clock time is set to 12 noon when the sun is at its zenith above the prime meridian. This is **Greenwich mean time** (GMT), now called **coordinated universal time** or **zulu time** (for zero meridian time). Since sun time changes by one hour for each 15° of longitude, the earth has been divided into time zones that are 15° of longitude wide. The time zones do not exactly follow lines of longitude; they follow political boundaries when necessary for the convenience of the people living in those zones (see fig. 1.14).

1.5 Earth: The Water Planet

Water on the Earth's Surface

As the earth and the other planets of the solar system cooled, the sun's energy gradually replaced the heat of planet formation in maintaining their surface temperatures. The earth developed a nearly circular orbital path at a mean distance of 149 million kilometers (149×10^6 km) or 93 million miles (93×10^6 mi) from the sun. (If you are unfamiliar with scientific notation to express very large numbers, see the Appendix.) Moving along this orbit the earth is 151×10^6 km (91×10^6 mi) from the sun in July and 146×10^6 km (88×10^6 mi) away in January, as shown in figure 1.15. At these distances from the sun, the earth's orbit keeps the annual heating and cooling cycle within moderate limits. The earth's mean surface temperature is about 16°C, which allows water to exist as a gas, as a liquid, and as a solid.

The rotation of the earth on its axis is also important in moderating temperature extremes. The earth completes one rotation turning from west to east in twenty-four hours. If the earth rotated more slowly, the side of the earth toward the sun would be exposed to the sun's energy for a longer period than it is at present and would become very hot, while the side in darkness would lose heat and become very cold. Temperature changes from day to night would be large. By contrast, a shorter period of rotation would decrease the present variation from day to night.

The earth's solar orbit, its rotation, and its blanket of atmospheric gases produce surface temperatures that allow the existence of liquid water. The atmosphere covering the earth's surface acts as a protective shield between the earth and the sun. Without it, solar heating would evaporate water at a much increased rate. Compare the earth's distance to the sun, its period of rotation, and the time required

Figure 1.14

Distribution of the world's time zones. Degrees of longitude are marked along the bottom of the diagram. Time zones are positive (west zones) or negative (east zones). Time at Greenwich is determined by adding the zone number to the local time.

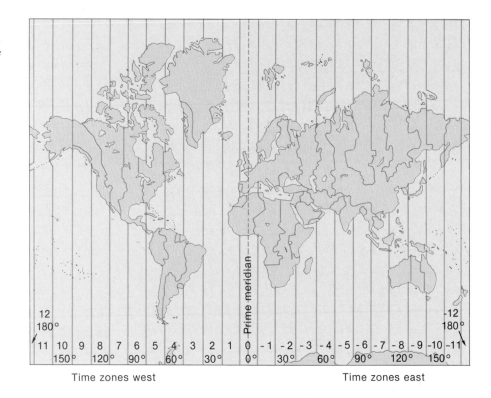

Time zones west   Time zones east

Figure 1.15

The earth's orbit around the sun is nearly circular. Note the change in distance between the earth and the sun during the year.

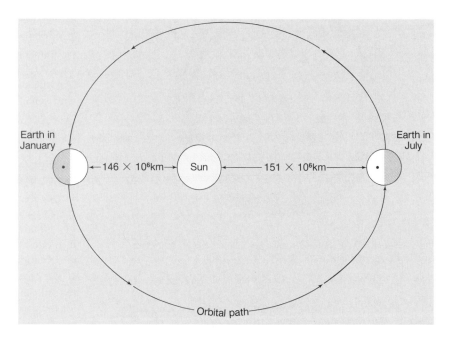

for the earth to complete one orbit of the sun with those of other planets, as shown in table 1.1. Note the surface temperatures of the earth as compared to those of other planets.

The amount of water on the earth's surface can be expressed in a number of ways. For example, the oceans cover 361 million square kilometers (361×10^6 km^2) or 139 million square miles (139×10^6 mi^2). These numbers are so large that they do not convey a clear idea of size; an easier concept is to remember that 71% of the earth's surface is covered by the world's oceans, and only 29% of the surface area is land above sea level.

The volume of water in the oceans is enormous: 1.37 billion cubic kilometers (1.37×10^9 km^3). Another way to express the oceanic volume is to think of a smooth sphere with exactly the same surface area as the earth (510×10^6 km^2 or 316×10^6 mi^2), uniformly covered with the water from the earth's oceans. The ocean water would be 2686 m (8800 ft) deep, a depth of about 1.7 miles. If the water from all other sources in the world were added, including water from the land and from the atmosphere, the depth would rise 56 m to 2742 m (9000 ft). When water volumes are considered as depths over a smooth sphere, they are

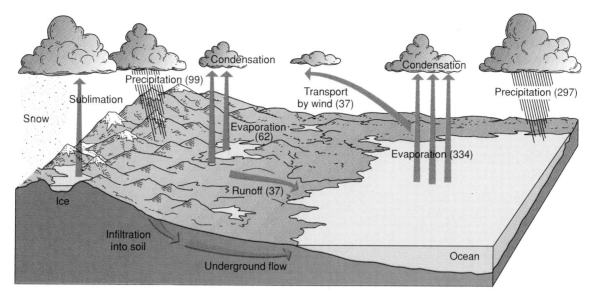

Figure 1.16

The hydrologic cycle and annual transfer rates. Snow and rain together equal precipitation. Sublimation, the direct transfer of ice to water vapor, is included under evaporation. Surface flow and underground flow are considered as land runoff. Annual transfer rates are in thousands of cubic kilometers (10^3 km^3).

Table 1.3
The Earth's Water Supply

Reservoir	Volume (km³)	% of total volume	Sphere depth (m)
atmospheric moisture expressed as water	15.3×10^3	0.001	0.03
rivers and lakes	510.0×10^3	0.036	1.0
groundwaters	$5,100.0 \times 10^3$	0.365	10.0
glacial and other land ice	$22,950.0 \times 10^3$	1.641	45.0
oceanic water and sea ice	$1,370,323.0 \times 10^3$	97.957	2,686.0
totals	$1,398,898.3 \times 10^3$	100	2,742.0

referred to as **sphere depths.** The ocean sphere depth is 2686 m, and the total water sphere depth of all the earth's water is 2742 m. This information is summarized in the last column of table 1.3.

The Hydrologic Cycle

The earth's water is found as a liquid in the oceans, rivers, lakes, and below the ground surface; it occurs as a solid in glaciers, snow packs, and sea ice; it takes the form of water droplets and gaseous water vapor in the atmosphere. The places in which water resides are called **reservoirs,** and each type of reservoir, when averaged over the entire earth, contains a fixed amount of water at any one instant. But water is constantly moving from one reservoir to another, as liquid water evaporates from the oceans into the air, icebergs melt in the oceans, rains fall on the land, and rivers flow back to the sea. This movement of water through the reservoirs is called the **hydrologic cycle** (fig. 1.16). Evaporation takes water from the surface of the ocean into the atmosphere; most of this water returns directly to the

sea, but air currents carry some water vapor over the land. Precipitation transfers this water to the land's surface, where it percolates into the soil; fills rivers, streams, and lakes; or remains for longer periods as snow and ice in some areas. Melting snow and ice, rivers, groundwater, and land runoff move the water back to the oceans to complete the cycle and maintain the oceans' volume. For a comparison of the water stored in the earth's reservoirs, see table 1.3 again.

In local areas excess water may evaporate from the land and excess precipitation may occur over the sea, but on a worldwide average, the cycle operates with a net removal of ocean water by evaporation, a net gain of water on land as a result of precipitation, and a return of the excess water on land to the sea by rivers and land drainage. The properties of climate zones are principally determined by the earth's surface temperature and evaporation-precipitation patterns: the moist, hot equatorial regions; the dry, hot subtropic deserts; the cool, moist temperate areas; and the cold, dry polar zones. Differences in these properties, coupled with the movement of air between the climate

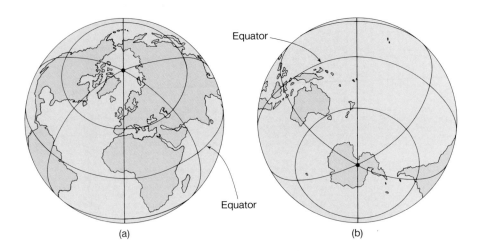

Figure 1.17
Continents and oceans are not distributed
uniformly over the earth. The Northern
Hemisphere (a) contains most of the land; the
Southern Hemisphere (b) is mainly water.

Equator

Equator

(a)

(b)

zones, moves water through the hydrologic cycle at different rates from one reservoir to another. The transfer of water between the atmosphere and the oceans alters the salt content of the oceans' surface water, and, with the seasonal and latitudinal changes in surface temperature, determines many of the characteristics of the world's oceans. These characteristics will be explored in chapters 6 to 9.

Reservoirs and Residence Time

Because the total amount of water on earth is nearly constant, the hydrologic cycle must maintain a balance between the addition and removal of water from the earth's water reservoirs. The rate of removal of water from a reservoir must equal the rate of addition to it, for if the balance is disturbed one reservoir gains at the expense of another. The time that the water spends in any one reservoir is called its **residence time.** Large reservoirs have long residence times because it takes a long time to move all the water in that reservoir through the hydrologic cycle, while the water in small reservoirs can be replaced comparatively quickly, so that they have short residence times. The size of the reservoir also determines how it reacts to changes in the rate at which the water is gained or lost. Large reservoirs show little effect from small rate changes, while small reservoirs may alter substantially when exposed to the same variation. For example, if the ocean volume decreased by 6 1/2% and this volume of water were added to the land ice, it would produce a 400% increase in the present volume of land ice but only a 250 m (820 ft) drop in sea level. This example reflects the changes that have occurred on earth during the major ice ages.

About 396,000 km³ (94,644 mi³) of water move through the atmosphere each year. Since the atmosphere holds the equivalent of 15,300 km³ (3,580 mi³) of liquid water at any one time, a little arithmetic shows that the water in the atmosphere can be replaced twenty-six times each year. Atmospheric water has a very short residence time. The residence time for water in the other larger reservoirs is much longer. For example, it would take 37,027 years to evaporate and pass the water from all of the oceans through the atmosphere, to the land as precipitation, and return it to the oceans via rivers. Further study of water's

movement shows us that annually 334,000 km³ (79,826 mi³) are evaporated from the oceans while the land loses 62,000 km³ (14,818 mi³). When the water returns as precipitation, 297,000 km³ (70,983 mi³) are returned directly to the sea surface, and 99,000 km³ (23,660 mi³) return to the land. However, the excess gained by the land (37,000 km³) (8,843 mi³) flows back to the oceans in the rivers, streams, and groundwater of the world (see fig. 1.16).

Distribution of Land and Water

To understand the present distribution of land and water on the earth, consider the earth when it is viewed from the north (see fig. 1.17a) and from the south (fig. 1.17b). About 70% of the earth's landmasses are in the Northern Hemisphere, and most of this land lies in the middle latitudes. The Southern Hemisphere is the water hemisphere, with its land located mostly in the tropic latitudes and in the polar region. The details of this land-water-latitude distribution are presented in figure 1.18.

The Oceans

Oceanographers view the world's oceans as three fingers—the Atlantic, Pacific, and Indian oceans—stretching up from a common source around Antarctica. The Arctic Ocean is considered an extension of the North Atlantic, and the Antarctic Ocean may be considered as the Southern Ocean at latitudes greater than 50°S or it may be divided along chosen lines of longitude to become the southern portions of the Atlantic, Pacific, and Indian oceans (see fig. 1.19). Each of these three oceans has its own characteristic surface area, volume, and mean (or average) depth. The Pacific Ocean has more surface area, a larger volume, and a greater mean depth than either the Atlantic or the Indian oceans. The Atlantic Ocean is the shallowest ocean and has the greatest number of shallow adjacent seas, such as the Arctic Ocean, Gulf of Mexico, Caribbean Sea, and Mediterranean Sea. The Indian Ocean is a Southern Hemisphere ocean; it is the smallest ocean in terms of area, but it is quite deep. The volume of the Pacific Ocean is over twice that of either the Atlantic Ocean or the Indian Ocean. Use table 1.4 to compare the three oceans.

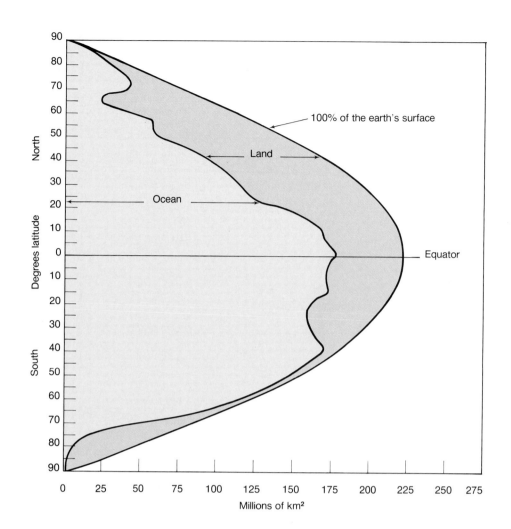

Figure 1.18

Distribution of land and ocean by latitude. In the Northern Hemisphere, middle latitude areas of land and ocean are nearly equal. Land is almost absent at the same latitudes in the Southern Hemisphere. The areas are calculated on the basis of 5° latitude intervals.

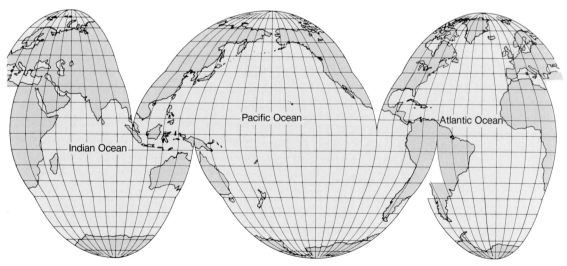

Figure 1.19

The world's three major oceans.

Table 1.4
Ocean Areas and Volumes versus Depth

Depth (m)	Atlantic (mean depth 3332 m)		Pacific (mean depth 4028 m)		Indian (mean depth 3897 m)		All oceans (mean depth 3795 m)	
	Area[1]	Volume[2]	Area	Volume	Area	Volume	Area	Volume
0	106.4	354.7	179.7	723.7	74.9	291.9	361.0	1370.3
1000	84.7	259.1	164.0	550.5	69.4	219.7	318.1	1029.3
2000	79.1	177.2	156.9	388.8	66.9	151.6	302.9	717.6
3000	69.7	102.8	147.5	236.6	61.3	87.5	278.5	426.9
4000	50.0	43.0	114.3	105.7	43.4	35.1	207.7	183.8
5000	22.6	6.7	51.0	23.1	14.8	6.0	88.4	35.8
6000	0.64	0	3.2	0.2	0.3	0.1	4.14	0.3
7000	0	0	0.36	0	0	0	0.36	0

[1] All areas are given in 10^6 km^2.
[2] All volumes are given in 10^6 km^3.

A view of an ocean from above is a view of 100% of that ocean's surface area at a depth of zero meters. It is also a view of 100% of the volume of water contained in that ocean. If successive layers of water 1000 m (3280 ft) thick were removed, then each view would show an ocean of reduced area and volume. This sequential removal would allow us to see how the ocean water is distributed with depth. A shallow ocean would reduce rapidly in both area and volume as the layers were removed, while a deep ocean would change more slowly. This effect is shown in table 1.4 for each of the three principal oceans and for all the oceans combined.

When the topmost 1000-m layer of water is removed, the Atlantic Ocean undergoes a large change in both volume and area compared to the Pacific and Indian oceans under the same conditions. Areas at given depths can be used to calculate percent changes in areas and volumes. As an example, for the Atlantic Ocean, the change in area from 0 to 1000 m ($[106.4 - 84.7] \times 10^6$ km^2) divided by the total area at 0 m (106.4×10^6 km^2) and multiplied by 100 shows that 20.4% of the seafloor of the Atlantic falls within this depth range. This indicates that this ocean has extensive shallow regions. These shallow regions and the absence of any appreciable depths greater than 6000 m combine to produce the relatively shallow mean depth of the Atlantic as compared to the Pacific and Indian oceans. The largest change in seabed area for all oceans occurs when the layers between 4000 m and 6000 m are removed. This change indicates that the most commonly occurring depths in the oceans are found between these two values.

The Pacific and Indian oceans have much less area, respectively, between 0 m and 1000 m (3280 ft) changing only 8.8% and 7.3%. This lack of shallow-water area and the presence of depths greater than 6000 m (19,680 ft) combine to give both of these oceans mean depths greater than that of the Atlantic.

The Hypsographic Curve

Another method used by oceanographers to depict land-water relationships is shown in figure 1.20. This graph of depth or elevation versus area is called a **hypsographic curve.** Find the line indicating sea level and note that the elevation of land above sea level is given in meters along the left margin; the depth below sea level is given in meters along the right margin. The scale across the top of the figure indicates total earth area in 10^8 km^2. The scales along the bottom of the figure indicate percentages of the earth's surface area; note that the curve crosses sea level at the 29% mark, showing that 29% of the earth's surface is above sea level and 71% is below sea level. The lower of the two scales gives land and ocean areas as separate percentages. Referring to the land area percentage scale, note that only 20% of all land areas are at elevations above 2 km. The ocean area scale shows that approximately 85% of the ocean area is below 2 km in depth. Compare this value with the data in table 1.4. The hypsographic curve helps us to see not only that our earth is 71% covered with water but that the areas whose depths are well below the sea surface are much greater than the areas whose elevations are well above it; there are basins beneath the sea that are about four times greater in area than the area of land in mountains above sea level. Mount Everest, the highest land peak, reaches 8.84 km (5.49 mi) above sea level, while the ocean's deepest trench descends 11.02 km (6.84 mi) below sea level.

Since the hypsographic curve is constructed as a plot of area versus height, an area of the diagram shows volume, because volume is the product of area times height. The mean elevation of the land is 840 m (2750 ft), and the entire land volume above sea level fits within a box 840 m high, covering 29% of the earth's surface. The mean depth of the ocean is 3795 m (12,400 ft). In this case the ocean volume is the mean ocean depth times 71% of the earth's surface area.

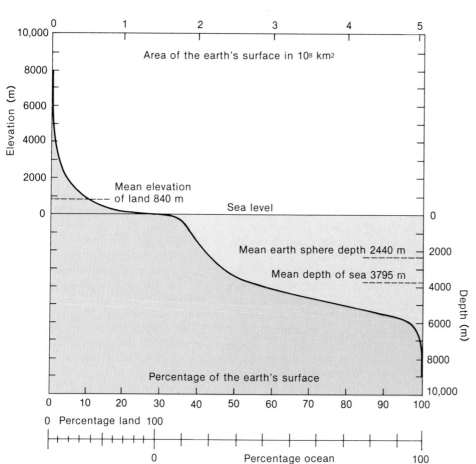

Figure 1.20
The hypsographic curve displays the area of the earth's surface at elevations above and below sea level.

Refer back to the discussion of sphere depths as a device for expressing water volumes and remember that, because humans are land dwellers, their reference line is sea level. But if we are considering the hypsographic curve for the total earth, a mean earth elevation becomes more suitable for describing the location of the earth's solid surface; this value is called the **mean earth sphere depth.** This level is 2440 m (8003 ft) below the present sea level, as shown in figure 1.20. This value represents the level at which the volume of crust above and water volume below the earth sphere depth are equal. To arrive at the mean earth sphere depth, think of moving all the elevated land and some of the sea bottom down into the ocean depressions until the earth is a perfectly smooth ball. When this is accomplished the mean earth sphere depth reference level would stand at the 2440-m (8003-ft) depth. Such an operation would also result in the displacement of seawater upward, until a new sea level 246 m (807 ft) above the present sea level was reached. This depth of water is the **mean ocean sphere depth,** 2686 m (8810 ft). It is the depth that the oceans would have if the volume of ocean water were spread over an area equal to the total earth area rather than over 71% of the earth's area. The oceans swallow the land, with only a comparatively small rise in water level, emphasizing again that earth is the water planet (or the ocean planet) and not the land planet.

1.6 Practical Considerations: Modern Navigational Techniques

Modern navigators still use chronometers and wait for clear skies to "shoot" the sun or stars with a sextant to determine positions at sea, but such measurements are used primarily to check their modern electronic navigational equipment. Chronometers are now calibrated, or reset, by broadcast time signals. When the vessels are near land, **radar** (radio detecting and ranging), gives an accurate line-of-sight picture of the shoreline and shows the position of vessels relative to the shore or other vessels. The image on the radar screen is formed by a pulse of radiation energy that is sent out by a transmitter, reflected from an object, returned to the antenna, and then displayed on a screen. **Loran** (long-range navigation) can be used farther out at sea. This electronic timing device is used to measure the difference in arrival time to the ship of radio signals from two pairs of land stations. The position of the ship may then be plotted on a chart that shows the time delay lines for signals from these station pairs. Recent improvements in loran include receivers with built-in computers, which the navigator can program with the latitude and longitude of the desired destination. The loran receiver monitors the signals from the

Because the great area of the world's oceans makes it impossible to equip enough research vessels to study more than a small area of one ocean at one time, and because what happens in any one area of an ocean is dependent on processes at work in other parts of the world's oceans, oceanographers have needed the ability to study the oceans as a total system.

If a satellite is set in high orbit about the earth's equatorial plane approximately 41,625 kilometers (22,500 nautical miles) above the earth, it will remain centered above one point on the earth's surface. This is a geosynchronous satellite that can be used to monitor changes at a single location. A satellite orbiting in a plane inclined to the earth's equatorial plane, at a speed allowing it to circle the earth every ninety minutes, and with an observational band of 22 1/2° of longitude, will have a much lower orbit and cover a new band of earth surface each time the satellite crosses the equator. Complete earth surface coverage will be accomplished in twelve hours within the north-south limits of the satellite's orbit.

If the satellite is placed in a low polar orbit (925 km or 500 nm) above the earth with an orbital period of ninety minutes and if its sensors cover a band 22 1/2° longitude-wide at the equator, an area of 100×10^6 km^2 (29.2×10^6 nm^2) is observed during each ninety-minute period. Under these conditions each data point covers a relatively large area and a fuzzy average portrayal of the total earth's surface is achieved in twelve hours. If the sensors on the satellite are restricted to a narrower path of observation, a much more detailed picture can be obtained. These conditions require more orbits and a longer time to achieve full earth coverage.

Oceanographers are using satellites with specialized sensors and measuring devices to provide total ocean surveillance and data on a global scale. Satellites are either passive, recording information from the earth such as radiation energy, or active, sending out radar or laser signals that echo back to the satellite. These data are sent back to earth in the form of radio signals that are recorded, processed, and analyzed by computer to re-create the satellite's observations. In the U.S. satellites are the responsibility of some private companies, the military services, the National Oceanographic and Atmospheric Administration (NOAA), and the National Aeronautic and Space Administration (NASA); some are intelligence gatherers; some are communication links for global telephone, radio, and television connections; some are navigation aids (see section 1.6); some gather data for earth sciences research. The latter are becoming more and more important to oceanographers as they work to understand the interactions between the oceans and the atmosphere.

The first earth-observing satellite was the TIROS weather satellite (Box fig. 1.1a) launched in 1960. TIROS has been followed by a large number of satellites launched by numerous countries and for many purposes. NASA's NIMBUS series of satellites was designed to observe the earth's atmosphere and surface features. NIMBUS-7 (Box fig. 1.1b), launched in 1978, gathered data that allowed detection of changes in the earth's surface temperatures and the atmosphere's chemical composition. It monitored back radiation and reflection from the earth's surface and carried a coastal zone color scanner to detect back radiation from the chlorophyll of sea and land plants. Although this scanner failed in 1986, it gave oceanographers their first global look at plant productivity in the oceans; its data continues to provide information for analysis and interpretation.

SEASAT (Box fig. 1.1c), a specialized oceanographic satellite, was launched in June 1978 and operated until October of that year. This short-lived satellite carried a radar altimeter capable of measuring the distance between the satellite and the sea surface with an accuracy of about 5 cm (2 in) over a narrow observational width. The altimeter measured both wave heights and sea-surface elevation. Radar was used to measure the reflected scattering patterns of wind waves, from which wind speed and direction were calculated. Radiometers recorded radiation from the earth's surface, allowing the measurement of sea-surface temperatures, atmospheric water vapor content, wind speed, and sea ice cover. SEASAT gave the first large-area coverage of oceanic winds and produced a picture of sea-surface elevations that showed the effect of local changes in the earth's gravity caused by variations in density of the earth's crust and the topography of the seafloor (see fig. 2.16).

GEOSAT (Box fig. 1.1d), launched in 1985, was designed to collect high-resolution data for military purposes. Its orbit was changed to replace SEASAT and from 1986 to 1990 monitored sea level topography, surface winds and waves, local gravity changes, and abrupt boundaries between water types. GEOSAT detected month-to-month changes in sea level that agreed within 4 cm (1.6 in) of tide gauge records. Like SEASAT, GEOSAT used the scattering of radar signals to estimate surface wind speeds and directions from wave heights (see fig. 7.13). It also detected major currents and boundaries between different types of water, based on water level changes. Ocean scientists have used GEOSAT data to detect the movement of surface water masses related to El Niño (see chapter 7).

At present a large number of cooperative satellite efforts are under way. The European Remote Sensing (ERS-1) satellite and a Japanese Resources Satellite (JERS-1) were launched in 1992. They will be joined by the Canadian Radarsat. Each of these satellites uses radar systems with different frequencies to allow more-accurate detection of sea-surface topography and wave patterns. Another satellite launched in 1992 was the TOPEX/ Poseidon (Box fig. 1.1e), a NASA-French cooperative effort, with a three-year mission of gathering information of the global circulation of the world's oceans, the interaction between atmosphere and ocean surface, and the ocean tides. A global ocean research program such as the Joint Global Ocean Flux Study (JGOFS) (discussed in the Prologue, The Present and the Future and in the chapter 8 Box) requires satellite support to follow the movement of carbon dioxide between the atmosphere and the oceans. A specialized satellite (SeaWIFS or sea wide-field sensor) designed for this project is scheduled for launch in 1993. SEASTAR is another satellite scheduled for 1993 launch; it carries a wide-field sensor for wavelengths of light chosen to provide improved data on the minute plant life of the near surface waters and is also associated with gaining a better understanding of the earth's abilities to cycle carbon dioxide.

NASA's Earth Observing System (EOS) program, to be launched between 1998 and 2002, is a long-range program using a series of three satellites, each with a five-year working life, to monitor the role of clouds, radiation, water vapor, and precipitation; productivity of the oceans;

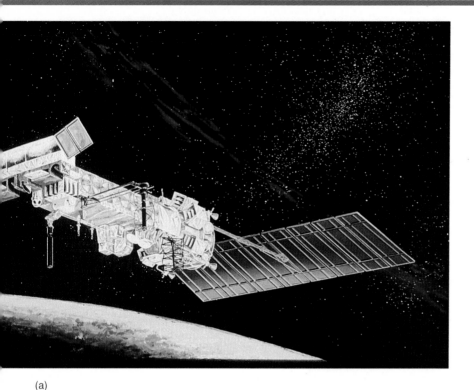

(a)

(b)

(c)

(d)   (e)

Box figure 1.1

U.S. satellites with ocean remote-sensing systems. (a) TIROS is one of a series of polar-orbiting meteorological satellites able to measure sea-surface temperatures to 1°C and 1 km resolution; it carries the international Search and Rescue (SAR) emergency beacon system. (b) NIMBUS-7 monitors surface temperatures, wind speed, and sea ice. (c) SEASAT failed after 104 days of operation. During this period it monitored sea-surface topography, wave heights, and sea-surface wind velocities and measured sea-surface temperatures, atmospheric water vapor, and sea ice distribution. (d) GEOSAT used radar altimetry to determine the topography of the ocean. (e) TOPEX is designed to determine ocean circulation patterns.

gas exchange between the oceans and the atmosphere; and the role of polar ice. New satellites and new programs mean vast amounts of new data being collected. Improved storage facilities for this data, its management, and its distribution are necessary if it is to be available to researchers. A new data system is being developed in conjunction with the EOS project; it is expected to serve up to 10,000 scientists around the world.

These data and their interpretation will help us to find important pieces of the puzzle that oceanographers and other earth scientists face as they work to develop a better understanding of the links between atmosphere and oceans. As we struggle to forecast global environmental cycles in our efforts to maintain a healthy global environment, satellite oceanography becomes more and more important. It is costly, but it is the only way to achieve a global view of the earth, its oceans, and its atmosphere.

station pairs on land and directly reads out the course to be sailed and the distance to the desired destination. The computer can also monitor the signals and continuously calculate the latitude and longitude, allowing the navigator to know the ship's position at all times.

The **satellite navigation system** is a more sophisticated navigational aid. In this system a satellite orbiting the earth emits a coded signal of a precise frequency, which is picked up by a receiver on the ship. The shipboard receiver follows the frequency shift as the satellite passes and determines the exact instant in time at which the frequency is correct. At this instant the ship's path and the satellite's orbit are at right angles. Given this information, a computer programmed with the satellite's orbital properties can determine the ship's position to within 30 meters (100 feet) or less.

A more versatile and accurate method of finding one's position uses the U.S. Navstar Global Position System (GPS). By late 1993 this system expects to have twenty-four orbiting satellites operating in six different orbits 12,500 miles above the earth. These satellites emit continuous radio signals that pinpoint the position of GPS receivers within a few feet, including latitude, longitude, and altitude. The system can be used to determine the distance between receiving stations, and these data are then used to determine earth positions within a few centimeters' accuracy. This system allows mapping of the earth's surface on land and across large bodies of water; it will be tied into signals from radio telescope networks that receive radiation signals from space, allowing very accurate positioning of the satellites. Once this is accomplished these GPS satellites will be able to sense variations in the earth's surface elevations associated with storms, tides, currents, earth movements, and changes in glacial ice volume and sea level.

The U.S. GPS system is similar to a global position system, GLONASS, employed by the former Soviet Union. The increasing cooperation between the United States and Russia is pointing to the development of receivers that will accept both systems, allowing increased positioning accuracy, better earth coverage, and independent verifications of results.

The navigator has been able to transfer the position information relayed by the satellite to a chart where the vessel's progress is checked over time, but new electronic aids are providing more automation and more flexibility. Shipboard computers can now store an electronic atlas that includes surface charts and seafloor bathymetry. The ship's position is tracked and its position determined continuously. Oceanographic vessels use devices that draw charts showing the vessel's changing position and at the same time conduct a seafloor survey and keep track of water measurements (such as temperature and salt content) made automatically as the ship moves along. Today's ocean scientists are able to return more and more accurately to the same place at sea if further measurements are needed. They also have the ability to evaluate data as they are being taken, allowing them to change the vessel's sampling pattern to insure that the best measurements are made to satisfy a project's research requirements.

Summary

The solar system began as a rotating cloud of gas. A series of events produced nine planets orbiting the sun, each planet having different characteristics. Over approximately 1 1/2 billion years the earth heated, cooled, changed, and collected a gaseous atmosphere and an accumulation of liquid water.

Reliable age dates for earth rocks, meteorites, and moon samples are obtained by radiometric dating. The accepted age of the earth is 4.6 billion years. Geologic time is used to express the time scale of the earth's history.

The distance between the earth and the sun, the earth's orbit, its period of rotation, and its atmosphere protect the earth from extreme temperature change and water loss. Because it rotates, the earth's shape is not perfectly symmetrical. Its exterior is relatively smooth. Natural time periods (the year, day, and month) are based on the motions of the sun, earth, and moon. Because of the tilt of the earth's axis as it orbits the sun, the sun moves annually between 23 1/2°N and 23 1/2°S, producing the seasons.

Latitude and longitude are used to form a grid system for the location of positions on the earth's surface. Different types of map and chart projections have been developed to show the earth's features on a flat surface. All of these projections distort the earth's features to some extent. Bathymetric and physiographic charts and maps use elevation and depth contours to depict the earth's topography.

In order to determine longitudinal position, time must be measured accurately. This need required the development of the seagoing clocks for celestial navigation.

The earth is the water planet. Of the earth's surface, 71% is covered by its oceans. There is a fixed amount of water on earth. Evaporation and precipitation move the water through the reservoirs of the hydrologic cycle. Water's residence time varies in each reservoir and depends on the volume of the reservoir and the rate of addition and removal of the water.

The earth's Northern Hemisphere is the land hemisphere; the Southern Hemisphere is the water hemisphere. The earth has three large oceans extending north from the Southern Ocean. Each has a characteristic surface area, volume, and mean depth. The hypsographic curve is used to show land-water relationships of depth, elevation, area, and volume. It is also used to determine mean land elevation, mean ocean depth, earth sphere depth, and ocean sphere depth.

Modern navigational techniques make use of radar, radio signals, computers, and satellites. A satellite network is allowing more accurate position readings and is also mapping storms, tides, sea level and properties of surface waters.

Key Terms

radiometric dating
isotope
half-life
vertebrates
Tropic of Cancer
summer solstice
Arctic Circle
Antarctic Circle
autumnal equinox
Tropic of Capricorn
winter solstice
vernal equinox
lunar month
solar day
sidereal day
latitude
longitude
parallel
equator
meridian
prime meridian
international date line
great circle

nautical mile
projection
Mercator projection
contour
topography
bathymetry
physiographic maps
Polaris
zenith
chronometer
Greenwich mean time
coordinated universal time
zulu time
sphere depth
reservoir
hydrologic cycle
residence time
hypsographic curve
mean earth sphere depth
mean ocean sphere depth
radar
loran
satellite navigation system

Study Questions

1. How and why have estimates of the age of the earth changed over the past few hundred years? Do you think the present estimate of the earth's age will change in the future?
2. Describe the distribution of water and land on the earth.
3. Why does the earth's average surface temperature differ from the surface temperature of other planets in the solar system?
4. Why is the twilight period at sunset shorter at low latitudes than it is at high latitudes?
5. The route of a ship sailing a constant compass course on a Mercator projection is indicated by a straight line that cuts all longitude lines at the same angle. This is a rhumb line. Discuss how this line appears (a) on a polar conic projection, (b) on a globe, and (c) on a tangent plane projection centered on the polar axis.
6. Discuss how the hypsographic curve is used to determine the mean depth and sphere depth of the oceans.
7. Why are the Arctic and Antarctic circles displaced from the poles by 23 1/2°, so that they are located at 66 1/2°N and 66 1/2°S?
8. What are some advantages of using satellites for oceanographic research? Are there any disadvantages?
9. How will the seasons change over a calendar year at each of these latitudes: (a) 10°N; (b) 70°N; (c) 30°S? Make a simple diagram for each latitude to show why the seasonal pattern occurs.
10. Explain why the earth sustains a wide variety of life-forms while the other planets of our solar system do not.
11. Trace several possible routes for a water molecule moving between a mountain lake and an ocean. In which reservoirs would the molecule spend the greatest amount of time and in which the least?
12. Use an atlas to find the appropriate latitudes and longitudes for each of the following:
    (a) St. John, Newfoundland and London, England;
    (b) Cape Town, South Africa and Melbourne, Australia;
    (c) Anchorage, Alaska and Moscow, Russia;
    (d) Straits of Gibraltar, Strait of Magellan, Straits of Florida;
    (e) Galápagos Islands, Tristan da Cunha, and Reykjavík, Iceland.
13. Although latitude and longitude were used on very early charts, navigators continued to use charts with many compass direction lines (portolano type) well into the seventeenth century. Why did they do this?
14. If the lunar month were used as the length of a month, what would happen to the calendar year relative to the sun?
15. At what 1000-m depth interval in the world's oceans would the greatest change in ocean area occur? Use the hypsographic curve to determine your answer.

Study Problems

1. Determine the distance between two locations: 110°W, 38.5°N and 110°W, 45.0°N. Express this distance in nautical miles and in kilometers.
2. The contour interval on a bathymetric chart is constant and equal to 100 m. Graph the slope of the seafloor across four evenly spaced contour lines if the distance between the first line and the fourth line is 2.5 km.
3. A plane leaves Tokyo, Japan, on June 6, at 0800 hours local Tokyo time and flies for nine hours, landing in San Francisco, California. What is the local time and date of arrival in San Francisco?
4. Show that the annual net evaporative loss of water from the world's oceans equals the annual net gain of water by precipitation on the land. Why does the ocean volume not decrease?
5. Show that the Atlantic Ocean has a larger percentage of its volume in the top layer (0–1000 m) than does the Pacific or Indian Ocean. Explain the significance of this difference.
6. Use the volume of the oceans and the earth's area to determine the sphere depth of the oceans.

Suggested Readings

Baker, J. D., and W. S. Wilson, 1986. Spaceborne Observations in Support of Earth Science. *Oceanus* 29 (4): 76–85.

Bowditch, N. 1984 ed. *American Practical Navigator,* vol. 1. U.S. Defense Mapping Agency Hydrographic Center, Washington, D.C. 1414 pp. (History of navigation, chart projections, and navigation aids are covered in chapters 1–5 and 41–46.)

Gore, R. 1985. The Planets, Between Fire and Ice. *National Geographic* 167 (1): 4–51.

Maranto, G. 1991. Way Above Sea Level. *Sea Frontiers* 37 (4): 16–23.

McClintock, J. 1987. Remote Sensing, Adding to Our Knowledge of Oceans and Earth. *Sea Frontiers* 33 (2): 105–13.

Ryan, P. R., ed. Fall 1981. *Oceanus* 24 (3). (Issue devoted to oceanography from space.)

Wagner, J. K. 1991. *Introduction to the Solar System.* Saunders College Publishing, Philadelphia. 522 pp. (General astronomy text.)

2

The Seafloor

2.1 Bathymetry of the Seafloor
  The Continental Margin
  The Ocean Basin Floor
  The Ridges and Rises
  The Trenches
2.2 Measuring the Depths
2.3 Sediments
  Sediment Sources

| Box: Bathymetrics: New Techniques for Surveying the Seafloor |

  Biogenous Oozes
  Red Clay
  Patterns of Deposit
  Particle Size
  Rates of Deposit
  Sampling Methods
2.4 Practical Considerations: Seabed Resources
  Sand and Gravel
  Phosphorite
  Sulfur
  Coal
  Oil and Gas
  Manganese Nodules
  Sulfide Mineral Deposits
  Laws and Treaties
Summary
Key Terms
Study Questions
Study Problems
Suggested Readings

The fog continued through the night, with a very light breeze, before which we ran to the eastward, literally feeling our way along. The lead was hove every two hours and the gradual change from black mud to sand, showed that we were approaching Nantucket South Shoals. On Monday morning, the increased depth and deep blue color of the water, and the mixture of shells and white sand which we brought up, upon sounding, showed that we were in the channel, and nearing George's; accordingly, the ship's head was put directly to the northward, and we stood on, with perfect confidence in the soundings, though we had not taken an observation for two days, nor seen land; and the difference of an eighth of a mile out of the way might put us ashore. Throughout the day a provokingly light wind prevailed, and at eight o'clock, a small fishing schooner, which we passed, told us we were nearly abreast of Chatham lights. Just before midnight, a light land-breeze sprang up, which carried us well along; and at four o'clock, thinking ourselves to the northward of Race Point, we hauled upon the wind and stood into the bay, north-north-west, for Boston light, and commenced firing guns for a pilot.

Richard Henry Dana, Jr.,
from *Two Years Before the Mast*

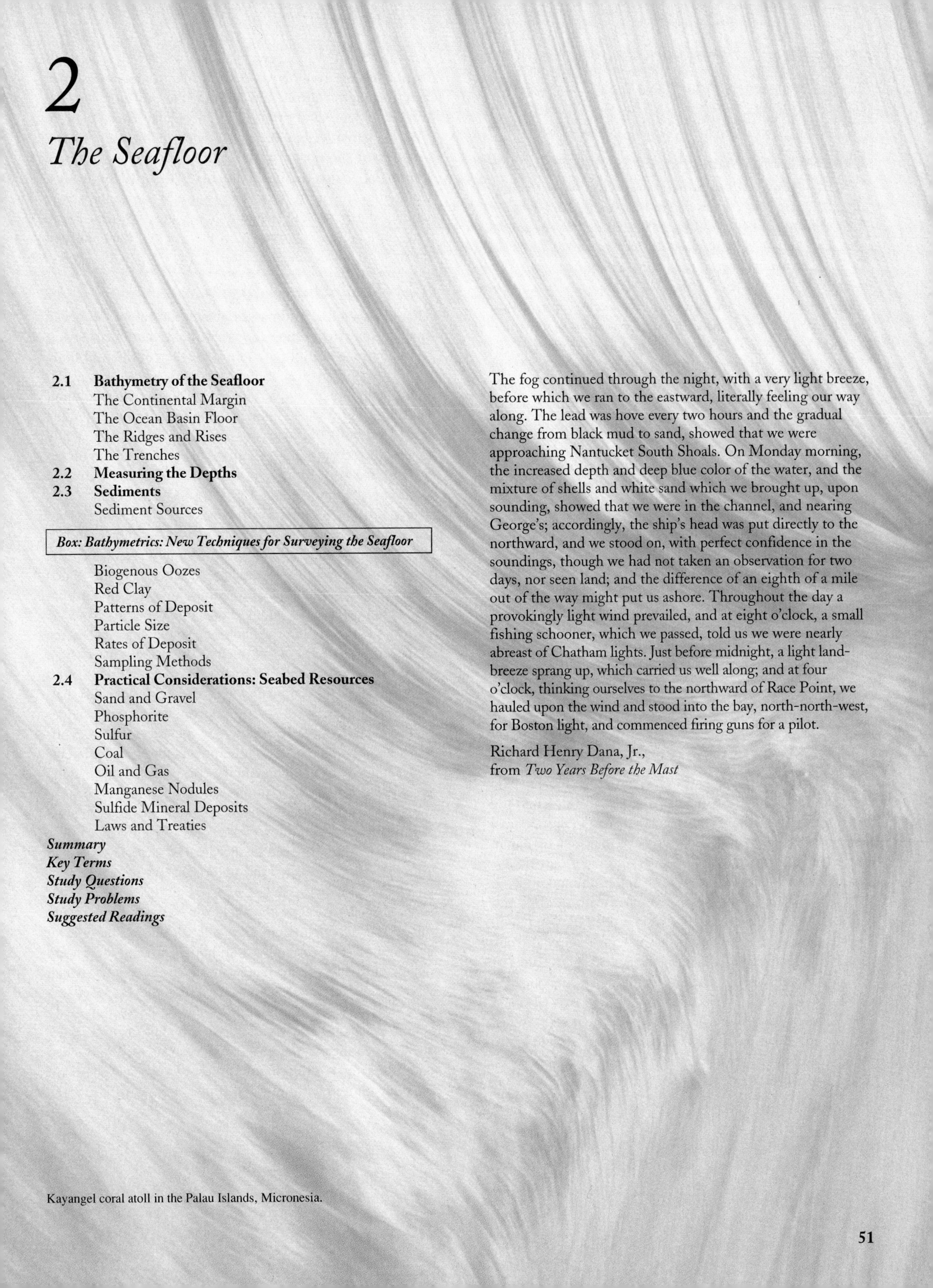

Kayangel coral atoll in the Palau Islands, Micronesia.

T   *he seafloor was an unknown environment to early mariners and the first curious scientists. They believed that the oceans were large basins or depressions in the earth's crust, but they did not conceive that these basins held features that were as magnificent as the major mountain chains, deep valleys, and great canyons of the land. As maps were created in greater detail and as ocean travel and commerce increased, it became essential to measure depths and record seafloor features in the shallower regions to maintain safe travel for sailing ships. The secrets of the deeper oceanic areas continued to lie hidden by water too deep to probe, and the bottom features of these areas remained unknown.*

It was not until the 1950s that improvements in technology made it relatively easy to map and sample the seafloor; the large numbers of vessels taking part in surveys accumulated sufficient data to give detail to this hidden terrain. What was found beneath the surface of the sea created more questions than answers and produced a demand for more and varied measurements to describe and explain the features of this portion of the earth's crust. This chapter surveys the world's ocean floors and discusses their present geography and geology.

Studies of the seabed also give us useful knowledge of mineral resources that we can use to supplement the dwindling supplies found on land. We need to know where deposits of minerals are located and how to remove them economically and with a minimal impact. We need information so that we can make wise and thoughtful decisions before irreversible steps are taken.

2.1   Bathymetry of the Seafloor

The land below the sea surface is as rugged as any land above it. The Grand Canyon, the Rocky Mountains, the desert mesas in the Southwest, and the Great Plains all have their undersea counterparts. In fact, the undersea mountain ranges are longer, the valley floors are wider and flatter, and the canyons are often deeper than those found on land. Features of land topography, such as mountains and canyons, are continually and actively eroded by wind, rain, and ice and are affected by changes in temperature and rock chemistry. Undersea mountains and canyons are features of the ocean's bathymetry. The erosion of undersea mountains and canyons is slow and occurs only by way of waves and currents along the shore and on steep underwater slopes. Beneath the surface the water dissolves away certain materials, changing the rock chemistry, but the features, once formed, retain their shape and are only slowly modified as

they are covered by a constant rain of particles, or **sediments,** falling from above. Movements of the earth's crust may displace features and fracture the seafloor, and the weight of some underwater volcanoes may cause them to sink, or **subside,** but the appearance of the bathymetric features of the ocean basins and seafloor has remained much the same through the last 100 million years. Computer-drawn profiles of crustal elevations across the United States and the Atlantic Ocean are shown in figure 2.1. Compare the topography of the Rocky Mountains and the undersea peaks shown in this figure.

The Continental Margin

The edges of the landmasses at present below the ocean surface and the steep slopes of these landmasses that descend to the seafloor are known as the **continental margin.** The continental margin is made up of the continental shelf, shelf break, slope, and rise. The **continental shelf** lies at the edge of the continent. Continental shelves are the nearly flat borders of varying widths that slope very gently toward the ocean basins. Shelf widths average about 65 km (40 mi) and vary from only a few tens of meters across in some areas to more than 100 km (60 mi) in others. The average depth of the seaward edge of the continental shelf is considered to be 130 m (430 ft).

The distribution of the world's continental shelves is shown in figure 2.2. The width of the shelf is often related to the slope of the adjacent land, being wide along low-lying land and narrow along mountainous coasts. Note the narrow shelf along the west coast of South America and the great wide expanses of continental shelf along the eastern and northern coasts of North America, Siberia, and Scandinavia. The continental shelves are geologically part of the continents; processes producing continental shelves include the storm waves that erode the edges of the continents and, in some cases, natural dams that trap sediments between the offshore dam and the coast. Such dams include reefs, volcanic barriers, upthrust rock, and salt domes. In the Gulf of Mexico, sediments from the land are trapped behind folds and domes of the seafloor containing salt deposits. Sediments along the northeast coast of North America are trapped behind upturned rock near the outer edges of the shelves (see fig. 2.3).

During past ages, the shelves have been covered and uncovered by fluctuations in sea level. During the glacial epoch of the Pleistocene (2.5 million years ago) there were a number of short-term changes in sea level, some of which were greater than 120 m (400 ft). When the sea level was low, erosion deepened valleys, waves eroded previously submerged land, and rivers left sediments far out on the shelf. When the glacial ice melted, these areas were flooded and sediments built up in areas closer to the new shore. At present, although submerged, these areas still show the

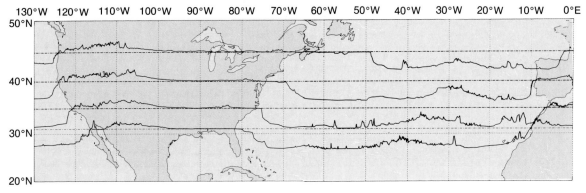

Figure 2.1

Computer-drawn topographic profiles from the west coast of Europe and Africa to the Pacific Ocean. The elevations and depths above and below 0 meters are shown along a line of latitude by using the latitude line as zero elevation. For example, the depth at 40°N and 60°W is 5040 m (16,531 ft). The vertical scale has been extended about one hundred times the horizontal scale. If both the horizonal and vertical scales were kept the same, a vertical elevation change of 5000 m would measure only 0.05 mm (0.002 in).

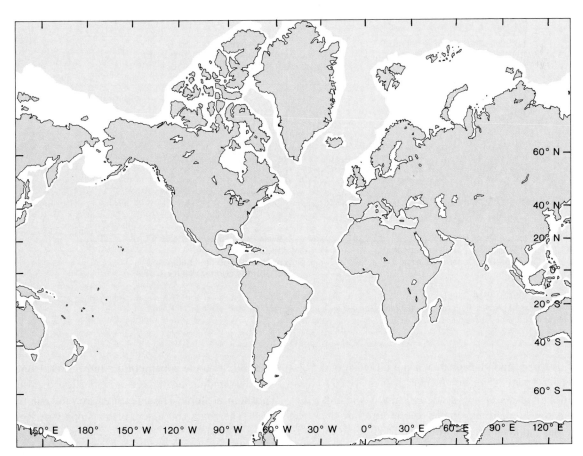

Figure 2.2

Distribution of the world's continental shelves. The seaward edges of these shelves are at an average depth of approximately 130 meters (430 ft).

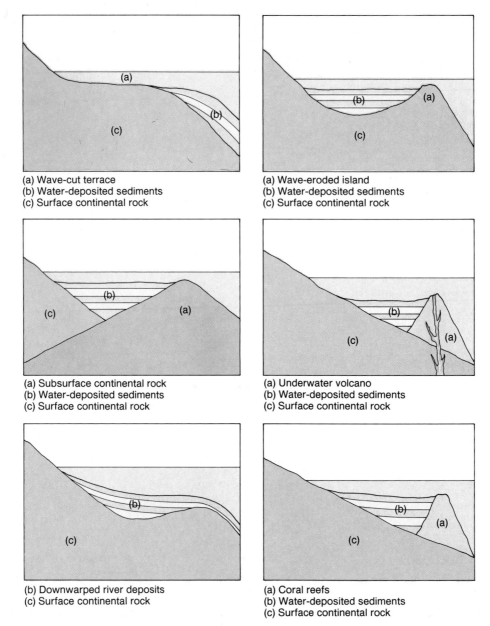

(a) Wave-cut terrace
(b) Water-deposited sediments
(c) Surface continental rock

(a) Wave-eroded island
(b) Water-deposited sediments
(c) Surface continental rock

(a) Subsurface continental rock
(b) Water-deposited sediments
(c) Surface continental rock

(a) Underwater volcano
(b) Water-deposited sediments
(c) Surface continental rock

(b) Downwarped river deposits
(c) Surface continental rock

(a) Coral reefs
(b) Water-deposited sediments
(c) Surface continental rock

Figure 2.3

Examples of how continental shelves are formed by trapping land-derived sediments at the edge of the continental landmasses.

scars of old riverbeds and glaciers they acquired when part of the landmass. Today, some continental shelves are covered with thick deposits of silt, sand, and mud sediments derived from the land; for example, the mouths of the Mississippi and Amazon rivers, where large amounts of such sediments are deposited annually.

Other shelves are bare of sediments, such as where the fast-moving Florida Current sweeps the tip of Florida, carrying the sediments northward to the deeper water of the Atlantic Ocean. The boundary of the continental shelf on the ocean side is determined by an abrupt change in slope and a rapid increase in depth. This change in slope is

referred to as the **continental shelf break,** while the steep slope extending to the ocean basin floor is known as the **continental slope.** These features are shown in figure 2.4.

The angle and extent of the slope vary from place to place. The slope may be short and steep (for example, the depth may increase rapidly from 200 m (650 ft) to 3000 m (10,000 ft), as in fig. 2.4), or it may drop as far as 8000 m (26,000 ft) into a great deep seafloor depression or trench (for example, off the west coast of South America, where the narrow continental shelf is bordered by the Peru-Chile Trench). The continental slope may show rocky outcroppings, and it is relatively bare of sediments because of its steepness.

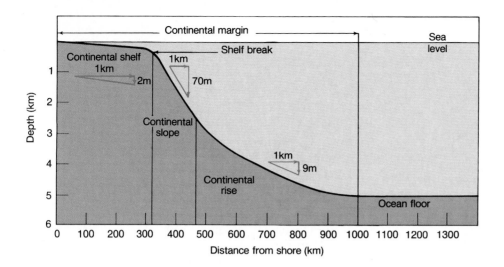

Figure 2.4
A typical profile of the continental margin. Notice both the vertical and horizontal extent of each subdivision. The slope is indicated for the continental shelf, slope, and rise. The vertical exaggeration is one hundred times greater than the horizontal scale.

Table 2.1
Major Submarine Canyons of the World

Pacific Ocean	Atlantic Ocean	Indian Ocean
Tokyo Canyon	Oceanographer Canyon	*Indus Canyon
Bering Canyon	*Hudson Canyon	*Ganges Canyon
*Columbia (or Astoria) Canyon	Wilmington Canyon	
Juan de Fuca Canyon	Norfolk Canyon	
Monterey Canyon	*Congo Canyon	
Arguello Canyon	*São Francisco Canyon	
Scripps Canyon	*Mississippi Canyon	
Coronados Canyon		

*Canyon has present-day relationship to a river. Other canyons are related to faults, inactive riverbeds, and drainage systems active at previous low sea levels.

The most outstanding features found on the continental slopes are **submarine canyons** (see table 2.1 for a list of major submarine canyons). These canyons sometimes extend up, into, and across the continental shelf. A submarine canyon is steep-sided and has a V-shaped cross section, with tributaries similar to those of river-cut canyons on land. Figure 2.5a shows the Monterey and Carmel canyons off the coast of California. Figure 2.5a is a bathymetric chart; figure 2.5b compares the profile of the Monterey Canyon with the profile of the Grand Canyon of the Colorado River. A submarine canyon is also shown in figures 1.9 and 1.10.

Many of these submarine canyons are associated with existing river systems on land and were apparently cut into the shelf during periods of low sea level, when the glaciers advanced and the rivers flowed across the continental shelves. Ripple marks on the floor of the submerged canyons, and sediments fanning out at the ends of the canyons, suggest that they have been formed by moving flows of sediment and water called **turbidity currents.** Caused by earthquakes or the overloading of sediments on steep slopes, turbidity currents are fast-moving avalanches of mud, sand, and water that flow down the slope, eroding and picking up sediment as they gain speed. In this way the currents erode the slope and excavate the submarine canyon. As the flow reaches the bottom it slows and spreads, and the sediments settle. Because of their speed and turbulence, such currents can transport large quantities of mixed materials. The settling process produces graded deposits of coarse materials overlaid by successive layers of decreasing particle size. These graded deposits are called **turbidites.** Figure 2.6 shows a turbidite preserved in rock. Such large and occasional currents have never been directly observed, although similar but smaller and more continuous flows, such as sand falls, have been observed and photographed (see fig. 2.7).

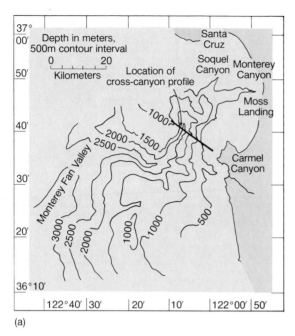

(a)

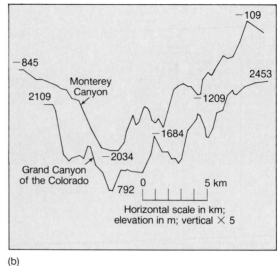

(b)

Figure 2.5

(a) Depth contours depict the Monterey and Carmel canyons off the California coast as they cut across the continental slope and up into the continental shelf. (b) Cross-canyon profile of the Monterey Canyon shown in (a). Compare this profile with that of the Grand Canyon drawn to the same scale.

Research on turbidity currents began with laboratory experiments in the 1930s. These experiments were based on earlier observations of the silty Rhone River water moving along the bottom of Lake Geneva in Switzerland. Later analysis of a 1929 earthquake that broke the transatlantic telephone and telegraph cables on the continental slope and rise off the Grand Banks of Newfoundland showed a pattern of rapid cable breaks high on the continental slope, followed by a sequence of downslope breaks after the earthquake. These breaks were calculated to have been caused by a turbidity current that ran for 700 km (420 mi) at speeds of 40–55 km (25–35 mi) per hour. Later samples taken from the area showed a series of graded sediments at the end of the current's path. Searches of cable company records showed similar patterns of cable breaks in other parts of the world.

At present, researchers are developing computer and laboratory models to predict a turbidity current's speed, depth, and sediment content and the distance it will travel. One of the first models is being applied to the Scripps submarine canyon in California.

Canyons with U-shaped cross sections and flat bottoms have been found off rivers that supply large quantities of sediments, such as the Ganges River. Continental shelves also show numerous gullies, probably formed by loose sediments moving down the slope.

Figure 2.6

This beach cliff shows a series of ancient turbidite flows that have been uplifted and then exposed by wave erosion.

Figure 2.7
Sand fall in San Lucas submarine canyon.

At the base of the steep continental slope there may be a gentle slope formed by the accumulation of sediment. This portion of the seafloor is the **continental rise;** it is a region of sediment deposit by turbidity currents, underwater landslides, and any other processes that carry sands, muds, and silt down the continental slope. The continental rise is a conspicuous feature in the Atlantic and Indian oceans and around the Antarctic continent. Few continental rises occur in the Pacific Ocean, where great seafloor trenches are often located at the base of the continental slope. Refer back to figure 2.4 to see the relationship of the continental rise to the continental slope.

The Ocean Basin Floor

The true oceanic features of the seafloor occur seaward of the continental margin. The deep seafloor, between 4000 and 6000 m (13,000–20,000 ft), covers more of the earth's surface (30%) than do the continents (29%). In many places the ocean basin floor is a flat plain extending seaward from the base of the continental slope. It is flatter than any plain on land and is known as the **abyssal plain.** Abyssal plains are formed from sediments covering the irregular topography of the seafloor, the sediments fall from the surface and are deposited by turbidity currents.

Abyssal hills and **seamounts** are scattered across the seafloor in all the oceans. Abyssal hills are less than 1000 m (3300 ft) high, and seamounts are steep-sided volcanoes rising abruptly and sometimes piercing the surface to become islands. These features are shown in figure 2.8. Abyssal hills are probably the earth's most common topographic feature. They are found over 50% of the Atlantic seafloor and about 80% of the Pacific floor; they are also abundant in the Indian Ocean. Most abyssal hills are probably volcanic while some may have been formed by other movements of the seafloor. Submerged, flat-topped seamounts, known as **guyots,** are found most often in the Pacific Ocean; a guyot is also shown in figure 2.8. These guyots are 1000 to 1700 m (3300–5600 ft) below the surface, with many at the 1300 m (4300 ft) depth. Many guyots show the remains of shallow marine coral reefs and evidence of wave erosion at their summits. This indicates that at one time they were surface features and that their flat tops are the result of wind and wave erosion. They have since subsided owing to their weight, the accumulated sediment load bearing down on the oceanic crust, and crustal motion. They have also been submerged by rising sea level during periods in which land ice melted.

In the warm waters of the Atlantic, Pacific, and Indian oceans, coral reefs and coral islands are formed in association with seamounts. Reef-building coral is a warm-water animal that requires a place of attachment and grows in intimate association with a single-celled plant. It is therefore confined to sunlit, shallow, tropic waters. When a seamount pierces the sea surface to form an island, it provides a base

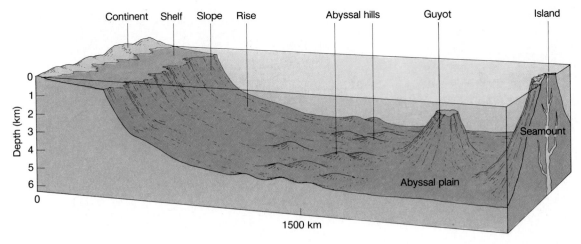

Figure 2.8
An idealized portion of ocean basin floor with abyssal hills, a guyot, and a seamount on the abyssal plain. Seamounts and guyots are known to be volcanic in origin.

on which the coral can grow. The coral grows to form a **fringing reef** around the island. If the seamount sinks or subsides slowly enough, the coral continues to grow upward at a rate that is not exceeded by the rising water, and a **barrier reef** with a lagoon between the reef and the island is formed. If the process continues, eventually the seamount disappears below the surface and the coral reef is left as a ring, or **atoll.** This process is illustrated in figure 2.9.

The steps required to form an atoll were suggested by Charles Darwin, based on the observations he made during the voyage of the *Beagle* from 1831 to 1836. Darwin's ideas have been proved to be substantially correct by more recent expeditions that drilled through the debris on the lagoon floor and found the basalt peak of the seamount that once protruded above the sea surface.

The Ridges and Rises

The most remarkable features of the ocean basin floor are the midocean **ridge and rise systems.** This series of great continuous underwater volcanic mountain ranges stretches for 65,000 km (40,000 mi) around the world and runs through every ocean. See figure 2.10 for a bird's-eye view of the system. These ridge systems are about 1000 km (600 mi) wide and 1000–2000 m (3500–7000 ft) high. If the slopes of these mountain ranges are steep they are referred to as ridges (such as the Mid-Atlantic Ridge and the Mid-Indian Ridge); if the slopes are more gentle they are called rises (such as the East Pacific Rise). Along some portions of the system's crest is a central **rift valley,** 15–50 km (9–30 mi) wide and 500–1500 m (1500–5000 ft) deep. The rift valley is volcanically very active and bordered by rugged rift mountains. Steep-sided fracture zones known as **transform faults** run perpendicular to the ridges and rises,

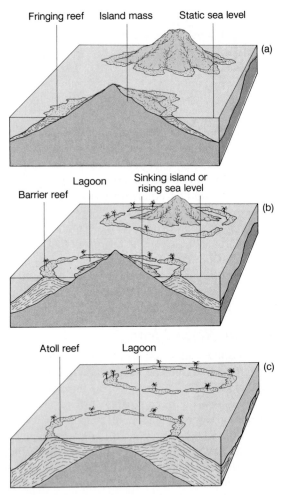

Figure 2.9
Types of coral reefs and the steps in the formation of a coral atoll are shown in profile: (a) fringing reef; (b) barrier reef; and (c) atoll reef.

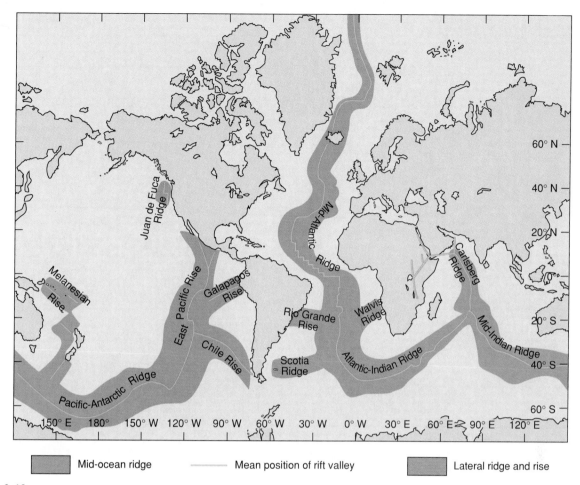

▨ Mid-ocean ridge	— Mean position of rift valley	▨ Lateral ridge and rise

Figure 2.10

The midocean ridge and rise system with its central rift valley.

connecting offset sections of the midocean ridges. These features are seen in the physiographic chart of the seafloor in figure 2.13.

The midocean ridges and rises separate the ocean basins into a series of sub-basins, which are shown in figure 2.11 and listed in table 2.2. The deep waters within these smaller basins are isolated from each other by the height of the ridges separating them. A low ridge allows good communication of deep water between basins, while a high, continuous ridge effectively isolates deep water in one basin from that in another. For example, deep water in the Angola Basin, located in the bight of the west coast of Africa, is cut off from the Brazil Basin to the west by the Mid-Atlantic Ridge and from the Cape Basin to the south by the Walvis Ridge.

The Trenches

Just as the Mid-Atlantic Ridge is the dominant feature of the floor of the Atlantic Ocean, the narrow, steep-sided, deep-ocean **trenches** characterize the Pacific (see fig. 2.12).

Some of these trenches are located on the seaward side of chains of volcanic islands known as **island arcs.** For example, the Japan-Kuril Trench, the Aleutian Trench, the Philippine Trench, and the deepest of all ocean trenches, the Mariana Trench, are all associated with island arc systems. The Challenger Deep, a portion of the Mariana Trench, has a depth of 11,020 m (36,000 ft), making it the deepest known spot in all the oceans. The longest of the trenches is the Peru-Chile Trench, stretching 5900 km (3700 mi) down the west side of South America. To the north, the Middle America Trench borders Central America. The Peru-Chile and Middle America trenches are not associated with volcanic islands but are bordered by land volcanoes. In the Indian Ocean the great Java Trench runs for 4500 km (2800 mi) along Indonesia, while in the Atlantic there are only two comparatively short trenches: the Puerto Rico-Cayman Trench and the South Sandwich Trench, both associated with chains of volcanic islands. To view the bathymetry of the ocean floor as it is known to exist today, see figure 2.13.

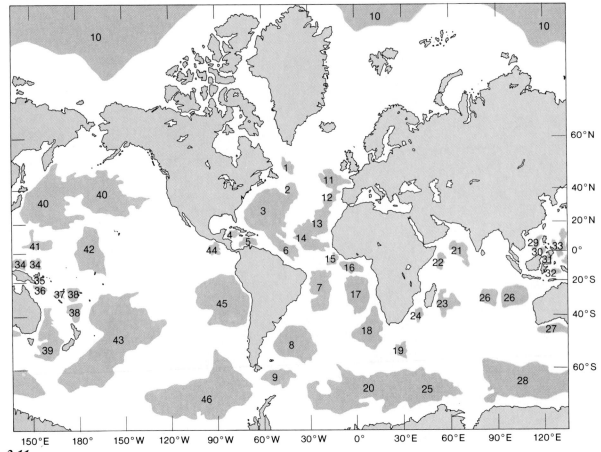

Figure 2.11

Major ocean basins of the world. Numbers are keyed to the basins listed in
table 2.2.

Table 2.2			
Major Ocean Basins of the World[1]			
Ocean	**Western side**	**Central**	**Eastern side**
Atlantic	1. Labrador Basin		10. North Polar Basin
	2. Newfoundland Basin		11. West Europe Basin
	3. North American Basin		12. Iberia Basin
	4. Western Caribbean Basin		13. Canaries Basin
	5. Eastern Caribbean Basin		14. Cape Verde Basin
	6. Guiana Basin		15. Sierra Leone Basin
	7. Brazil Basin		16. Guinea Basin
	8. Argentina Basin		17. Angola Basin
	9. South Antilles Basin		18. Cape Basin
			19. Agulhas Basin
			20. Atlantic-Indian Antarctic Basin
Indian	21. Arabian Basin		26. India-Australia Basin
	22. Somali Basin		27. South Australia Basin
	23. Mascarenes Basin		28. Eastern Indian Antarctic Basin
	24. Madagascar Basin		29. South China Basin
	25. Atlantic-Indian Antarctic Basin		30. Sulu Basin
			31. Celebes Basin
			32. Banda Basin
Pacific	33. Philippines Basin	40. North Pacific Basin	44. Guatemala Basin
	34. Caroline Basin	41. Mariana Basin	45. Peru Basin
	35. Solomon Basin	42. Central Pacific Basin	46. Pacific Antarctic Basin
	36. Coral Basin	43. South Pacific Basin	
	37. New Hebrides Basin		
	38. Fiji Basin		
	39. East Australia Basin		

[1]The numbers are keyed to areas marked in figure 2.11.

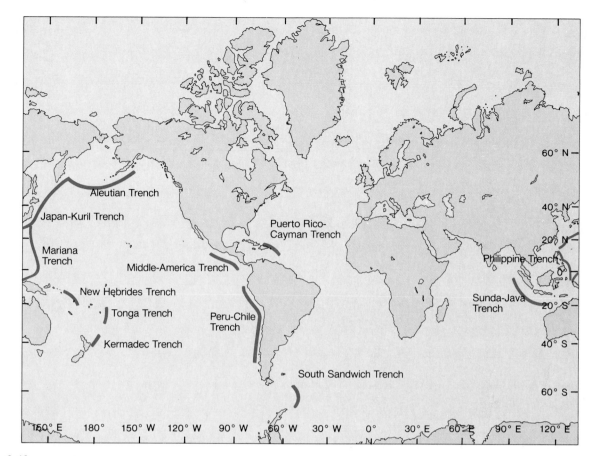

Figure 2.12
Major ocean trenches of the world.

2.2   Measuring the Depths

The discovery of the deep ocean's bathymetric features and the accumulation of accurate depth measurements in the deep sea are both accomplishments of recent technology. Early ocean explorers were not concerned with knowing how deep the ocean was; they were interested in how shallow the water was in the path of their ships. There were no practical reasons to measure the great depths; only the curiosity of the explorer and scientist required satisfaction.

Early mariners made their **soundings,** or depth measurements, using a hemp line or rope marked in equal distances (usually **fathoms,** which was the length between a person's fully outstretched hands, standardized at six feet) with a greased lead weight at the end. The change in line tension when the weight touched bottom indicated depth, and the particles from the bottom adhering to the grease confirmed the contact and brought a bottom sample to the surface. This method was quite satisfactory in shallow water, and the experienced captain used the properties of the bottom sample to aid in navigation, particularly at night or in heavy fog. In deep water, however, the weight of the hemp line was so great that it was difficult to sense when the lead weight touched the bottom. The sediments adher-

ing to the grease could still confirm a touch, but there was no way of knowing how much slack line lay on the bottom. For this reason the deeper areas measured by this technique were often thought to be greater than their real depth.

Later, piano wire with a cannonball attached to one end was used in deep water. The heavy weight of the ball compared to the weight of the wire made it easier to sense the bottom, but the time (eight to ten hours) to winch the wire out and in and the effort consumed for each measurement were so great that by 1895 only about 7000 depth measurements had been made in water greater than 2000 m and only 550 measurements had been made of depths greater than 9000 m over all the world's oceans.

It was not until the 1920s, when acoustic sounding equipment was invented, that deep-sea depth measurements became routine. The **echo sounder,** or **depth recorder,** which measures the time required for a sound pulse to leave the surface vessel, reflect off the bottom, and return, allows continuous measurements to be made easily and quickly when the ship is under way. The behavior of sound in seawater and its uses as an oceanographic tool are discussed in chapter 4. A trace from a depth recorder is shown in figure 2.14.

Figure 2.13

A physiographic chart of the world's oceans.

From *Floor of the Oceans,* Bruce C. Heezen and Marie Tharp, 1975, copyright © by
Marie Tharp 1980. Reproduced by permission of Marie Tharp, 1 Washington Ave.,
South Nyack, NY 10960.

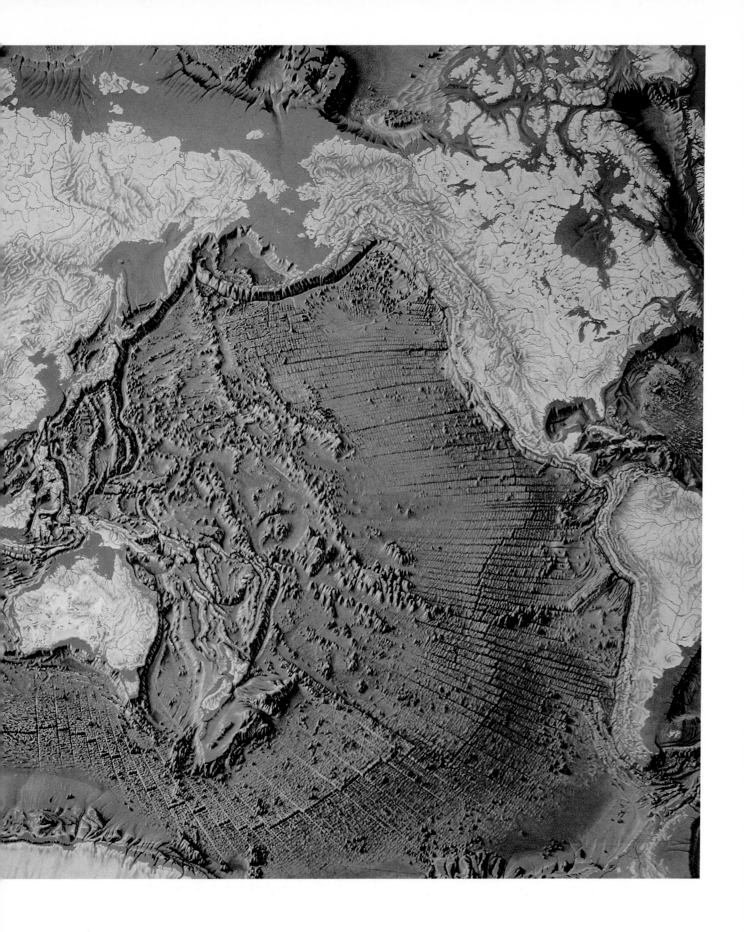

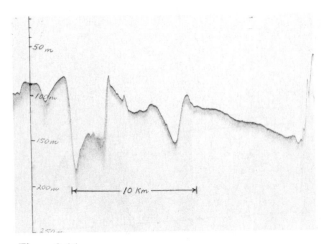

Figure 2.14

Depth recorder trace. A sound pulse reflected from the ocean floor traces a depth profile as the ship sails a steady course.

Figure 2.15

This internally recording television camera is placed on the seafloor, where it automatically photographs events until it is retrieved by the researchers.

In 1925, the German *Meteor* expedition made the first large-scale use of an echo sounder on a deep-sea oceanographic research cruise. After this expedition, depth measurements gradually accumulated at an ever-increasing rate. As the acoustic equipment improved and was used more frequently, knowledge of the ocean floor's bathymetry expanded and improved, culminating in the 1950s with the detailed mapping of the midocean ridge system. Today, data from precision depth recorders, underwater photography, and television and direct observation by submersibles are used together to produce even more detailed knowledge of the bathymetry of the seafloor (see fig. 2.15).

A new method of sensing seafloor features is based on satellite measurements of the ocean's irregular surface. These irregularities are related to changes in local gravity due to the uneven seafloor. The sea surface may be depressed as much as 60 m (200 ft) over ocean trenches and may bulge as much as 5 m (15 ft) over seamounts and ridges. See figure 2.16.

2.3 Sediments

The margins of the continents and the ocean basin floors receive a continuous supply of particles from many sources. Whether these particles have their origin in living organisms, the land, the atmosphere, or the sea itself, they are called sediment when they accumulate on the seafloor. Thick deposits of sediment are found near the continental margins, while the deep seafloor receives a constant but slow accumulation of sediment.

Oceanographers study the rate at which sediments accumulate, the distribution of sediments over the sea bottom, their sources and abundance, their chemistry, and the history they record in layer after layer as they slowly but continuously accumulate on the ocean floors. In order to describe and catalogue the sediments, geological oceanographers have classified the sediments by their source, place of deposit, chemistry, particle size, age, and color.

Sediment Sources

Classification by source relates the sediments to their origins from rocks, organisms, seawater, and space. Sediments derived from rock are **lithogenous sediments.** Rock at or above sea level is weathered by wind, water, and continual freezing and thawing. The resulting particles are transported to the oceans by water, wind, gravity, and ice. Most of the lithogenous sediments are deposited on the continental margins; smaller amounts are carried to the deep sea. Windblown dust from arid areas of the continents and ash from active volcanoes are additional sources of lithogenous particles.

Sediments formed from the remains of living organisms are called **biogenous sediments** and may include shell and coral fragments as well as the hard skeletal parts of single-celled plants and animals that live in the surface waters. Biogenous sediments are discussed in more detail later in this chapter.

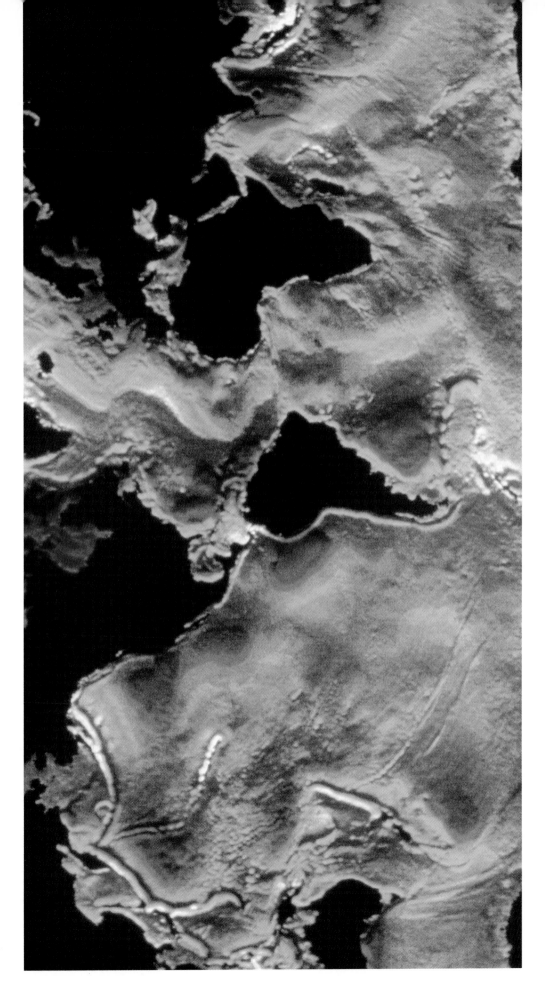

Figure 2.16
Sea-surface topography reflects
seafloor bathymetry. Seventy
days' worth of data collected by
the short-lived SEASAT satellite
were used to produce this
computer-generated chart.
Marked are (1) the Mid-Atlantic
Ridge, (2 and 3) trenches along
the west and northwest margins
of the Pacific, (4) the Hawaiian
Island-Emperor seamount
system, (5) the Louisville Ridge
of the South Pacific, (6) fracture
zones produced by stresses acting
on the earth's crustal plates, and
(7) low-amplitude bumps that
may mark upwellings of molten
rock circulating in the earth's
interior.

Visualizing the seafloor began with single soundings made with a lead line and continued with simple echo sounders and contour maps drawn by hand. In one hour an individual with a lead line could take 20 measurements in water 10 m (33 ft) deep and in four hours only one measurement in water 4000 m (132,000 ft) deep; the echo sounder allowed 36,000 measurements to be made each hour in 10 m of water and 680 in the same time in 4000 m of water. Today's multibeam sound systems can take 293,000 measurements per hour in 10 m of water and 20,000 measurements in 4000 m. Advances in multibeam sound system technology and improved computer graphics, combined with satellite navigation for precise positioning, are opening dramatic new windows to the seafloor.

A single sound beam device releases a cone of sound; as the depth of the water increases the area of the seafloor from which the echo is reflected also increases. Depth is averaged over the "footprint" of the sound beam, and therefore seafloor features smaller than the footprint are difficult to detect and detail is reduced. Two new technologies using multiple sound devices are being used to produce detailed, high-resolution seafloor maps: (1) side-scan acoustical imaging and (2) swath bathymetry.

Side scanner measurements can be made either from a surface vessel or from a submerged system towed behind a vessel. If the ship is pitching and rolling, the path of the sound beam from a surface vessel will be displaced from its intended direction, resulting in inaccurate data. A towed system is below the depth of surface waves and winds; it is also closer to the seafloor, allowing the use of a conical sound beam that produces a smaller sound footprint on the seafloor. The smaller footprint increases the detail that is imaged but decreases the scanned area for each cone of sound. To increase the area surveyed, multiple sound beams are sent out obliquely on either side of the sound device; no image is obtained from directly under the side scanner.

Side-scan acoustical images are the product of the reflectivity of the seafloor materials and the angle at which the

Box figure 2.1

The image of a plane on the seafloor obtained by side-scan sonar. Note the shadow generated when sound was not returned from the seabed.

sound beams strike the seafloor. Changes in the reflection of the sound come from the irregularities and the changing properties of the bottom being scanned. The sides of a seamount, a fault, and other objects with strong topographic relief act as good reflectors. Side-scan acoustical imaging also works very well to detect sunken ships, planes, or other structures, because the reflecting surfaces of these structures are at an angle to the seafloor, and their acoustical properties are very different from those of the seafloor. The object's shape is accompanied by an acoustical shadow (seen behind the plane in the Box fig. 2.1) that provides strong image definition and indicates elevation above the seafloor.

Multiple side-scan acoustical beams can also be used to produce detailed bathymetric images of the seafloor. Very slight differences in the times that the echoes return to the side scanner's multiple receivers can be used to calculate the angle of the returning sound beam and therefore the depth from which it is reflected. Another technique determines depth and produces images by analyzing the sound interference patterns between the outgoing sound beams and their returning echoes.

Swath bathymetry is accomplished by mounting a series (as many as 59) of sound sources along the length of a vessel's hull. Multiple sound receivers are mounted in the other direction, across the hull. Both sound sources and receivers are directed vertically downward. The outgoing cones of sound overlap and combine to produce focused signals over a small seafloor area directly below and at right angles to the vessel's track. The length of the area returning echoes is about twice the depth of the water; this is the swath length. The width of each swath in the direction of the vessel's motion is determined by the sound cone angle. The receivers of this system sense the returning echo's angle to measure depth; the reflecting properties of the seafloor are not used.

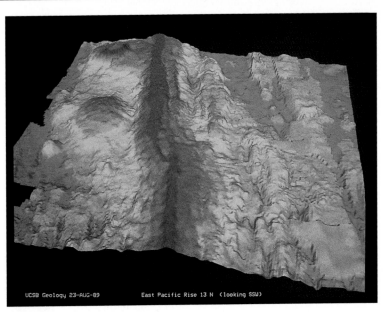

Box figure 2.2
Three-dimensional, computer processed image of a section of the East Pacific Rise. Data for this image was obtained by using the Sea Beam echo sounder.

To produce a bathymetric survey chart the returning echo data from either side-scan acoustical systems or swath bathymetry, plus the location of the vessel calculated by satellite using GPS control (see chapter 1) and the known depth and position of the towed system are integrated by computer. Large computer driven plotters produce charts showing latitude and longitude, depth, and depth contours as the data is collected. With systems such as these it is possible to produce three-dimensional images of the seafloor from any angle and with the perspective required for any particular task; see Box figure 2.2.

In certain areas sediments are produced in seawater by chemical processes. These sediments are called **hydrogenous sediments** and include **carbonates** (limestone-type deposits), **phosphorites** (phosphorus in phosphate form in crusts and nodules), and **manganese nodules.** The manganese nodules are deposits of manganese, iron, copper, cobalt, and nickel in the form of variously shaped nodules scattered across the ocean floor and lying on top of the other sediments, as shown in figure 2.17. Nodules develop very slowly (1–10 mm per million years), accumulating layer upon layer, often around a hard skeletal piece such as a shark's tooth or the ear bone of a whale. The nodules form in areas of slow sediment accumulation or rapid bottom currents. If there is a rapid accumulation of other sediments they either do not form or are buried quickly by the constant rain of particles from above.

Salt deposits occur when a high rate of evaporation removes most of the water and leaves a very salty brine in shallow areas. Chemical reactions occur in the brine and salts are deposited on the bottom. In such a process carbonate salts are formed first, followed by sulfate salts, and then rock salt.

Outer space is an intriguing source of ocean sediment. Particles from space constantly bombard the earth's surface; the particles are very small, stay in suspension for a very long time, and most are thought to dissolve as they slowly sink to the seafloor. These iron-rich **cosmogenous sediments** are found in small amounts in all oceans scattered on the surface of the other sediments. The pattern of related cosmic materials indicates the direction of the particle shower that supplied them. The particles become very hot as they pass through the earth's atmosphere and partially melt; this melting gives the particles a characteristic rounded or teardrop shape. Cosmic bodies can disintegrate and melt surface materials as they strike the earth. Their impact can cause a splash of melted particles that spray outward and produce splash-form **tektites** (see fig. 2.18). Microtektites are found on the ocean floor and on land.

Biogenous Oozes

Deep-sea biogenous sediments are composed almost entirely of the hard outer coverings of single-celled organisms (fig. 2.19). Their chemical composition is either calcareous (calcium carbonate or shell material) or siliceous (clear and hard). If fine sediments from the deep sea are more than 30% fine biogenous sediment by weight, then the sediment is known as an **ooze**—a **calcareous ooze** or a **siliceous ooze.**

The distribution of calcareous and siliceous oozes in the sediments is related to the supply of organisms, the dissolving abilities of the water through which they fall, water depth, and dilution with other sediment types. In the Atlantic Ocean, the waters from the surface down to a depth of about 4000 m (13,000 ft) are saturated with calcite, a form of calcium carbonate, and the calcareous materials do not dissolve in these calcite-rich waters. Below 4000 m (13,000 ft) the water is undersaturated with calcite, and calcareous materials dissolve readily. In the North Pacific Ocean, water saturation with calcite only occurs from the surface to a few hundred meters depth. This means that calcite-type calcareous remains dissolve very little as they fall through the upper 4000 m (13,000 ft) of the Atlantic and the upper 500 m (1600 ft) of the northern Pacific. The South Pacific has water saturated with calcite down to about 2500 m (8000 ft). Below these depths the higher carbon dioxide content of aged water makes the water more acidic and capable of dissolving the easily broken down calcite particles.

Calcite solubility has an unusual temperature dependence: calcite is more soluble in cold water than in warm water. This causes the low temperature surface water at higher latitudes to have an increased ability to dissolve calcite carbonates, further limiting the accumulation of carbonates in the underlying sediments. This cold surface water effect coupled with the water's changing carbonate-dissolving ability with depth and age produces the observed distribution of carbonate sediments. Calcareous deposits are found at temperate and tropical latitudes in shallower areas of the seafloor such as the Caribbean Sea, on elevated ridge systems, and in coastal regions.

The waters of both oceans are undersaturated for biogenic silica over their entire depth. The siliceous remains of marine organisms are resistant to dissolution and break down slowly, even though the entire water column through which they fall is undersaturated. The siliceous remains of organisms are found at all depths because of their slow dissolution rates at all temperatures. In regions where there is a high supply rate of siliceous biogenic remains, and where calcareous remains are more readily dissolved, the siliceous content of the accumulating sediments is high. The sediments surrounding Antarctica are an example.

(a)

(b)

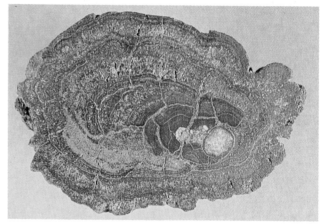

(c)

Figure 2.17

(a) Manganese nodules photographed under artificial light on the seafloor. (b) A box core sample, photographed on deck in natural light, showing manganese nodules resting on red clay. (c) A cross section of a manganese nodule showing concentric layers of formation.

Figure 2.18
Tektites collected from Southeast Asia.

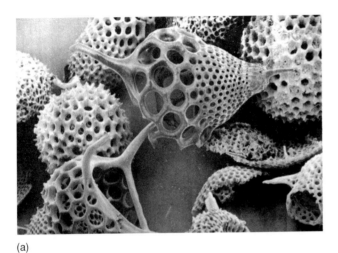

(a)

(b)

Figure 2.19
Photomicrographs of biogenous sediments: (a) radiolarians; (b) diatoms and coccolithophores. The small disks are detached coccoliths.

Siliceous oozes have one principal source in colder and temperate latitudes: single-celled plants called **diatoms** that produce the **diatomaceous ooze** found around the Antarctic and in a band across the North Pacific. (See fig. 2.19b for a picture of diatom skeletons.) Because the diatoms are plants, they require sunlight and fertilizers, or **nutrients,** for growth. The sunlight is available at the ocean's surface; the nutrients are produced by the decomposition of all plant and animal life in the ocean, and these nutrients are liberated in the deeper water as the decomposition takes place. Only at certain locations are these nutrients returned to the surface by a large-scale upward flow of deeper water. Where this upward flow occurs, the sunlight combines with the nutrients to provide the conditions needed for high levels of plant production. Production is discussed in chapter 14. Because diatoms reproduce rapidly in cold waters, large populations are found in the areas that combine light, nutrients, and suitable temperatures: the waters of the North Pacific and Antarctic oceans. See figure 2.20 to trace the distribution of diatomaceous oozes.

Siliceous sediments are also found in equatorial latitudes (again, see fig. 2.20). These sediments are not produced by diatoms because the water at these latitudes is too warm. Here, the principal source is a single-celled, amoeba-like animal, the **radiolarian,** which produces a siliceous outer shell that is often covered with long spines. These delicate, glasslike external coverings may accumulate in sufficient quantity to produce **radiolarian ooze** in tropic areas. Figure 2.19a shows radiolarian sediments at high magnification.

Calcareous oozes are formed at the shallower oceanic depths from the remains of other small animals and plants, with worldwide distribution. Look once more at figure 2.20 to locate the deposits of calcareous oozes that are often found on the major ridge systems. Both the microscopic **foraminifera,** another single-celled relative of the amoeba, and the **pteropods,** small, graceful, swimming snails, contribute their shells to these sediments. Minute, single-celled plants, **coccolithophores,** covered with calcareous plates called coccoliths, also play a role in the formation of these bottom deposits. Each of these is named for its principal constituent: foraminifera ooze, pteropod ooze, and coccolithophore ooze.

Red Clay

On the deep-ocean floor, far away from the land and under water that holds little marine life, a very fine lithogenous material accumulates very slowly. This fine rock powder may remain suspended in the water for many years; it is believed to have been blown out to sea by wind or to have been swept out of the atmosphere by rain. This fine sedi-

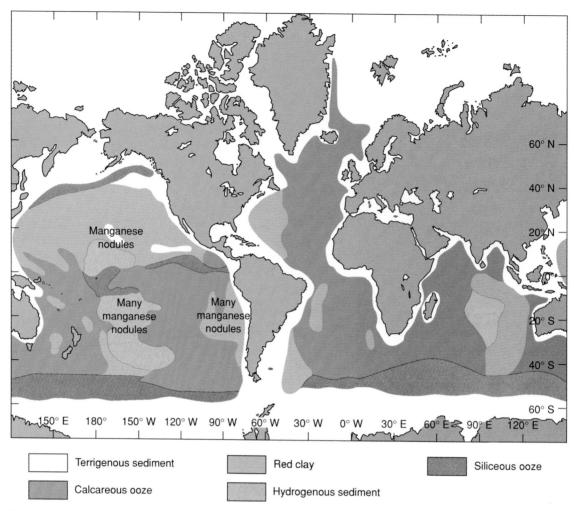

Figure 2.20

Distribution of the principal sediment types on the deep seafloor. Sediments are usually a mixture but are named for their major component.

ment is known variously as **red clay, brown clay,** or **brown mud.** The red-brown color is due to the oxidizing (or rusting) of the iron present in the sediment. Refer back to fig. 2.17b for a view of red clay. Although traces of red clay are found throughout the oceans, it only becomes a substantial component of the sediment in areas where sediments from other sources do not mask its presence, and in the deep basin areas of the Pacific Ocean.

Patterns of Deposit

Ocean scientists classify sediments by the area in which they are deposited as well as by their source. **Terrigenous sediments** are derived from land and found close to their land source (for example, silt, sand, gravel, wood chips, and sewage sludge). **Neritic sediments,** or coastal sediments, come from land and sea sources and are found deposited under the shallow waters of the continental shelf. **Pelagic sediments** are deep-sea sediments that have been deposited far from the direct influence of land. Areas of sediment deposits are shown in figure 2.21.

The patterns formed by the sediments on the seafloor reflect both distance from their source and processes that control the rates at which they are produced, transported, and deposited. Seventy-five percent of marine sediments

are terrigenous; the majority of these terrigenous sediments are deposited on the continental margins. The terrigenous sediments of coastal regions are primarily lithogenous, supplied by rivers and wave erosion along the coasts. Coarse sediments are concentrated close to their sources in high-energy environments, for example beaches with swift currents and breaking waves. The waves and currents move quite large rock particles in the shore zone, but these larger particles settle out quickly. Finer particles are held in suspension and carried farther away from their source. This results in a gradation by particle size: coarse particles close to shore and to their source, with finer and finer particles predominating as the distance from the source increases.

Finer sediments are deposited in low-energy environments, offshore away from the currents and waves or in quiet bays and estuaries. In higher latitudes, deposits of rock and gravel carried along by glaciers are found in coastal environments, while in equatorial latitudes fine sediments predominate and are considered to be products of large rivers, heavy rainfall, and loose surface soils.

At the present time, most of the land-derived sediments are accumulating off the world's river mouths and in estuaries. Estuaries and river deltas serve as sediment traps, preventing these terrigenous sediments from reaching the

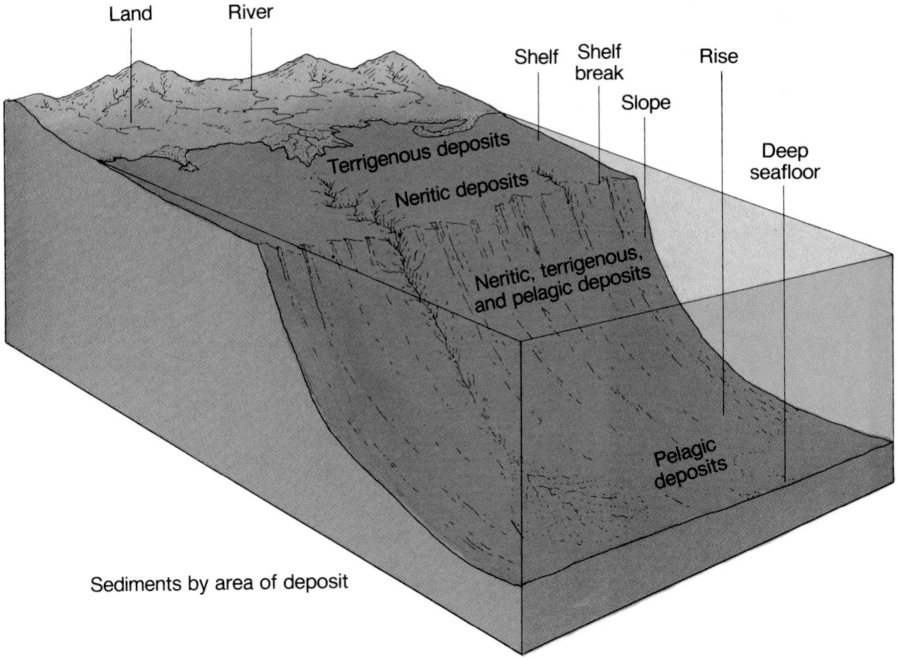

Figure 2.21

Classification of sediments by location of deposit. The distribution pattern is in part controlled by the sediments' proximity to source and rate of supply.

Land River

Shelf Shelf break Rise

Slope

Deep seafloor

Terrigenous deposits

Neritic deposits

Neritic, terrigenous, and pelagic deposits

Pelagic deposits

Sediments by area of deposit

Figure 2.22

A deep-sea sediment core obtained by the drilling ship *Glomar Challenger*. Note the layering of the sediments.

deep seafloor in such places as the Chesapeake and Delaware bay systems along the North Atlantic coast, in the Georgia Strait of British Columbia, and in California's San Francisco Bay along the west coast. If sediments are supplied to a delta faster than they can be retained, the sediments will move across the shelf into the deeper water environments. This is currently the case with the sediments of the Mississippi River (see the box in chapter 12). Much of the total thickness of sediments on the outer continental shelf is sediment laid down during the ice ages when the sea level was lower; for example, Georges Bank southeast of Cape Cod. Little is currently being added to these outer regions of the continental shelf.

The rapid accumulation of sediments on the continental shelf results in large, unstable, steep-sided deposits that may slump, sending a flow of terrigenous sediment moving rapidly down the continental slope in a turbidity current (see the previous discussion of turbidity currents). Turbidity currents move coarse materials of land origin a long way out to sea; in doing so they distort the general deep-sea sediment pattern. Near shore, the spring flooding of the rivers alternates with periods of low river discharge in summer and fall. The floods bring large quantities of sediment to the coastal waters, and a record of the contribution of this flooding is seen in the layering of the sediments. Sudden contributions of sediment material from the collapse of a cliff or the eruption of a volcano are seen in the sediment pattern as specific additions of large quantities of sand or ash.

Oceanic sediments are laid down in layers characterized by changes in color, particle size, and kinds of particles (see fig. 2.22). Layering occurs if the sediment type varies over time or if the rate at which a sediment type is supplied varies. Seasonal variations and the patterns of long

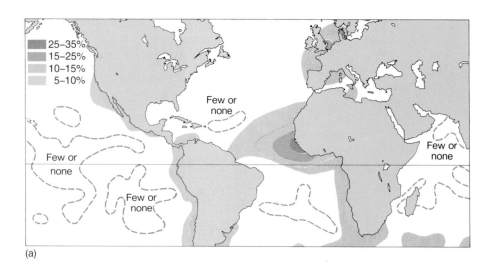

(a)

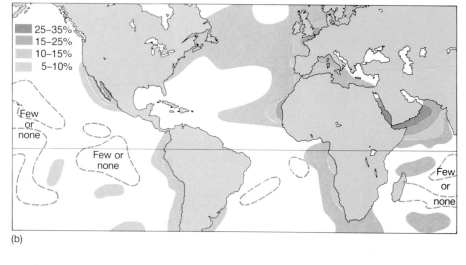

(b)

Figure 2.23

Frequency of haze as a result of airborne
dust during the Northern Hemisphere's
(a) winter and (b) summer. Values are given
in percentages of total observations.

and short growing seasons among marine life can also be
determined from the properties and thicknesses of the layers
of biogenous material. Over long periods of geologic time,
climatic changes such as the ice ages have altered the biological populations that produce sediment and have left their
record in the sediment layers.

Along the continental margins, biogenous sediments
are diluted by the large amounts of lithogenous sediment
washing from the land. When marine life is abundant in
coastal areas, biogenous sediments are formed from shells
and corals. In the more homogeneous environment of the
deep sea, biogenous sediments make up the majority of the
pelagic deposits. There is little dilution with terrigenous
materials, and few environmental changes disturb the bottom deposits. Seafloor conditions may stay relatively
unchanged for long periods of time in contrast to the cycles
of climate change, which cause processes to fluctuate along
the continental margins. Calcareous oozes are found where
the production of organisms is high, dilution by other sediments is small, and depths are less than 4000 m (13,000 ft).
See the areas including the midocean ridges and the
warmer shallower areas of the South Pacific (refer back to
fig. 2.20). Siliceous oozes cover the deep seafloor beneath
the colder surface waters of 50–60° N and S latitude, and in

equatorial regions where cold deeper water is brought to
the surface by vertical circulation processes. Deep basin
areas of the Pacific have extensive deposits of red clay
(again, refer back to fig. 2.20).

Large rock particles of land origin are also moved out
to sea by a process known as **rafting.** Glaciers carry sand,
gravel, and rocks with them as they cut through the earth's
crust. If the glacier reaches the sea, parts of it break off and
fall into the water as icebergs. The icebergs are carried
away from land by the currents and winds, and as the ice
melts, the rocks and gravel that were frozen in the ice sink
to the seafloor, far from their original source. Sometimes
large, brown seaweeds known as kelp, which grow attached
to rocks in coastal areas, are dislodged by storm waves. The
kelp may have enough buoyancy to float away, carrying
with it the rock to which it is attached. When the plant dies
or sinks, the rock is deposited on the ocean floor at some
distance from its origin. The deposit of larger rocks by this
process of rafting is an irregular and isolated process.

In some parts of the world the wind is an effective
agent for the movement of lithogenous materials out to sea.
Figure 2.23 indicates the frequency with which winds carry
airborne dust, or haze, out to sea. Winds blowing offshore
from the Sahara Desert or other arid regions transfer sand

Figure 2.24
Screens used for sorting sediment samples into different size ranges. Examples of materials sorted into phi (φ) sizes.

Table 2.3

Classification of Sediments by Particle Size

Descriptive name		Diameter ranges (mm)	φ sizes
gravel	boulder	> 256	< –8
	cobble	64–256	–6 to –8
	pebble	4–64	–2 to –6
	granule	2–4	–1 to –2
sand	very coarse sand	1–2	0 to –1
	coarse sand	0.5–1	1 to 0
	medium sand	0.25–0.5	2 to 1
	fine sand	0.125–0.25	3 to 2
	very fine sand	0.0625–0.125	4 to 3
mud	silt	0.0039–0.0625	8 to 4
	clay	< 0.0039	> 8

Geologists use a phi or φ size scale, based on the negative of the power of 2, to describe these size ranges. For example, if particle diameter is 64 mm or 2^6, then its phi size is –6; a fine silt with grains 1/128th millimeter, or 2^{-7} mm has a phi size of 7. When a sediment sample is collected it is dried and shaken through a series of screens that have woven mesh of decreasing size (fig. 2.24). Material that passes through one screen but not the next is said to have a diameter size range between the two phi sizes of the screens. The geological oceanographer determines the percentage by weight of sediment in each size category in a sediment sample. This process effectively identifies the sediment and allows comparisons to be made with deposits at other locations; it also shows any changes in the sediments at a location sampled over a period of time. Figure 2.25 shows how sediments appear when they are mixtures of different particle sizes.

If a sediment contains particles of mixed sizes, it is said to be poorly sorted (fig. 2.25a). But if the smaller and finer particles are carried away by turbulence and current flow before it is laid down, the remaining sediment becomes a well-sorted, coarse deposit. Figure 2.25b shows the fewer size categories in the graded series that are typical of a well-sorted sample. The washed-out finer material may then be deposited as a well-sorted fine deposit in another location. Size frequency graphs of the type shown in figure 2.25 made from deposit samples are compared with graphs from sediment sources. Such comparisons indicate to geologists the net result of the processes that acted on the sediment before they are deposited on the seafloor.

particles directly from land to sea, sometimes 1000 km (600 mi) or more offshore. A similar process can occur between sand dunes and coastal waters.

Particle Size

Sediments may be classified by particle size; see table 2.3. Familiar terms such as gravel, sand, mud, and pebble are defined by the specific size ranges of their particles.

Rates of Deposit

The rates at which sediments accumulate vary enormously, owing to the natural variability of the processes already discussed. An average accumulation value for the deep oceans is 0.5–1.0 cm (.2–.4 in)/1000 years, while at the ocean margins the rates are considerably greater and more varied. In river estuaries the rate may be more than 800,000 cm (315,000 in)/1000 years (800 cm (315 in)/year); each year the rivers of Asia, such as the Ganges, the Yangtse, the Yellow, and the Brahmaputra, contribute more than one-quarter of the world's land-derived marine sediments. In quiet bays, the rate may be 500 cm (197 in)/1000 years, while on the continental shelves and slopes values of 10–40 cm (3.9–15.7 in)/1000 years are appropriate, with the flat continental shelves receiving the larger amounts. Many of the sediments covering the continental shelves away from river mouths are sediments laid down by processes that no longer exist at that location. Such sediments are called **relict sediments** and represent conditions that existed three thousand to seven thousand years ago or longer, when the sea level was lower than it is today owing to the accumulation of water in ice caps and glaciers.

Although deep-sea sedimentation rates are excessively slow, there has been plenty of time during geologic history to accumulate the average deep-sea sediment thickness of approximately 500 to 600 m (1600–2000 ft). At a rate of 0.5 cm (.2 in)/1000 years, it takes only 100 million years to accumulate 500 m (1600 ft) of sediment, and the oldest seafloor crustal rocks are known to have existed for twice as long, or 200 million years. Very small particles sink extremely slowly (for example, a particle 0.001 mm (.0004 in) in diameter will take about 100 years to sink 3000 m (10,000 ft)), and scientists have puzzled over the close correlation in distribution between the particles found in the surface waters and those found on the seafloor. If particles sank at such slow rates, they would be carried very long distances from their source by ocean currents before reaching the bottom. Therefore, some mechanisms must be working to aggregate the tiny particles into larger particles, increasing the sinking rate, in order to explain how the particle distribution on the ocean floor could be so similar to that found in the overlying surface waters. Scientists have observed that small particles often attract each other owing to their electrical charges. This attraction forms larger particles that sink more rapidly. This process is important in the formation of the abundant sediment deposits in river deltas. Also, when small organisms ingest these tiny particles, the particles are expelled with the organism's wastes as larger fecal pellets. The fecal pellets themselves may be microscopic, but they are large enough to reach the seafloor in ten to fifteen days. Once the pellets are deposited, the breaking down of the organic portion of the pellets liberates the small particles.

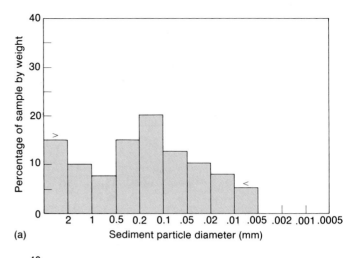

(a)

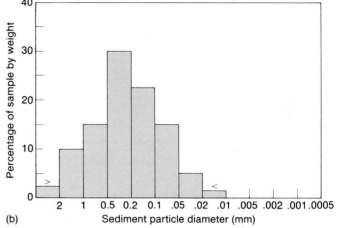

(b)

Figure 2.25

(a) A poorly sorted sample. Particles fall into a wide variety of size ranges in approximately equal amounts. (b) A well-sorted sample. One size range predominates in a limited distribution of sizes.

Sampling Methods

To analyze sediments the geological oceanographer must have an actual bottom sample to examine. A variety of devices have been developed to take a sample from the seafloor and return it to the laboratory for analysis. **Dredges** are net or wire baskets that are dragged across a bottom to collect loose bulk material, surface rocks, and shells in a somewhat haphazard manner (see fig. 2.26). **Grab samplers** are hinged devices that are spring- or weight-loaded to snap shut when the sampler strikes the bottom. See figure 2.27 for examples of this type of device. Grab samplers sample surface sediments from a fixed area of the seafloor from a single known location.

A **corer** is essentially a hollow pipe with a sharp cutting end. The free-falling pipe is forced down into the sediments by its weight or, for longer cores, by a piston device that enables water pressure to help drive the core barrel into the sediment; coring devices are shown in figure 2.28.

Figure 2.26

A geological dredge with heavy-duty chain bag. The dredge picks up loose rock and breaks off fragments from rock outcroppings on the seafloor.

The result is a cylinder of mud, usually 1 to 20 m (3.3–65.6 ft) long, that contains undisturbed sediment layers (refer back to fig. 2.22). Box corers are used when a large and nearly undisturbed sample of surface sediment is needed. These corers drive a rectangular metal box into the sediment, with doors that close over the bottom before the sample is retrieved. Longer cores may be obtained by drilling. For a discussion of the highly sophisticated drilling techniques used by the research vessel *JOIDES Resolution,* see chapter 3.

Figure 2.27

Grab samplers: Vanveen (left) and orange peel (right), both in open positions. Grabs take sediment samples.

<u>**2.4**</u> **Practical Considerations: Seabed Resources**

Long ago, people began to exploit the materials of the seabed. The ancient Greeks extended their lead and zinc mines under the sea, medieval Scottish miners followed veins of coal under the Firth of Forth, and more recently, coal has been mined from undersea veins off Japan, Turkey, and Canada. As technology has developed, and as people have become concerned about the depletion of onshore mineral reserves, interest in seabed minerals and mining has grown. During the past twenty-five years a reasonable but still incomplete inventory of these minerals and materials has been taken. The current situation for a number of these materials will be summarized and reviewed in this section. Keep in mind that each potential deep-sea source is in competition with an onshore supply. Whether the seabed source will be developed depends largely on international markets, needs for strategic materials, and whether offshore production costs can compete with onshore costs.

Sand and Gravel

The largest superficial seafloor mining operation is for sand and gravel, used in cement and concrete for buildings, for landfills, and to construct roads and artificial beaches. The technology and cost required to mine sand and gravel in shallow water differ very little from land operations. This is a high-bulk, low-cost material tied to the economics of transport and the distance to market. Annual world production is approximately 1.2 billion metric tons; the reported potential reserve is more than 800 billion metric tons. The United Kingdom and Japan each take 20% of their total annual sand and gravel requirements from the seafloor.

Sand and gravel mining is the only significant seabed mining done by the United States at this time. It is estimated that the United States has a reserve of 450 billion tons of

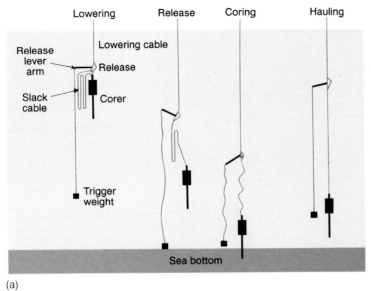

(a)

(c)

(d)

(e)

(b)

Figure 2.28

(a)The Phleger corer is a free-fall gravity corer. The weights help to drive the core barrel into the soft sediments. Inside the corer is a plastic liner. The sediment core is removed from the corer by removing the plastic tube, which is capped to form a storage container for the core. (b) A sketch of a piston corer in operation. The corer is allowed to fall freely to the sea bottom. The action of the piston moving up the core barrel owing to the tension on the cable then allows water pressure to force the core barrel into the sediments. (c) Loading a piston corer with weights to prepare it for use. (d) A gravity corer ready to be lowered. (e) A box corer is used to obtain large, undisturbed seafloor surface samples.

sand off its northeast coast, and large deposits of gravel along Georges Bank of New England and also in the area off New York City. Along the coasts of Louisiana, Texas, and Florida, shell deposits are mined for use in the lime and cement industries, as a source of calcium oxide used to remove magnesium from seawater, and, when crushed, as a gravel substitute along roads and highways.

Sands are mined as a source of calcium carbonate throughout the Bahamas Islands, which have an estimated reserve of 100 billion metric tons. Coral sands are mined in Fiji, in Hawaii, and along the U.S. Gulf Coast. Other coastal sands contain iron, tin, uranium, platinum, gold, and diamonds. The "tin belt" stretches for 3000 km (1800 mi) from northern Thailand and western Malaysia to Indonesia. Here, sediments rich in tin have been dredged for hundreds of years and supply more than 1% of the world's market. Iron-rich sediments are dredged in Japan, where the reserve of iron in shallow coastal waters is estimated at 36 million tons. The United States, Australia, and South Africa recover platinum from some sands, and gold is found in river delta sediments along Alaska, Oregon, Chile, South Africa, and Australia. Diamonds, like gold, are found in sediments washed down the rivers in some areas of Africa and Australia. Muds bearing copper, zinc, lead, and silver also occur on the continental slopes, but they lie too deep for exploitation, considering the present demand and their market value.

Phosphorite

Phosphorite, which can be mined to produce the phosphates needed for fertilizers, is found in shallow waters as phosphorite muds and sands containing 12% to 18% phosphate, and as nodules on the continental shelf and slope. The nodules contain about 30% phosphate, and large deposits are known to exist off Florida, California, Mexico, Peru, Australia, Japan, and northwestern and southern Africa. Recently a substantial source of phosphorite appears to have been located in Onslow Bay, North Carolina. Eight beds have been located and five are thought to be economically valuable; they have been estimated to contain 3 billion metric tons of phosphate concentrates.

The world's ocean reserve of phosphorite is estimated at about 50 billion tons. Readily available land reserves are not in short supply, but most of the world's land reserves are controlled by relatively few nations. Therefore, political considerations may make these marine deposits attractive as mining ventures for some countries. No commercial phosphorite mining occurs at present in the oceans.

Sulfur

In the Gulf of Mexico, sulfur is mined by injecting high-pressure steam to melt the sulfur, which is then pumped ashore to processing plants. Today, the cheaper and easier recovery of sulfur waste from pollution-control equipment has replaced all but one of the mining operations, and it is thought that the remaining mining operation may be closed down completely by the year 2000. Sulfur is necessary for the production of sulfuric acid, which is needed in nearly all industrial processes. If sulfur must be mined again, millions of tons of sulfur reserves are known to exist in the Gulf of Mexico and the Mediterranean Sea.

Coal

Coal was produced by land vegetation in tropic climates, but changes in sea level and land geography over geologic time have caused some coal deposits to become submerged. Coal deposits under the seafloor are mined when the coal is present in sufficient quantity and quality to make the operation worthwhile. In Japan, the undersea coal deposits are reached by shafts that stretch under the sea from the land or descend from artificial islands.

Oil and Gas

Oil and gas represent more than 90% of mineral value presently taken from the sea. 1988 figures show that 10.8% of U.S. oil and 24.6% of U.S. gas production came from offshore areas. Major offshore oil fields are found in the Gulf of Mexico, the Persian Gulf, and the North Sea, and off the northern coast of Australia, the southern coast of California, and the coasts of the Arctic Ocean. Three new large fields are opening in deeper water in the 1990s: Hibernia off northeastern Newfoundland, Auger in 850 m (2800 ft) of water off Louisiana, and Heidrun in 350 m (1150 ft) off Norway in the North Sea. Heidrun's development cost is about $3 billion. At the present time many U.S. companies are finding it more profitable to drill for oil and gas in foreign waters and are moving their rigs to the waters of the North Sea, West Africa, and Brazil.

Bringing the offshore oil fields into production has required the development of massive drilling platforms and specialized equipment to withstand heavy seas and fierce storms, and to allow drilling and development of wellheads at great depth (see fig. 2.29). Although the cost of drilling and equipping an offshore well is three to four times greater than that of a similar venture on land, the large size of the deposits allows offshore ventures to compete successfully. The gas and oil potential of the much greater depths of the ocean floor is still unknown, but the deeper the water in which the drilling must be done, the higher the cost. Exxon is studying a project that would require drilling in 1400 m (4600 ft), more than twice the depth of its current deepest well. Even though legal restraints, environmental concerns, and worldwide political uncertainties will continue to contribute to the slow development of offshore deposits, petroleum exploration and development will undoubtedly continue to be the main focus of ocean mining in the near future.

The new methods and equipment used and developed for deep-sea oceanographic drilling and research have provided the prototypes for new generations of deep-sea

Figure 2.29
The Chevron oil-drilling platform *Hermosa* off the California coast.

commercial drilling systems. Even though legal restraints, environmental concerns, and worldwide political uncertainties will continue to contribute to the slow development of offshore deposits, petroleum exploration and development will undoubtedly continue to be the main focus of ocean mining in the near future.

Manganese Nodules

Manganese nodules are hydrogenous, pelagic deposits found scattered across the world's deep-ocean floors, with particular concentrations in the red clay regions of the northeast Pacific (refer back to fig. 2.17). The nodules were discovered by the *Challenger* expedition (1873–76) and in the last twenty-five years have been the focus of intense research and development of mining and extraction techniques by large mining corporations and multinational consortia (see fig. 2.30). The mineral content varies from place to place, but the nodules in some areas contain 30% manganese, 1% copper, 1.25% nickel, and 0.25% cobalt, much higher concentrations than are usually found in land ores. Although the nodules grow at an estimated rate of 0.1 mm/1000 years, they are already present in vast quantities, and an estimated 16 million additional tons accumulate each year.

Figure 2.30
The *Deep Sea Miner II,* a converted ore carrier, is used to make deep-ocean mining tests.

Since the 1960s, large multinational consortia have spent over $600 million to locate the highest nodule concentrations and develop technologies for their collection. However, their expectations of rapid development have been disappointed. In the 1980s some of these consortia had withdrawn completely while others were dormant. The primary reason for the lack of development of this industry is the presently depressed international market in metals. Another reason involves the history of ownership of the pelagic nodules (see the section in this chapter on Laws and Treaties).

Progress in the 1990s has continued to be slow and mostly outside the U.S. An organization of twelve South Pacific Island nations (Cook Islands, Federated States of Micronesia, Fiji, Guam, Keribati, Marshall Islands, Papua New Guinea, Solomon Islands, Tonga, Tuvalu, Vanuatu, and Western New Guinea) with Australia and New Zealand as associate members have supported over twenty-five deep-ocean survey cruises. In 1991 three of the U.S. consortia holding licenses for exploration under U.S. laws applied for three-year extensions, and the final environmental impact statement for leasing areas rich in cobalt crusts off Hawaii and Johnston Island has been approved, but no follow-up action has been taken. These crusts have four to five times the cobalt content of manganese nodules. India has applied to the United Nations to develop manganese nodule deposits in the Indian Ocean and is working to advance mining systems and processing plant designs. Japan continues its research interests.

Sulfide Mineral Deposits

Expeditions to the rift valleys of the East Pacific Rise near the Gulf of California, the Galápagos Ridge off Ecuador, and the Juan de Fuca and Gorda ridges off the northwestern U.S. have found deposits of minerals combined with sulfur to form sulfides of zinc, iron, copper, and possibly silver, molybdenum, lead, chromium, gold, and platinum. Molten material from beneath the earth's crust rises along the rift valleys, fracturing and heating the rock. Seawater percolates into and through the fractured rock, forming mineral-rich hot solutions. When these solutions rise from the cracks and cool, the metallic sulfides precipitate to the seafloor. Deposits may be tens of meters thick and hundreds of meters long. Too little is presently known about these deposits for us to know whether they might be of economic importance at some future date. No practical technology exists to sample or retrieve them at this time, and like the manganese nodules, these deposits are found outside national economic zones, which presents ownership problems; see the next section.

In the 1960s metallic sulfide muds were discovered in the Red Sea. Deposits of mud 100 m (330 ft) thick were found in small basins at depths of 1900 to 2200 m (6200–7200 ft). High amounts of iron, zinc, and copper and smaller amounts of silver and gold were found. The salty brines over these muds were hundreds of times richer in some metals than normal seawater. In 1985, a pilot mining and processing operation was set up under a Saudi-Sudanese joint commission.

Laws and Treaties

Because of the potential value of deep-sea minerals, specifically manganese nodules, and because the nodules are found in international waters, outside the usual two-hundred-mile economic zones of coastal nations, the developing nations of the world feel they have as much claim to this wealth as those countries that are presently technically able to retrieve the nodules. The developing nations want access to the mining technology and a share in the profits. For nearly ten years, the United Nations' Law of the Sea Conferences worked to produce a treaty to regulate deep-ocean exploitation, including mining. The Law of the Sea Treaty was completed in April, 1982. The treaty recognizes the nodules as the heritage of all humankind, to be regulated by a U.N. seabed authority that would license private companies to mine in tandem with a U.N. company. The quantities removed would be limited, and the profits would be shared.

The United States chose not to sign the treaty. In 1984 the United States, Belgium, France, West Germany, Italy, Japan, and the Netherlands signed a separate Provisional Understanding Regarding Deep Seabed Matters, and, under this provisional understanding, four international consortia have been awarded exploration licenses by the United States, West Germany, and the United Kingdom. By 1991 forty-five countries had ratified the Law of the Sea Treaty. Tempers have cooled as the years have passed, and the U.N., realizing that it needs U.S. support to ensure international cooperation for the exploration and exploitation of ocean resources, is making gestures to include the U.S. in working groups to resolve mining conflicts. Some U.S. officials believe that movement toward an acceptable treaty can occur within the next three to four years. In any case, it is unlikely that any deep-sea mining will occur until well into the twenty-first century. The high costs of sea mining, low metal prices, and still-undeveloped land sources combine to make rapid commercialization unlikely under any legal system.

Because there may be rich deposits of minerals and oil in Antarctica and its surrounding seas and because both national and private corporations are interested in surveying these areas for possible future mining and drilling, a multinational meeting in June 1988 produced an agreement to regulate mining exploration in and around that continent. This treaty must be ratified by sixteen of the twenty nations that signed the 1959 Antarctica Treaty that banned all military activity and permitted scientific research. According to the 1988 treaty, no activities will be permitted if they will cause "significant changes" in atmospheric, terrestrial, or marine environments. Since no one really knows for sure how much mineral and oil wealth may be under Antarctica's ice and snow, the extent of future operations is also unknown.

In the United States, there is no offshore mining except for the sand and gravel operations in the state-owned waters of a few coastal states. Mining for hard minerals, which include manganese nodules, cobalt crusts, metallic sulfides, and phosphorites, would be carried out within U.S. waters under provisions of U.S. domestic laws. Under these laws the Department of the Interior is authorized to issue leases for mineral exploration and development on the continental shelf. No leasing or regulatory program has yet been developed, but federal and state task forces are currently working to develop programs for cobalt-rich crusts in the Hawaiian Islands, metallic sulfides on the Gorda Ridge off the coast of Oregon and northern California, and phosphorites off North Carolina.

Summary

The bathymetric features of the ocean floor are as rugged as the topographic features of the land but erode more slowly. The continental margin includes the continental shelf, slope, and rise. The continental shelf break is located at the change in steepness between the continental shelf and the continental slope. Submarine canyons are major features of the continental slope and, in some cases, the continental shelf. Some canyons are associated with rivers; others are believed to have been cut by turbidity currents. Turbidity currents deposit graded sediments known as turbidites.

The ocean basin floor is a flat abyssal plain, but it is interrupted by scattered abyssal hills, volcanic seamounts, and flat-topped guyots. In warm shallow water, corals have grown up around the seamounts to form fringing reefs. A barrier reef is formed when a seamount subsides while the coral grows. An atoll results when the seamount's peak is fully submerged. Great, continuous volcanic mountain ranges, called midocean ridges and rises, extend through all the oceans and separate the ocean floor into basins. Rift valleys run along the ridge and rise crest; fracture zones run perpendicular to the ridge system. Trenches are long depressions in the ocean floor that are associated with island arcs; they are found mainly in the Pacific Ocean.

Depth soundings were made first with a hand line, then with wire, and, since the 1920s, with echo sounders.

Sediment classifications are based on their sources, areas of deposit, and particle size. Sediments formed from rock are lithogenous, those from living organisms are biogenous, those from seawater are hydrogenous, and cosmogenous sediments are particles from outside the atmosphere. Sediments arising from and deposited close to land are terrigenous; those arising from processes over the continental shelf are neritic; and pelagic sediments are found on the deep seafloor, away from the influence of land. Red clay is a pelagic, lithogenous sediment that accumulates slowly. Sediments made up of 30% or more biogenous material are oozes. Siliceous oozes are insoluble; calcareous particles are soluble at depths below 4000 m.

Patterns of sediment deposit result from the distance from their source, the abundance of living forms contributing their remains, the seasonal variations in river flow, the waves and currents including turbidity currents, the variability in land sources, the prevailing winds, and sometimes rafting.

Coarse sediments are concentrated close to shore; finer sediments are found in quiet offshore or nearshore environments. Terrigenous sediments are found mainly along coastal margins; most deep-sea sediments come from biogenous sources. Mixtures of particle sizes reveal the processes that formed the deposit. Sediment layers provide clues to ancient climate patterns.

In general, sedimentation rates are slowest in the deep sea and greatest near the continents. Relict sediments were deposited under conditions that no longer exist. Mechanisms that increase the sinking rates of particles include clumping and incorporation of sediment particles into the larger fecal pellets of small marine organisms.

Sediments are sampled with dredges, grabs, corers, and by drilling.

Seabed resources include sand and gravel used in construction and landfills. Sands and muds that are rich in mineral ores are mined. Phosphorite nodules have potential as fertilizer. Oil and gas are the most valuable of all seabed resources. Manganese nodules are rich in copper, nickel, and cobalt; they are present on the ocean floor in huge numbers. The status of manganese-nodule mining is clouded by disputes over international law with reference to mining claims and shared technology. Sulfide mineral deposits have been discovered along rift valleys; their economic importance is unknown.

Key Terms

sediment	hydrogenous sediment
subsidence	carbonate
continental margin	phosphorite
continental shelf	manganese nodule
continental shelf break	cosmogenous sediment
continental slope	tektite
submarine canyon	ooze
turbidity current	calcareous ooze
turbidite	siliceous ooze
continental rise	diatom
abyssal plain	diatomaceous ooze
abyssal hill	nutrient
seamount	radiolarian
guyot	radiolarian ooze
fringing reef	foraminifera
barrier reef	pteropod
atoll	coccolithophore
ridge and rise system	red clay
rift valley	brown clay
transform fault	brown mud
trench	terrigenous sediment
island arc	neritic sediment
soundings	pelagic sediment
fathom	rafting
echo sounder	relict sediment
depth recorder	dredge
lithogenous sediment	grab sampler
biogenous sediment	corer

Study Questions

1. Are calcareous oozes more common in the southern Pacific Ocean or the northern Pacific Ocean? Why is this so?
2. What is a turbidity current? Where would you expect a turbidity current to occur? How does the structure of a sediment deposit left by a turbidity current differ from other sediment deposits?
3. List the four basic sediment types classified by source. Where is each sediment type most likely to be found?
4. Discuss the probability of commercial development and exploitation of deep-sea mineral resources during the next twenty-five years.
5. Imagine that you are in a submersible on the ocean bottom. You leave New York and travel across the North Atlantic to Spain. Draw a simple ocean-bottom profile showing each major bathymetric feature you see as you move across the ocean. Name each feature. Do the same for the South Pacific between the coast of Chile and the west coast of Australia. Compare the two profiles. Did your depth scale differ from your horizontal scale? How much?
6. Which of the sampling devices discussed in this chapter takes a relatively undisturbed bottom sample? Why is this important to a marine geologist?
7. What is the continental margin? What pattern of sediment deposit would you expect to find associated with it? What processes produce these patterns of deposit?
8. What combination of factors is required to form a coral atoll?
9. What is a relict sediment? Where would you be likely to find such a deposit and why would you find it in that place?
10. Describe several ways in which a continental shelf may be formed.
11. Describe methods used to recover sediment samples from the seafloor. Discuss the advantages and disadvantages of each method.
12. What is the average depth of the oceans in meters, in miles, in fathoms?
13. How is particle size used in understanding the pattern of ocean floor deposits?
14. What are the implications for the marine environment as exploitation of seabed resources continues? Consider mining and drilling on continental shelves, in Antarctic waters, in the open ocean.
15. The Grand Banks is an extensive, relatively shallow area southeast of Newfoundland, Canada and off the U.S. northeast coast. Why are there large boulders scattered across this area so far from shore?

Study Problems

1. If underwater cables are spaced 14 km apart on the seafloor, and if monitoring equipment shows that they break in sequence from shallow to deeper water at fifteen-minute time intervals, what can you determine about the event causing the breaks?

2. If the average concentration of suspended sediment in the water is 1 g/m^3 and the volume of water in a harbor is 158 km^3, what is the average residence time of sediment in the water of this harbor? The daily sediment supply rate averages 1×10^7 kg.

3. In how many years will each of the particles listed in the following table reach the seafloor if the particles fall through 4000 m of seawater? All the particles are derived from land rock of the same density (2.8 g/cm^3). Settling rate is calculated from Stokes's law:

$$V \text{ cm/sec} = 2.62 \times 10^4 r^2$$

where r is the radius expressed in centimeters. (2.62×10^4 contains gravity, viscosity of water, the difference between the density of the particle and the density of water, and a constant for particle shape.) How does the settling rate change if the diameter remains constant but the density of a particle changes?

Particle Type	Particle Diameter (mm)	Settling Rate (V) (cm/sec)
granule	4	1048
sand	0.25	4.09
clay	0.004	10.48×10^{-4}

4. Assuming a constant sedimentation rate of 0.4 cm/1000 years, how thick will the sediments be in a portion of an ocean basin where the underlying crust is 130 million years old?

Suggested Readings

Geology of the Seafloor

Emery, K. O. 1969. The Continental Shelves. In *Ocean Science,* readings from *Scientific American* 221 (3): 32–44 (1977).

Menard, H. W. 1969. The Deep-Sea Floor. In *Ocean Science,* readings from *Scientific American* 221 (3): 55–64 (1977).

Montgomery, C. 1993. *Physical Geology,* 3d ed. Wm. C. Brown Communications, Dubuque, Ia. 454 pp. (General geology text.)

Open University. 1989. *The Ocean Basins: Their Structure and Evolution.* Pergamon Press, Oxford, England; Open University, Milton Keynes, England. 171 pp.

Open University. 1989. *Ocean Chemistry and Deep Sea Sediments.* Pergamon Press, Oxford, England; Open University, Milton Keynes, England. 120 pp.

Rice, A. L. 1991. Finding Bottom. *Sea Frontiers* 37 (2): 28–33.

Shepard, F. P. 1973. *Submarine Geology,* 3d ed. Harper & Row, New York. 517 pp. (Classic text in geological oceanography.)

Stephenson, A. G. 1970. Hydrographic Surveying. In *Oceanography, Contemporary Readings in Ocean Sciences,* 2d ed., Pirie, R. G., ed. (1977). Oxford Univ., New York. pp. 34–38.

Seabed Resources

Blissenbach, E., and Z. Nawab. 1982. Metalliferous Sediments of the Seabed: The Atlantic II-Deep Deposits of the Red Sea. In *Ocean Yearbook 3,* Borgese, E., and N. Ginsburg, eds. Univ. of Chicago, Chicago. pp. 77–104.

Broadus, J. M. 1987. Seabed Materials. *Science* 235 (4791): 853–60. (Review of seabed resources.)

Cruickshank, M. J. 1991. Ocean Mining: For the Future a Good Omen. *Sea Technology* 32 (1): 38–39.

Dubs, M. 1986. Minerals of the Deep Sea: Politics and Economics in Conflict. In *Ocean Yearbook 6,* Borgese, E. M., and N. Ginsburg, eds. Univ. of Chicago, Chicago. pp. 55–83.

Edmond, J. M., and K. Von Damm. 1983. Hot Springs on the Ocean Floor. *Scientific American* 248 (4): 78–93.

Ellers, F. S. 1982. Advanced Offshore Oil Platforms. *Scientific American* 246 (4): 39–49.

Koski, R. A., W. R. Normark, J. L. Morton, and J. R. Delaney. 1982. Metal Sulfide Deposits on the Juan de Fuca Ridge. *Oceanus* 25 (3): 43–48.

Mottl, M. J. 1980. Submarine Hydrothermal Ore Deposits. *Oceanus* 23 (2): 18–27.

Pagano, S. 1991. Offshore Drilling, Production—New Waves of Technology. *Sea Technology* 32 (4): 19–22.

Ryan, P. R., ed. Fall, 1983. *Oceanus* 26 (3). (Issue devoted to offshore oil and gas.)

Sea Technology 33 (1), 1992. (Annual review and forecast of oil, gas, and mining activities.)

3

The Not-So-Rigid Earth

3.1 **The Interior of the Earth**
Internal Layers
Evidence for the Internal Structure
3.2 **The Lithosphere**
The Layers
Isostasy
3.3 **The Movement of the Continents**
History of a Theory
Evidence for a New Theory
Evidence for Crustal Motion
3.4 **Plate Tectonics**
Plate Boundaries
Rifting
Subduction
Continental Margins
Driving Forces for Plate Motion
3.5 **The Motion of the Plates**
Rates of Motion
Hot Spots
Polar Wandering Curves
3.6 **The History of the Continents**
The Breakup of Pangaea
Before Pangaea
Terranes
3.7 **Current Research**
Drilling Plans
Project FAMOUS
Hydrothermal Vents

Box: Megaplumes

3.8 **Practical Considerations: Ocean Waste Management**
Summary
Key Terms
Study Questions
Study Problems
Suggested Readings

We shall not cease from exploration
And the end of all our exploring
Will be to arrive where we started
And know the place for the first time.
Through the unknown, remembered gate
When the last of earth left to discover
Is that which was the beginning;
At the source of the longest river
The voice of a hidden waterfall ...
Not known, because not looked for,
But heard, half heard, in the stillness
Between two waves of the sea.

T. S. Eliot,
from "Little Gidding," *The Four Quartets*

Lava flow from Kilauea volcano meets the sea, Hawaii.

M *any features of our planet have presented scientists with contradictions and puzzles. For example, the remains of warm-water coral reefs are found off the coast of the British Isles; marine fossils occur high in the Alps and the Himalayas; and coal deposits that were formed in warm, tropic climates are found in northern Europe, Siberia, and northeast North America. At the same time, similar patterns were observed in widely scattered places but for no known reason. The volcanoes known as the ring of fire border the east and west coasts of the Pacific Ocean, and deep ocean trenches are found adjacent to long island arcs, such as the Aleutian Islands and Trench, the Mariana Islands and Trench, and the Japanese Islands and Japan-Kuril Trench. No single coherent theory explained all these features, until new technology and new scientific discoveries in the 1950s and early 1960s combined to trigger a complete reexamination of the earth's history.*

In this chapter we explore the history of an idea and the evidence that allowed this idea to become, first, theory and, at last, generally accepted fact. During the last forty years our understanding of our planet has changed completely; new tools and new knowledge have allowed us to gain insights into the past, appreciate the present, and even look forward into the future.

3.1   The Interior of the Earth

Internal Layers

Living as we do on the surface of the outer crust of the earth, we cannot directly observe the interior of the planet. However, scientists have been able to learn a great deal about the structure and composition of the earth's interior using indirect methods.

If the earth were cut and a piece removed, the layers would appear as in figure 3.1. At the planet's center is an **inner core,** its radius 1070 km (665 mi). It is solid and very dense because it exists under extremely high pressure; it is also magnetized, rich in iron and nickel, and very hot (5000°C). The inner core is surrounded by a transition zone about 700 km (435 mi) thick, which is in turn surrounded by a 1700-km (1056-mi) thick layer of liquid material; together they form the **outer core.** The liquid material of the outer core is similar in composition to the inner core but cooler (4000°C). The next layer, the **mantle,** contains the largest mass of material of any of the layers (about 70% of the earth's volume). This layer is 2835 km (1761 mi) thick; it is less dense than the core, still cooler (1500° to 3000°C), and is composed of magnesium-iron silicates. The outer part of the mantle is thought to be rigid, while the inner region is deformable and flows slowly over the deeper mantle. The earth's outermost layer is the cold, rigid, thin

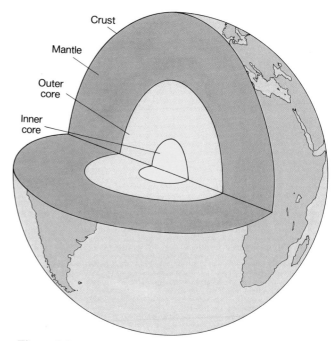

Figure 3.1

The layered structure of the earth.

(10–65 km) (6–40 mi), familiar surface, or **crust.** See figure 3.1 and table 3.1 for a comparison of these features and their properties.

Evidence for the Internal Structure

The evidence that has allowed scientists to build this model of the earth's interior has come from many sources. Using the earth's spherical shape, its mean (or average) radius, and its mass, it is possible to determine the average density of the earth; 5.52 g/cm^3. **Density** is defined as mass per unit volume and is usually given in grams per cubic centimeter, written g/cm^3. This calculated density is considerably greater than the average density of the earth's crust found by direct measurement, which is approximately 2.8 g/cm^3. The high average density value, requires that the material below the crust have a much greater density. Since the earth wobbles only very slightly as it rotates and since the acceleration due to gravity over the earth's surface is quite uniform, the earth's mass must be distributed uniformly about the earth's center as a series of concentric layers. Based on gravity, density, and the earth's dimensions, it is possible to calculate the pressures within the earth and also the temperatures that can be reached under these pressures; and, because there is a magnetic field around the earth, the earth's core must include materials that produce magnetic fields.

Another clue is furnished by meteorites that occasionally hit the earth and are considered to be the remains of unknown planets. More than half of the meteorites that have been found are "stony" silicate or rocky lumps, another large group is made mainly of iron, nickel, and other metals, and a few are "stony-iron" with metal inclusions. Radiometric dating of meteorites gives a maximum age of

Table 3.1
Layers of the Earth

Layer	Depth (km)	Thickness (km)	State	Composition	Density (g/cm^3)	Temperature (°C)
Crust	10–65					
continental		55	solid	magnesium-aluminum silicates	2.8	– 40–1000
oceanic		10	solid	magnesium-iron silicates	3.0	1000–1500
Mantle	65–2900	2835	solid and mobile	magnesium-iron silicates	4.5	1500–3000
Outer core	2900–5300	2400	liquid	iron, nickel	11.5	4000
Inner core	5300–6370	1070	solid	iron, nickel	13.0	5000

4.6 billion years, the same as the age of the solar system. These fragments allow us to directly analyze the density, chemistry, and mineralogy of the nickel-iron cores and stony shells of bodies that we believe to have a similar composition to that of the earth.

A major tool used to understand the earth's interior is the **seismic wave.** A seismic wave is an underground earth shock wave, or vibration, that is produced by an earthquake or underground explosion. All over the surface of the earth geologists and geophysicists monitor seismic recording stations, which pick up the vibrations caused by earthquakes, volcanic eruptions, landslides, and deliberately caused detonations.

Two basic kinds of seismic waves are associated with earth movements and travel through the earth's interior. These are pressure waves, or **P-waves,** which oscillate in the same direction as they move (similar to sound waves), and shear waves, or **S-waves,** which oscillate at right angles to their direction of movement (similar to a plucked string). These waves are shown in figure 3.2. P-waves are able to move through solids and liquids. S-waves cannot pass through a liquid, so they cannot pass through the outer core of the earth.

As the seismic waves move through one earth layer and into another, their speeds of travel change and the waves bend or **refract** as shown in figure 3.3a and b. The paths taken by the seismic waves as they pass through the earth provide information about the dimensions, structure, and physical properties of each of the internal layers. Because of the paths followed by the S-waves (shown in fig. 3.3b) we know that the outer core is liquid. The P-waves (fig. 3.3a) are refracted, but not stopped, by changes in density as they pass through the layers. In both diagrams (fig. 3.3a and 3.3b), shadow zones are formed where waves are absent. Measurements of travel time and wave direction confirm that there are abrupt changes of speed and direction

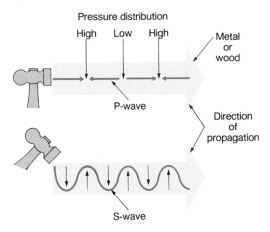

Figure 3.2

Waves passing through a solid are of two types: pressure waves, or P-waves (top), and shear waves, or S-waves (bottom).

at certain depths corresponding to the interior layers. If the interior of the earth had uniform properties, the waves would follow straight lines and their speed would not change, as shown in figure 3.3c.

Data from an increasingly densely spaced and sophisticated array of seismic-recording stations and the computer capacity to analyze the travel time of the thousands of seismic waves caused by the world's earthquakes are being used to produce three-dimensional maps of the changing physical properties of the lithosphere, asthenosphere, and the remaining mantle. This process, known as **seismic tomography,** is giving us a more detailed description of the earth's interior layers and is demonstrating that these layers are less homogeneous than had been thought.

Changes in the speed of P-waves as they pass around and through the areas of the midocean ridges are being used to develop three-dimensional images of the structure beneath the ridge systems. The P-waves define location, size, and physical state of mantle features. Features in the

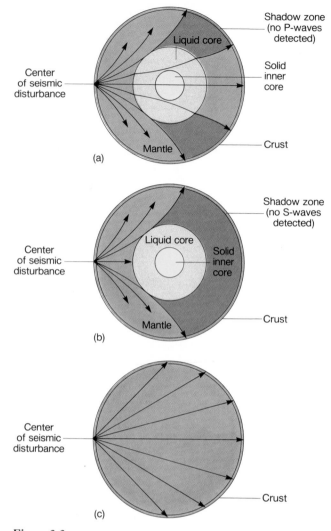

Figure 3.3

Movement of seismic waves through the earth. (a) Refraction of P-waves and shadow zones produced by the earth's interior structure. (b) Refraction of S-waves and shadow zones produced by the earth's interior structure. (c) No refraction and no shadow zones occur in an earth with a uniform structure.

upper oceanic crust have sizes measured in hundreds of meters; in magma chambers of the ridges the features are in thousands of meters or kilometers, and in areas of mantle upwelling beneath the ridges the measurements are in tens or hundreds of kilometers. Repeated observations of changes in space and time provide a picture of the dynamic processes occurring in this area of the mantle. Changes in the behavior of seismic waves at deeper mantle depths provide data about the physical properties of these areas. The new data indicates that the core is not smooth, but has major peaks and valleys over its surface. These features extend as much as 11 km (7 mi) above and below the mean surface of the outer core. At this time, it is thought that the regions above a peak are areas where the mantle has excess heat and mantle material rises toward the crust, drawing the core upward. Cooler, more viscous mantle material sinks to cause depressions in the core's surface. It is likely that

these peaks and valleys last only as long as it takes a rising plume to lose its excess heat and sink back toward the core, perhaps a hundred million years.

Three-dimensional tomographic images of the mantle indicate that there are masses of material unlike the surrounding mantle suspended in this layer. It has been proposed that these masses are slabs of crust that have sunk unmelted and have been turned down into the lower mantle. These cold, dense crustal slabs are modified at their boundaries to create the local vulcanism associated with these zones, and their high density also increases gravity values around such zones. These slabs may sink to great depths in the mantle approaching the outer core of the earth, appearing as shadowy forms sensed by tomography. As these crustal slabs gradually become heated they become less dense, rise again to the bottom of the outer layer, remelt, and form magma pools that feed regions of vulcanism.

Seismic tomography is combined with model studies to try to understand how plumes of mantle material form. Using layered fluids of different densities and viscosities, researchers build rising plume models with a lower-density fluid underlying a more dense fluid. The structure of these model plumes, their distribution, and the speeds at which they ascend are controlled by the differences in density and viscosity of both liquids and the thickness of the overlying layer. Continuing research will eventually give us a more complete understanding of the processes at work in the earth's interior and a better window into the heart of our planet.

3.2   The Lithosphere

The Layers

The large continental landmasses are formed primarily from **granite**-type rock, which has a high content of aluminum and magnesium silicate; quartz and feldspar are the principal mineral components. The crustal rock layer lying under the ocean is **basalt**-type rock, which is lower in silica and higher in iron and magnesium. Continental crust has a density of 2.8 g/cm^3 and is approximately 55 km (34 mi) thick, while oceanic crust is only about 10 km (6 mi) thick and has a density of 3.0 g/cm^3.

The boundary between the crust and the mantle is the Mohorovicic discontinuity, named for its discoverer, and usually called the **Moho**. The Moho is a boundary at which there is a sudden change in the chemical composition and the speed of seismic waves. At one time some thought that the Moho was the structure at which the earth's rigid crust moved relative to the mantle. Current research emphasizes the function or behavior of the crust and upper mantle rather than the structures that might be detectable and places the zone of movement within the mantle, from 70 km (43 mi) below the ocean crust to 150 km (93 mi) below the continental crust. The mantle just below the crust is rigid, solidified, basalt-type rock, fused to the crust but at the same time separated from it by the Moho. This rigid layer of crust and upper mantle is the **lithosphere.** Seismic

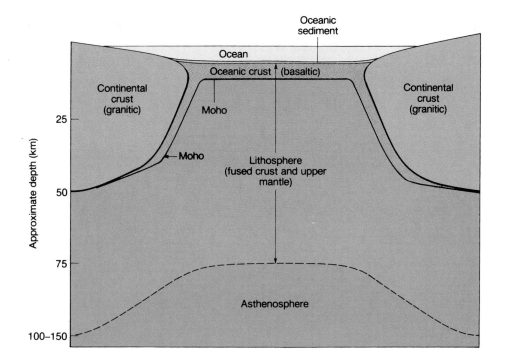

Figure 3.4
The lithosphere is formed from the fusion of crust and upper mantle. Notice that the Moho is relatively close to the earth's surface under the ocean's basaltic crust but is depressed with the mantle under the granite continents.

waves pass through it at high speeds, implying that the lithosphere is both strong and rigid. The subregion of the mantle extending about 250 km (155 mi) below the lithosphere is the **asthenosphere;** this region of the mantle passes seismic waves more slowly and is considered to be partly melted, flowing slowly when stressed. The lithosphere is less dense than the asthenosphere; both continental and oceanic lithosphere float on the asthenosphere. The lithosphere and asthenosphere are shown in figure 3.4.

Isostasy

The distribution of elevated continents and depressed ocean basins over the crust requires keeping a balance between the internal pressures under the land blocks and those under the ocean basins. This is the principle of **isostasy.** The balance is possible because the greater thickness of low-density granitic crust in the continental regions is compensated for by the elevated higher-density mantle material under the oceans (see fig. 3.5). The situation is often compared to the floating of an iceberg. The top of the iceberg is above the sea surface, supported by the buoyancy of the displaced water below the surface. The deeper the ice extends below the surface, the higher the iceberg reaches above the water. The less-dense continental land blocks float on the more-dense mantle in the same way, with most of the continental volume below sea level.

In other words, the asthenosphere offers buoyant resistance to a section of lithosphere sagging under the weight of a mountain range. Because the lithosphere is cooler, it is more rigid and stronger than the asthenosphere. If a thick section of lithosphere has a large area and it is mechanically strong, it depresses the asthenosphere only slightly and is able to support a mountain range such as the Himalayas or the Alps. In other places where the crust is

fractured and mechanically weak, growing continents and mountains must penetrate deeper into the mantle in order to provide buoyancy and keep the continental mass from sinking; the Andes are a mountain range with roots deep in the mantle.

If material is removed from or added to the continents, isostatic adjustment will occur. For example, parts of North America and Scandinavia continue to rise as the continents readjust to the lost weight of the ice sheets that receded at the close of the last ice age ten thousand years ago. Newly formed volcanoes protruding above the sea surface as islands often subside as their weight depresses the oceanic crust, and they sink back under the sea. The outer mantle gradually changes its shape in response to weight changes in the overlying, more-rigid crust.

3.3 The Movement of the Continents

History of a Theory

As world maps became complete and more accurate, observant individuals were intrigued by the shapes of the continents on either side of the Atlantic Ocean. The possible "fit" of the bulge of South America into the bight of Africa was noted by the English scholar and philosopher Francis Bacon (1561–1626), the French naturalist George Buffon (1707–88), the German scientist and explorer Alexander von Humboldt (1769–1859), and others. In the 1850s the idea was expressed that the Atlantic Ocean had been created by a separation of two landmasses during some unexplained cataclysmic event early in the history of the earth. Thirty years later it was suggested that a portion of the earth's continental crust had been torn away to form the

Figure 3.5

Isostasy. Columns of crustal material are unequal in height and density but generate the same pressure at the same depth within the mantle.

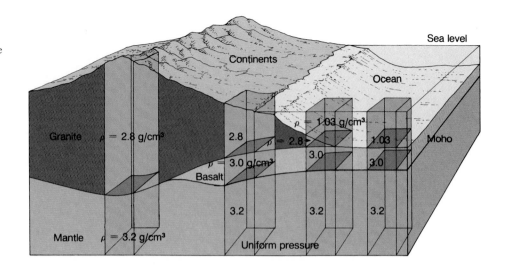

Figure 3.6

Pangaea 200 million years before the present. Wegener's supercontinent is composed of the two subcontinents, Laurasia and Gondwanaland.

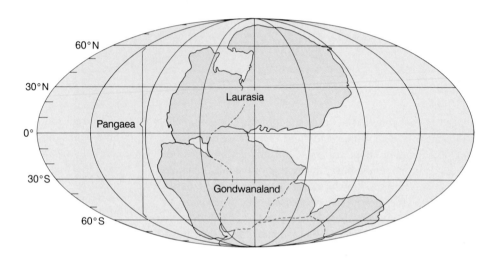

moon, creating the Pacific Ocean and triggering the opening of the Atlantic. As scientific studies of the earth's crust continued, patterns of rock formation, fossil distribution, and mountain range placement began to show even greater similarities between the lands now separated by the Atlantic Ocean. In a series of volumes published between 1885 and 1909, an Austrian geologist, Edward Suess, proposed that the southern continents had been joined into a single continent he called **Gondwanaland.** He assumed that isostatic changes had allowed portions of the continents to sink and create the oceans between the continents. This idea was known as the subsidence theory of separation. At the beginning of this century, Alfred L. Wegener and Frank B. Taylor independently proposed that the continents were slowly drifting about the earth's surface. Taylor soon lost interest, but Wegener, a German meteorologist, astronomer, and Arctic explorer, continued to pursue this concept until his death in 1930.

Wegener's theory of **continental drift** proposed the existence of a single supercontinent he called **Pangaea** (see fig. 3.6). He thought that forces arising from the rotation of the earth began Pangaea's breakup. First, the northern portion composed of North America and Eurasia, which he called **Laurasia,** separated from the southern portion

formed from Africa, South America, India, Australia, and Antarctica, for which he retained the earlier name Gondwanaland. Laurasia and Gondwanaland are shown in figure 3.6. The continents as we know them today then gradually separated and moved to their present positions. Wegener based his ideas on the geographic fit of the continents and the way in which some of their older mountain ranges and rock formations appeared to relate to each other when the landmasses were assembled to form Pangaea. He also noted that fossils more than 150 million years old collected on different continents were remarkably similar, implying the ability of land organisms to move freely from one landmass to another. Fossils from different places dated after this period showed quite different forms, suggesting that the continents and their evolving populations had separated from one another.

Wegener's theory provoked considerable debate in the decade of the 1920s, but most geologists agreed that it was not possible to move the continental rock masses through the rigid basaltic crust of the ocean basins. There was no mechanism to cause the drift, and the theory was not regarded very seriously in the scientific community; it became a footnote in geology textbooks.

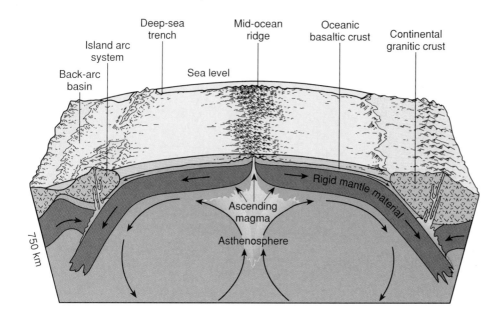

Deep-sea
trench
Island arc
system
Back-arc
basin
Mid-ocean
ridge
Sea level
Oceanic
basaltic crust
Continental
granitic crust
Rigid mantle material
Ascending
magma
Asthenosphere
750 km

Figure 3.7

Seafloor spreading creates new crust at
midocean ridges and loses old crust in deep-
sea trenches. This process is shown being
driven by convection cells in the
asthenosphere.

Evidence for a New Theory

Armed with new sophisticated instruments and the technology developed during World War II, earth scientists returned to a study of the earth's crust in the 1950s. For the first time, scientists were able to examine the floor of the deep ocean in detail and explore the world's largest mountain range running through the oceans. Refer back to figures 2.10 and 2.13 to trace this remarkable and continuous underwater feature. The theories current at the time of its discovery had not predicted the existence of such an extensive midocean mountain range and could not explain it, but a new theory was advanced that made the old idea of continental drift seem plausible.

In the early 1960s H. H. Hess of Princeton University proposed that deep within the earth's mantle there are currents of low-density molten material heated by the earth's natural radioactivity. When these upward-moving currents of mantle material reach the lithosphere they move along under it, cooling as they do so until they become cool enough and dense enough to sink down toward the core again. These patterns of moving mantle material are called **convection cells** (see fig. 3.7). The convection cells are generally believed to occur in the asthenosphere, although there is no definite evidence that the convection cells are confined to this layer. Some scientists believe that they extend throughout the mantle.

If the upward-moving mantle material, or **magma,** breaks through the lithosphere of the seafloor instead of continuing to flow underneath it, underwater volcanoes are produced and a mountain range forms along the crack in the crust. As the cooling magma oozes out along the underwater mountain range, or ridge system, it becomes lava, cools, hardens, and is added to the earth's surface as new oceanic-type basaltic crust. If new crust is being produced in this manner, a mechanism is needed to remove old crust since there is no measurable change in the size of the earth. The great, deep trenches of the Pacific were proposed as

areas where the older lithosphere dips down and disappears back into the earth's interior, eventually to be recycled into the mantle and the convection cell system. Figure 3.7 shows this process of producing new lithosphere at the ridges and losing old lithosphere at the trenches in a system driven by the motion of convection cells.

Although some of the ascending molten material breaks through the crust and solidifies, most of the rising material is turned aside under the rigid lithosphere and moves away toward the descending sides of the convection cells, dragging pieces of the lithosphere with it. This lateral movement of the crust produces **seafloor spreading** (see fig. 3.7). Areas in which new crust is formed above rising magma are **spreading centers;** areas of descending older crustal material are **subduction zones.** The seafloor spreading mechanism provides the forces to cause continental drift. The continents are not moving through the basalt of the seafloor; instead, the continents are being carried as passengers on the lithosphere similar to boxes on a conveyor belt.

Evidence for Crustal Motion

Additional evidence was needed to support the idea of seafloor spreading, and as oceanographers, geologists, and geophysicists explored the earth's crust on land and under the oceans, evidence began to accumulate.

Earthquakes were known to be distributed around the earth in narrow and distinct zones. These zones were found to correspond to the areas along the ridges (or spreading centers) and the trenches (or subduction zones). Compare figure 3.8 with figures 2.10, 2.12, and 2.13.

Researchers sank probes into the seafloor to measure the heat from the interior of the earth moving through the crust. The measured heat flow shows a pattern with a high degree of variability, even over closely spaced intervals. In part, this variation in the data is blamed on seawater seeping down through porous or fractured portions of the crust at

Figure 3.8

The world's major earthquake belts coincide with midocean ridge systems and the deep-ocean trenches.

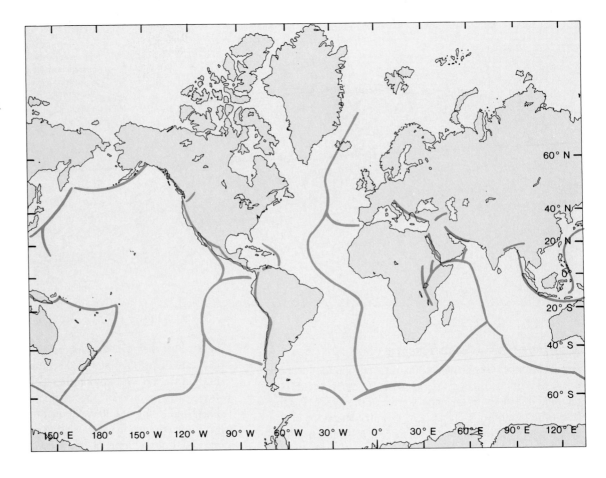

Figure 3.9

Heat flow through the Pacific Ocean floor. Values are shown against age of crust and distance from the ridge crest.

Data from J.G. Sclater and J. Crowe, "On the Variability of Oceanic Heat Flow Average." In *Journal of Geographical Research, 8* (17 June 1976), p. 3004. American Geophysical Union, Washington, D.C.

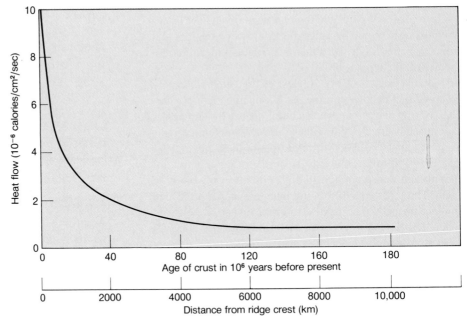

one location along a ridge system and rising as heated water at another. This circulation does not occur over regions of the seafloor that are sealed by a thick layer of sediment. In sediment-sealed areas, the measured heat flow shows a regular pattern; here it is highest in the vicinity of the midocean ridges over the ascending portion of the mantle convection cell (see fig. 3.9).

Radiometric dating of the age of rocks from the land and from the seafloor shows that the oldest rocks from the oceanic crust are only abut 200 million years old, while the rocks from the land are much older. The seafloor formed by convection-cell processes at the spreading centers is young and short-lived, for it is lost at the subduction zones, where it plunges back down into the asthenosphere.

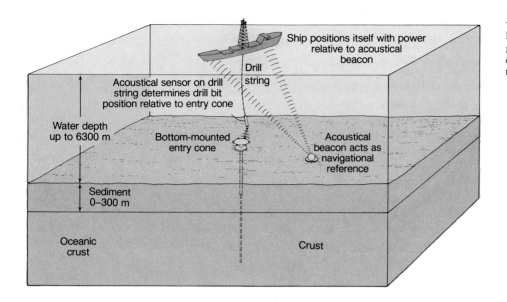

Figure 3.10

Deep-ocean drilling technique. Acoustical guidance systems are used to maneuver the drilling ship over the bore hole and to guide the drill string back into the bore hole.

Cores were drilled down through the sediments that cover the ocean bottom and into the ocean floor. This process required the development of a new technology and a new kind of ship. In the late summer of 1968 the specially constructed drilling ship *Glomar Challenger* was used for a series of studies. This ship, 122 m (400 ft) long with a beam of 20 m (65 ft) and a draft of 8 m (27 ft), displaces 10,500 tons when loaded. Its specialized bow and stern thrusters and its propulsion system respond automatically to computer-controlled navigation, using acoustic beacons on the seafloor. This allows the ship to remain for long periods of time in a nearly fixed position over a drill site in water too deep to anchor. A sonar guidance system enables it to replace drill bits and reenter the same bore holes in water about 6000 m (20,000 ft) deep. This method is illustrated in figure 3.10.

In 1983 the *Glomar Challenger* was retired, after logging 600,000 km (375,000 mi), drilling 1092 holes at 624 drill sites, and recovering a total of 96 km (58 miles) of deep-sea cores for study. A new deep-sea drilling program with a new drill vessel, the *JOIDES Resolution,* began in 1985. The *JOIDES Resolution* is about the same size as the *Glomar Challenger,* complete with the world's most sophisticated, state-of-the-art scientific and drilling equipment (see fig. 3.11). Information on the current research being conducted by the *JOIDES Resolution* will be found in the Current Research section later in this chapter.

The cores taken by the *Glomar Challenger* provided much of the factual data needed to establish the existence of seafloor spreading. No ocean crust older than 180 million years old was found, and sediment age and thickness were shown to increase with distance from the ocean ridge system (see fig. 3.12). Note that the sediments closest to the ridge system are thin over the new crust that has not had long to accumulate its sediment load. The crust farther away from the ridge system is older and is more heavily loaded with sediments.

Although each of these pieces of evidence fits the theory that the earth's crust produced at the ridge system is new and young, the most elegant proof for seafloor spreading came from a study of the magnetic evidence locked into the oceans' floors.

The earth's familiar north and south geographic poles at 90°N and 90°S latitude mark the axis about which it rotates. The earth also behaves as if it had a giant bar magnet embedded in its interior. The magnetic North Pole is located in the Hudson Bay area, and the magnetic South Pole is directly opposite, in the South Pacific Ocean. Like any magnet, the earth has a magnetic field, and we can visualize the earth's magnetic field as lines of force surrounding the planet. The lines of force converge and dip toward the earth at the magnetic poles and are parallel to the earth at the magnetic equator. At other positions on the earth's surface the lines of force have both horizontal and vertical components.

When materials that can be magnetized are heated in a magnetic field they become magnetized as they cool, with their north and south magnetic poles lined up with the poles of the magnetic field. In the same way, as molten volcanic material cools and solidifies, its iron-bearing minerals become magnetized and permanently aligned with both the horizontal and vertical components of the earth's magnetic field (see fig. 3.13).

Research on age-dated layers of volcanic rock found on land shows that the polarity, or north-south orientation, of the earth's magnetic field reverses for varying periods of geologic time. This means that at different times in the earth's history the present north and south magnetic poles have changed places. Each time a layer of volcanic material solidified, it trapped the magnetic orientation and polarity of the time period in which it occurred. The dating and testing of samples taken through a series of volcanic layers have allowed scientists to build a calendar of these events (see fig. 3.14). No one knows what triggers these **polar reversals,** but during a reversal the earth's magnetic field gradually collapses, removing this barrier to cosmic

Figure 3.11
The deep-sea drilling vessel, *JOIDES Resolution*.

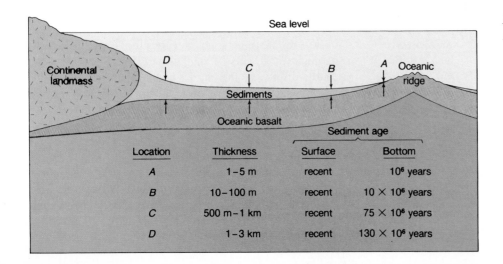

Figure 3.12
Age and thickness of seafloor sediments.

Location	Thickness	Sediment age	
		Surface	Bottom
A	1–5 m	recent	10^6 years
B	10–100 m	recent	10×10^6 years
C	500 m–1 km	recent	75×10^6 years
D	1–3 km	recent	130×10^6 years

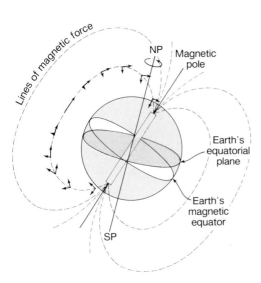

Figure 3.13
Lines of magnetic force surround the earth and converge at the magnetic poles. The lines of force have components that are parallel and perpendicular to the earth's surface. The vertical components are not present at the earth's magnetic equator and are at their maximum at the magnetic poles, where the horizontal components disappear. A compass orients its needle with the horizontal component to indicate the direction to the magnetic pole. NP and SP indicate the present north and south geographic poles.

radiation, and recent work has suggested a correlation between such periods and a decrease in populations of delicate, single-celled organisms that live in the surface layers of the ocean. Nearly 170 reversals have been identified during the last 76 million years, and our present magnetic orientation has existed for 710,000 years. How long our current polarity will last we do not know.

When magnetometers were towed over the seafloor by scientists from the Scripps Institution of Oceanography in the early 1960s, the resulting maps revealed a pattern of magnetic stripes that ran parallel with the midocean ridge (see fig. 3.15). Their significance was not understood until 1963, when F. J. Vine and D. H. Matthews of Cambridge

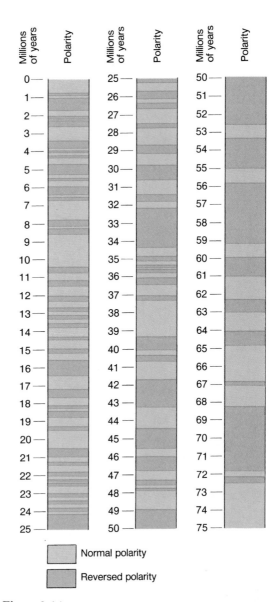

Normal polarity

Reversed polarity

Figure 3.14
Polarity reversal time scale during the Cenozoic era. Time is given in millions of years.

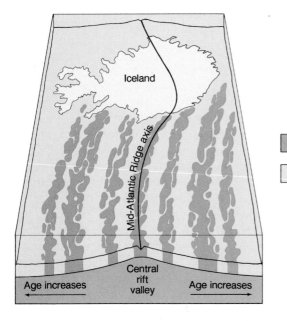

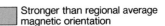

Stronger than regional average magnetic orientation

Weaker than average magnetic orientation (reversal)

Figure 3.15
Reversals in the earth's magnetic polarity cause the symmetrically striped pattern centered on the Mid-Atlantic Ridge. The age of the seafloor increases with the distance from the ridge. The spreading rate along the Mid-Atlantic Ridge is about one centimeter per year.

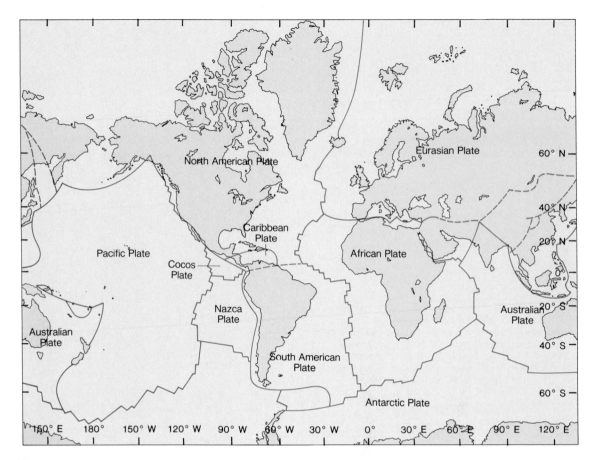

Figure 3.16
Major lithospheric plates of the world.

University proposed that these stripes represented a recording of the polar reversals of the vertical component of the earth's magnetic field, frozen into the seafloor. As the molten basalt rose along the crack of the ridge system and solidified, it locked in the direction of the prevailing magnetic field. Seafloor spreading moved this material off on either side of the ridge, to be replaced by more molten materials. Each time the earth's magnetic field reversed, the direction of the magnetic field was recorded in the new crust. Vine and Matthews proposed that if such were the case, there should be a symmetric pattern of magnetic stripes centered at the ridges and becoming older away from the ridges. The polarity and age of these stripes should correspond to the same magnetic field changes found in the dated layers on land. Experimental work confirmed their ideas and thus dramatically confirmed seafloor spreading. The present ocean basins are not old but new, created by seafloor spreading during the past 200 million years, or the last 5% of the earth's history.

3.4 Plate Tectonics

When the theory of continental drift was joined with the idea of seafloor spreading, it led to the formation of a single concept of crustal movement: **plate tectonics.** The earth's lithosphere is seen as a series of rigid plates outlined by the

major earthquake belts of the world (refer back to fig. 3.8). These earthquake belts are the plate boundaries and coincide with the trenches, ridges, and faults discussed in the next section. Seven major lithospheric plates are recognized—the Pacific, Eurasian, African, Australian, North American, South American, and Antarctic—as well as numerous smaller plates off South and Central America, in the Mediterranean area, and along the northwest United States (see fig. 3.16).

Plate Boundaries

Sections of lithosphere are known as plates; each plate is made up of continental and/or oceanic crust and the attached rigid mantle material. The plate boundaries are trenches, ridges, or faults. The direction of plate motion at each boundary depends on whether lithospheric plates move together, move apart, or slide past each other. Plate boundaries move apart at the midocean ridges and create new lithosphere; these are known as **divergent plate boundaries.** At the trenches, plates move toward each other, destroying old lithosphere at subduction zones and forming **convergent plate boundaries.** Plates move past each other along **faults,** where there is a break in the rocky crust with displacement of one side relative to the other. There are many types of faults; in some the movement is vertical, in some horizontal, and in some oblique. Close

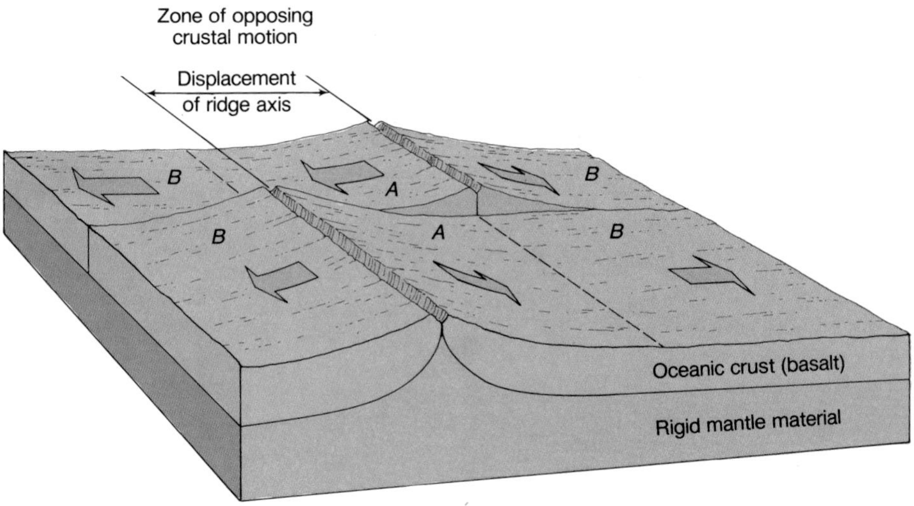

Zone of opposing
crustal motion

Displacement
of ridge axis

Oceanic crust (basalt)

Rigid mantle material

Figure 3.17
Relative motion and displacement of the
crust along a transform fault. Motion at *A* is
in opposite directions. Motion at *B* is in the
same direction.

inspection of the midocean ridge and rise system shows sections of the ridge which are offset or displaced laterally from each other along a special kind of fault called a **transform fault.** The opposite sides of a transform fault are two different plates that are moving in opposite directions (see area *A* in fig. 3.17). This movement creates a fault zone that is active and may be the seat of frequent and severe but shallow earthquakes. Where this same fault line extends outside both ridge axes, the crustal plates are moving in the same direction, and differential motion along the fault is much reduced (see area *B* in fig. 3.17). These regions of the transform faults are quiet, but the faults can still be detected.

The lateral motion of the ridge along these transverse faults produces sharp vertical displacements called **escarpments** across the width of the ridge. These are regions where there are sudden changes in the depth of the ocean. The escarpments and transform faults are boundaries where one piece of crust moves relative to another. Crust is neither created nor destroyed along these faults.

The San Andreas Fault, an excellent example of one of the larger transform faults, extends from the northern end of the East Pacific Rise up through the Gulf of California, to the San Francisco Bay region, and out to sea to the north (see fig. 3.18). It terminates at another north-south ridge system, the Gorda and Juan de Fuca ridges. The San Andreas Fault system allows crustal sections on either side to move relative to each other. The land on the Pacific side, including the coastal area from San Francisco to the tip of Baja California, is actively moving northward past the land on the east side of the fault. The motion is not uniform; when stress accumulates there is a sudden movement and displacement along the fault, causing severe earthquakes. Movement along this fault system caused the famous 1906 and recent October 1989 earthquakes in the San Francisco Bay area.

The reason for the many transform faults that are found associated with the ridge system is not completely understood. These faults probably develop because the crust is subject to different degrees of rotational motion

depending on latitude. Changes in the direction of crustal motion, due to variations in the strength or location of convection cells or due to collisions between sections of crust, could also result in transform faults.

Rifting

Rift zones, or spreading zones, are regions where the lithosphere splits, separates, and is forced apart as new crustal material intrudes into the crack or rift. The major rift zones are the midocean ridge and rise systems (return to fig. 3.8), but rifting is not limited to oceanic areas. It occurs on land and, in the past, has been responsible for breaking up landmasses.

The lithosphere under the continents is 100–150 km (60–90 mi) thick, but under the oceans it is thin, 10–100 km (6–60 mi), where new at the rift zones; and thicker, 100–150 km (60–90 mi), where older and farther away from the rift. To crack the lithosphere under the continents some mechanism must thin and apply tension to the lithospheric plate. How this mechanism operates is not fully known. The crust may be under tension because of subduction occurring at the plate margin, stretching and thinning the plate to allow magma to rise, weakening the plate so that it cracks under tension and produces a fault zone into which magma eventually flows (see fig. 3.19).

Another possibility requires a more active role for the mantle processes and appears to be a better explanation of crustal rifting. Rising magma in a mantle convection cell may heat the overlying crust and bow it upward. This weakens the crust and results in localized faulting that allows blocks of crust to drop downward, as seen in the Great Rift Valley of Africa that stretches from Mozambique to Ethiopia. Such a sunken rift zone is called a **graben.** Continuing convection cell motion forces the plates apart, thinning the crust and deepening the fault. Gradually the magma penetrates into the crack and widens the rift. If the rift deepens sufficiently, seawater may enter. The Red Sea today is at this intermediate stage, with both oceanic-type basalt and subsided blocks of continental-type

Figure 3.18

(a) (*inset*) The San Andreas Fault runs north-south from the Gulf of California through the San Francisco peninsula. (b) The sunken portion of the fault south of San Francisco and west of Palo Alto is used to store water for the city's use.

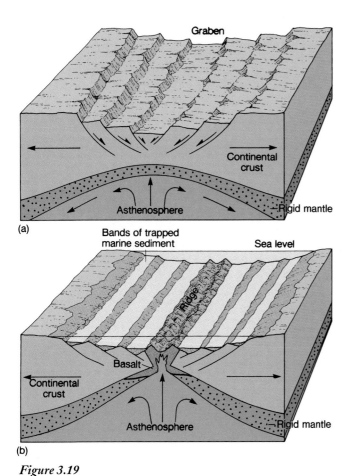

Figure 3.19

(a) Rifting is thought to happen when magma rises, causing tension and stretching of the overlying plate, resulting in a graben. (b) As the spreading continues, the fault deepens and cracks, allowing magma to penetrate and eventually form a ridge. Marine sediments accumulate in the fault, smoothing the seafloor.

crust present on the floor of the rift. As the lithospheric plate cools and moves away from the rift, magma becomes attached to the lower side of the lithosphere, causing the plate to gradually thicken away from the rift. As the crustal plate continues to thin at the rift, magma wells into the rift at an increasing rate and the landmass is gradually separated into two parts, with a low-lying region of oceanic basalt-type crust between the two sections. A new ocean basin is formed as well as a new ridge system (see fig. 3.19).

Subduction

As one oceanic plate collides with and overrides another, the overriding plate edge scrapes, folds, and slices off seafloor sediments, some of which are carried down into the subduction zone. When the descending plate slides past the overriding one, the friction may be great enough to melt the upper part of the downturned slab, including the wet sediments it is carrying. These melted sediments rise upward to form belts of volcanoes along the trench. When such volcanoes are separated from the continent they often form island arc systems (see fig. 3.20a). Island arcs may also be compressed against a landmass, or volcanoes may

arise along the border of the landmass. Examples of the latter are the California coast range, parts of the Alps, and the Andes.

When two continents collide at a subduction zone, the two landmasses resist sinking because their density is lower than the mantle density. Instead, they are crumpled and deformed at the colliding edges, and continental material from one plate may override the continental material of the other plate, producing a large thickness of continent. The sediments and the less-dense crust of the intervening seafloor may also be compressed together and piled upward; the Himalayas and parts of the Alps are examples of mountain ranges made in this way. Old marine sediments and fossils including the limestone remains of coral reefs are found on the summits of peaks in these ranges. This situation is illustrated in figure 3.20b.

If a plate carrying a continent and one carrying an ocean collide, the continental crust is less dense while the oceanic crust is closer to the density of the asthenosphere and is more easily forced down into the mantle. When the oceanic crust melts it moves upward through crustal fractures adjacent to the subduction zone. The volcanic action that occurs under these conditions is distinctly different from the volcanic action associated with spreading centers. Volcanic eruptions associated with recycled oceanic crust and silica-rich, water-saturated sediments are explosive, producing large amounts of ash, gas, and steam. Volcanic action of this type forms **andesite** volcanoes, named for the type of silicic magma erupted.

Andesite volcanoes are common to the Andes and the Cascade Mountains of the Pacific Northwest, including Mount St. Helens (see fig. 3.21). These volcanoes form the island arc chains of the Aleutians, Japan, the Philippines, and Malaysia. By contrast island formation at hot spots is more often a quiet, smooth outpouring of thick, flowing magma; for example, the Icelandic and Hawaiian volcanoes. See the section on hot spots in this chapter.

Continental Margins

As a plate carrying a continent moves away from a spreading center, the continental margin closest to the midocean ridge is known as a **trailing margin** or **trailing edge.** Basalt-type rock of the seafloor adheres to the trailing edge, and the aging continent edge slowly subsides as the spreading lithosphere thickens, cools, and contracts. A trailing edge is eroded by the waves and the currents; it may also be modified by shell or coral reefs. These continental margins accumulate an orderly series of sediment deposits eroded from the continents, to a depth of about 3 km (1.8 mi). Turbidity currents move the sediments down the continental slope to the continental rise, where thick deposits of graded sediments build on top of the oceanic crust. The trailing edges of continental margins are passive and not greatly modified by tectonic processes; they are often extensive, broad, and shallow with thick sedimentary deposits.

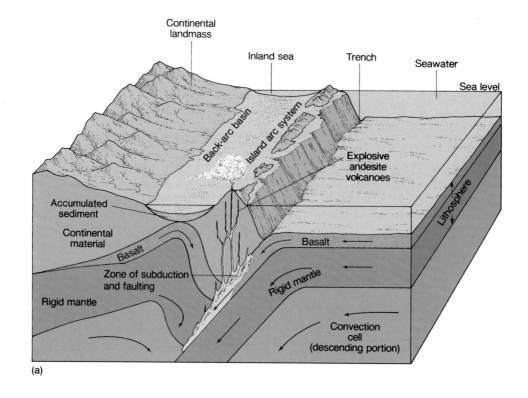

(a)

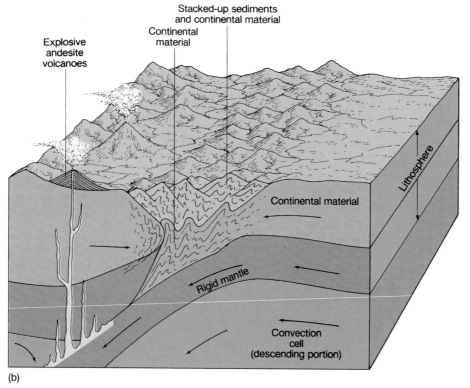

(b)

Figure 3.20

Subduction processes produce (a) island arc systems, volcanic activity, and (b) additions to a landmass and explosive volcanoes.

Figure 3.21
Mount St. Helens erupted violently on May 18, 1980. The mountain lost nearly 4.1 cubic kilometers from its once symmetrical summit, reducing its elevation from 2950 meters to 2550 meters. The force of the lateral blast blew down forests over a 594-square-kilometer area. Huge mud floes of glacial meltwater and ash flowed down the mountain.

The edge of a continent that is moving toward a deep-sea trench or subduction zone is known as a **leading margin** or **leading edge.** When a leading edge reaches a subduction zone or fault boundary, it is modified by tectonic processes. Although these continental margins receive sediments from the land, the sediments move downslope to deeper water, into trenches or adjacent basins. Thick sediment deposits do not accumulate on leading margins, and narrow continental shelves result.

Driving Forces for Plate Motion

The mechanism that drives the plates apart is still not fully understood. The plates could be forced apart at the ridges by the formation of new crust and then pushed downward at the trenches, or the subduction process could be caused by the weight of the thick, dense, older crust and its sediment load sinking into the mantle and dragging the remainder of the

plate with it. This latter mechanism would result in tension cracks through which the mantle material escapes upward to form the ridges. In this theory the weight of the dense, cold, descending slab of lithosphere pulls the trailing portion of the plate, and the friction between the moving plate and the asthenosphere helps to drive the convection cell rather than the reverse.

New computer models of this "slab pull" theory lead some researchers to believe that the cold ocean plates can drive plate tectonics without the forces of seafloor spreading. The situation is probably more complex, combining both mechanisms. Other factors that may be involved include the shape of the earth, changes in the weight of plates resulting from land erosion or the accumulation of sediments, the rate at which magma wells up into the ridges, and the thickening of the older lithosphere.

3.5   The Motion of the Plates

The directions and the rates at which pieces of the earth's crust move have varied over time. Although we do not understand why such directions have sometimes changed, we are able to trace back these movements and follow the paths taken by the continents as they have moved around the earth. Island chains, fixed centers of volcanic activity, and magnetic rocks help us trace some of these movements.

Rates of Motion

Spreading centers, seafloor spreading, and subduction zones provide the mechanisms for moving the continents. Acting like a conveyor belt, the seafloor moves away from the ridge system at rates between 1 and 17 cm (0.5–6 in)/year, and the old crust disappears into the trenches at a comparable rate. Although the rates are small by every-day standards, they produce large changes over geologic time. For example, at the rate of 1.5 cm (0.6 in)/year it takes 100,000 years to move 1.6 km, or 1 mile; therefore, in the 200 million years since the breakup of Pangaea the crust could move 3200 km, or 2000 miles, which is more than half the distance between Africa and South America. Spreading rates vary in magnitude and with time. A ridge system with a steep profile (the Mid-Atlantic Ridge) has a slower spreading rate than a ridge system with less-steep sides (the East Pacific Rise). Spreading rates are estimated at 3 cm (1 in)/year for the Mid-Atlantic Ridge and at 6–17 cm (2–6 in)/year for the East Pacific Rise. The largest spreading velocity known is 18.3 cm (7 in)/year off the west coast of South America. Keep in mind that the process of spreading does not occur smoothly and continuously but goes on in fits and starts, with varying time periods between occurrences.

Although seafloor spreading can be observed directly only with great expense and difficulty at sea, there is one place where many of the processes can be seen on land: in Iceland. Iceland is the only large island lying across a midocean ridge and rift zone, and here many of the processes of seafloor spreading can be seen and measured on land. Spreading in Iceland occurs at rates similar to those found at the crest of the Mid-Atlantic Ridge. Northeastern Iceland had been quiet for one hundred years, until volcanic activity began in 1975; in six years this activity widened by 5 m (17 ft) an 80-km (50-mi)-long stretch of the ridge's rift zone. Over one hundred years this spreading rate is 5 cm (2 in)/year, which is within the range of 1–17 cm (0.5–6 in)/year.

Recent investigations of ancient granitic rocks, formed during the Archean eon and found exposed in West Greenland, eastern Labrador, Wyoming, Western Australia, and southern Africa, indicate that 3.5 billion years ago crustal plates existed and moved granitic continental blocks at an average rate of about 1.7 cm (0.67 in)/year, again within today's average rates of plate motion.

Hot Spots

Scattered around the earth are approximately forty fixed areas of isolated volcanic activity known as **hot spots.** They are found under continents and oceans, in the center of plates, and at the midocean ridges. These hot spots periodically channel hot material to the surface from deep within the mantle. At these sites, a plume of mantle material may force its way through the lithosphere and form a volcanic peak or seamount directly above. If the hot spot does not break through, it may produce a broad swelling of the ocean floor or the continent that will subside as the crust moves over the magma source. Hot spots resupply the asthenosphere, which is constantly cooling and becoming attached to the base of the lithosphere, thus thickening the crust. Some people believe the breakup of Pangaea began when the supercontinent came to rest above a chain of hot spots.

Hot spots channel heat to the surface from deep within the mantle. Hot spot plumes of magma are not uniform; they differ in chemistry, suggesting that they come from different depths, and it has been suggested that their discharge rates may also vary. Hot spots may fade away and new ones may form. The life span of a typical hot spot appears to be about 100 million years. Although their positions may change slightly, they tend to remain relatively stable in comparison to the plates, and therefore are useful in tracing plate motions. As the oceanic crust moves over a hot spot, successive eruptions can produce a linear series of peaks or seamounts. In such a series, the youngest peak is above the hot plume, and the seamounts increase in age as the distance from the hot spot increases (see fig. 3.22). For example, in the islands and seamounts of the Hawaiian Islands system, the island of Hawaii with its active volcanoes is presently located over the hot spot. The newest volcanic seamount in the series is Loihi, found in 1981 45 km (28 mi) east of Hawaii's southernmost tip and rising 2450 m (8000 ft) above the seafloor but still under water.

The land features of the island of Hawaii show little erosion, for it is comparatively young. To the west are the islands of Maui, Oahu, and Kauai, which have been displaced by the moving crust. Kauai's canyons and cliffs are the result of erosion over the longer period of time that it has been exposed to the winds and rains. Although these four islands are the most familiar, other islands and atolls attributed to the same hot spot stretch farther west across the Pacific. They are peaks of eroded and subsided seamounts formed by the same hot spot. When the seamounts in tropic areas sank slowly, corals grew upward and coral atolls resulted.

West of Midway Island, the chain of peaks changes direction and stretches to the north, indicating that the crust over the hot spot moved in a different direction some 40 million years ago. This line of peaks is the Emperor Seamount Chain, volcanic peaks that once were above the sea surface as islands but have since eroded and subsided over time, resulting in many flat-topped guyots 1000 m (0.6 mi) below the surface. The northern end of the Emperor

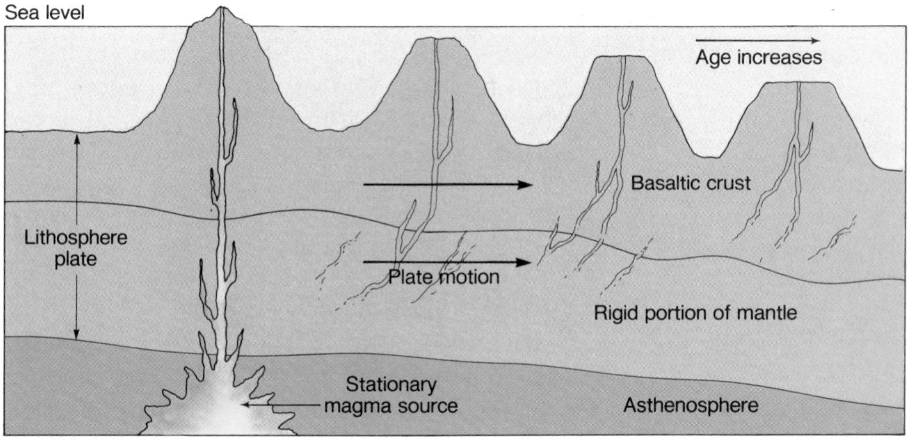

Sea level

Age increases

Basaltic crust

Lithosphere
plate

Plate motion

Rigid portion of mantle

Stationary
magma source

Asthenosphere

Figure 3.22
Chains of islands are produced when a crustal plate moves over a stationary hot spot.

Seamount Chain is estimated to be 75 million years old, while Midway Island itself may be 25 million years old. Refer to figure 2.13 to follow this seamount chain. Notice that the peaks formed by the hot spot get older in the direction in which the plate is moving. The plate is presently moving westward; Midway Island is northwest of the main Hawaiian Islands, and it is also older.

It is possible to check the rate at which the plate is moving by using the distance between seamounts in conjunction with radiometric dating. For example, the distance between the islands of Midway and Hawaii is 2700 km (1700 mi). Midway was an active volcano 25 million years ago, when it was located above the hot spot currently occupied by Hawaii. In other words, Midway has moved 2700 km (1700 mi) in 25 million years, or 11 cm (4 in)/year.

There is another hot spot at 37° 27′S, in the center of the South Atlantic, which is marked by the active volcanic island of Tristan da Cunha. Volcanic activity at this more slowly diverging plate boundary produces seamounts that can be carried either to the east or to the west, depending on the side of the spreading center on which the seamount was formed. The hot spot produces a continuous series of seamounts very close together, forming a **transverse ridge:** the Walvis Ridge to the east, between the Mid-Atlantic Ridge and Africa, and the Rio Grande Rise to the west, between the Mid-Atlantic Ridge and South America. Again, refer to figures 2.10 and 2.13.

If a hot spot is located at a spreading center, the flow of material to the surface is intensified. The crust may thicken and form a platform. Iceland is an extreme example of this process in which the crust has become so thick that it stands above sea level.

Seamount chains, plateaus, and swellings of the seafloor, all products of hot spots, are being used to trace the motion of the earth's plates over known hot spot locations. Recently, hot spot tracks have been used to reconstruct the opening of the Atlantic and Indian oceans.

Polar Wandering Curves

Magnetic rocks of different ages found on the same continent point to different magnetic pole positions. Records of the angular relationship between the magnetic pole and the

Figure 3.23
The positions of the North Pole millions of years before present (mybp), judged by the magnetic orientation and the age of the rocks of both North America and Eurasia. The divergence of the two paths indicates the landmasses of North America and Eurasia have been displaced from each other as the Atlantic Ocean opened.

geographic pole are locked in these rocks. If the direction of magnetic north is plotted, and if it is assumed that the continent has remained in a fixed position through time, it appears that the earth's magnetic poles have wandered away from their present positions relative to the earth's north-south axis. The path plotted in this way for any continent or region is known as a **polar wandering curve** (see fig. 3.23). Because rocks of the same age from two different continents on different plates point to two different locations for the magnetic pole, and because all evidence points to the magnetic poles remaining close to the geographic poles, even though the magnetic polarity has reversed through time, it is the continents that have moved and not the poles. If the magnetic poles remain close to the geographic poles and the continents move and rotate, then polar wandering curves can be used to track the movements of the continents with time.

When the polar wandering curves formed from the changes in the magnetic orientation of North America and northern Europe are superimposed, these two landmasses are seen to reconstruct the formation of the two-lobed supercontinent proposed by Wegener, with Laurasia to the north, Gondwanaland to the south, and the Sea of Tethys in between. Based on this method, the present landmasses join nicely at the edges of the continental blocks, at a boundary on the continental slope that is about 2000 m below the present sea level. When the landmasses are moved together, major ancient mountain and fault systems spanning several of our modern continents join; for example, the fault through the Caledonian Mountains of Scotland joins the Cabot Fault extending from Newfoundland to Boston. Also, countries of western Europe and Great Britain that are at present located in temperate and high latitudes were once found in the equatorial zone, which explains their fossil coral reefs and desert-type sand deposits.

Current research leads some scientists to believe that the assumption that the rotational polar axis has not varied its location over time may not be correct. The rotating earth possesses tremendous inertia, which keeps it spinning on its axis. However, as the lithospheric plates migrate about on the earth's surface, the earth may tend to roll about its center slightly, trying to adjust so that the majority of the earth's crustal mass is centered about the equator. If this is so, the location of the points marking the earth's rotational axis relative to the crust may also shift.

3.6   The History of the Continents

The Breakup of Pangaea

About 200 million years ago Pangaea began to break apart. Figure 3.24 traces the plate movements that led to the configuration of the earth's continents and oceans as we know them today. Laurasia (North America and Eurasia) moved away from Gondwanaland, and the South Indian Ocean began to form (fig. 3.24a). India drifted away from Antarctica to move northward toward Asia and its eventual collision with the Asian mainland. Water flooded the spreading center between Africa and South America 135 million years ago; North America and Eurasia were still firmly attached, as were Australia and Antarctica (fig. 3.24c). South America separated from Africa 65 million years ago; Africa moved northward, and the Mediterranean Sea was created. India continued to move toward Asia. Australia began to separate from Antarctica, and Madagascar split off from Africa (fig. 3.24d). During the last 16 million years, North America and Eurasia separated, creating the North Atlantic Ocean, and Greenland moved away from Europe. North and South America united, and

Australia at last became free of Antarctica. India reached Asia, and because the continental landmass of India was less dense than the basaltic ocean crust, India was not subducted but crumpled up against Asia, resulting in the Himalayas. A similar situation occurred when Italy became detached from Africa and moved across the Mediterranean to collide with Europe and produce the Alps. Twenty million years ago, Arabia moved away from Africa to form the Gulf of Aden and the new and still opening Red Sea.

Before Pangaea

Since the earth is about 4.6 billion years old, there is no reason to believe that Pangaea and its breakup into today's continental configuration was the first or only time during which the seafloor spread or the continents changed position and recombined. Scientists are now searching for new and more extensive evidence to indicate the position of the landmasses before the breakup of Pangaea. Because there is no record of pre-Pangaean relationships in the present 200-million-year-old oceanic crusts, scientists must depend on evidence from the continents, where the oldest rocks are found.

Radiometric dating and magnetic and mineral analyses indicate that ancient granitic formations are common to all landmasses; they have been present in crustal plates for more than 3 billion years. Ancient mountain ranges located in the interiors of today's continents are evidence of collisions between old plates; seismic evidence has been used to mark old plate edges below the Ural and Appalachian mountains. The magnetic and fossil evidence frozen into the land rocks during the 350 million years prior to Pangaea have been used to propose a series of pre-Pangaea plate movements during the Paleozoic period, which began about 600 million years ago.

Since the earth's polar axis (and therefore its magnetic poles) have held relatively constant positions through time, the magnetic data found in the ancient land rocks (called paleomagnetic data) are the basis for the positioning of the continents during the Paleozoic era. A fixed polar axis also means the earth's climate zones have remained fixed in their latitudinal positions as well. Therefore, any portion of the earth's crust drifting through these climate zones will undergo environmental changes. For example, deposits of mineral salts indicate that an area was once a hot, dry desert, while coal deposits were formed in warm, freshwater swamps; in order for either deposit to have occurred, the land must have been at a latitude favorable to such climatic conditions. Present-day geographic and climatological data are combined with paleomagnetic evidence and paleoclimatology to move the continents back through time. In reconstructing their positions it is assumed that pieces of the earth's crust remained in a climatic zone long enough to collect such a record. The length of the required period varies between 5 million and 15 million years, which is quite reasonable when we consider that the rate of plate movement is 2–8 cm (1–3 in)/year.

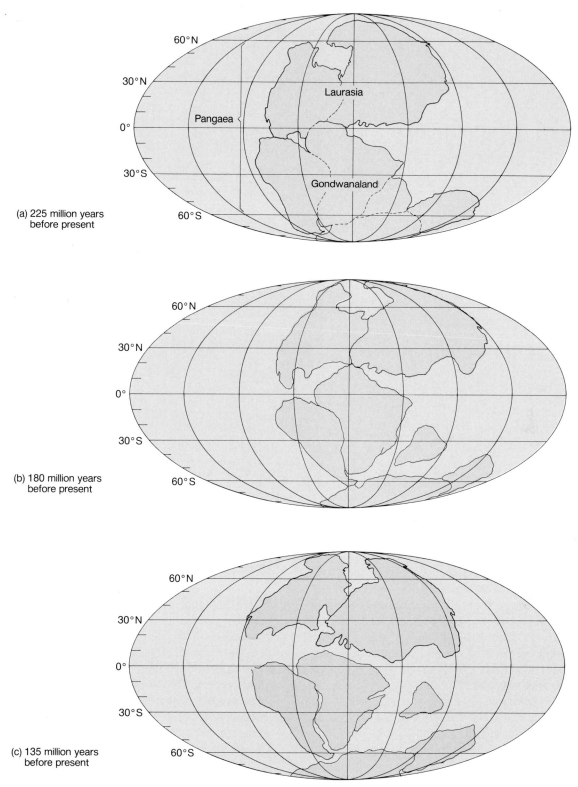

(a) 225 million years
before present

(b) 180 million years
before present

(c) 135 million years
before present

Figure 3.24
The movements of the continents from Pangaea to the present.

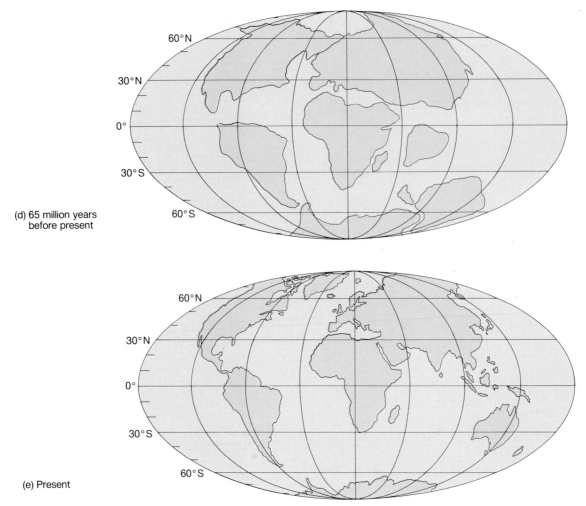

(d) 65 million years before present

(e) Present

Figure 3.24
continued

Six major continents are recognized from the Paleozoic era: Gondwana (Africa, South America, India, Australia, Antarctica), Baltica (Scandinavia), Laurussia (North America), Siberia, China, and Kazakhstania. Approximately 555 million to 540 million years ago (fig. 3.25a) these landmasses were strung along the earth's equator; there was no land above 60°N and 60°S, and the polar regions were wide expanses of ocean. Figure 3.25b shows Gondwana and Baltica shifting to the east and south 490 to 475 million years ago. Gondwana's southward motion carried it across the South Pole 435 million to 430 million years ago (fig. 3.25c), and then north again (fig. 3.25d) on the opposite side of the earth, reaching a position that would eventually make it a part of Pangaea about 350 million years ago (fig. 3.25e). This movement completely reversed the north-south orientation of modern Africa and South America. Their present-day southern tips pointed to the north at the beginning of the journey, but as the landmass drifted across the South Pole to the other side of the world, they pointed to the south, as we know them today. About 310 million years ago (fig. 3.25f) the ocean between Gondwana and the other continental fragments began to close. Siberia moved from low to high latitudes and merged

with Kazakhstania, while China moved westward (fig. 3.25g). The assembly of Pangaea 260 million to 250 million years ago resulted from a series of collisions that formed great mountain belts that are still detectable.

Because the tectonic processes are continuous, there probably has never been a stable geography of the earth. The modern configuration is no more stable than in the Paleozoic era. They are both steps in an ever-changing pattern. We have followed this changing pattern of the earth's landmasses from wide dispersal 550 million years ago, to a single joined continental mass 250 million years ago, and then through another period of dispersal to the present time.

It has been suggested that this is an orderly, cyclic, 500-million-year pattern powered by the heat from radioactive decay in the earth's interior. Although the production of heat is continuous, this model proposes that the heat is released in periodic bursts that are related to the movement of the continents. As the continents shift, coalesce, and compress, the sea level drops. When sufficient heat accumulates under the continental mass, the rifting process begins unstitching the continents, and as the continents move apart, they cool and subside, then the sea level rises and covers their border lands.

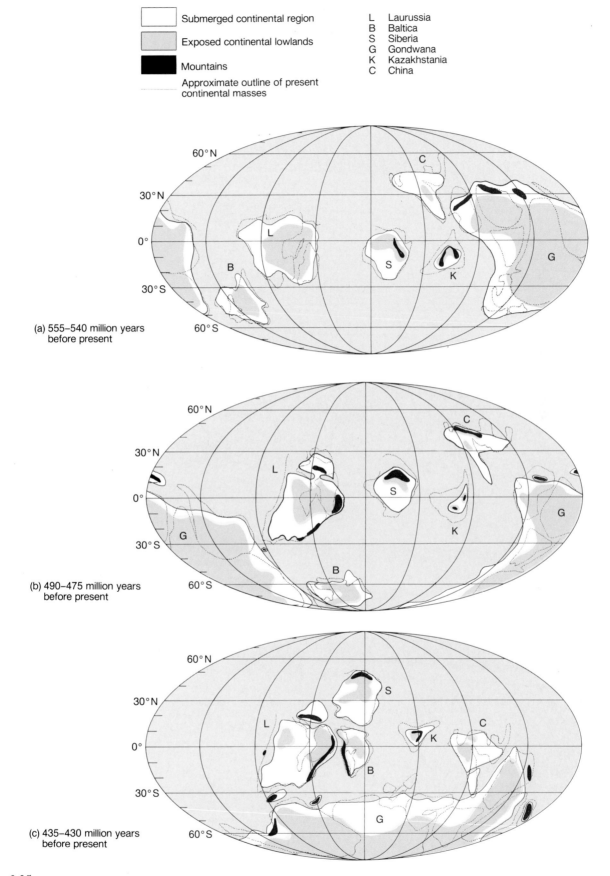

Legend:
- Submerged continental region
- Exposed continental lowlands
- Mountains
- Approximate outline of present continental masses

L Laurussia
B Baltica
S Siberia
G Gondwana
K Kazakhstania
C China

(a) 555–540 million years before present

(b) 490–475 million years before present

(c) 435–430 million years before present

Figure 3.25

Movements of landmasses before Pangaea.

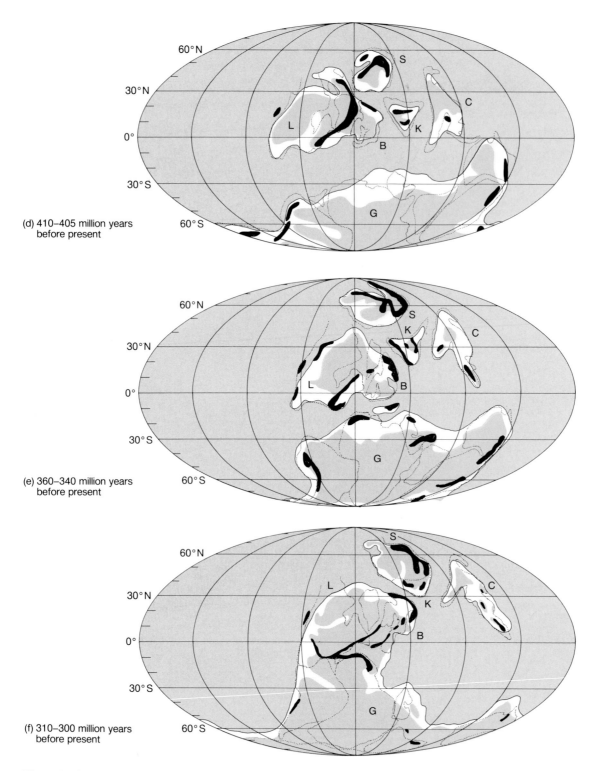

(d) 410–405 million years
before present

(e) 360–340 million years
before present

(f) 310–300 million years
before present

Figure 3.25

continued

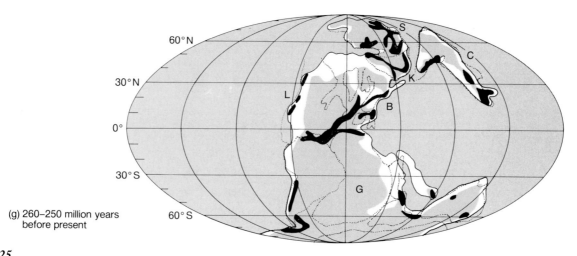

60°N

30°N

0°

30°S

60°S

S

C

K

L

B

G

(g) 260–250 million years
before present

Figure 3.25
continued

Using this model we see the Atlantic Ocean opening and closing as the continents move apart and then reassemble over the 500-million-year cycle, while the Pacific Ocean boundaries, which have remained more stable, approximate the wide, ancient hemispheric ocean of that period. We can even look forward in time to see the Mediterranean Sea being closed and lost as Africa continues to move northward, and the Atlantic and Indian oceans continuing to expand while the Pacific Ocean narrows as North and South America collide with Asia. Australia will continue to move northward, eventually colliding with Eurasia, while Los Angeles and coastal southern California will pass San Francisco on their way toward the Aleutian Trench. This is drama on the grandest scale, never to be seen by any one individual, yet it can be appreciated by all.

Terranes

Studies of ancient crustal rocks from the interior of North America indicate that the core of the continent was assembled about 1.8 billion years ago from several large pieces of granitic crust more than 3 billion years old. It appears that four or five large pieces of crust called **cratons** collided and joined over a 100-million-year period. Mixtures of sediments from the craton margins and the deep seafloor, the remains of ancient island arc systems, and other small pieces of continental crust were trapped between the colliding blocks of continent. Radiometric dating has identified boundaries between North American cratons at 1.78 billion and 1.65 billion years old.

Geological studies of Alaska show that the characteristics of one area do not necessarily relate to the history, age, structure, or mineral composition of an adjacent area. Alaska is made up of crustal fragments, some showing the characteristic homogeneity and higher density of oceanic

crust and others showing the highly varied mineral content and lower density of continental crust. Smaller crustal fragments associated with craton margins, bounded by faults, and with a history distinct from adjoining crustal fragments are known as **terranes.** See figure 3.26 for a map of Alaskan terranes.

Terranes are often elongate, as if produced by an island arc system that has collided with a craton, or by faulting, which has cut off a sliver of continent similar to the land west of California's San Andreas Fault. The terranes of Alaska appear to have arrived from the south, rotated clockwise, and faulted and stretched as the Pacific and North American plates moved relative to each other. A terrane known as Wrangellia originated at or to the south of the equator, moved northward, and crashed into western North America about 70 million years ago. Faulting has spread its fragments throughout eastern Oregon, Vancouver Island, the Queen Charlotte Islands, and the Wrangell mountains of southeast Alaska.

India is considered a single giant terrane by some; north of the Himalayas are terranes that predate the arrival of India. The area of North America west of Montana, south into New Mexico, and north into Canada and Alaska appears to be an assemblage of terranes that have been modified by faulting and vulcanism. In Oregon some volcanic peaks appear to be basalt seamounts that have moved landward from relatively short distances offshore. Rock formations found in the San Francisco area are typical of rock formed in the South Pacific. Fossils collected between Virginia and Georgia indicate that a long section up and down the East Coast was formed somewhere adjacent to an island arc system; the fossils point to a past European connection. See figure 3.26.

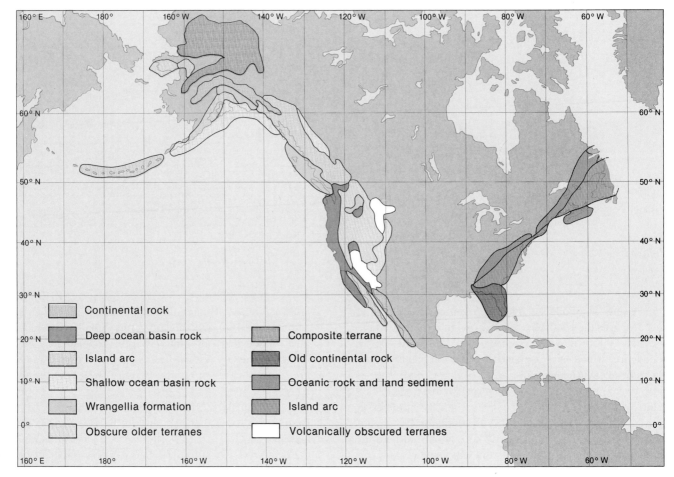

Figure 3.26
Examples of terranes of the coasts of North America.

3.7 Current Research

Drilling Plans

Scientists are exploring the ocean bottom with increasingly sophisticated technology as they continue to search for more evidence and a better understanding of present and past events. The new Ocean Drilling Program, guided by the Joint Oceanographic Institutions for Deep Earth Sampling (JOIDES) and using the *JOIDES Resolution* (fig. 3.11) operated by Texas A&M University, plans a ten-year international program of scientific ocean drilling. Objectives of the new program include improving our understanding of the nature and evolution of oceanic crust, investigating fossils and the sediments that cover them to create a record of the changing conditions that have occurred on this planet, searching for evidence of global environmental changes, and developing a better understanding of mantle-crust interactions and tectonic processes by recovering rock cores from the seafloor. The ability to drill into the exposed rock of the seafloor will also help us to understand how water circulating through the crust alters the crust's chemical structure and produces

the rich mineral deposits associated with divergent plate boundaries (see the section on sulfide deposits in Practical Considerations in chapter 2).

The *JOIDES Resolution* has drilled in the regions of Baffin Bay and the Labrador Sea, seeking to determine the tectonic evolution of these two basins. In Caribbean waters, off the island arc of the Lesser Antilles, where the edge of the North American Plate moves under the Caribbean Plate, cores have been made of the sediments along this plate boundary. Another plate boundary under investigation is along the coast of Peru, where the Pacific plate slides underneath South America. Recently, the *JOIDES Resolution* drilled in the Antarctic, working to understand the formation of the great ice sheets covering the Antarctic region. In 1988 drilling in the southern Indian Ocean on the Kerguelen Plateau discovered fragments of fossil wood and soils that indicate that this area was once above sea level, a barren land that with time developed a lush vegetation and then sank below the ocean to its present depth of 800 m (2600 ft). Through 1991 the ship operated in the Pacific Ocean, exploring the Lua basin near Hawaii. It then moved to the eastern tropical Pacific to explore the ridge systems and an area known as the Hess Deep. In 1992 *JOIDES Resolution* explored a triple-plate junction off Chile and the

Figure 3.27
The submersible *Alvin* being lowered for a dive.

atolls and guyots between Hawaii and Guam. It also plans a series of drill holes across the North Pacific, ending in the Cascadia basin off the coast of Washington state.

Thousands of meters of core are taken at the drill sites, and it takes many years to complete the analyses of these cores. However, the results from each cruise are eagerly anticipated, for they help us to build a more precise and more detailed understanding of earth processes and the development of our planet.

Project FAMOUS

The first descent into the rift valley of the midocean ridge occurred in 1973. The previous year more than twenty U.S., French, Canadian, and British oceanographic vessels mapped a small section of the Mid-Atlantic Ridge. In the summer of 1973 in an area centered at 36° 50′ N, southwest of the Azores, French oceanographers aboard the submersible *Archimede* made the first visual study of a rift valley, the area where new crust is produced. The following summer the French submersible *Cyana* and the U.S. submersible *Alvin* (fig. 3.27) joined the dive program. This effort was known as the French-American Mid-Ocean Undersea Study, or Project FAMOUS.

The submersibles descended nearly 3000 m (9800 ft) to the floor of the huge rift valley. In a series of 50 dives they took more than 50,000 photographs and made 100 hours of television recording, as well as recovering 150

rock samples and 70 water samples from the bottom. The submersibles found an enormous variety of rocks, topography, and sediments. The cracks and fissures running parallel to the direction of the rift indicated a continuous spreading with episodic, localized volcanic eruptions occurring in the central rift area.

Hydrothermal Vents

The Galápagos Rift, lying between the East Pacific Rise and the South American mainland, 964 km (599 mi) west of Ecuador, has also been the focus of much study. Measurements of water temperature and chemistry taken in 1972 showed active circulation of seawater through newly formed oceanic crust, which appeared to be the cause of plumes of hot water rising from **hydrothermal vents** along the rift. In 1976 the area's seafloor was mapped in detail in preparation for submersible exploration.

In 1977 and 1979, expeditions from Woods Hole Oceanographic Institution (WHOI) used the submersible *Alvin* to locate and explore the vent areas. The *Alvin* carries two scientists and a pilot. It is equipped with cameras inside and out, as well as baskets for samples taken with its mechanical claws (see fig. 3.28). It dives at a speed of 2 knots to a depth of 3000 m (10,000 ft). It requires three hours for the round trip, and *Alvin* spends four to five hours on the bottom. The underwater camera assembly called ANGUS (Acoustically Navigated Underwater Survey

Figure 3.28
Alvin's mechanical arm collects a sample.

their energy source has been removed. Further discussion of these communities, the organisms, and their food supply is found in chapter 17.

Along a section of the Galápagos Rift, the basaltic oceanic crust is new, so new that it shows little sediment cover or chemical change. A spreading rate of 3.5 cm (1.4 in)/year occurs across the rift, while the sedimentation rate is only about 5 cm (2 in)/1000 years. Some of the basalt shows no sediment, indicating its formation within the last 100 years. At the outer edges of the rift valley, about 500 m (1640 ft) from the central rift, sediments are thick enough to cover the new lava, indicating an age of about 10,000 years.

Water issuing from these vents had temperatures of 17°C, compared to 2°C for the surrounding seawater. The hot water mixes with the cold bottom water to produce shimmering, upward-flowing streams rich in silica, barium, lithium, manganese, hydrogen sulfide, and sulfur. Rocks in the vicinity of the vents are coated with chemical deposits rich in metals precipitated out of the vent waters.

Also in 1979 the Mexican-French-American Riviera Submersible Experiment (RISE) research program investigated ocean floor hot springs near the tip of Baja California. Two thousand meters (6000 ft) down, *Alvin* found mounds and 20-m (65-ft)-high chimney-shaped vents ejecting hot (350°C) black streams of particles containing lead, cobalt, zinc, silver, and other minerals. Here, too, the vents are surrounded with rich animal life.

Since these initial discoveries, expeditions have continued to search for hydrothermal vents along the spreading centers of the world's plates. In 1981 researchers first photographed a small section of a rift valley in the ridge system along the small Juan de Fuca Plate off the coast of Oregon. The *JOIDES Resolution* identified vents along the Mid-Atlantic Ridge in 1986, and that same year additional sites on the Gorda Ridge of the Juan de Fuca Plate were surveyed.

During one dive on the Juan de Fuca Ridge an electronic camera carried on the *Alvin* photographed light coming from seafloor vents; see the photo series in figure 3.29. On another dive in this area the *Alvin's* cameras recorded the features of the massive sulfide structures associated with vent areas; see figure 3.30. The large flanges that protrude from these structures trap beneath them hot vent water (350°C) that can be monitored without having to penetrate the vents and interfere with their natural processes.

At present two major U.S. programs are associated with identifying and understanding processes along the midocean ridges: the National Science Foundation's Ridge Inter-Disciplinary Global Experiments (RIDGE) and NOAA's VENTS program. The VENTS program is exploring the source and strength of hydrothermal discharges as well as the time intervals between discharges. This program is also interested in the pathways followed by the materials issuing from the vents, and its researchers are setting up monitors to discover the impact of vent discharges on the chemistry of the water. Researchers from the VENTS program discovered the megaplume described in this chapter's box.

System) was slowly towed above the ocean floor to pinpoint the vent area, and the *Alvin* made twenty-four dives to the 2500-m (8000-ft) deep rift.

The most surprising discovery of these expeditions was the unexpected and perplexing presence of large communities of animals so far removed from surface food sources. Clams, mussels, limpets, tube worms, and crabs were viewed, photographed, and collected. New species were identified; giant tube worms and clams with red blood and flesh similar to beef were collected for laboratory analysis. The presence of so many large animals at such depths immediately brought up the question of their source of food. Such dense communities could not be supported by the fall of organic matter from the sea surface. Instead of being dependent on the sun to provide energy for plant life to produce organic matter to be used as food, these ocean-bottom populations rely on the hydrogen sulfide and particulate sulfur in the hot vent water, which is utilized by bacteria to provide the energy necessary for life. The other vent animals then feed on the bacteria, which were calculated to reach concentrations as high as 0.1 to 1 gram per liter of vent water. No sunlight is necessary and no food is needed from the surface; the populations are sustained by the vents themselves. In areas where the vents have become inactive the animal communities have died, as

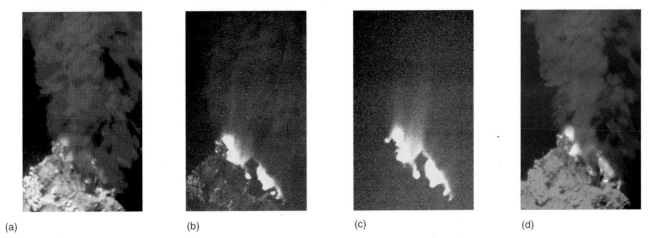

(a)   (b)   (c)   (d)

Figure 3.29

(a) A digital image of vent light taken with an electronic still camera and illuminated by a thallium iodide lamp mounted on the outside of *Alvin*. (b) A digital image of vent light taken with an electronic still camera and the light of a flashlight shining from inside *Alvin*. (c) A digital image of vent light taken with an electronic still camera and vent radiation. (d) False color image produced from figures a, b, and c by passing different wavelengths of light through these images.

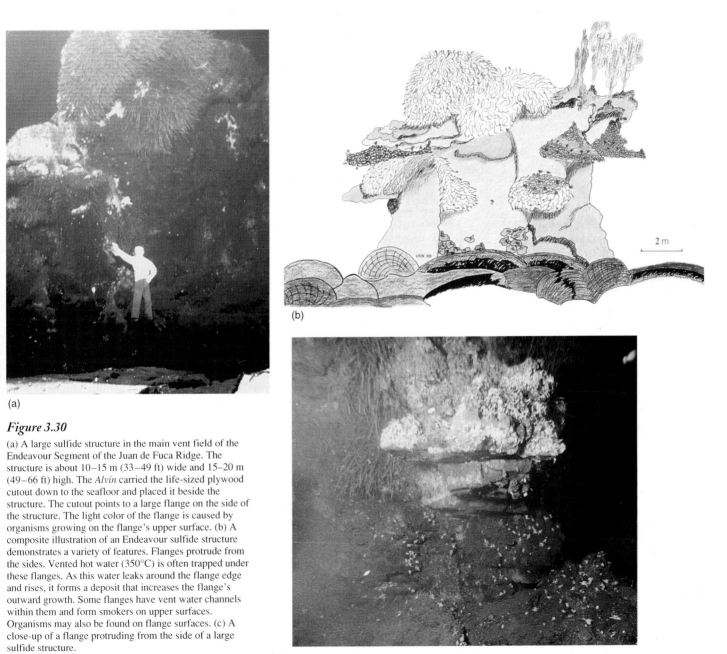

(a)

(b)

(c)

Figure 3.30

(a) A large sulfide structure in the main vent field of the Endeavour Segment of the Juan de Fuca Ridge. The structure is about 10–15 m (33–49 ft) wide and 15–20 m (49–66 ft) high. The *Alvin* carried the life-sized plywood cutout down to the seafloor and placed it beside the structure. The cutout points to a large flange on the side of the structure. The light color of the flange is caused by organisms growing on the flange's upper surface. (b) A composite illustration of an Endeavour sulfide structure demonstrates a variety of features. Flanges protrude from the sides. Vented hot water (350°C) is often trapped under these flanges. As this water leaks around the flange edge and rises, it forms a deposit that increases the flange's outward growth. Some flanges have vent water channels within them and form smokers on upper surfaces. Organisms may also be found on flange surfaces. (c) A close-up of a flange protruding from the side of a large sulfide structure.

Line illustration by Véronique Robigou-Nelson, University of Washington.

Megaplumes

On August 17, 1986, Dr. Edward Baker and Gary Massoth from the Pacific Marine Environmental Laboratory of the National Oceanographic and Atmospheric Administration (NOAA) were making water property measurements aboard the NOAA research vessel *Discoverer.* They were 450 km (260 mi) off the coast of Oregon, just north of the southern end of the Juan de Fuca Ridge. The Juan de Fuca Ridge and the Gorda Ridge to the south were being explored for hydrothermal vents; oceanographers in the VENTS program were testing the idea that hot water from a vent field would rise 100–300 m (300–900 ft) above the ridge and spread out much like chimney smoke.

During their survey of the north end of the Cleft segment of the ridge, heat sensors indicated they were above a large mass of warm water, 0.25°C above the temperature of the surrounding water and 1300–2000 m (4000–6500 ft) below the surface. They were able to determine that the warm-water mass was about 700 m (2300 ft) thick and 20 km (12 mi) in diameter; its top was about 1000 m (3300 ft) above the ridge. The volume of water and its height above the ridge were quite unexpected, and the large warm-water mass was christened a **megaplume,** Box figure 3.1.

Data analysis has established three important facts about this first observed megaplume: (1) chemical analysis of the water showed that its source was a hydrothermal vent, (2) the presence of crystals that are formed only during the initial mixing of hydrothermal fluids with seawater showed that it had been formed very recently, and (3) its volume and its symmetry indicated that it had been formed by a brief but massive discharge rather than a continuous venting of hot water. It is likely that the megaplume formed over a great crack in the seafloor with an area larger than any known hydrothermal vent.

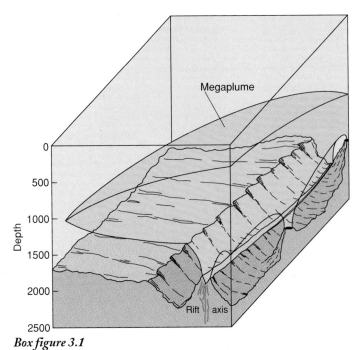

Box figure 3.1

A large volume of heated water with a high sulfur mineral content is formed by a rapid outpouring of water from vents located along a seafloor rift system.

Because of the megaplume's known symmetrical boundaries, its heat content could be calculated. Its volume of over 130 km^3 (20.4 nmi^3) contained about 2×10^{16} calories, or enough energy to provide New York City with electricity for a year. The original volume of fluid released from the seafloor was calculated as 100 million cubic meters at a temperature of 350°C. To gather information to link the megaplume to volcanic activity, bathymetric maps were made in 1987 and 1989 and compared to older surveys. The 1987 and 1989 surveys showed a new series of lava mounds stretching for 16 km (10 mi) and running through the center of the megaplume area.

In 1987 a smaller, older plume was observed, but it was not known how far it had drifted from its source. The discovery of these plumes has raised more questions than have been answered. How did the large volume of water needed to produce the megaplume seep into the seafloor crust? How was the water stored, and what was the mechanism that released it? Was this a special event or are such plumes common? Most basically, what can this phenomenon teach us about (1) the relationship between the earth's crust and mantle, and (2) the deep ocean's chemical balance?

Further understanding requires a long-term observational program of the seafloor in this area. The NOAA VENTS program included the 1991–92 installation of sensors to monitor temperature and current velocity above the vent field, seismic recorders on the seafloor, and acoustical beacons installed on either side of the rift zone to measure seafloor spreading. The data from these instruments will require analysis and interpretation before future experiments are planned.

The RIDGE program is studying the formation of transform faults, the extension of the ridges, and the formation of discontinuities and overlapping ridge segments. These processes are associated with the ways in which magma is supplied to the ridges. Researchers are developing computer models of magma activity to account for changing rates of migrating magma. In 1991 the *Alvin* made dives on the East Pacific Rise as a part of the RIDGE program and discovered an area of fresh lava. The animals in this vent area were recently dead. The new magma flow may have been associated with a series of earthquakes recorded in the area during the previous month. The program is installing sampling devices, recording cameras, and seismic event detectors in areas of interest to discover how such events evolve over time.

3.8 Practical Considerations: Ocean Waste Management

Using the sea as a dump for trash and garbage is a common practice in many countries. Pipelines discharge municipal wastewater, sometimes untreated, into coastal waters; dredged sediments from ports and waterways are commonly dumped at sea, and probably more than 25% of the mass of all material dumped at sea is dredged material. One of the major methods of industrial waste disposal is dumping at sea, and large amounts of industrial wastes are discharged into coastal seas. Nearly half the world population lives in coastal areas, and the numbers of these people will increase as the world population increases by more than 3 billion in the next 60–65 years.

Industrialization and economic growth have increased our amounts of garbage and have also changed its characteristics. Aluminum and plastics are displacing traditional materials such as paper, steel, and glass. Many modern consumer products contain toxic substances such as heavy metals (lead, mercury, cadmium), household cleaners, solvents, and pesticides. Estimates of world wastes produced each year are shown in table 3.2. Current waste management practices are not sufficient to handle our present problems; landfill areas are being overwhelmed, and much of their contents is increasingly recognized as a threat to the earth's soil and freshwater reservoirs.

In 1991 a group of scientists held a workshop at the Woods Hole Oceanographic Institution to explore the scientific research required to assess the potential of the deep ocean as an option for future waste management. They argued that controlled disposal in some ocean environments presents less risk to life than current and projected land disposal with its associated groundwater contamination or incineration accompanied by atmospheric pollution. They pointed out that the abyssal plain of selected midocean areas supports only sparse populations, and the slowly moving water above it stays at the bottom for thousands of years before it returns to the surface. In addition they believe that the deep-ocean floor far from shore may be a more appropriate disposal site than coastal landfills that

Table 3.2
Estimate of World Wastes Produced per Year*

Type	Millions of metric tons	%
Municipal solid waste	1,500	36.2
Dry sewage sludge**	65	1.6
Dredged material	1,075	26.0
Industrial wastes	1,500	36.2
TOTAL	4,140	100.0

*Source: D. Spencer, *An Abyssal Ocean Option for Waste Management*, 1991. Woods Hole Oceanographic Institution, Woods Hole, Ma.
**Produced by treatment plants.

pollute inshore waters and degrade beaches. They proposed a large-scale experiment to assess environmental effects, depositing one million tons of sewage sludge per year for ten years on the floor of the Hatteras Abyssal Plain in the Atlantic Ocean.

The 1972 U.S. Marine Protection, Research and Sanctuaries Act (commonly called the Ocean Dumping Act) prohibits dumping of all types of materials into ocean waters unless a permit is obtained from the Environmental Protection Agency (EPA). Permits for dredged material are obtained from the U.S. Corps of Engineers. In 1988 the Ocean Dumping Act was amended to prevent all ocean dumping of sewage sludge by 1991. New federal guidelines for the open ocean disposal of dredged materials were issued in 1992. Some feel that any experiments with ocean waste disposal will have to proceed by affiliation with other countries.

After each modern war, obsolete military hardware and munitions have been disposed of at sea. The North Atlantic was the dumping ground for toxic gases confiscated from Germany at the end of World War II, and the bays and lagoons of many South Pacific islands were used for the disposal of jeeps, tanks, bombs, and other items. After the United States left Vietnam, there was a similar unloading of vehicles and explosives in the waters around Southeast Asia.

Highly radioactive nuclear waste and hazardous chemical waste products are usually stored underground or in aboveground repositories. To avoid possible contamination of land areas and freshwater supplies, it has been suggested that these long-lived wastes should be disposed of away from the continents in the deep-sea subduction zones where the waste products will sink back into the mantle and eliminate the problem of long-term hazardous waste storage. No experiments involving these wastes are being planned at present.

The deep-ocean basins are the lowest places in the earth's crust, and all the products of society and nature ultimately work their way down to these basins. Should we use them and speed up this process? How do we handle our growing waste production as our populations increase?

Summary

The earth is made up of a series of concentric layers: the crust, the mantle, the liquid outer core, and the solid inner core. The evidence for this internal structure comes indirectly from studies of the earth's dimensions, density, rotation, gravity, and magnetic field and of the remains of meteorites. It also comes from the ways in which seismic waves change speed and direction as they move through the earth. Seismic tomography is being used to describe the earth's interior layers.

Continental crust is formed from granite-type rock, which is less dense than oceanic crust formed of basalt. The top of the mantle is fused to the crust to form the rigid lithosphere. The lithosphere floats on the deformable upper mantle, or asthenosphere. The pressures underneath the elevated continents and depressed ocean basins are kept in balance by vertical adjustments of the crust and mantle, a process known as isostasy.

Alfred Wegener's theory of drifting continents was based on the geographic fit of the continents and the similarity of fossils collected on different continents. His ideas were ignored until the discovery of the midocean ridge system and until the proposal of convection cells in the asthenosphere led to the concept of seafloor spreading. New lithosphere is formed at the ridges, or spreading centers. Old lithospheric material descends into trenches at subduction zones. Seafloor spreading is the mechanism of continental drift. Evidence for lithospheric motion includes the match of earthquake zones to spreading centers and subduction zones, the greater heat flow along ridges, age measurement of seafloor rocks, age and thickness measurements of sediment from deep-sea cores, and the magnetic stripes in the seafloor on either side of the ridge system. Rates of seafloor spreading average 1 to 10 cm/year. Plate tectonics is the unifying concept of lithospheric motion. Plates are made up of continental and oceanic lithosphere bounded by ridges, trenches, and faults.

Rift zones separate ocean basins, and in the past have separated landmasses and produced new ocean basins. Subduction produces island arcs, mountain ranges, earthquakes, and volcanic activity. Trailing continent margins move away from spreading centers. Leading continent margins move toward subduction zones. Hot spots and polar wandering curves are used to trace plate motions. The exact mechanism that drives the plates is unknown.

Plate movements traced over the last 225 million years show the breakup of Pangaea. Paleozoic plate movements in the 225 million years prior to Pangaea have recently been estimated using climatological evidence. Much of North America appears to be made up of continental fragments or terranes around a core continent or craton.

Deep-sea drilling programs are being conducted to investigate tectonic processes, the evolution of oceanic crust, and the sediment records of past conditions. Submersibles have been used to explore the rift valley of the Mid-Atlantic Ridge and the hydrothermal vents and animal communities found along the Galápagos Rift, Baja California, and the Oregon coast.

Dumping waste into ocean waters has been a common practice. Because of the world's increasing quantities of solid waste, the deep-ocean plains are being considered for waste disposal.

Key Terms

inner core	seafloor spreading
outer core	spreading center
mantle	subduction zone
crust	polar reversal
density	plate tectonics
seismic wave	divergent plate boundary
P-waves	convergent plate boundary
S-waves	fault
refraction	transform fault
seismic tomography	escarpment
granite	rift zone
basalt	graben
Moho	andesite
lithosphere	trailing margin (edge)
asthenosphere	leading margin (edge)
isostasy	hot spot
Gondwanaland	transverse ridge
continental drift	polar wandering curve
Pangaea	craton
Laurasia	terrane
convection cell	hydrothermal vent
magma	megaplume

Study Questions

1. What is meant by the term "polar wandering"? Have the magnetic poles actually wandered?
2. Describe the three types of plate boundaries. What processes take place at each type of boundary? In what direction do the plates move at each boundary?
3. What mechanisms have been proposed to account for plate motion?
4. What is the difference between the leading edge and the trailing edge of a continent? A divergent plate boundary and a convergent plate boundary?
5. If the ability of the oceanic crust to transmit heat were uniform, the rate of heat flow through the ocean floor would depend only on the temperature change across the oceanic crust. Under such a condition, how do the heat flow measurements in figure 3.9 indicate the presence of ascending convection cells in the asthenosphere?
6. If the polar wandering curves for North America and Europe are made to coincide, how will these continents move relative to each other?
7. Using the techniques and reasoning employed to discover the properties of the interior of the earth, explain how you would determine what is inside a sealed box (for example, measuring the box, weighing it, spinning

it, balancing it on different axes, sampling its exterior). What clue would each of these measurements give you to the contents of the box?

8. What had to be learned about the earth to allow acceptance of Alfred Wegener's ideas?

9. Why does a newly formed midocean volcanic island gradually subside?

10. Explain the formation and symmetry of the magnetic stripes found on either side of the midocean ridge system. What is their significance when the magnetic information is correlated with the age of the crust?

11. Under what conditions will a subduction zone form a mountain range? An island arc system? Why do volcanoes associated with subduction zones usually erupt more explosively than midocean volcanoes associated with hot spots and spreading centers?

12. On an outline map of the world draw in (a) earthquake belts, (b) midocean ridges, and (c) trenches. Relate your map to figure 3.16. What do you conclude?

13. How have recent advances in tomography modified our ideas of the earth's internal layers as shown in figure 3.1?

14. Describe the configuration of the continents prior to Pangaea. According to the cyclic theory, what configuration might be present in another 225 million years?

15. What is a terrane? What role do terranes play in our understanding of today's continents?

Study Problems

1. If a plate moves away from a spreading center at the rate of 5 cm per year, what is the displacement of a landmass carried by that plate after 180×10^6 years?

2. Magnetic stripes with the same magnetic orientation are measured on either side of a ridge crest. The stripe on the west side of the ridge is displaced 11 km from the crest; the stripe on the east side is displaced 9 km from the crest. The age of the rock in both stripes is 4×10^5 years. Calculate the average spreading rate at this ridge.

3. If the north end of the Emperor Seamount Chain is 75 million years old and Midway Island is 25 million years old, what was the rate of movement of the Pacific Plate during the period of the seamount chain's creation? (You will have to use an atlas to determine the distance between the north end of the Emperor Seamount Chain and Midway Island.) What can you deduce about the past direction of the Pacific Plate's movement compared to its present direction of motion from the orientation of the islands and seamounts from Hawaii to Midway and from Midway to the north end of the Emperor Seamount Chain?

Suggested Readings

Interior of the Earth

Heppenheimer, T. A. 1988. The Sum of Its Parts. *Mosaic* 19 (3/4): 38–51.

Montgomery, C. 1993. *Physical Geology* (3rd ed.). Wm. C. Brown, Dubuque, Ia. 454 pp. (General geology text.)

Spindel, R.C., and P.F. Worcester. 1990. Ocean Acoustic Tomography. *Scientific American* 263 (4):94–99.

Toomey, D.R. 1991/92. Tomographic Imaging of Spreading Centers. *Oceanus* 34 (4): 92–99.

The Moving Continents

Baker, E. T. 1991/92. Megaplumes. *Oceanus* 34 (4): 84–91.

Ballard, R. 1975. Dive into the Great Rift. *National Geographic* 147 (5): 604–15. (Project FAMOUS and photographs from *Alvin*.)

Bambach, R. K., C. R. Scotese, and A. M. Ziegler. 1980. Before Pangaea: The Geographies of the Paleozoic World. *American Scientist* 68 (1): 26–38.

Bloxham, J., and D. Coubbins. 1989. The Evolution of the Earth's Magnetic Field. *Scientific American* 261 (6): 68–75.

Bonatti, E. 1987. The Rifting of Continents. *Scientific American* 256 (3): 97–103.

Brimhall, G. 1991. The Genesis of Ores. *Scientific American* 264 (5): 84–91.

Continents Adrift and Continents Aground. Readings from *Scientific American,* 1976. Freeman, San Francisco. 230 pp.

Decker, R., and B. Decker. 1981. The Eruptions of Mount St. Helens. *Scientific American* 244 (3): 68–80.

Golden, F. 1991. Birth of the Caribbean. *Sea Frontiers* 37 (5): 20–25.

Gore, R. 1985. Our Restless Planet Earth. *National Geographic* 168 (2): 142–81.

Hoffman, K. A. 1988. Ancient Magnetic Reversals: Clues to the Geodynamo. *Scientific American* 258 (5): 76–83.

Howell, D. G. 1985. Terranes. *Scientific American* 253 (5): 116–25.

Jones, D. L., A. Cox, P. Coney, and M. Beck. 1982. The Growth of Western North America. *Scientific American* 247 (6): 70–84.

Lonsdale, P., and C. Small. 1991/92. Ridges and Rises: A Global View. *Oceanus* 34 (4): 26–35.

Macdonald, K. C., and P. J. Fox. 1990. The Mid-Ocean Ridge. *Scientific American* 262 (6): 72–79.

Moores, E., ed. 1990. *Shaping the Earth—Tectonics of Continents and Oceans*. W. H. Freeman, New York. 206 pp. (Readings from *Scientific American*.)

Murphy, J. B., and R. D. Nance. 1992. Mountain Belts and the Supercontinent Cycle. *Scientific American* 266 (4): 84–91.

Mutter, J. C. 1986. Seismic Images of Plate Boundaries. *Scientific American* 254 (2): 66–75.

Nance, R. D., T. R. Worsley, J. B. Moody. 1988. The Supercontinent Cycle. *Scientific American* 259 (1): 72–79.

Open University. 1989. *The Ocean Basins: Their Structure and Evolution*. Pergamon Press, Oxford, England; Open University, Milton Keynes, England. 171 pp.

Rabinowitz, P. D., L. E. Garrison, and A. W. Meyer. 1989. Four Years of Scientific Deep Ocean Drilling. *Sea Technology* 30 (6): 35–40.

Rabinowitz, P. D., S. Herrig, and K. Riedel. 1986. Ocean Drilling Program Altering Our Perception of Earth. *Oceanus* 29 (3): 36–41. (Report on the first nine voyages of the *JOIDES Resolution* drill ship.)

Rona, P. 1986. Mineral Deposits from Seafloor Hot Springs. *Scientific American* 254 (1): 84–92.

Tivey, M. K. 1991/92. Hydrothermal Vent Systems. *Oceanus* 34 (4): 68–74.

Vink, G. E., W. Morgan, and P. Vogt. 1985. The Earth's Hot Spots. *Scientific American* 252 (4): 50–57.

4

The Properties of Water

4.1   The Water Molecule
4.2   Changes of State
4.3   Heat Capacity
4.4   Cohesion, Surface Tension, and Viscosity
4.5   Compressibility
4.6   Density
          The Effect of Temperature
          The Effect of Salt
          The Effect of Pressure
4.7   Dissolving Ability
4.8   Transmission of Energy
          Heat
          Light
          Sound
4.9   Practical Considerations: Ice and Fog
          Sea Ice
          Icebergs
          Fog

Box: The Heard Island Experiment

Summary
Key Terms
Study Questions
Study Problems
Suggested Readings

A sudden fog-drift muffled the ocean,
A throbbing of engines moved in it,
At length, a stone's throw out, between the rocks and the vapor,
One by one moved shadows
Out of the mystery, shadows, fishing-boats, trailing each other
Following the cliff for guidance,
Holding a difficult path between the peril of the sea-fog,
And the foam on the shore granite.
One by one, trailing their leader, six crept by me,
Out of the vapor and into it.
The throb of their engines subdued by the fog, patient and cautious,
Coasting all around the peninsula
Back to the buoys in Monterey harbor. A flight of pelicans
Is nothing lovelier to look at;
The flight of the planets is nothing nobler; all the arts lose virtue
Against the essential reality
Of creatures going about their business among the equally
Earnest elements of nature.

Robinson Jeffers,
Boats in a Fog

Sea ice in Antarctica.

Water is the most common of substances on our earth, yet it is uncommon in many of its properties. Water is a unique liquid. It makes life possible, and its properties largely determine the characteristics of the oceans, the atmosphere, and the land. In order to understand the oceans, it is necessary to examine water as a substance and to learn something of its physical and chemical characteristics. In this chapter we learn about the structure of the water molecule and explore the properties of water.

4.1   The Water Molecule

About a billion years after the formation of the earth and as a consequence of the reorganization of the planet described in chapter 1, water released from the earth's interior and carried to the surface as water vapor condensed to its liquid form. The properties of water have excited scientists for over two thousand years. The early Greek philosophers (500 B.C.) counted four basic elements from which they believed all else was made: fire, earth, air, and water. More than 1800 years later the English scientist Henry Cavendish, in 1783, determined that water was not a simple element but a substance made up of hydrogen and oxygen. Shortly afterward another Englishman, Sir Humphrey Davey, discovered that the correct formula for water was two parts hydrogen to one part oxygen, or H_2O.

The properties that make water such a special, useful, and essential substance are the result of its molecular structure. The water molecule is deceptively simple, made up of three atoms: two hydrogen atoms and one oxygen atom (H_2O). The positively charged (+) hydrogen atoms and negatively charged (−) oxygen atom attract one another, but the two positively charged hydrogen atoms repel each other. The result is a V-shaped molecule with the hydrogen atoms on one side and the oxygen atom at the other. The central core, or nucleus, of the oxygen atom forms an angle of about 105° with the nuclei of the two hydrogen atoms (see fig. 4.1a).

The three atoms are linked by **covalent bonds.** Two hydrogen atoms each form a covalent bond with a single oxygen atom; each bond is formed by the sharing of two electrons between the two atoms. The single negatively charged electron of the hydrogen atom is shared with the oxygen atom, which contributes one of its negatively charged electrons to the hydrogen atom (fig. 4.1b). Each V-shaped molecule of water has a neutral charge, but the negatively charged electrons within the molecule are distributed unequally, staying closer to the oxygen atom and giving this end a slightly negative charge. The hydrogen

end of the molecule carries a slightly positive charge. The opposite sides of the water molecule have opposite charges, and the molecule is termed a **polar molecule.**

Because of the molecules' polarity they attract each other, and bonds form between the positively charged side of one water molecule and the negatively charged side of another water molecule (see fig. 4.1c). These bonds are known as **hydrogen bonds.** Each water molecule can establish hydrogen bonds with four other water molecules. Any single hydrogen bond is relatively weak (less than one-tenth the strength of the covalent bonds between the hydrogen and oxygen atoms), but as one hydrogen bond is broken, another is formed. As a result, the water molecules cling together. Water is not a typical liquid; compared to other liquids, it has extraordinary properties that are the consequence of attractions among water molecules.

4.2   Changes of State

Water exists on the earth in three physical states: as a solid, a liquid, and a gas (fig. 4.2). When it is a solid we refer to it as ice, and when it is a gas we call it water vapor. Pure water ice melts at 0°C and pure water boils at 100°C at standard atmospheric pressure. Pure water is defined as fresh water without suspended particles or dissolved substances, including gases. Standard atmospheric pressure is equal to 76 cm of mercury and is discussed in chapter 7. If you are unfamiliar with the Celsius temperature scale, see the Appendix.

Because of the bonding between the water molecules, it takes a great deal of energy to separate them from the water's surface to form vapor and to separate them from ice to form water. Heat is a form of energy and can be measured in **calories.** One calorie is the amount of heat needed to raise 1 gram of water 1°C. One thousand of these calories is equivalent to 1 Calorie, kilocalorie (K-calorie), or food calorie. Do not confuse heat measured in calories and temperature measured in degrees. Heat is energy, while temperature measures only the gain or loss of heat that occurs in a body as heat energy is added or removed. If two cups of water and two quarts of water at the same temperature are placed on the same size stove burners at the same time, it will take longer to boil the larger amount because more energy is needed to bring the larger quantity to a boil. Both amounts of water boil at the same temperature; the same temperature change occurs in both cases, but the amount of heat required to make that change is different.

When pure water makes any of the changes among liquid, solid, and gas, it is said to change its state. Changes of state are due to the addition or loss of heat. When enough heat is added to solid water (or ice) to raise its temperature above its freezing point, it melts to form liquid water. When heat is added to liquid water, the temperature of the water rises and some of the water evaporates to form water vapor. When heat is removed from water vapor and the temperature falls below the **dew point,** or temperature

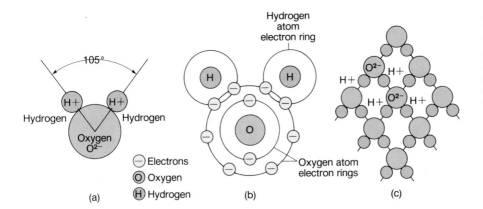

Figure 4.1

The water molecule. (a) The angle at which the hydrogen atoms bond to the oxygen atom results in a polar molecule. (b) The hydrogen atoms share electrons with the outer ring of the oxygen atom. (c) The positive and negative charges allow each water molecule to form hydrogen bonds with other water molecules.

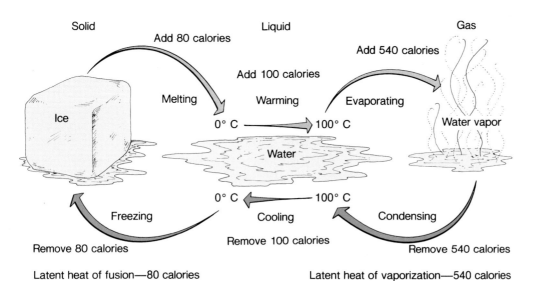

Figure 4.2

Heat energy must be added to convert a gram of ice to liquid water and to convert liquid water to water vapor. The same quantity of heat must be removed to reverse the process.

of water vapor saturation, the water vapor condenses to liquid. When liquid water loses heat and its temperature is lowered to its freezing point, ice is formed.

To change pure water from its solid state (ice) to liquid water at 0°C requires the addition of 80 calories for each gram of ice. There is no change in temperature; there is a change in the physical state of the water. The reverse of this process is required to change liquid water to ice. For each gram of liquid water that becomes ice, 80 calories of heat must be removed at 0°C. The heat necessary to change the state of water between solid and liquid is known as the **latent heat of fusion.** This addition or loss of heat takes time in nature. A lake does not freeze immediately, even though the surface water temperature is 0°C, nor does it thaw on the first warm day. Time is required to remove or add the heat needed for the change of state.

One gram of liquid water requires 1 calorie of heat to raise its temperature 1°C. Therefore, 100 calories are needed to raise the temperature of 1 g of water from 0° to 100°C. In comparison, the 80 calories of heat required per gram to convert ice to and from liquid water at 0°C with no change in temperature is relatively large. Ice is a very stable form of water; it takes a lot of heat to melt it and a large loss of heat to form it.

The change of state between liquid water and water vapor requires 540 calories of heat to convert 1 g of water to water vapor at 100°C. When 1 g of water vapor condenses and returns to the liquid state, 540 calories of heat are liberated. Again, there is no change in temperature; there is only a change in the water's physical state. The heat needed for a change between the liquid and vapor states is the **latent heat of vaporization.**

The energy required to change the state of water from solid to liquid to vapor is shown in figure 4.2. Both the latent heat of fusion and the latent heat of vaporization are presented in this diagram, as well as the calories needed for each change of state and the temperatures at which these changes occur. Water converts from the liquid to the vapor state at temperatures other than 100°C; for example, rain puddles evaporate and clothes dry on the clothesline. This type of change requires slightly more heat to convert the liquid water to a gas at these lower temperatures (see table 4.1).

As water is evaporated from the world's lakes, streams, and oceans, and is then returned as precipitation, heat is being removed from the earth's surface and liberated into the atmosphere. This heat energy is a major source of the energy used to power the earth's weather systems.

Table 4.1
Energy Required to Convert Water to Water Vapor
(Latent Heat of Vaporization)

Temperature of water (°C)	Calories required (per g of water)
0°	596.0
10°	590.8
20°	585.6
30°	580.4
40°	575.2
50°	568.5
60°	563.2
70°	557.5
80°	551.7
90°	545.8
100°	539.5
110°	532.9
120°	525.7

Under certain conditions it is possible (1) to cool liquid water below 0°C and keep it as a liquid, (2) to change ice directly to a gas, a process known as **sublimation,** or (3) to boil water at temperatures below 100°C. Cooling below 0°C is accomplished in the laboratory by controlled cooling of pure water to produce supercooled water. Sublimation is seen in nature when snow or ice evaporates directly under very cold and dry conditions, as on the high desert lands of Utah and Arizona. Anyone living at high altitudes knows that potatoes must be boiled longer and that brewed coffee is cooler there than at sea level because of the decrease in atmospheric pressure and the lower boiling temperature of the water.

The previously described behavior of water is explained by considering processes occurring between water molecules. At the molecular level the addition of heat energy increases the speed with which molecules move while the loss of heat energy decreases their rate of motion. Heat energy is required to break hydrogen bonds between water molecules, and heat is released when hydrogen bonds are formed. The addition of heat to water causes a relatively small change in the temperature of the water, because much of the heat energy is used to disrupt the hydrogen bonds. These bonds must be broken before the molecules are able to move more rapidly. When heat is removed from water, the molecules slow, many additional hydrogen bonds are formed, and considerable energy is released as heat, which prevents any rapid drop in temperature.

Water molecules stay close together because they are attracted to each other. If the molecules are moving fast enough, they overcome their attractions and leave the liquid,

entering the air as a gas. Some of the most rapidly moving molecules move into the air even at low temperatures. The greater the addition of heat, the greater the average energy of motion of the molecules, and more water molecules leave the liquid more quickly. Relatively large amounts of heat are needed to evaporate water because hydrogen bonds must first be broken. In the same way, large amounts of heat must be extracted from water to form the hydrogen bonds required to freeze water. On earth, natural temperatures required for boiling are rare and those for freezing are infrequent; therefore, liquid water is abundant.

The addition of salt to water changes its boiling and freezing points. The boiling temperature is raised and the freezing temperature is lowered. The amount of change is controlled by the amount of salt added. The change in the boiling temperature is of little consequence since seawater does not normally reach such high temperatures in nature, but the lowering of the temperature of the freezing point is important in the formation of sea ice. Seawater freezes at about −2°C.

4.3 Heat Capacity

Of all the naturally occurring earth materials, water changes its temperature the least for the addition or removal of a given amount of heat. The ability of a substance to give up or take in a given amount of heat and undergo large or small changes in temperature is a measure of the substance's **heat capacity.** The heat capacity of water is very high compared with that of the land and the atmosphere. For example, summer temperatures in the Libyan desert reach 50°C and temperatures in the Antarctic drop down to −50°C, for a world range of 100°C. Ocean temperatures vary from a nearly constant high of approximately 28°C in the equatorial areas to a low of −2°C in Antarctic waters, for a world range of 30°C. The water in a lake changes its temperature very little between noon and midnight, while the adjacent land and air temperature changes are large during this same time. The high heat capacity of water as well as the ability of water to redistribute heat over depth allow the world's lakes and oceans to change temperature slowly, helping to keep the earth's surface temperature more stable and consequently more livable.

When the heat capacity of a substance is divided by the heat capacity of pure water the resulting ratio is termed the **specific heat** of the substance. The specific heat of a substance has no units but is numerically equal to heat capacity. The heat capacities of some common materials are given in table 4.2. When salt is added to pure water the changes in heat capacity, latent heat of fusion, and latent heat of vaporization are small. Heat capacity and specific heat, like changes of state, are related to the hydrogen bonding between water molecules. Review the discussion of the topic in the preceding section.

Table 4.2 Heat Capacity of Common Materials	
Material	**Heat capacity (calories/g/°C)**
Acetone	0.51
Aluminum	0.22
Ammonia	1.13
Copper	0.09
Grain alcohol	0.23
Lead	0.03
Mercury	0.03
Silver	0.06
Water	1.00

4.4 Cohesion, Surface Tension, and Viscosity

In its liquid state the bonds between water molecules are quite fragile; they form, break, and re-form with great frequency. Each bond lasts only a few trillionths of a second. However, at any instant a substantial percentage of all water molecules are bonded to their neighbors. Therefore, water has more structure than other liquids. Collectively, the hydrogen bonds hold water together; this is known as **cohesion.**

Cohesion is related to **surface tension,** which is a measure of how difficult it is to stretch or break the surface of a liquid. At the surface between air and water, water molecules arrange themselves in an ordered system, hydrogen-bonded to each other laterally and to the water molecules beneath. This forms a weak elastic membrane that can be demonstrated by filling a water glass carefully; the water can be made to brim above the top of the glass but not overflow. A steel needle can be floated on water; insects like the water strider walk about on the surface of lakes and streams. All of these things are possible because water has a high surface tension. This property is important in the early formation of waves. A gentle breeze stretches and wrinkles the smooth water surface, enabling the wind to get a better grip on the water and add more energy to the sea surface.

The addition of salt to pure water increases the surface tension. Decreasing the water temperature also increases the surface tension, while increasing the water temperature decreases it.

Liquid water pours and stirs easily. It has little resistance to motion or internal friction. This property is called **viscosity.** Water has a low viscosity when compared to motor oil, paint, or syrup. Viscosity is affected by temperature. Consider pancake syrup. When it is stored in the refrigerator it becomes thick and slow to pour. It has a high viscosity. When the syrup is returned to room temperature or heated, it becomes thin and runny; it has a low viscosity. The same is true of water, but the change in viscosity is much less, and it is not noticeable under normal temperature changes. Surface water at the equator is warmer and therefore less viscous than surface water in the Arctic. Minute microscopic organisms find it easier to float in the more viscous polar waters; their tropical-water cousins have adapted to the less-viscous water by developing spines and frilly appendages to help keep them afloat. The addition of salt to pure water increases the viscosity of the water, but the change is small.

4.5 Compressibility

Pure water is a nearly incompressible substance. The same is true of seawater. Pressure in the oceans increases with depth. For every 10 m in depth, the pressure increases by about 1 atmosphere. One atmosphere is equal to the pressure of a column of mercury 76 cm (30 in) high; in English units, 1 atm equals 14.7 pounds per square inch (lb/in^2). Pressure in the deepest ocean trenches, 11,000 m (36,000 ft) deep, is about 1100 atm. These great pressures have only a small effect on the volume of the oceans because of the slight compressibility of the water. A cubic centimeter of seawater at the surface will lose only 1.7% of its volume if it is lowered to 4000 m (13,000 ft), with a pressure of 400 atm. The average pressure acting on the total world's ocean volume results in the reduction of ocean depth by about 37 m (121 ft). In other words, if the ocean water were truly incompressible, sea level would stand about 37 m (121 ft) higher than it does at present. The pressure effect is small enough to be ignored in most instances, except when very accurate determination of density is required. Further information on units of pressure is found in the Appendix.

4.6 Density

Density is defined as mass per unit volume of a substance and is usually measured in grams per cubic centimeter, or g/cm^3. Pure water has a density of 1.00 g/cm^3 at 4°C. Thus a cube of pure water that is 1 cm high, 1 cm wide, and 1 cm deep has a mass of 1 g. The density of seawater is greater than the density of pure water at the same temperature, because seawater contains dissolved salts. At 4°C the density of seawater of average salinity is 1.0278 g/cm^3. The densities of other substances may be considered in the same way (see table 4.3). Less-dense substances will float on more-dense liquids (for example, oil on water, dry pine wood on water, and alcohol on oil). Note from table 4.3 that ocean water is more dense than fresh water; therefore fresh water floats on salt water.

Table 4.3	
Densities of Common Materials	
Material	**Density (g/cm³)**
Ice (pure) 0°C	0.917
Water (pure) 0°C	0.99987
Water (pure) 3.98°C	1.0000
Water (pure) 20°C	0.99823
White pine wood	0.35–0.50
Olive oil 15°C	0.918
Ethyl alcohol 0°C	0.791
Seawater 4°C, 35⁰/₀₀	1.0278
Steel	7.60–7.80
Lead	11.347
Mercury	13.6

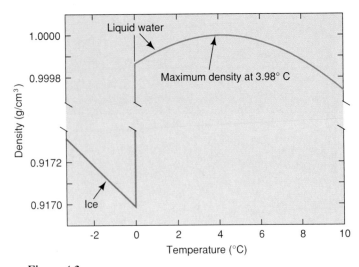

Figure 4.3
The density of pure water reaches its maximum at 3.98°C. Pure water is free of dissolved gases.

The Effect of Temperature

Water density is very sensitive to temperature changes. When water is heated, energy is added and the water molecules move apart; therefore the mass per cubic centimeter becomes less because there are fewer water molecules per cubic centimeter. For this reason, the density of warm water is less than that of cold water, so that warm water floats on cold water. When water is cooled it loses heat energy, and the water molecules slow down and come closer together; there are then more water molecules, or a greater mass, per cubic centimeter. Because cold water is more dense than warm water, it sinks below the warm water. To see these changes try the following experiment. Chill a small quantity of water in the refrigerator; mix it with a little dye such as ink or food coloring to make it easy to see. Allow the water from the tap to run hot and fill a glass half full of the hot water. Now slowly and carefully pour the colored water down the side of the glass on top of the hot water and watch what happens.

In pure fresh water, molecules move closer and closer together as water is cooled to 4°C. At this temperature fresh water reaches its greatest density, 1 g/cm³. As the temperature of the water drops below 4°C some water molecules start to form a crystal lattice, each molecule bonded to a maximum of four other molecules. Water begins to freeze when the water molecules no longer move enough to break hydrogen bonds with neighboring molecules. The hydrogen bonds keep the molecules far enough apart to make ice about 10% less dense than liquid water at 4°C. At 3°–0°C the water is less dense than at 4°C (see fig. 4.3) and therefore stays at the surface, where ice is formed with further cooling. As the ice crystals form, the water molecules become widely spaced, resulting in fewer water molecules per cubic centimeter. Therefore ice is less dense than water, and it floats on water.

When ice absorbs enough heat to increase the temperature above 0°C, the hydrogen bonds between molecules are disrupted, and the crystal lattice collapses. The ice melts and the molecules move closer together. Above 4°C, the water begins to expand again, due to the increasing space between the faster-moving molecules. If enough heat is added, the energy level of the molecules increases until they can overcome the attraction between one another and the vapor pressure of the air at the surface of the water. The molecules can then escape, or evaporate, into the air as water vapor. Since water vapor is less dense than the mixture of gases that form the atmosphere, a mixture of water vapor and dry air is less dense than dry air alone at the same temperature and pressure.

The Effect of Salt

When salts are dissolved in pure water the density of the water increases because the salts have a greater density than water. In other words, they have more mass per cubic centimeter. The average density of seawater is approximately 1.0278 g/cm³, in comparison to 1 g/cm³ for fresh water at 4°C. Therefore fresh water floats on salt water. To see this happen, take two small quantities of fresh water; add dye to one and some salt to the other. Carefully and slowly add the salt water to the dyed fresh water and watch what happens.

If seawater contains less than 24.7 g of salt per kilogram of seawater, then the seawater, although it is more dense, will behave much like fresh water. Seawater of this concentration will reach its maximum density before it freezes. Cooling surface water with a salt content of less than 24.7 g/kg causes the surface water to increase in density and sink. This process continues with surface cooling until the temperature of maximum density is reached. Further cooling causes the low-salinity surface water to become less dense and remain at the surface. At a salt content of 24.7 g/kg, the freezing point and the maximum density of

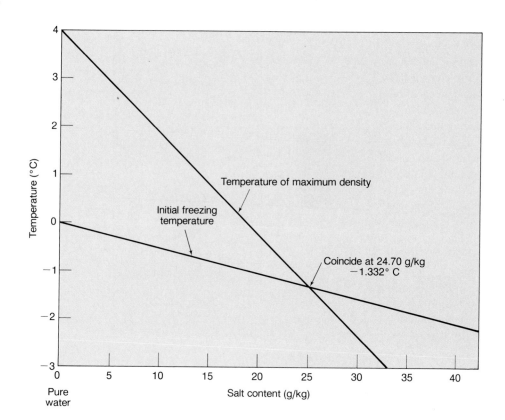

seawater coincide at −1.332°C. If the salt content is greater than 24.7 g/kg, freezing occurs before the maximum density is reached. This relationship is shown graphically in figure 4.4. Open ocean water generally has an average salt content of 35 g/kg; therefore cooled surface water sinks continuously as its density increases.

The Effect of Pressure

Density is also affected by pressure, and although the effect is small it is taken into consideration by research oceanographers. Increasing the pressure increases the density of the water by reducing the volume occupied by a fixed mass of seawater. The reduction in volume results because the water and salt molecules are being crowded together by the applied pressure. The small reduction in volume is the compressibility of the water, which was discussed previously.

4.7 Dissolving Ability

Water is an extremely good solvent. More substances—solids, liquids, and gases—dissolve in water than in any other common liquid. Water has been called the universal solvent because of its exceptionally good dissolving ability for a wide range of materials.

The excellent dissolving ability of water is also related to its molecular structure. For example, common table salt dissolves in water. Common salt is chemically sodium chloride (NaCl); each molecule is made up of one atom of sodium and one atom of chlorine. When the salt is in its natural crystal form, the positively charged sodium atoms and the

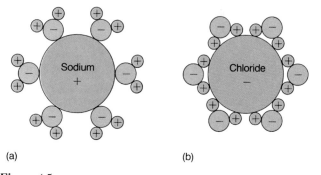

(a)   (b)

Figure 4.5
Salts dissolve in water because the polarity of the water molecule keeps positive ions separated from negative ions. (a) Sodium ions are surrounded by water molecules with their negatively charged portion attracted to the positive ion. (b) Chloride ions are surrounded by water molecules with their positively charged portion attracted to the negative ion.

negatively charged chlorine atoms are held together in an alternating pattern by their bonds, resulting from the strong attraction force between the two differently charged atoms. When sodium chloride is placed in water these bonds break, and the resulting charged atoms are called **ions.** The positive sodium ions are attracted to the negatively charged oxygen side of the water molecules, while the negative chloride ions are attracted to the positively charged hydrogen side. The salt atoms are surrounded and separated by water molecules (see fig. 4.5a and 4.5b). This ability of water molecules to separate compounds into their ions makes water an excellent solvent.

For millions of years the rainwater washing over the land has been dissolving materials out of the land and carrying them to the sea via rivers, streams, and groundwater seepage. In this way much of the oceans' salt content has been supplied. Once these dissolved substances reach the ocean basins there is no similar continuous mechanism to return them to the land. The water is recycled to the land by oceanic evaporation followed by precipitation on land to repeat this process, but the salts are trapped in the sea and its bottom sediments. Some land salt deposits are the result of geologic uplift processes bringing seafloor deposits above the ocean surface. Others are the remnants of ancient shallow seas that became isolated over geologic time; the water evaporated, leaving the salt deposits behind.

4.8 Transmission of Energy

Fresh water and salt water both transmit energy. The energy may be in the form of light, heat, or sound. Since the principles of transmission are the same in fresh water and salt water, our discussion will center on energy transmission in the oceans.

Heat

There are three ways in which heat energy may be transmitted: by **conduction,** by **convection,** and by **radiation.** Conduction is a molecular process. When heat is applied at one location the molecules move faster due to the addition of energy; gradually this more rapid molecular motion passes on to the adjacent molecules and the heat spreads. For example, if the bowl of a metal spoon is placed in a hot liquid, the handle soon becomes hot. Heat has been conducted from the part of the spoon that is in contact with the heat source to the handle. Metals are excellent conductors; water is a poor conductor and transmits heat slowly in this way.

Convection is a density-driven process in which fluid material moves and carries heat to a new location. In older heating systems, hot air is supplied through vents at floor level; the hot air rises because it is less dense, and the air carries the heat with it. When this air is cooled it drops down again because it is then more dense. In these energy-conscious days, ceiling fans are being used to force the less dense hot air down, to keep the room warm at the human level rather than accumulating excess heat at the ceiling. Water behaves in the same way, rising when it is heated from below and its density decreases; sinking when it is cooled at its surface and its density is increased.

Radiation is the direct transmission of heat from the energy source. Switch on an incandescent light bulb and feel from a foot away the instantaneous heat released. This is radiant energy. The sun provides the earth with radiant energy, and the surface waters of the earth are warmed by this solar radiation. Unlike conduction and convection, which require a medium for the transfer of heat, radiated heat can be transferred through the vacuum of space.

Consider all three processes with respect to a container of water. The surface of the water receives solar radiation; some of this radiation is reflected from the surface and adds

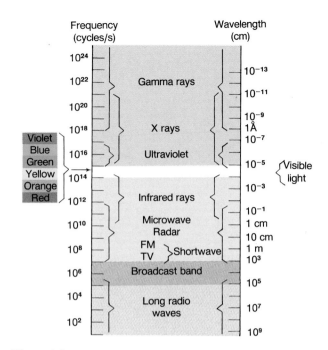

Figure 4.6
The electromagnetic spectrum.

no heat to the water, while some of the radiation is absorbed and raises the temperature of the surface water. Since warming the surface water makes it less dense, the warm water floats at the surface. The only way to transmit the heat downward is molecular conduction, unless the container is stirred. Stirring requires that mechanical energy be added from another source. Therefore heating a body of water from the top is not an efficient method of uniformly warming all the water in the container. Compare this situation with that of a similar container heated from below. In this case, the downward molecular conduction of heat is replaced by convection; as the water at the bottom gains heat, decreases in density, and rises, it takes the heat with it and distributes it throughout the water volume. As this warmed water rises it is replaced by colder water, which is warmed in turn. This is a much more rapid and efficient method of distributing heat, since the water circulates on its own without the need for stirring.

The oceans are heated from above by solar radiation, which is absorbed in the upper surface layers of water. The heat gained at the surface of the ocean is transmitted slowly downward by the natural turbulent stirring action of the wind and the currents, as well as by the slower molecular conduction. The oceans' surface water heated by the sun then loses heat to warm the atmosphere above it. Since the heating of the atmosphere is done from below, the heat is efficiently transferred upward by the convection of the air.

Light

The light striking the ocean surface is one of the many forms of **electromagnetic radiation** the earth receives from the sun. The full range of this radiation may be seen in the **electromagnetic spectrum,** shown in figure 4.6. Note that visible light occupies a very narrow segment of

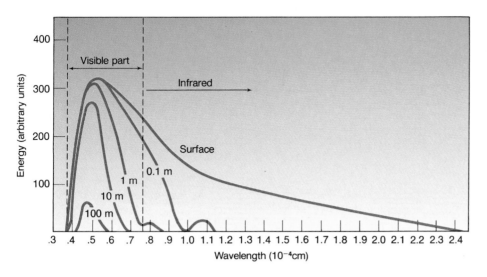

Figure 4.7

Total available solar energy in the sea (shown by the areas under the curves) decreases as depth increases. The longer, red wavelengths are absorbed first. The color peak shifts to the shorter, blue wavelengths as the depth increases.

From Sverdrup/Johnson/Fleming, *The Oceans,* ©1942, renewed 1970, p. 105. Adapted by permission of Prentice-Hall, Inc., Englewood Cliffs, New Jersey.

the spectrum. Wavelengths shorter than visible light include ultraviolet light, X rays, and gamma rays. Wavelengths longer than visible light include infrared (heat) waves, microwaves, TV, FM, and AM radio waves, all of which are used in modern technology. Visible light may be broken down into the familiar spectrum of the rainbow: red, orange, yellow, green, blue, and violet. Each color represents a range of wavelengths; the longest wavelengths are at the red end of the spectrum, and the shortest are at the blue-violet end. When combined, these wavelengths produce white light.

Seawater transmits only the visible light portion of the electromagnetic spectrum. About 60% of the entering light energy is absorbed in the first meter, and about 80% is gone after 10 m (33 ft). Only 1% of the total light available at the surface is left in the clearest water below 150 m (500 ft). No light penetrates below 1000 m (3300 ft). Not all wavelengths of visible light are transmitted equally (see fig. 4.7). The long wavelengths at the red end of the spectrum are absorbed rapidly within the upper 10 m (33 ft); the shorter wavelengths of blue-green light are transmitted to the greater depths.

Since color perception is due to the reflection back to our eyes of wavelengths of a particular color, the oceans usually appear blue-green because these wavelengths are being absorbed the least and are available to be reflected and scattered. Because all of the wavelengths, or colors, are present to illuminate objects in shallow waters, objects are seen in their natural colors at the surface. Objects in deeper waters usually appear dark in color because they are illuminated by blue light. Coastal waters may appear green or even yellow, brown, or red, while deep, open ocean water appears blue. Coastal waters usually contain silt from the rivers and the land, as well as large numbers of microscopic organisms; the color of these particles and organisms is revealed by the light entering the water, so that the color of the water is changed. Open-ocean water beyond the influence of land is often clear and blue, which indicates that it is nearly empty of both living and nonliving suspended materials. The clearer the water, the less suspended matter present, and the deeper the light penetration.

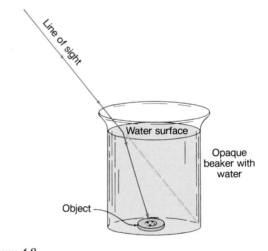

Figure 4.8

Objects that are not directly in the individual's line of sight can be seen in water due to the refraction of the light rays. The refraction is caused by the decreased speed of light in water.

When light passes from the air into the water it is bent, or **refracted,** because its speed is faster in air than in water. Because the light rays bend as they enter water from the air, objects seen through the water's surface are not where we believe them to be (see fig. 4.8). Refraction is affected slightly by changes in salinity, temperature, and pressure.

As light passes through the water it is **absorbed** and **scattered** by suspended particles, including silt, single-celled organisms, and the water and salt molecules. It is also absorbed by plants, to be used in their life processes. This decrease in the intensity of light over distance is known as **attenuation.** The clearer the water, the greater the light penetration, and the smaller the attenuation.

Oceanographers use a light meter in a watertight case to measure the intensity of light, first at the sea surface and then at increasing depths. Light meters (fig. 4.9) measure total available light, or they may be equipped with filters to measure only specific wavelengths. The measure of light attenuation with depth allows scientists to estimate the depth at which available light limits plant growth. This

Figure 4.9

An immersion light meter is used to determine the percentage of surface solar radiation at depth. The photo cell on the left is placed on deck to measure solar radiation at the sea surface. The solar cell on the right is lowered in the sea.

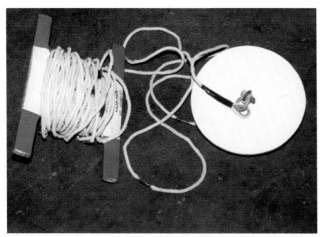

Figure 4.10

A Secchi disk. The line is marked with tape at meter intervals.

information is then used to determine the ability of any ocean area to support the plant populations required to feed other living creatures. This ability is of considerable importance to people who control the length of fishing seasons and the size of catches.

The simplest way of measuring light attenuation in surface water is to use a **Secchi disk** (fig. 4.10). This is a white disk about 30 cm (12 in) in diameter, which is lowered on a line to the depth at which it just disappears from view. The measure of this depth can be used to determine the average attenuation of light. Advantages of a Secchi disk include never needing adjustment, low cost, never leaking, and never requiring replacement of electronic components. In waters rich with living organisms or suspended silt, the Secchi disk may disappear from view at depths of 1 to 3 m (3–10 ft) while in the open ocean it may be visible down to 20 to 30 m (65–100 ft).

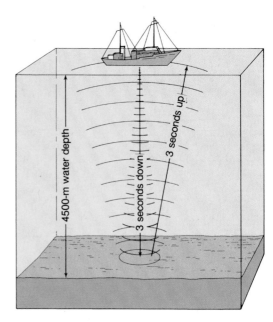

Figure 4.11

Traveling at an average speed of 1500 meters per second, a sound pulse leaves the ship, travels downward, strikes the bottom, and returns. In 4500 meters of water, the sound requires three seconds to reach the bottom and three seconds to return.

Sound

The sea is a noisy place, because sound is transmitted both farther and faster in seawater than it is in air. Waves break, fish grunt and blow bubbles, crabs snap their claws, and whales whistle and sing. Sound travels in seawater at an average velocity of 1500 m (5000 ft)/sec as compared with 334 m (1100 ft)/sec in dry air at 20°C. The speed of sound in seawater increases with increasing temperature, pressure, and salt content, and decreases with decreasing temperature, pressure, and salt content. High-frequency sounds do not travel as far as low-frequency sounds due to the friction between water molecules, which absorb energy rapidly as the high-frequency sound waves pass through the water.

Because water transmits sound, sound can be used to locate the objects producing the sounds; because sound is reflected back after striking an object, it can be used to find objects, sense their shape, and determine their distance away from the sound's source. If a sound signal is sent into the water and the time required for the return of the reflected sound, or echo, is measured accurately, the distance to the object may be determined. For example, if 6 seconds elapse between the outgoing sound pulse and its return after reflection, the sound has taken 3 seconds to travel to the object and 3 seconds to return. Since sound travels at an average speed of 1500 m/sec in water, the object is 4500 m away, as shown in figure 4.11.

Echo sounding to determine depth uses a narrow sound beam that is directed vertically to the sea bottom. The sound waves pass through nearly horizontal layers of water in which salt content, temperature, and pressure change. The speed of the sound waves continuously changes as they pass from layer to layer, until they reach

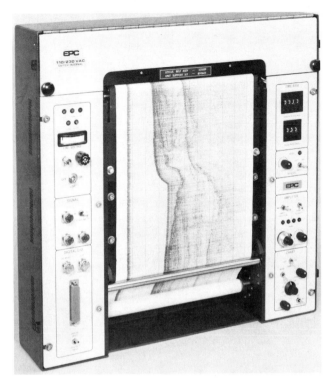

Figure 4.12
A precision depth recorder displaying bottom and subbottom profiles.

the bottom and are reflected back to the ship. Little refraction, or bending, of the sound beam occurs, because the path of the sound beam is perpendicular to the water layers. Echo sounders, or **depth recorders,** are used by all modern vessels to measure the depth of water beneath the ship. Oceanographic vessels record the depths on a chart, producing a continuous reading of depth as the vessel moves along its course. The **precision depth recorder (PDR)** on an oceanographic research vessel uses a narrow sound beam to give detailed and continuous traces of the bottom while the ship is in motion, as shown in figure 4.12.

Geologists can detect the properties of the seafloor by studying the echo charts, because some seafloor materials reflect back a stronger signal than other materials. If high-intensity sound pulses are transmitted, some sound energy can penetrate the seafloor and reflect back from layers in the sediments. In this case, the echo chart displays the layering of the seabed materials; see figure 4.12.

Porpoises and whales use sound in water in the same way that bats use sound in air. The animal produces a sound, which travels outward until it reaches an object from which it is reflected. The animal is able to judge the direction from which the sound returns, the distance to the reflecting object, and the properties of that object. Human technology has produced an underwater location system called **sonar** (sound navigation and ranging), which uses sound in a similar way. Sonar technicians direct a sound beam through the water, searching for targets that return echoes. They are then able to determine the distance and direction of the target. An electronic screen is used to display the direction of sound beams relative to the

sending vessel. If a reflective target is found, the screen also portrays the distance to the target, using the time difference between the sending of the sound pulse and the return of the echo. After much training and practice, technicians are able to distinguish between the echoes produced by a whale, a school of fish, or a submarine. It is even possible to determine the type and class of a vessel merely by listening to the sounds produced by its propellers and engines. See chapter 2 for a discussion of multiple sonar devices being used in seafloor mapping.

However, the target may not be at the depth, distance, and angle indicated, because the sound beam may have been bent, or refracted, as it moved through water layers of differing densities (see fig. 4.13a and 4.13c). Figure 4.13b and 4.13d illustrates the refraction of sound beams and the formation of **sound shadow zones,** areas of the ocean into which sound does not penetrate. Sound beams bend toward regions in which sound travels more slowly and away from regions in which sound waves travel more rapidly.

In order to interpret the returning echo and to determine distance and depth correctly, the sonar operator must have information about the properties of the water through which the sound passes. Governments and their navies have conducted intensive research in the field of underwater sound, for in wartime naval operations the winner in an encounter is often determined by who has the best understanding of the water's properties and its effects on underwater sound. Survival for a surface vessel depends on its accuracy in locating a submarine's position, while the submarine must remain at the correct depth and distance from the sonar detector to remain invisible in the shadow zone.

At about 1000 m (3300 ft) the combination of salt content, temperature, and pressure creates a zone of minimum velocity for sound, the **sofar** (sound fixing and ranging) **channel** (see fig. 4.14a). Sound waves produced in the sofar channel do not escape from it unless they are directed outward at a sharp angle. Instead, the majority of the sound energy bounces back and forth along the channel for great distances (see fig. 4.14b). Test explosions set off in this channel near Australia have produced sound heard as far off as Bermuda. It has been suggested that listening stations be established around the world to monitor the sofar channel. Vessels and aircraft in distress could drop small depth charges programmed to explode at the sofar depth and their location could be established by using the difference in arrival time of the sound at listening stations.

Depth recorders can be confused by still another underwater phenomenon. Large numbers of small organisms, including fish, move toward the surface during the night and sink down to greater depths during the day. These organisms form a layer known as the **deep scattering layer (DSL).** This layer reflects a portion of the sound beam energy and creates the image of a false bottom on the depth recorder trace.

Other echoes are returned from mid-depths by fish swimming in schools or by large individual fish. Echo sounders are designed and marketed as "fish finders," and those who fish learn to distinguish fish school reflections.

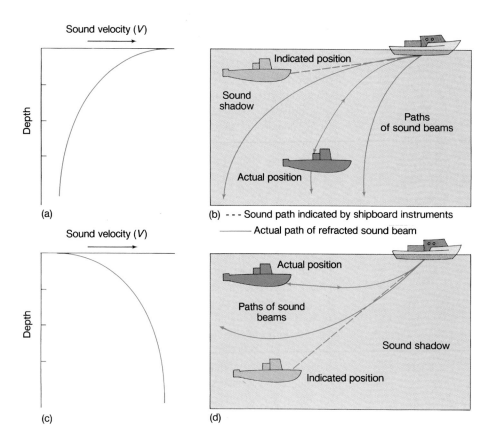

(a)

(b) - - - Sound path indicated by shipboard instruments
—— Actual path of refracted sound beam

(c)

(d)

Figure 4.13
Sound waves change velocity and refract as they travel at an angle through water layers of different densities. The angle at which the sound beam leaves the ship indicates a target in the indicated, or ghost, position. To determine the actual position of the target the degree of refraction and change in velocity with depth must be known.

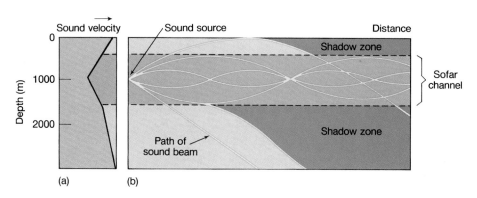

Figure 4.14
(a) The temperature, salinity, and pressure variation with depth combine to produce a minimum sound velocity at about 1000 meters. (b) Sound generated at this depth is trapped in a layer known as the sofar channel.

The depths of the echoes give them the depth to operate their nets, while their location and the appearance of the echo pattern inform them of the fish species.

Note the similarities between the behavior of sound and light in water. Both represent energy in wave form and both are absorbed, reflected, and refracted.

4.9 Practical Considerations: Ice and Fog

Ice and fog are forms of water that are extremely hazardous to ocean navigation and transportation. At polar latitudes in the Northern Hemisphere, ice prevents the routine use of shipping routes for much of the year. **Sea ice** driven by winds and currents hampers oil drilling, and polar research

is also difficult under these conditions. A knowledge of sea ice is required for operation on or below the sea surface at high latitudes in either hemisphere. Fog restricts visibility. Although electronic eyes (radar) have been developed to helps ships under way in foggy seas, safety is compromised, collisions are not infrequent, and many small-boat sailors lose their way. The processes that produce ice and fog at sea are discussed here.

Sea Ice

Ice is present year-round at all latitudes if the elevation is high enough to keep the average earth surface temperature below the freezing point of water. This elevation varies between sea level at the poles and approximately 5000 m (16,400 ft) at the equator. On land, ice and snow are the result of low temperatures and precipitation. In the sea or in

(a)

(b)

(c)

Figure 4.15

(a) Pancake ice, an early stage of ice formation. (b) A pressure ridge formed by colliding floes. (c) Winds and currents move sea ice, opening leads and forming pressure ridges.

lakes, precipitation is not necessary; the ice is formed when the temperature of the surface water goes below the freezing point of water.

Sea ice is formed at polar latitudes because of the extremely low air and water temperatures. As the seawater begins to freeze, the surface water becomes dull, and clouds of ice crystals are formed. As the number of ice crystals increases, a layer of slush forms, covering the ocean in a thin sheet of ice. Sheets of new sea ice are broken into "pancakes" by waves and wind (see fig. 4.15a). As the freezing continues the pancakes move about, unite, and form floes (see fig. 4.15b). Ice floes move with the currents and the wind, collide with each other, and form ridges and hummocks. Some floes shift constantly, breaking apart and freezing together in response to winds and water motion. Others remain anchored to the landmass. These stationary floes are called **fast ice.** As freezing continues, the ice increases in thickness. Snow on the surface may also freeze and contribute to the flat, floating ice masses.

The dissolved salts of the seawater do not fit into the ice crystal structure and are left below in the surface water. This saltier water is very cold and dense, and so it sinks to be replaced by less salty water, which in turn cools to its freezing point. As the ice is formed, some seawater is trapped in the voids between the ice crystals. If the ice forms slowly, most of the trapped seawater drains out and escapes; if the ice forms quickly at subfreezing temperatures, more salt water is trapped. As time passes and the ice continues to age, the salt water slowly escapes through the ice, and eventually the ice becomes fresh enough to drink when melted.

In one season ice about 2 m (6.5 ft) thick forms at the ocean's surface. The thickness of the ice is limited by the necessity of cooling the ice surface and extracting sufficient heat through the ice from the underlying water to form new ice beneath the existing layer. The latent heat of fusion must be removed from the water beneath the ice, and it can only be extracted by the process of conduction through the

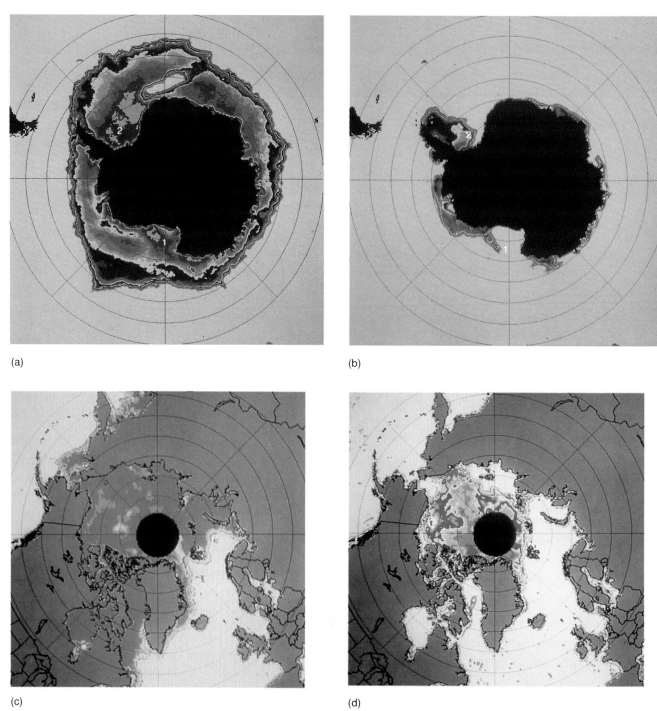

(a)

(b)

(c)

(d)

Figure 4.16

Sea ice around the Arctic and Antarctic. Data from NASA's *NIMBUS-5* satellite was used to construct seasonal maps of sea ice coverage at high latitudes. This information is used to help develop a better understanding of global climate. The Antarctic ice extends more than 100 km from the land into the Ross (1) and Weddell (2) seas during the winter (a) and is much reduced during the summer (b). The Arctic Ocean is filled with ice in the winter (c) but by the next fall the ice cover is much less (d). Short-term motions of the ice boundaries can be calculated from repeated satellite images. These show that the ice floes move up to 50 km per day. Purple indicates high ice concentration and blue indicates open water (in a and b). Yellow areas are open water in c and d.

initial ice layer, a slow process at polar temperatures. Snow accumulating on the ice surface acts as an insulator that also retards the extraction of heat from the water below.

Sea ice is seasonal around the bays and shores of parts of the northeast United States, Canada, Russia, Scandinavia, and Alaska. The ice exists year-round in the central Arctic and around Antarctica. Sea ice covers the entire Arctic Ocean in the northern winter and pushes far out to sea around Antarctica during the Southern Hemisphere's winter (see fig. 4.16). Waves, currents, winds, and tides fracture the edges of the ice and cause floes to collide and pile into irregular masses to form pressure ridges (see fig. 4.15c). In summer the ice melts back along its edges. In areas where the ice does not melt entirely during the year, new ice is added each winter, and a thickness of 3 to 5 m (10–16 ft) can accumulate.

(a)

(b)

Figure 4.17

(a) A castle berg drifting in the Atlantic Ocean just north of the Grand Banks. (b) A tabular berg in the South Atlantic Ocean. Its approximate length is 500 m (1600 ft).

Remember that sea ice is a product of the salt water; it is not to be confused with the freshwater glacial **land ice** that covers Greenland and the Antarctic continent. Pieces of the land ice sometimes break off and fall into the sea. These pieces of freshwater ice are called **icebergs.**

Icebergs

Icebergs are massive, irregular in shape, and float with only about 12% of their mass above the sea surface (see fig. 4.17). They are formed by glaciers, large bodies of ice that begin inland in the snows of central Greenland, Antarctica, and Alaska, and inch their way toward the sea. The forward movement, the melting at the base of the glacier where it meets the ocean, and the waves and wind action cause blocks of ice to break off and float out to sea. The production of icebergs by a glacier is called calving, and it occurs mainly during the summer months.

The icebergs produced in the Arctic drift south with the currents as far as New England and the busy shipping lanes of the North Atlantic. It was one of these icebergs that

Figure 4.18
Iceberg B-9 as observed by Landsat 4, Nov. 28, 1987, about 4 to 6 weeks after detachment from the Ross Ice shelf, Antarctica.

sank the *Titanic*. On its maiden voyage the great luxury liner struck an iceberg and sank with a loss of 1517 lives on the night of April 14, 1912. The ice had most probably become detached from a glacier in Greenland and then floated south. Following the sinking, the United States began an iceberg patrol to warn vessels of the location of icebergs. The patrol became an international effort in 1914 and continues to the present.

Icebergs produced in the Antarctic usually stay close to the polar continent, caught by the circling currents, although they have been known to reach latitudes of 40°S to 50°S. Alaskan icebergs are usually released in narrow channels and semienclosed bays; they do not usually escape into the open ocean. Rather than being a hazard to shipping, Alaskan icebergs have become a tourist attraction. Cruise ships visit Glacier Bay, Alaska, to show their passengers the glaciers and icebergs.

Icebergs from valley glaciers are relatively high and narrow, with above-water shapes resembling towers and battlements; these are called **castle bergs.** Icebergs from broad, flat, continental ice sheets are usually much larger than castle bergs. These huge, flat icebergs are called **tabular bergs** and form the ice islands that are used as bases for

polar research. Some of these icebergs have been large enough to accommodate aircraft runways, housing, and research facilities. Such an ice island may be used for years before it eventually breaks up.

In late 1987, a large tabular berg known as B-9, 155 km (96 mi) long and 230 m (755 ft) thick, with a surface area about the size of Long Island, New York, broke from Antarctica and drifted 2000 km (1250 mi) along the Antarctic coast. See figure 4.18. This berg was so large and so deep that its drift was the product of subsurface currents; conventionally sized tabular bergs drift with the wind. It drifted clockwise and to the west in the Ross Sea for two years before it grounded and broke into three pieces. B-9 was tracked by plane and satellite; its drift path helped increase our knowledge of Ross Sea currents.

The calving of such a spectacularly large tabular berg is infrequent. The seaward movement of glacial ice is about 1 meter (3 ft) per day; a berg of B-9's size contains about 70 years of glacial advance. Formation of such a large berg is controlled by the spacing of the major crevasses that produce weak spots along which a berg separates from the glacier. The developing rift along which B-9 broke away was first identified in aerial photos taken between 1965 and 1971.

Figure 4.19
Advective fog obscures the Golden Gate Bridge at the entrance to San Francisco Bay.

Large, floating masses of ice have attracted considerable interest as sources of fresh water. The first of a series of meetings on this possibility was hosted in 1978 by Arab interests. Discussions at the meetings centered on the feasibility of towing icebergs from the Antarctic to the Red Sea. Among the techniques necessary would be those of detaching the ice, shaping it for the long miles of transport, and designing a bridle for towing. Although the expense would be enormous and much of the ice would melt as it was towed into equatorial latitudes, the need for water in the Arab countries is so great that it is believed that sufficient ice would remain after the journey to make the project worthwhile.

An Alaskan company harvests up to approximately 1000 tons of ice per month from icebergs floating in Alaskan waters. The company then sells the ice to a Japanese firm for processing into gourmet ice cubes. The harvesting company has obtained permits from the Alaskan Department of Natural Resources subject to limitations on quantity and place of harvesting.

Fog

The ability of water to remain as water vapor is a function of temperature: warm air can hold more moisture than cold air. When water vapor in the air is cooled, it condenses around any small particles in the air, forming liquid droplets. Large accumulations of water droplets form clouds, and when clouds form close to the ground they are called **fog.** Fog at sea is a navigation hazard to mariners, but the ocean scientist is interested in fog and in predicting its occurrence because of the role fog plays in transferring water and heat between the atmosphere and the ocean surface.

There are three basic types of fog, each formed by a different process. The most common fog at sea is **advective fog,** forming when warm air saturated with water vapor moves over colder water. This advective fog hugs the sea surface like a blanket. For example, when the warm, moist air over the Gulf Stream flows over the cold water of the Labrador Current, the famous fogs of the Grand Banks result. In northern California, Oregon, and Washington, when offshore warm, moist air moves over the cold coastal waters, particularly in summer, a coastal fog results, giving moisture to the redwood forests and grassy cliffs and cooling the city of San Francisco (see fig. 4.19).

Streamers of **sea smoke** rising from the sea surface on a cold winter day are a spectacular sight. Dry, cold air from the land or from the polar ice pack moves out over the warmer water, and the water warms the air above it. This warmed air picks up water vapor from the sea surface and begins to rise rapidly. As it rises it is cooled below the dew point, and rising ribbons of fog are formed. When sea smoke is formed the vertical convection results in high evaporation from the water's surface, whereas the advective fog returns moisture to the earth's surface by condensation.

Radiative fog is the result of warm days and cold nights. The earth's surface warms and cools, and so does the air above it. If the air holds enough moisture during the day, it condenses as the earth cools at night, forming the low-lying, thick, white fogs that are often found in river valleys and occasionally in bays and inlets along the coast. These fogs usually disappear during the morning as the sun gradually warms the air, changing the water droplets back to water vapor.

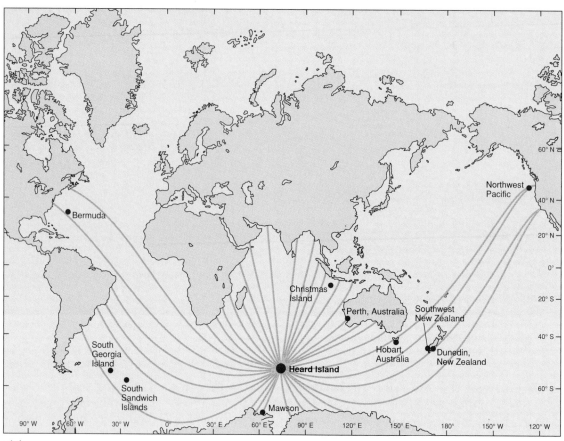

Box figure 4.1

Sound from the Heard Island experiment travels through the oceans in the sofar channel. Some listening stations are indicated.

Acoustic tomography uses variations in the speed of low-frequency sound waves traveling through seawater to create a three-dimensional image of an area. Walter Munk of Scripps Institution of Oceanography has developed a program to use acoustical tomography in a world-wide experiment to determine whether the world's oceans are changing temperature in response to global warming trends. If the oceans are warming, detectable world-wide changes in sound speed should occur, because sound waves travel faster in warm water than in cold water.

In January 1991 the first ocean-scale tests were conducted in the South Indian Ocean near Heard Island, a location offering clear pathways for sound waves across all the oceans of the world; see

Box figure 4.1. A sound generator was placed at a depth of 150 meters in the sofar channel. High-energy sound pulses were generated every third hour for 10 to 12 days at a frequency chosen to be distinguishable from the miscellaneous signals of the world's power grids. About three-and-a-half hours after transmission the sound pulses were picked up with underwater microphones or hydrophones at San Francisco, Bermuda, Ascension Island, Tasmania, Goa, Christmas Island, Mawson Station Antarctica, Sao Paulo, Capetown, Kerguelen Island, New Zealand, Fiji, Florida, and other locations. Comparing the arrival time of the sound to the time of its generation allows the calculation of the sound's travel time that is then related to the properties of the

seawater along its sofar channel travel path. Calculating the travel time in this way allows averaging of sound speed changes over large volumes of water.

The precision of travel time measurements is about 1 millisecond over a path 1000 km (660 mi) long. Repeated tests are needed to filter out small changes in speed due to short-term seasonal variations in water properties so that sound speed from any experiment can be compared with speeds measured in other experiments in future years. Scientists expect to detect sound speed changes with a sensitivity of 0.015 m/sec; this will allow detection of average seawater temperature changes as small as 0.003°C per year. In two to three years sufficient data will be available to determine whether this technique yields statistically significant information to describe changes in average ocean temperature. If it is successful it will be followed by a five- to ten-year larger-scale study using multiple sound sources and an increased number of listening posts.

When the initial Heard Island tests were proposed, questions were raised regarding the impact of high-intensity sound waves on marine animals, especially marine mammals. There was concern that animals using sound for communication could be adversely affected. An impact evaluation was made, and it was concluded that confinement of the sound energy to the sofar channel would minimize any large-scale effect; provisions were also made to conduct the test at times when marine mammals were not in the vicinity of the sound source. A permit was granted by the Marine Mammal Protection Agency of NOAA, and the study of marine mammal behavior in the vicinity of the sound source has become a secondary test objective.

Future plans are to use multiple sound sources to improve detection of changes in seawater properties; this will allow three-dimensional imaging of the ocean's salinity, temperature, and current structure. See the box, New Ways to Measure the Oceans, in chapter 6 for more information on this process.

The earth's water, whether found as a liquid filling our lakes, streams, and oceans; as a solid in the land glaciers, snowpacks, and sea ice; or as a gas in the atmosphere, is a remarkable substance. The properties of this water as well as its abundance give the earth its habitable characteristics and make it unique in our solar system.

Summary

A water molecule is made up of two positively charged atoms of hydrogen and one negatively charged atom of oxygen. The molecule has a specific shape with oppositely charged sides; it is a polar molecule. Because of the distribution of the charges, water molecules interact with each other by forming hydrogen bonds between molecules. As a result of its structure, the water molecule is very stable. The structure of the water molecule is responsible for the properties of water.

Water exists as a solid, a liquid, and a gas. Changes from one state to another require the addition or extraction of heat energy. Water has a high heat capacity; it is able to take in or give up large quantities of heat with a small change in temperature. The surface tension of water, which is related to the cohesion between water molecules at the surface, is high. The viscosity of water is a measure of its internal friction; it is primarily affected by temperature. Water is nearly incompressible. Pressure increases 1 atm for every 10 m of depth in the oceans.

The density, or mass per unit volume, of water increases with a decrease in temperature and an increase in salt content. The effect of pressure on density is small. Less-dense water floats on more-dense water. Pure water reaches its maximum density at 4°C. Open-ocean water does not reach its maximum density before freezing. Water vapor is less dense than air; a mixture of water vapor and air is less dense than dry air.

The ability of water to dissolve substances is exceptionally good. River water dissolves the salts from the land and carries the salts to the sea.

Seawater transmits energy as light, heat, and sound. The long red wavelengths of light are lost primarily in the first 10 m (33 ft); only the shorter wavelengths of blue-green light penetrate to depths of 150 m (500 ft) or more. Light passing into water is refracted. Attenuation, or the decrease in light over distance, is the result of absorption and scattering by the water and by particles suspended in the water. Light attenuation is measured by light meters and with a Secchi disk.

The sea surface layer is heated by solar radiation, and heat is transmitted downward by conduction. This is an inefficient process compared to convection, which transfers heat from the earth's surface to the atmosphere.

Sound travels farther and faster in water than in air. Its speed is affected by the temperature, pressure, and salt content of the water. Echo sounders are used to measure the depth of water, and sonar is used to locate objects. Sound is refracted as it passes at an angle through the density layers, and sound shadows are formed. The sofar channel, in which sound travels for long distances, is the result of salt content, temperature, and pressure in the oceans. The deep scattering layer, formed by small animals moving toward the surface at night and away from it during the day, reflects portions of a sound beam and creates a false bottom on a depth recorder trace.

Sea ice is formed in the extreme cold of polar latitudes. The process is self-insulating, so that large thicknesses of ice do not form each winter. Icebergs are formed from glaciers that break off into the sea.

Fog occurs when water vapor condenses to form liquid droplets. Three types of fog occur: advective fog occurs when warm, water-saturated air passes over cold water; sea smoke occurs when dry, cold air moves over warm water; and radiative fog occurs when warm, moist air is cooled at night.

Key Terms

covalent bond	surface tension
polar molecule	viscosity
hydrogen bond	density
calorie	ions
dew point	conduction
latent heat of fusion	convection
latent heat of vaporization	radiation
sublimation	electromagnetic
heat capacity	radiation
specific heat	electromagnetic
cohesion	spectrum

refraction
absorption
scattering
attenuation
Secchi disk
echo sounding
depth recorder
precision depth recorder
  (PDR)
sonar
sound shadow zone
sofar channel

deep scattering layer (DSL)
sea ice
fast ice
land ice
iceberg
castle berg
tabular berg
fog
advective fog
sea smoke
radiative fog

Study Questions

1. Discuss the structure of the water molecule.
2. What happens to the arrangement of water molecules when water freezes?
3. What happens to the sea level of the oceans if all the sea ice melts? Assume that the ocean area does not change and ocean temperature stays the same.
4. Compare the heat capacity of the oceans with that of the land and the atmosphere. Compare the climate of the west and east coasts of the United States with that of the central and midwestern states. How is climate related to heat capacity and the direction of the prevailing wind?
5. How do the properties of the water molecule affect (a) surface tension and (b) the dissolving ability of water?
6. Explain why heating the atmosphere from below is an efficient way to distribute heat in the atmosphere, while heating the oceans from above is an inefficient way to distribute heat in the oceans.
7. If there were no scattering of light by seawater, what color would the oceans appear? Why?
8. Both advective fog and sea smoke transfer heat between the atmosphere and the oceans. How does the rate and direction of heat transfer differ in each case? Why?
9. How are the amounts of heat energy lost and gained when water changes its state related to the bonding and motion of the water molecules?
10. If the substances in table 4.2 are exposed to the same quantity of heat energy, which liquid and which solid undergo the greatest temperature change?
11. Explain the effect of increasing and decreasing the salinity, temperature, and pressure on the density of seawater.
12. How are light and sound in seawater affected by changes in the water's density?
13. Where do icebergs come from? How are they formed?
14. Explain differences between the terms sonar and sofar.
15. Why might one expect to find that traces of every known naturally occurring substance are dissolved in seawater?

Study Problems

1. If surface cooling and evaporation remove heat from the surface of a deep lake at the rate of 800×10^3 calories per hour, and if the surface temperature of the lake is $0°C$, at what rate must deep water be pumped to the surface of the lake to prevent the formation of ice?
2. Determine the freezing point of seawater at 32 g/kg salt content. Use figure 4.4.
3. If an echo sounder measures the depth of the water as 3500 m, but the instrument is shown to have a timing error of ± 0.001 second over the total time period of the measurement, what is the error in the depth measurement?
4. If water from a given area of sea surface evaporates at the rate of 1 metric ton per day at $20°C$, at what rate is heat being added to the atmosphere above this surface area?
5. If heat is removed from a kilogram of water at a constant rate that lowers the temperature of the water from $25°C$ to $0°C$ in thirty minutes, how long will it take to convert $0°C$ water to ice at this rate of heat removal?

Suggested Readings

Alper, J. 1991. Munk's Hypothesis. *Sea Frontiers* 37 (3): 38–41.
Bowditch, N. 1984 ed. *American Practical Navigator*, vol. 1. U. S. Defense Mapping Agency Hydrographic Center, Washington, D.C. 1414 pp. (Soundings are covered in chapter 28 and sound in chapter 35; ice and weather in chapters 36 and 39.)
Georges, T. M. 1992. Taking the Ocean's Temperature with Sound. *The World & I* July: 282–89.
The Heard Island Experiment. 1991. *Oceanus* 34 (1): 6–8.
Jacobs, S. 1992. The Voyage of Iceberg B-9. *American Scientist* 80 (1): 32–42.
MacLeish, W. H., ed. Spring 1977. *Oceanus* 20 (2). (Issue includes articles on uses of sound in navigation, warfare, animal behavior, and seismics.)
Neumann, G., and W. J. Pierson. 1966. *Principles of Physical Oceanography*. Prentice-Hall, Englewood Cliffs, N.J. 545 pp.
Open University. 1989. *Seawater: Its Composition, Properties and Behavior*. Pergamon Press, Oxford, England; Open University, Milton Keynes, England. 165 pp.
Stefanick, T. 1988. The Nonacoustic Detection of Submarines. *Scientific American* 258 (3): 41–47.
Stewart, W. K. 1991. High Resolution Optical and Acoustic Remote Sensing for Underwater Exploration. *Oceanus* 34 (1): 10–22.

5
The Salt Water

5.1 **The Salts**
Ocean Salinities
Dissolved Salts
Sources of Salt
Regulating the Salt Balance
Residence Time
Constant Proportions
Determining Salinity
5.2 **The Gases**
Distribution with Depth
Carbon Dioxide as a Buffer

Box:	*The Messages in Polar Ice*

The Carbon Dioxide Cycle
The Oxygen Balance
Measuring the Gases
5.3 **Other Substances**
Nutrients

Box:	*The Effect of Sunlight on Seawater*

Organics
5.4 **Practical Considerations: Salt and Water**
Chemical Resources
Desalination
Summary
Key Terms
Study Questions
Study Problems
Suggested Readings

And truly, though we were at sea, there was much to behold and wonder at, to me, who was on my first voyage. What most amazed me was the sight of the great ocean itself, for we were out of sight of land. All round us, on both sides of the ship, ahead and astern, nothing was to be seen but water—water—water; not a single glimpse of green shore, not the smallest island, or speck of moss anywhere. Never did I realize till now what the ocean was: how grand and majestic, how solitary, and boundless, and beautiful and blue; for that day it gave no tokens of squalls or hurricanes, such as I had heard my father tell of, nor could I imagine how anything that seemed so playful and placid could be lashed into rage, and troubled into rolling avalanches of foam, and great cascades of waves, such as I saw in the end.

Herman Melville,
from *Redburn*

Receding waves leave sea foam along a flat, sandy beach in California.

S eawater is salt water, and historically seawater has been valued for its salt. Until recently salt was enormously important as a food preservative, and at one time salt formed the basis for a major commercial trade. Today, although salt is still extracted from seawater, it is the water that has become increasingly valuable in many areas of the world. Seawater is also much more than salt water. Seawater is a complex mixture containing dissolved gases, nutrient substances, and organic molecules as well as salts.

In this chapter we investigate this seawater mixture, and we explore the physical, chemical, and biological processes that regulate the mixture. We follow these processes as they cycle dissolved substances and water from land and atmosphere to ocean and ocean floor and back to land and atmosphere.

5.1 The Salts

Ocean Salinities

In the major ocean basins, 3.5% of the weight of seawater, on the average, is dissolved salt and 96.5% is water, so that a typical 1000-g, or 1-kg, sample of seawater is made up of 965 g of water and 35 g of salt. Oceanographers measure the salt content of ocean water, or **salinity,** in grams of salt per kilogram of seawater (g/kg), or parts per thousand (‰). The average ocean salinity is approximately 35‰.

The salinity of ocean surface water is associated with latitude. Latitudinal variations in evaporation and precipitation, as well as freezing, thawing, and freshwater runoff from the land, affect the relative amount of salt in seawater. The relationship between evaporation, precipitation, and

surface salinity is shown in figure 5.1. Notice the low surface salinities in the cool and rainy 40°–50°N and S latitude belts, high evaporation rates and high surface salinities in the desert belts of the world centered on 25°N and S, and low surface salinities again in the warm but rainy tropics centered at 5°N. Sea-surface salinities during the Northern Hemisphere summer are shown in figure 5.2.

In coastal areas of high precipitation and river inflow, surface salinities fall below the average. For example, during periods of high flow, the water of the Columbia River lowers the Pacific Ocean's surface salinity to less than 25‰ as far as 35 km (20 mi) at sea. Also, sailors have dipped up water fresh enough to drink from the ocean surface 50 miles from the mouth of the Amazon River. In subtropic regions of high evaporation and low freshwater input, the surface salinities of nearly landlocked seas are well above the average: 40 to 42‰ in the Red Sea and the Persian Gulf and 38 to 39‰ in the Mediterranean Sea. In the open ocean at these same latitudes the surface salinity is closer to 36.5‰. Surface salinities change seasonally in polar areas, where the surface water forms sea ice in winter, leaving behind the salt and raising the salinity of the water under the ice. In summer the water returns as a freshwater layer when the sea ice melts. Deep-water samples from the mid-latitudes are usually slightly less salty than the surface waters because the deep water is formed at the surface in high latitudes at times of high precipitation. The formation of these deep-water types will be further described in chapter 6.

Dissolved Salts

When salts are added to water, the salts dissolve. The salt crystals dissociate (or break apart) into **ions,** which are positively or negatively charged atoms or groups of atoms. An ion with a positive charge is a **cation;** an ion with a negative charge is an **anion.** The salts in seawater are mostly present in their ionic forms as anions and cations. For example, table salt (or sodium chloride) is the most common salt in seawater. When sodium chloride is added to

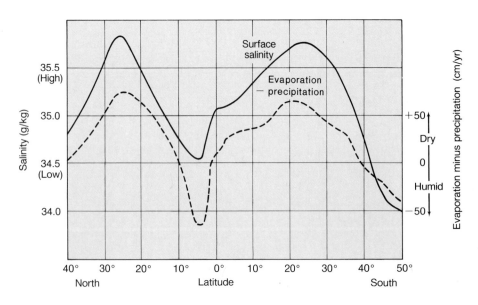

Figure 5.1

Midocean average surface salinity values match the average changes in evaporation minus precipitation values that occur with latitude.

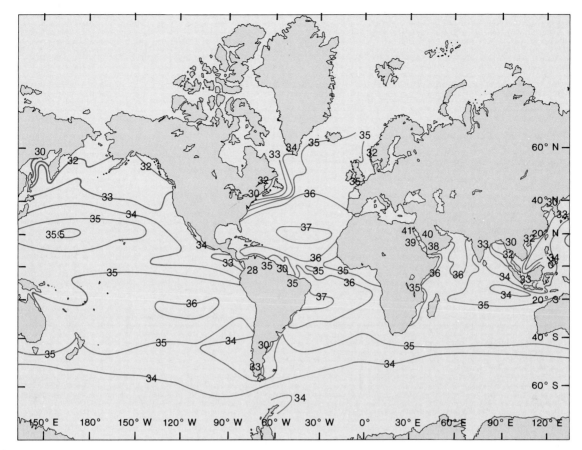

Figure 5.2
Average sea-surface salinities in the Northern Hemisphere summer, given in parts per thousand (‰).

water the bonds between the atoms break, and positively charged sodium ions and negatively charged chloride ions are formed. This reaction can be written as:

sodium chloride → sodium ion + chloride ion,

or it can be written with chemical abbreviations as:

$NaCl \rightarrow Na^+ + Cl^-$.

Refer back to section 4.7 and figure 4.5 for a discussion of the dissolving of salts in water.

Six ions make up more than 99% of the salts dissolved in seawater. Four of these ions are cations: sodium (Na^+), magnesium (Mg^{2+}), calcium (Ca^{2+}), and potassium (K^+); and two are anions: chloride (Cl^-) and sulfate (SO_4^{2-}). Table 5.1 lists these six ions and five more. The ions are arranged in order of the amount present in seawater. The ions listed in table 5.1 are known as the **major constituents** of seawater. Note that sodium and chloride ions account for 86% of the salt ions present in seawater.

All the other elements dissolved in seawater are present in concentrations of less than one part per million and are called **trace elements** (see table 5.2). Some of these elements are important to organisms that are able to concentrate these ions in their bodies. For example, long before the presence of iodine could be determined chemically as a

trace element of seawater, it was known that shellfish and seaweeds were rich sources for this element. Seaweed has even been harvested commercially for iodine extraction.

Because the major constituents of seawater do not change their ratios to each other with changes in salinity, and because they are not generally removed or added to by living organisms, they are termed **conservative constituents.** Certain of the ions present in much smaller quantities, some dissolved gases, and assorted organic molecules and complexes do change their concentrations with biological and chemical processes that occur in the oceans; these are the **nonconservative constituents.**

Sources of Salt

The original sources of the seas' salts include the crust and the interior of the earth. The chemical composition of the earth's rocky crust can account for most of the positively charged ions found in seawater. Large quantities of cations are present in **igneous rocks** that are formed by the crystallization of molten magma from volcanic processes. The physical and chemical weathering of rock over time breaks it into small pieces and the rain dissolves out ions, which are carried to the sea by rivers. Anions are present in the earth's interior and may have been present in the earth's early atmosphere. Some anions may have been washed from the atmosphere by long periods of rainfall, but the

Table 5.1
Major Constituents of Seawater

Constituent	Symbol	g/kg in seawater[1]	Percentage by weight
chloride	Cl^-	19.35	55.07
sodium	Na^+	10.76	30.62 ⎫
sulfate	SO_4^{2-}	2.71	7.72 ⎪
magnesium	Mg^{2+}	1.29	3.68 ⎬ 99.36
calcium	Ca^{2+}	0.41	1.17 ⎪
potassium	K^+	0.39	1.10 ⎭
bicarbonate	HCO_3^-	0.14	0.40
bromide	Br^-	0.067	0.19
strontium	Sr^{2+}	0.008	0.02
boron	B^{3+}	0.004	0.01
fluoride	F^-	0.001	0.01
total			99.99

From Riley and Skirrow, *Chemical Oceanography*, Vol. 1, 2d ed. Copyright © 1975 Academic Press, Orlando, Florida. Reprinted by permission.

Table 5.2
Concentrations of Trace Elements in Seawater[1]

Element	Symbol	Concentration[2]	Element	Symbol	Concentration[2]
aluminum	Al	5.4×10^{-1}	manganese	Mn	3×10^{-2}
antimony	Sb	1.5×10^{-1}	mercury	Hg	1×10^{-3}
arsenic	As	1.7	molybdenum	Mo	1.1×10^1
barium	Ba	1.37×10^1	nickel	Ni	5×10^{-1}
bismuth	Bi	$\leq 4.2 \times 10^{-5}$	niobium	Nb	$\leq (4.6 \times 10^{-3})$
cadmium	Cd	8×10^{-2}	protactinium	Pa	5×10^{-8}
cerium	Ce	2.8×10^{-3}	radium	Ra	7×10^{-8}
cesium	Cs	2.9×10^{-1}	rubidium	Rb	1.2×10^2
chromium	Cr	2×10^{-1}	scandium	Sc	6.7×10^{-4}
cobalt	Co	1×10^{-3}	selenium	Se	1.3×10^{-1}
copper	Cu	2.5×10^{-1}	silver	Ag	2.7×10^{-3}
gallium	Ga	2×10^{-2}	thallium	Tl	1.2×10^{-2}
germanium	Ge	5.1×10^{-3}	thorium	Th	(1×10^{-2})
gold	Au	4.9×10^{-3}	tin	Sn	5×10^{-4}
indium	In	1×10^{-4}	titanium	Ti	$< (9.6 \times 10^{-1})$
iodine	I	5×10^1	tungsten	W	9×10^{-2}
iron	Fe	6×10^{-2}	uranium	U	3.2
lanthium	La	4.2×10^{-3}	vanadium	V	1.58
lead	Pb	2.1×10^{-3}	yttrium	Y	1.3×10^{-2}
lithium	Li	1.7×10^2	zinc	Zn	4×10^{-1}
			rare earths		$(0.5–3.0) \times 10^{-3}$

[1]Nutrients and dissolved gases are not included.
[2]Parts per billion or $\mu g/kg$.
Parentheses indicate uncertainty about concentration.
From Riley and Chester, *Chemical Oceanography*, Vol. 8. Copyright © 1983 Academic Press, Orlando, Florida. Reprinted by permission.

more likely source of most of the anions is thought to have been the mantle. During the formation of the earth, described in chapter 1, gases from the mantle are believed to have carried anions into the newly forming oceans.

Anions are released during volcanic eruptions as gases (for example, hydrogen sulfide, sulfur, and chlorine); they dissolve in rainwater or river water and are carried to the oceans as Cl^- (chloride) and SO_4^{2-} (sulfate). These anions are also dissolved directly in the water formed during a volcanic eruption. River water carrying chloride ions and sulfur-containing ions is acidic and erodes and dissolves the rock over which it flows, helping to liberate the cations. Tests show that the most abundant ions in today's rivers (table 5.3) are the least abundant ions in ocean water. This difference results in part because the rivers have previously removed the most easily dissolved land salts, and they are now carrying the less-soluble salts. Exceptions to this pattern are found in rivers used for irrigation. These rivers are flowing through soils that have not lost much of their salt content because of the arid conditions in these areas. The river water is frequently used several times as it passes through a number of irrigation projects on its way downstream. This usage causes the river water to become increasingly salty and unfit for irrigation purposes toward the end of its run. This situation has produced years of continuous conflict between the United States and Mexico over the waters of the Colorado and Rio Grande rivers, which irrigate much of the agricultural land of the U.S. desert southwest before becoming available to Mexico.

In addition, we know that hot water vents located on the seafloor supply chemicals to the ocean water and also remove them. Hydrothermal activity found at the midocean ridges and associated with hot spots and seamount formation may play an important role in stabilizing the ocean's composition. When hot magma is introduced into the cold crust, it cracks and becomes permeable. The pressure exerted by the water above the seafloor is high (1 atmosphere for every 10 meters of water depth), and the water is forced into cracks and voids where it is heated to very high temperatures without boiling. The salinity of seawater entering the hydrothermal system is relatively constant worldwide, but the water emerging from hydrothermal vents has a salinity about one-third to more than double its original value. The process controlling this salinity change and its role in the ocean's total salt balance are unclear at this time.

Regulating the Salt Balance

The total amount of dissolved material in the world's oceans is calculated to be 5×10^{22} g for ocean water of 36‰ salinity. Each year the runoff from land adds another 2.5×10^{15} g, or 0.000005% of total ocean salt. We know from the age of older marine sedimentary rocks that the oceans have been present on the earth for about 3.5 billion years, and chemical and geologic evidence leads researchers to believe that the composition of the oceans has been the same for about the last 1.5 billion years. If we assume that the rivers have been flowing to the sea at the same rate over the last 3.5 billion years, more dissolved

Table 5.3
Dissolved Salts in River Water

Ion	Symbol	Percentage by weight
carbonate	CO_3^{2-}	35.15
calcium	Ca^{2+}	20.39
sulfate	SO_4^{2-}	12.14
silicon dioxide (nonionic)	SiO_2	11.67
sodium	Na^+	5.79
chloride	Cl^-	5.68
magnesium	Mg^{2+}	3.41
oxides (nonionic)	$(Fe, Al)_2O_3$	2.75
potassium	K^+	2.12
nitrate	NO_3^-	0.90
total		100.00

Source: After C. W. Wolfe, et al., *Earth and Space Science*. D. C. Heath, Lexington, Mass., 1966. Reprinted by permission of D. C. Heath and Company.

material has been added by this process than is presently found in the sea. In order for the oceans to remain at the same salinity, the rate of addition of salt by rivers must be balanced by the removal of salt; input must balance output.

Salts are removed from seawater in a number of ways. Sea spray from the waves is blown ashore, depositing a film of salt on rocks, land, houses, and cars. This salt is later returned to the oceans by runoff from the land. During the earth's history, shallow arms of the sea have become isolated, the water has evaporated, and the salts have been left behind to become land deposits called **evaporites.** Salt ions can also react with each other to form insoluble products that precipitate and settle to the ocean floor. Biological processes concentrate salts, which are removed if the organisms are harvested. Organisms' excretion products trap ions, which are transferred to the sediments or returned to the seawater. Other biological processes remove Ca^{2+} (calcium) by incorporating it into shells, and silica is used to form the hard exterior coverings of certain plants and animals. These hard parts are accumulated in the sediments when the organisms die. For a discussion of the sediments produced by biological processes, refer back to chapter 2.

A chemical process known as **adsorption,** the adherence of ions and molecules onto a surface, serves to remove other ions and molecules from the seawater. In this process, tiny clay mineral particles, weathered from rock and brought to the oceans by the rivers, bind ions such as K^+ (potassium) and trace metals to their surfaces, carrying them downward and eventually incorporating them into the sediments. Strongly adsorbable ions replace weakly adsorbable ions in a process known as **ion exchange.** If the clay minerals and sediments adsorb and exchange with one ion more easily than with another, there will be a lower concentration of the more easily adsorbed ion in the seawater. For example, potassium is more easily adsorbed than sodium, and so K^+ is less concentrated than Na^+ in seawater. Nickel, cobalt, zinc, and copper are adsorbed on nodules that form on the ocean floor. The fecal pellets and skeletal remains of small organisms also act as adsorption surfaces. In all of these cases the ions are removed from the water and are transferred to the sediments.

The process of forming the new earth's crust at the ridge system of the deep-ocean floor (chapter 3) participates in the input and output of the ions in seawater. Where molten rock rises from the mantle into the crust, magma chambers are formed. These chambers are found mainly along plate boundaries but have also been identified under volcanic hot spots in the middle of plates. Cold water seeps down several kilometers through the fractured crust at a spreading center. The seawater becomes heated by flowing near the magma chamber. By convection, the heated water rises up through the crust and reacts chemically with the rocks of the crust. Magnesium ions (Mg^{2+}) are transferred from the water to form mineral deposits in the crust. At the same time, hydrogen ions (H^+) are released and the seawater solution becomes more acidic. Chemical reactions change sulfate (SO_4^{2-}) to sulfur and then to hydrogen sulfide (H_2S). The acidic water releases and dissolves metals from the crust including copper, iron, manganese, zinc, potassium, and calcium. It has been estimated that a volume of water equivalent to the entire mass of the oceans circulates through the crust at oceanic ridges every ten million years.

The most important process for the removal of most elements from seawater is still the adsorption of ions onto fine particles and their removal to the sediments. Ions deposited in the sediments are trapped there. They are not returned to or redissolved in the seawater; but geologic uplift processes elevate some sediments from the seafloor to positions above sea level. Erosion works to dissolve and wash these deposits back to the sea.

The processes described in this section are summarized in figure 5.3, which you can use to understand the input and output processes that occur in the oceans.

Residence Time

The relative abundance of the salts in the sea is due in part to the ease with which they are introduced from the earth's crust and in part to the rate at which they are removed from the seawater. Sodium is moderately abundant in fresh water, but because it reacts slowly with other substances in seawater, it remains dissolved in the ocean. Calcium is removed rapidly from seawater, forming limestone and the shells of marine organisms.

The average (or mean) time that a substance remains in solution in seawater is called its **residence time.** Aluminum, iron, and chromium ions have short residence times, in the hundreds of years. They react with other substances quickly and form insoluble mineral solids in the sediments. Sodium, potassium, and magnesium are very soluble and have long residence times, in the millions of years (see table 5.4).

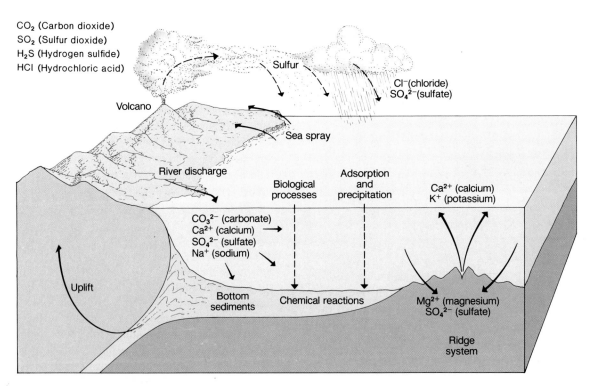

Figure 5.3
Processes that regulate the major constituents in seawater.

Table 5.4	
Approximate Residence Times of Ions in the Oceans	
Ion	**Time in years**
chloride	80 million
sodium	60 million
magnesium	10 million
sulfate	9 million
potassium	6 million
calcium	1 million
manganese	7 thousand
aluminum	1 hundred
iron	1 hundred

The residence time of an ion may be determined as follows:

1. Calculate the total mass of the ion in the oceans by multiplying the concentration of the ion per kilogram of seawater by the total mass of the oceans:

$$\text{Total amount ion (kg)} = \frac{\text{Ionic concentration (kg)}}{\text{Seawater (kg)}} \times \text{Ocean mass (kg)};$$

2. Calculate the input rate by multiplying the concentration of the ion in river water by the volume of river water added to the oceans each year (37,000 km^3/yr):

$$\text{Input rate (kg/yr)} = \frac{\text{Ionic concentration (kg)}}{\text{River water (km}^3)} \times \frac{\text{River discharge (km}^3)}{\text{Year}};$$

3. Since the ionic concentration of the oceans is nearly constant with time, the input of an ion equals its rate of loss to the sediments, or input equals output:

$$\text{Input (kg/yr)} = \text{Output (kg/yr)};$$

4. Residence time is calculated by dividing the total amount present from step 1 by the output, which is equal to the input as calculated in step 2:

$$\text{Residence time (years)} = \frac{\text{Total amount ion (kg)}}{\text{Input or Output rate (kg/yr)}}.$$

If other input sources of an ion are significant, they must also be considered.

Constant Proportions

Seawater is a well-mixed solution; currents and eddies in both surface and deep water, vertical mixing processes, and wave and tidal action have all helped to stir the oceans through geologic time. Because of this thorough mixing the ionic composition of open-ocean seawater is the same from place to place and from depth to depth. That is, the ratio of one major ion or seawater constituent to another remains the same. Whether the salinity is 40‰ or 28‰, the major ions exist in the same proportions. This concept was first suggested by the chemist Alexander Marcet in 1819. In 1865 another chemist, Georg Forchhammer, analyzed several hundred seawater samples and found that these constant proportions did hold true. During the world cruise of the *Challenger* Expedition (1872–76), seventy-seven water samples were collected from different depths and locations in the world's oceans. When chemist William Dittmar analyzed these samples, he verified Forchhammer's findings and Marcet's suggestion. These analyses led to the **principle of constant proportion** (or **constant composition**), still in use, which states that regardless of variations in salinity, the ratios between the amounts of major ions in open-ocean water are constant. Note that the principle applies to major ions in the open ocean; it does not apply along the shores, where rivers may bring in large quantities of dissolved substances or may reduce the salinity to very low values. The ratios of abundance of minor constituents vary, because some of these ions are closely related to the life cycles of living organisms. Populations of organisms remove ions during periods of growth and reproduction, temporarily reducing the amounts of these ions in solution. Later, the population declines, and decay processes return these ions to the seawater.

Determining Salinity

The great advantage of the major ions' being in constant proportion is that it allows us to determine salinity by measuring the concentration of one type of ion. Once the concentration of one major ion is known, then the concentration of all other major ions can be calculated from the known ratios.

Historically, the quantity of chloride ions present in a water sample has been measured to establish the salinity. This measurement is done quite easily and quickly by adding silver nitrate to a seawater sample. The silver combines with the chloride ion; so if the amount of silver required to react with the chloride ion in a known amount of seawater is measured, the amount of chloride is known. However, the silver also combines with bromine, iodine, and some other trace elements. The chloride concentration measured in this way is termed **chlorinity** (Cl‰); the chlorinity is measured in parts per thousand, or grams of chloride per kilograms of seawater. Chlorinity is defined as the quantity of silver required to remove all the **halogens** from 0.3285 kg of seawater. Halogens are members of a chemical family that includes chlorine, bromine, iodine, and fluorine, which react in similar ways. Chlorinity and salinity are related by the equation

$$\text{Salinity (‰)} = 1.80655 \times \text{Chlorinity (‰)},$$

or

$$\text{S‰} = 1.80655 \times \text{Cl‰}.$$

Figure 5.4

The salinometer
determines salinity
by measuring the
conductivity of the
seawater sample.

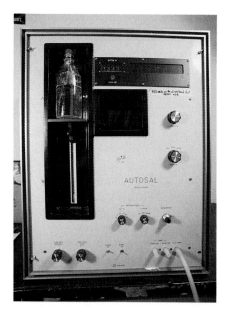

Once the chlorinity of a sample is determined, the concentration of any other major constituent is calculated by using the principle of constant proportion. This is a much faster and easier process than determining the concentration of each of the individual ions.

In order to assure that all determinations of chlorinity and salinity are done in exactly the same way in all oceanographic laboratories around the world, a standard method of analysis and a standard seawater reference is needed. A standard seawater is used to calibrate the silver nitrate solution used in these tests. Since the chemical procedures are all done in the same way and use the same standard seawater, salinity data that were determined in different laboratories can be directly compared.

Until 1975, the Hydrographic Laboratories in Copenhagen, Denmark, were responsible for the production of the standard (or normal) seawater (*eau de mer normale*). This standard seawater had an adjusted chlorinity of 19.4‰. It was sealed in glass vials and shipped to oceanographic laboratories around the world. In 1975, the production of standard seawater was taken over by the Institute of Oceanographic Services in Wormly, England. Today's standard seawater is adjusted to constant chlorinity and electrical conductance.

Because of the ions it contains, seawater conducts electricity; the more ions there are, the greater the conductance. This relationship makes it possible to determine salinity by using a salinity (or conductivity) bridge, often called a **salinometer.** The great advantage of the salinometer is that the measurements are made quickly and directly on the water sample with an electrical probe (see fig. 5.4). It is therefore not necessary to analyze a water sample chemically, except to test and calibrate the conductivity instruments. Because the conductivity of seawater is affected by both salinity and temperature, a conductivity instrument must correct for the temperature of the sample if it is calibrated to read directly in salinity units. Other direct measurement techniques are discussed in chapter 6.

5.2 The Gases

Gases move between the sea and the atmosphere at the sea surface. Atmospheric gases dissolve in the seawater and are distributed to all depths by mixing processes and currents. The most abundant gases in the atmosphere and in the ocean are nitrogen (N_2), oxygen (O_2), and carbon dioxide (CO_2). The percentage of each of these gases in the atmosphere and in seawater are given in table 5.5. Oxygen and carbon dioxide play important roles in the ocean because they are necessary to life, and biological activities modify their concentrations, at various depths. Nitrogen is not used directly by living organisms except for certain bacteria. Gases such as argon, helium, and neon are present in small amounts, but they are inert and do not interact with the ocean water, nor are they used by its inhabitants.

The amount of any gas that can be held in solution without causing the solution either to gain or to lose gas is the **saturation value.** The saturation value changes because it depends on the temperature, salinity, and pressure of the water. If the temperature or salinity decreases, the saturation value for the gas increases. If the pressure decreases, the saturation value decreases. In other words, colder water holds more dissolved gas than warmer water, less-salty water holds more gas than more-salty water, and water under more pressure holds more gas than water under less pressure.

Distribution with Depth

In the process known as **photosynthesis** plants produce oxygen and use carbon dioxide to form organic molecules. Plants are confined to the sea surface, the top 100 m (330 ft) of the water column, where there is sufficient light to energize the photosynthesis process. Therefore oxygen is produced in surface water and carbon dioxide is consumed there. By contrast, **respiration,** which breaks down organic substances to provide energy, requires oxygen and produces carbon dioxide. All living organisms respire to produce energy, and therefore respiration occurs at all depths. Photosynthesis and respiration are discussed in more detail in chapter 14. Decomposition, the bacterial breakdown of nonliving organic material, also requires oxygen and releases carbon dioxide. Decomposition becomes the more important factor in the removal of oxygen from seawater at greater depths because the densities of living populations diminish with depth.

The depth at which the rate of photosynthesis balances the rate of plant respiration is called the **compensation depth.** Above the compensation depth, plants produce oxygen at the expense of carbon dioxide; below it, carbon dioxide is produced at the expense of oxygen. Oxygen can only be added to the oceans at the surface, from exchange with the atmosphere or as a waste product of photosynthesis. Carbon dioxide also enters from the atmosphere at the surface, but it is available at all depths from respiration and decomposition.

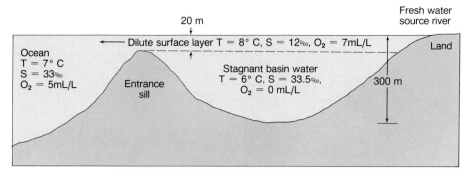

Figure 5.5

High-density ocean water is trapped behind the sill. The trapped water becomes anoxic due to continual respiration and decomposition.

Table 5.5 Abundance of Gases in Air and Seawater				
Gas	Symbol	Percentage by volume in atmosphere	Percentage by volume in surface seawater[1]	Percentage by volume in total oceans
nitrogen	N_2	78.03	48	11
oxygen	O_2	20.99	36	6
carbon dioxide	CO_2	0.03	15	83
argon, helium, neon, etc.	Ar, He, Ne	0.95	1	
totals		100.00	100	100

[1]Salinity = $36^0/_{00}$, temperature = $20°$ C.

Dissolved oxygen concentrations vary from 0 to 10 milliliters per liter of seawater. Very low or zero concentrations occur in the bottom waters of isolated deep basins, which have little or no exchange or replacement of water. Such an area can occur at the bottom of a trench, in a deep basin behind a shallow entrance sill (as in the Black Sea), or at the bottom of a deep fjord (300 to 400 m; 900–1300 ft). The deep water is trapped and becomes stagnant; respiration and decomposition use up the oxygen faster than the slow circulation to this depth can replace it. The bottom water becomes **anoxic,** or stripped of dissolved oxygen; **anaerobic** (or non-oxygen-using) bacteria live in such water. This condition is shown in figure 5.5.

At the surface, where oxygen is produced by the plants and added from the atmosphere, oxygen values are higher. Because oxygen is more soluble in cold water, more oxygen is found in surface waters at high latitudes than at lower latitudes. If the water is quiet, the nutrients and sunlight abundant, and a large population of plants present, oxygen values at the surface can rise above the equilibrium (or saturation) value, by 150% or more. This water is **supersaturated.** Wave action will tend to liberate oxygen to the atmosphere and return the condition to the 100% saturation state.

Carbon dioxide levels range between 45 and 54 milliliters per liter of seawater. The concentration is not substantially affected by biological processes, but tends to remain almost constant, controlled by temperature, salinity, and

pressure, because of the complex way in which carbon dioxide reacts with seawater. These reactions are discussed in the next section.

Figure 5.6 shows typical oxygen and carbon dioxide concentrations with depth. Use this diagram to review the relationships of these dissolved gases to the biological processes in the sea and to each other. At the surface, the concentration of oxygen is high due to the available atmospheric oxygen and to photosynthesis. Below the surface layer, oxygen decreases as animal respiration and the decomposition of organic material remove the oxygen; the **oxygen minimum** occurs at about 800 m (2600 ft). At depths greater than this, the rate of removal of oxygen decreases, because the population density of animals and the abundance of decaying organic matter has decreased. The supply of oxygen in water sinking from the surface increases the oxygen concentration above that found at the oxygen minimum. The carbon dioxide graph in figure 5.6 shows almost the direct opposite effect. The CO_2 concentration at the surface is low, because it is used in photosynthesis. Below the surface layer the concentration increases with depth as animal respiration and decomposition continually produce CO_2 and add it to the water.

Carbon Dioxide as a Buffer

When carbon dioxide dissolves in seawater, the CO_2 combines with the water to form carbonic acid (H_2CO_3). The carbonic acid dissociates to form bicarbonate (HCO_3^-) and

The Messages in Polar Ice

The chemical history of the earth's climate for the last 200,000 years is trapped in the earth's polar ice sheets. Scientists are going back in time by drilling ice cores in Greenland and Antarctica to discover connections between climate and atmospheric chemistry, and the impact of human activities on climate and atmospheric composition (see Box fig. 5.1a). The chemical composition of the ice preserves, layer by layer, a record of soluble substances (sodium, chloride, sulfate, and nitrate) and heavy metals (lead, cadmium, and zinc). Dust tells how stormy the atmosphere was and when volcanic eruptions occurred, and bubbles in the ice contain trapped air, tracing the natural changes of carbon dioxide and other gases through time.

At the summit of the Greenland ice cap the ice holds the longest, most continuous and detailed record of climate in the Northern Hemisphere. A collaborative European effort known as the Greenland Ice-core Project or GRIP began drilling in 1989, and in the summer of 1992 they reached the 2827-m (9300-ft) depth. Thirty-five kilometers (20 mi) away, the U.S. Greenland Ice Sheet Project or GISP 2 has spent every summer on the ice since 1988; it reached 2300 m (7550 ft) in June 1992 and the rock beneath the ice at 3000 m (9850 ft) at the end of the summer. A research tunnel carved out of solid snow extends away from the covered drill site (see Box fig. 5.1b). Here the cores are cut into sections and sliced lengthwise for initial sampling and measurement, Box figure 5.1c and d. Portions are packed into refrigerated containers and sent by air to cold storage facilities at the researchers' home laboratories for further, sophisticated study.

At the Vostok Antarctic research station, Russian scientists have drilled to 2546 m (8350 ft). In Antarctica there is a longer total time record over the same depth, but less ice is accumulated each year; therefore it is more difficult to date the layers accurately. It is important to compare the results from the two hemispheres, both for verification and to see if the two hemispheres reacted similarly.

When GRIP and GISP 2 reached the 3000-m depth, they recovered the climate record for the past 200,000 years. At present there is great interest in the time of the transition from the last glacial period, 11,000 years ago, because at this time the rate of temperature rise was similar to

(a)

(b)

(c)

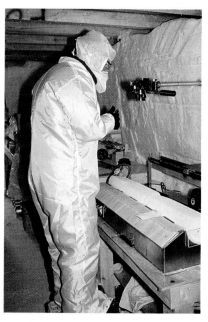

(d)

Box figure 5.1

(a) Scott-Amundsen Base, 90°S, November–December, 1982. Cores 10 cm (4 in) in diameter were drilled to 227 m (745 ft) with a mechanical drill. The ice at the bottom of the core is approximately 2200 years old. (b) Greenland Ice Sheet Project 2 (GISP 2) drilling station, summer 1990. The tunnel houses laboratories and leads to the drilling site. The boxes at the tunnel entrance hold core sections ready for shipment to the scientists' home labs. (c) A lengthwise slab is removed from a core section for analysis in the tunnel lab. (d) GISP 2, July, 1989. The layering of this ice core was inspected and photographed before it was analyzed for gas and particle content, chemistry, and conductivity. The core was 13 cm (5 in) in diameter.

that being predicted for the coming century. Researchers are working to determine how rapidly the last shift from glacial to interglacial occurred and how the properties of the atmosphere changed before and after the shift.

Scientists investigating global pollution have used data from the ice cores to compare the chemical composition of ice layers deposited before and after the

Industrial Revolution. Studies show that the concentration of sulfate derived from sulfuric acid has tripled and the concentration of nitrate derived from nitric acid has doubled in Greenland during the last century. Significant increases in the concentration of industrial heavy metals have also been detected, for example, the concentration of lead, a by-product of leaded-fuel combustion, increased dramatically with the world's increased dependence on gasoline.

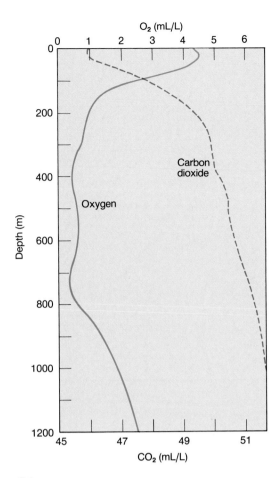

Figure 5.6
The distribution of O_2 and CO_2 with depth.

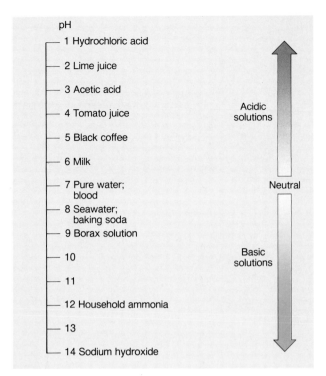

Figure 5.7
The pH scale.

a hydrogen ion (H^+) or carbonate (CO_3^{2-}) and two hydrogen ions ($2H^+$). The CO_2, H_2CO_3, HCO_3^-, and CO_3^{2-} exist in equilibrium with each other and with the H^+ ions, as shown in the following equation. The arrows indicate that the reactions both produce and take up hydrogen ions:

$$CO_2 + H_2O \leftrightharpoons H_2CO_3 \leftrightharpoons HCO_3^- + H^+ \text{ or } CO_3^{2-} + 2H^+.$$

This series of reactions allows CO_2 to play an additional and extremely important role in the ocean: it acts as a **buffer.** A buffer prevents sudden changes in the acidity or alkalinity of a solution. The buffering capacity of seawater is important to organisms requiring a steady **pH** for their life processes and to the chemistry of seawater, which is controlled, in part, by pH. The pH scale indicates acidity and alkalinity by measuring the concentration of hydrogen ions in a solution. A pH of 7 indicates neutrality, neither acid nor alkaline. The pH values between 1 and 6 are acid while the pH values between 8 and 14 are alkaline, or basic. The pH scale is presented in figure 5.7. The pH range of seawater is between 7.5 and 8.5; an average pH value for the world's oceans is approximately 7.8.

The concentration of hydrogen ions (H^+) in the seawater controls the acidity or alkalinity. An increase in hydrogen ions shifts the reaction series to the left, removing

them from solution. If the concentration of hydrogen ions decreases, the reaction moves from carbonic acid (H_2CO_3) to bicarbonate (HCO_3^-), liberating hydrogen ions. Therefore, the removal of CO_2 by photosynthesis or the addition of CO_2 by respiration or decay has little effect on the acid-alkaline balance of seawater.

The Carbon Dioxide Cycle

At present, the annual ocean uptake of carbon as carbon dioxide (CO_2) is calculated to be 2 billion metric tons per year. The rate at which the oceans absorb CO_2 is controlled by water temperature, pH, salinity, the chemistry of the ions (presence of calcium and carbonate ions), and biological processes, as well as mixing and circulation patterns of seawater which transfer CO_2 from the surface to the deep ocean (see fig. 5.8).

The transfer of carbon from CO_2 to organic molecules by photosynthesis results in the addition of CO_2 to the deep-ocean water, when the organic material sinks and decays. This process is often called the "biological pump," because biological processes pump carbon to the deep ocean. Enormous quantities of carbon as CO_2 transferred to the ocean bottom in this way are fixed in marine sediments as calcium carbonate ($CaCO_3$) (refer back to the Biogenous Oozes section in chapter 2). The rate at which this occurs is not fully understood. Cold, dense surface water sinking in subpolar regions also transfers CO_2 from the surface to deep water; these transfers are slow, requiring a few hundred to more than a thousand years. At other latitudes, deep water returns to the surface and provides a return source of

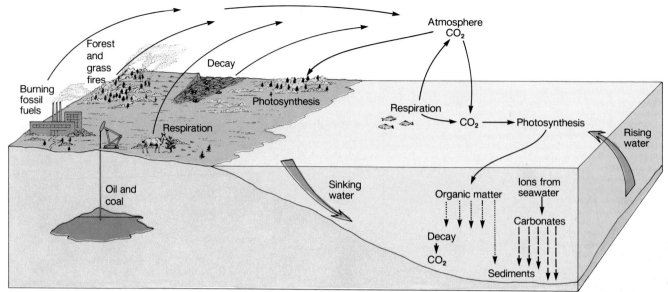

Figure 5.8

Major carbon dioxide pathways through the earth's environment.

atmospheric CO_2. These circulation patterns and the effect of the increased production of CO_2 by the burning of fossil fuels are discussed further in chapter 7.

The Oxygen Balance

The oceans also play a large role in regulating the oxygen balance in the earth's atmosphere. Photosynthesis in the oceans releases oxygen that is consumed by respiration and decay processes in the same way as on land. However, some organic matter is incorporated into the seafloor sediments, preventing the organic matter's decay and decomposition. Therefore, oxygen is not consumed to balance the oxygen produced in photosynthesis, and 300 million metric tons of excess oxygen should be released into the atmosphere each year, resulting in an increase of oxygen over time. This does not happen because the marine sediments are formed into rocks by the earth's geologic processes. Some of these rocks are uplifted onto land, and oxygen is eventually consumed in the weathering and oxidation of the materials in the rocks. This balances the atmosphere's oxygen budget. The mechanisms that link and control this process are not well understood.

Measuring the Gases

The amount of dissolved oxygen present in seawater samples can be measured by traditional chemical techniques in the laboratory. It is also possible to directly measure oxygen in the oceans, using specialized probes that send an electronic signal back to the ship or store the information in the testing unit. Carbon dioxide is rarely measured directly; rather, it is determined from the pH of the water, which can be measured directly.

Table 5.6	
Nutrients in Seawater	
Element	**Concentration** μg/kg[1]
nitrogen (N)	500
phosphorus (P)	70
silicon (Si)	3000

[1]Parts per billion.

5.3 Other Substances

Nutrients

Ions required for plant growth are known as **nutrients;** these are the fertilizers of the oceans. As on land, plants require nitrogen and phosphorus in the form of nitrate (NO_3^-) and phosphate (PO_4^{3-}) ions. A third nutrient required in the oceans is the silicate ion (SiO_4^-), which is needed to form the hard outer wall of the single-celled plants called diatoms and the skeletal parts of some protozoans. These three nutrients are among the dissolved substances brought to the sea by the rivers and land runoff. Despite their importance, they are present in very low concentrations (see table 5.6). The nutrients are removed from the water as the plant populations grow and reproduce and are then returned to the water as the organisms die and decay. Nutrients are cycled into the animal populations as the animals feed on the plants. These animals may be eaten

The Effect of Sunlight on Seawater

Understanding the many and complex chemical reactions caused by sunlight in the oceans' surface waters is a new area of study in marine chemistry. Photosynthesis is the best-known reaction involving sunlight in seawater, but researchers are finding an increasing number of other chemical reactions activated by the absorption of sunlight. When sunlight enters the surface waters, much of the radiation is absorbed by a variety of natural substances. These include living and nonliving particles, dissolved organic matter, and the water itself. Salt ions do not usually absorb sunlight. Some of the substances that are able to absorb the sun's energy become active and initiate a variety of other chemical reactions.

In some cases, the substance that absorbs the sunlight is changed as a direct result of the light energy. In other cases, the absorbing substance causes other substances that have not absorbed light to react. Both possibilities are shown in Box figure 5.2. An example of the first case is the nitrite ion (NO_2^-), which is changed by the absorption of sunlight to form activated forms of nitrous oxide (NO•) and hydroxide (OH•). The activated compounds then rapidly take part in other chemical reactions. This reaction may affect the availability of nitrogen to marine plants. The second case is illustrated by an organic compound (OC) that absorbs light energy, becomes energized (OC•), and transfers the energy to dissolved oxygen, producing a reactive, excited form of oxygen. The energized organic compound (OC•) is returned to its original form (OC). The excited oxygen passes its energy on to other compounds to be used in reactions thought to be important in the formation of organic molecules and the degradation of some pollutants. Other chemical reactions involving the sun's energy may influence the availability of trace metals (iron, manganese) to marine plants.

There is a great deal more to learn about these processes. The reactions are difficult to follow and to measure because concentrations of substances may be very low. NO• in the surface waters of the equatorial Pacific is only about 2×10^{-9} g/L, and molecules may have a very short lifetime (the half-life of reactive oxygen is 3×10^{-6} sec). Marine chemists are just beginning to realize the complexities of these processes and to understand the significance of sunlight's effect on seawater.

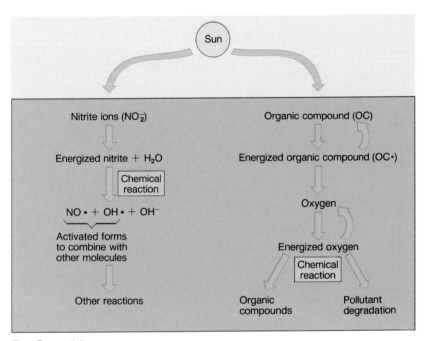

Box figure 5.2
The effect of sunlight on chemicals in seawater.

in their turn, and eventually the nutrients are returned to the oceans by way of death and bacterial decomposition. Excretory products from these animals are also added to the seawater, to be broken down and used again by a new generation of plants and animals. Nutrients do not maintain constant ratios in the way most major ions do. The cyclic nature of the nutrient pathways for nitrogen and phosphorus is discussed in chapter 14.

Organics

A wide variety of organic substances is present in seawater. Proteins, carbohydrates, lipids (or fats), vitamins, hormones, and their breakdown products are all present. Some are eventually reduced to their inorganic components; others are used directly by organisms and are incorporated into their systems. Another portion of the organic matter accumulates in the sediments, where over geologic time it slowly forms deposits of oil and gas. In the areas of the ocean that are high in plant and animal life, the surface layer may take on a green-yellow color due to the presence of so many organic substances, which are sometimes called "gelbstoff," or yellow substances.

5.4 Practical Considerations: Salt and Water

Chemical Resources

About 30% of the world's salt is extracted from seawater. In order to keep extraction costs low, the industrially produced energy required to remove the water is kept to a minimum. In warm, dry climates seawater is allowed to flow

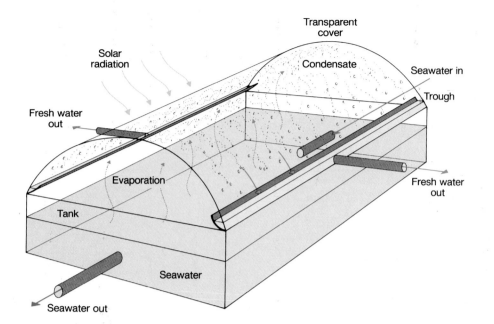

Figure 5.10

Solar energy is used to evaporate fresh water from seawater. Solar radiation penetrates the transparent cover of the still and causes evaporation of the seawater contained in the tank.

into shallow ponds and evaporate down to a concentrated brine solution. More seawater is added, and the process is repeated several times, until a dense brine is produced. Evaporation continues until a thick, white salt deposit is left on the bottom of the pond. The sea salt is collected and refined to produce table salt. This technique has been recently used in southern France, Puerto Rico, and California (see fig. 5.9).

In cold-climate areas, salt has been recovered by freezing the seawater in similar ponds. The ice that forms is nearly fresh; the salts are concentrated in the brine beneath the ice. The brine is removed and heated to remove the last of the water.

Of the world's supply of magnesium, 60% comes from the sea, and so does 70% of the bromine. There are vast amounts of dissolved minerals in the world's seawater, including 10 million tons of gold and 4 billion tons of uranium, but the concentration is very low (one part per billion or less). In the case of gold, no method has been devised in which the cost of extraction does not exceed the value of the recovered element. Japan, West Germany, and the United States have expressed an interest in the extraction of uranium. Japan set up a land-based test plant in 1986 to produce 10 kg (26 lb) of uranium per year by passing seawater over a uranium absorber. The system, including seawater pumps, absorber costs, and plant operation, has proved to be expensive, and the Japanese are looking at methods to use wave energy to run the system. (Some possible ways to harness wave energy are discussed in chapter 9.)

The dense, hot salt brines at the bottom of the Red Sea were discussed in chapter 2. These brines are estimated to contain minerals with values in the billions of dollars.

Desalination

Desalination is the process of obtaining fresh water from salt water. There are several possible desalination methods:

1. processes involving a change of state of the water—liquid to solid or liquid to vapor;
2. processes using a **semipermeable membrane—electrodialysis** and **reverse osmosis;** and
3. processes requiring ion exchange.

The simplest process involving a change of state is a solar still (fig. 5.10). In this process a pond of seawater is capped by a low plastic dome. Solar radiation penetrates the dome and evaporates the seawater. The evaporated water condenses on the undersurface of the dome and trickles down to be caught in a trough, where it accumulates and flows to a freshwater reservoir. The rate of production is slow, and a very large system is needed to supply the water requirements of even a small community.

When water is distilled by heating water to boiling, evaporation proceeds at a rapid rate and large quantities of fresh water are produced, but the energy requirement is high. If water is introduced to a chamber with a reduced air pressure, the boiling occurs at a much lower temperature and therefore uses less energy. Change of state by freezing can also be used to recover fresh water from seawater. The energy requirement is approximately one-sixth of that needed for evaporation, but the mechanical separation of the freshwater ice from the salt brine remains difficult.

Electrodialysis uses an electric field to transport ions out of solution and through semipermeable membranes. This technique works best in low-salinity (or brackish) water. Columns containing ion-exchange resins that extract

Figure 5.9 (facing page)

The southern end of San Francisco Bay has been diked into shallow ponds, where seawater is evaporated to obtain salt.

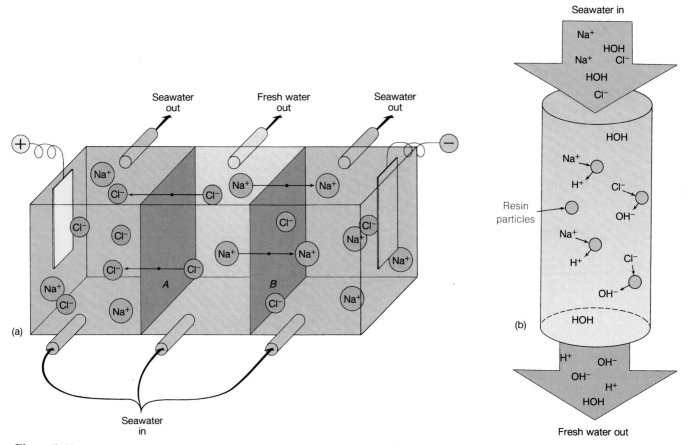

Figure 5.11

(a) Electrodialysis. A tank is separated into three compartments by membranes *A* and *B*. Membrane *A* passes only Cl⁻. Membrane *B* passes only Na⁺. Electrodes that are placed in the end compartments and are supplied with a direct-current voltage will cause the salt ions to migrate out of the center compartment. Fresh water is recovered from the center compartment; excess salt water is removed from each side compartment. (b) Ion-exchange column. Seawater passes through a column of resin particles that exchange H^+ for Na^+ and Cl^- for OH^- to produce HOH, or fresh water (H_2O).

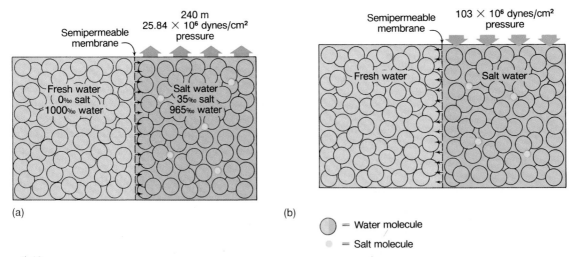

Figure 5.12

(a) Osmosis. Water moves from the freshwater side to the saltwater side. (b) Reverse osmosis. Water moves from the saltwater side to the freshwater side.

ions from salt water also work well with water of low salinity, but the resins need to be replaced periodically. Neither process produces the large volumes of low-cost fresh water that are required in some areas of the world. See figure 5.11a and b for diagrams of both processes.

Osmosis is the movement of water across a semipermeable membrane; the water moves from the side with the higher concentration of water molecules (or low salinity) to the side with the lower concentration of water molecules (or high salinity); this creates a higher pressure on the low water concentration (or high-salinity side of the membrane). See figure 5.12a. **Reverse osmosis** produces fresh water from seawater by applying pressure to seawater and forcing the water molecules through a semipermeable membrane, leaving behind the salt ions and other impurities. See figure 5.12b. This requires that the pressure applied to the seawater exceeds 24.5 atmospheres or 25.84×10^6 dynes/cm^2. It requires a pressure of about 101.5 atmospheres or 103×10^6 dynes/cm^2 to achieve a reasonable rate of freshwater production. The theoretical energy requirement is about one-half that needed for the evaporative process.

Reverse osmosis is the most rapidly growing form of desalination technology. As older evaporative desalination plants wear out, reverse osmosis plants are replacing them. Of Florida's 110 desalination plants 99% use reverse osmosis. Santa Barbara, California, faced with extreme drought conditions, has built the nation's largest reverse osmosis plant, capable of producing about 26 million liters (7 million gallons) per day. To the south, twenty-six miles off the coast from Los Angeles, Santa Catalina Island is using this method to produce about one-third of its potable water needs, 5×10^5 liters (132,000 gallons) per day, and is considering expansion. Reverse osmosis systems are a part of the U.S. Army's mobile water purification units; virtually all of the drinking water supplied to U.S. troops during the Persian Gulf War came from reverse osmosis. Small units powered by hand pumps are available to produce fresh water for drinking in emergency situations (for example, in lifeboats and rafts). Small commercial units are available for producing fresh water for small vessels and summer homes on islands with little fresh water.

At present nearly four thousand plants worldwide produce a total of about 13 billion liters (3.4 billion gallons) of potable water each day. About 60% of this capacity is located in Arab countries. The principal process used in these countries is multistage flash distillation, and the next most important process is reverse osmosis. Most flash distillation plants are associated with petroleum-powered electric plants that use high-temperature, high-pressure steam to drive their turbines. Condensate from the power plants at about 120°C is used to produce fresh water in a multistage, lower-pressure system that produces six to twenty times the amount of fresh water that would be obtained from a single-step distillation system. In addition there are twenty-two reverse osmosis plants in Saudi Arabia, with a total capacity of 230 million gallons of water per day.

In areas like Kuwait, Saudi Arabia, Morocco, Malta, Israel, the West Indies, California, and the Florida Keys, water is a limiting factor for population and industrial growth. The greatest drawback to the production of fresh water from seawater is the high cost, which is linked to the energy required. In the Arab countries water costs are low because fuel costs are low. Desalination plants in California and Texas produce fresh water at more than twice the cost of water from other sources. The total costs of producing potable water from seawater are on the order of $4 per 1000 gallons; this is cheap enough to use for drinking and for certain industrial applications but far too expensive for agriculture.

Summary

The average salinity of ocean water is 35‰. The salinity of the surface water changes with latitude and is affected by evaporation, precipitation, and the freezing and thawing of sea ice. Soluble salts are present as ions in seawater. Positive ions are cations; negative ions are anions. Six major constituent ions make up 99% of the salt in seawater. Trace elements that are present in very small quantities are particularly important to living organisms.

Most of the positively charged ions come from the weathering and erosion of the earth's crust and are added to the sea by rivers. Gases from volcanic eruptions are dissolved in river water as anions. Since the average salinity of the oceans remains constant, the salt gain must be balanced by the removal of salt; input must equal output. Salts are removed as sea spray, evaporites, and insoluble precipitates, as well as by biological reactions, adsorption, chemical reactions, and uplift processes. Seawater circulating through magma chambers in the earth's crust deposits dissolved metals and releases other chemicals in solution. The time that salts remain in solution, known as residence time, depends on their reactivity.

The proportion of one major ion to another remains the same for all open-ocean salinities. Ratios may vary in coastal areas and in association with biological processes.

Salinity is determined chemically by the amount of chloride ion present and by the conductivity and temperature of the water.

The saturation value of gases dissolved in seawater varies with salinity, temperature, and pressure. Carbon dioxide is added to seawater at the sea surface from the atmosphere and by respiration and decay processes at all depths; it is removed at the surface by photosynthesis. Oxygen is added only at the surface from the atmosphere and the photosynthetic process; it is depleted at all depths by respiration and decay. Seawater may become supersaturated with oxygen or it may become anoxic. Carbon dioxide levels tend to change little over depth. Carbon dioxide has the additional role of buffer in keeping the pH range of ocean water between 7.5 and 8.5. Large quantities of CO_2 are absorbed by the oceans. Biological processes pump carbon as carbon dioxide into deep water, where it is fixed in the marine sediments as calcium carbonate. Atmospheric oxygen is regulated by oceanic processes. The amount of oxygen present in seawater is measured chemically and electronically. Carbon dioxide content is determined from the pH of the water.

Nutrients include the nitrates, phosphates, and silicates required for plant growth. A wide variety of organic products is also present.

Salt, magnesium, and bromine are currently being commercially extracted from seawater. Direct extraction of other chemicals is neither economic nor practical at present. Fresh water is an important product of seawater. Desalination methods include change-of-state processes, movement of ions across semipermeable membranes, and ion exchange. The practicality of desalination is determined by cost and need. Reverse osmosis can produce fresh water, and small processing units are available.

Key Terms

salinity	salinometer
ion	saturation value
cation	photosynthesis
anion	respiration
major constituent	compensation depth
trace element	anoxic
conservative constituent	anaerobic
nonconservative constituent	supersaturation
igneous rock	oxygen minimum
evaporite	buffer
adsorption	pH
ion exchange	nutrients
residence time	desalination
principle of constant	semipermeable membrane
proportion (constant	electrodialysis
composition)	osmosis
chlorinity	reverse osmosis
halogens	

Study Questions

1. How is the ocean's salt balance regulated? Make a diagram showing inputs and outputs.
2. Explain the concept of the constant proportions of the major ions in seawater.
3. What is the least expensive method of desalination? Where is it most likely to be used?
4. How does the salinity of midocean surface water change with latitude? What processes produce these changes?
5. List the sources of the salts found in seawater. How is their input regulated?
6. Silicate is a nonconservative constituent of seawater and does not obey the principle of constant proportions. Explain why.
7. Compare the distributions of oxygen and carbon dioxide with depth in seawater. What processes are responsible for the distribution of each gas?
8. Explain the relationship between carbon dioxide and the pH of the oceans. Why is the buffering of seawater important?

9. Suggest several pathways a carbon dioxide molecule might follow as it is transferred from the atmosphere to the seafloor sediments.
10. Why does the addition of rain and river water to the oceans not decrease the ocean's overall salinity?
11. Why is the residence time in seawater different for different salts?
12. Explain reverse osmosis and how it produces fresh water from seawater.
13. What is the significance of the compensation depth to plant life?
14. Why does the concentration of dissolved substances in river water differ from the concentration of the same substances in seawater?
15. Why are nutrients considered to be nonconservative materials?

Study Problems

1. If there is 1.4×10^{21} kg of water in the oceans, what is the potential mass of NaCl, sodium chloride, in the oceans? Use table 5.1.
2. If the chloride ion (Cl^-) content of a seawater sample is 18.5‰, what is the concentration of magnesium in the same sample? Express your answer in g/kg.
3. Determine the residence time of calcium using the following information:
   Calcium ion present in seawater = 0.41 g/kg
   Seawater in the oceans = 1.4×10^{21} kg
   Calcium ion = 20.39% by weight of the average dissolved substances in river water
   Average salt content of river water = 0.001‰
   Annual river runoff = 3.7×10^{17} kg/yr.
4. How many kilograms of seawater would have to be processed to obtain 1 kg of gold? Use table 5.2.

Suggested Readings

The Seawater
Anderson, A. T. 1982. The Ocean Basins and Ocean Water. In *Ocean Yearbook 3,* Borgese, E., and N. Ginsburg, eds. Univ. of Chicago, Chicago. pp. 135–56. (The origin of seawater and the interaction of seawater and ocean crust.)

Berner, R. A., and A. C. Lasaga. 1989. Modeling the Geochemical Carbon Cycle. *Scientific American* 260 (3): 74–81.

Brown, N. 1991. The History of Salinometers and CTD Sensor Systems. *Oceanus* 34 (1): 61–66.

Edmond, J. M., and K. L. Von Damm. 1992. Hydrothermal Activity in the Deep Sea. *Oceanus* 35 (1): 76–81.

MacIntyre, F. 1970. Why the Sea Is Salt. In *Ocean Science,* readings from *Scientific American* 223 (5): 104–15 (1977).

Monastersky, R. 1991. Tales from Ice Time. *Science News* 140 (11): 168–72. (Ice cores in Greenland.)

Open University. 1989. *Ocean Chemistry and Deep Sea Sediments.* Pergamon Press, Oxford, England; Open University, Milton Keynes, England. 120 pp.

Open University. 1989. *Seawater: Its Composition, Properties and Behavior.* Pergamon Press, Oxford, England; Open University, Milton Keynes, England. 165 pp.

Riley, J. P., and R. Chester, eds. 1983. *Chemical Oceanography,* Vol. 8. Academic Press, New York. 398 pp.

Weiner, J. 1989. Glacier Bubbles Are Telling Us What Was in Ice Age Air. *Smithsonian* 20 (2): 78–87.

Chemical Resources

Anderson, A. T. 1982. The Ocean Basins and Ocean Water. In *Ocean Yearbook 3,* Borgese, E., and N. Ginsburg, eds. Univ. of Chicago, Chicago. pp. 135–56.

Blissenbach, E., and Z. Nawab. 1982. Metalliferous Sediments of the Seabed: The Atlantis II-Deep Deposits of the Red Sea. In *Ocean Yearbook 3,* Borgese, E., and N. Ginsburg, eds. Univ. of Chicago, Chicago. pp. 77–104.

Charlier, R. H. 1983. Water, Energy, and Nonliving Ocean Resources. In *Ocean Yearbook 4,* Borgese, E. M., and N. Ginsburg, eds. Univ. of Chicago, Chicago. pp. 75–120. (An overview of mineral resources.)

Hotta, H. 1987. Recovery of Uranium from Seawater. *Oceanus* 30 (1): 44–46.

Kent, P. 1980. *Minerals from the Marine Environment.* Edward Arnold, London. 88 pp.

Rona, P. 1986. Mineral Deposits from Seafloor Hot Springs. *Scientific American* 254 (1): 84–92.

Desalination

Friedman, R. 1990. Salt-free Water from the Sea. *Sea Frontiers* 36 (3): 48–54.

Levine, S. N., ed. 1968. *Selected Papers on Desalination and Ocean Technology.* Dover, New York. 437 pp.

Spiegler, K. S. 1977. *Salt Water Purification.* Plenum, New York. 189 pp.

6

Structure of the Oceans

6.1 **Heating and Cooling the Earth's Surface**
The Heat Budget
Annual Cycles of Solar Radiation
Heat Capacity
6.2 **Evaporation and Precipitation Patterns**
6.3 **Density Structure and Vertical Circulation**
Surface Processes
Changes with Depth
Density-Driven Circulation
Sigma-t
6.4 **Upwelling and Downwelling**
Chemical Tracers
6.5 **The Layered Oceans**
Structure of the Atlantic Ocean
Structure of the Pacific Ocean
Structure of the Indian Ocean
Mediterranean Sea and Red Sea Water
Comparing the Oceans
6.6 **T-S Curves**
T-S Curves and Water Masses
6.7 **Measurement Techniques**
Determining Water Properties
Temperature Measurements

Box: New Ways to Measure the Oceans

6.8 **Practical Considerations: Energy Sources**
Osmotic Pressure
Ocean Thermal Energy Conversion
Summary
Key Terms
Study Questions
Suggested Readings

The weeks passed. We saw no sign either of a ship or of drifting remains to show that there were other people in the world. The whole sea was ours, and, with all the gates of the horizon open, real peace and freedom were wafted down from the firmament itself.

It was as though the fresh salt tang in the air, and all the blue purity that surrounded us, had washed and cleansed both body and soul. To us on the raft the great problems of civilized man appeared false and illusory—like perverted products of the human mind. Only the elements mattered. And the elements seemed to ignore the little raft. Or perhaps they accepted it as a natural object, which did not break the harmony of the sea but adapted itself to current and sea like bird and fish. Instead of being a fearsome enemy, flinging itself at us, the elements had become a reliable friend which steadily and surely helped us onward. While wind and waves pushed and propelled, the ocean current lay under us and pulled, straight toward our goal.

Thor Heyerdahl,
from *Kon-Tiki*

Rain falling over the ocean.

*H*idden below its surface is the ocean's structure. If we could remove a slice of ocean water in the same way we might cut a slice of cake, we would find that, like a cake, the ocean is a layered system. The layers are invisible to us, but they can be detected by measuring the changing salt content and temperature and by calculating the density of the water from the surface to the ocean floor. This layered structure is a dynamic response to processes that occur at the surface: the gain and loss of heat, the evaporation and addition of water, the freezing and thawing of ice, and the movement of water in response to wind. These surface processes produce a series of horizontally moving layers of water, as well as local areas of vertical motion. In this chapter, we will study both the surface processes and their below-the-surface results in order to understand why the ocean is structured in this way and how the structure is maintained. We will also explore the ways in which oceanographers gather data about this layered system, and we will survey the possibilities of extracting useful energy from it.

6.1 Heating and Cooling the Earth's Surface

The intensity of solar radiation per unit area of earth surface is greatest at the equator, moderate in the middle latitudes, and least at the poles. If the earth had no atmosphere, the intensity of solar radiation available on a surface at right angles to the sun's rays would be 2 calories per square centimeter per minute ($cal/cm^2/min$); this value is called the **solar constant.** The solar constant is approached only at latitudes between 23 1/2° N (the Tropic of Cancer) and 23 1/2° S (the Tropic of Capricorn), because only between these latitudes does sunlight strike the earth at right angles. At all other latitudes the earth's surface is inclined to the sun's rays at a greater angle, because of the earth's spherical shape. Compare the angles at which the sun's rays strike the earth in figure 6.1. Because the sun is so far away from the earth, its rays are parallel when they reach the earth. Where the rays strike the earth at right angles, the same amount of radiant energy strikes each unit area. But as the angle at which the sun's rays strike the earth increases, the unit areas receive less energy. This difference is also shown in figure 6.1.

When the sun stands directly above the equator at noon, during either the vernal or the autumnal equinox, the radiation value is about 1.6 $cal/cm^2/min$. This value is less than the solar constant, because the earth's atmosphere stands between the earth's surface and the incoming solar radiation, and the atmosphere absorbs a portion of the sun's energy. Atmospheric absorption causes the solar radiation per unit of surface area to decrease with increasing latitude in each hemisphere. The greater the latitude, the longer the

distance through the atmosphere the sun's rays must travel (see fig. 6.1). The combined effects of atmospheric path length and inclination of the earth's surface cause the earth to receive more solar heat in the tropics, less at the temperate latitudes, and the least at the poles.

The Heat Budget

In order to maintain its long-term mean temperature pattern, the earth must lose heat as well as gain it. On the world average, the earth and its atmosphere must reradiate as much heat back to space as it receives from the sun. These gains and losses in heat can be represented as a **heat budget.** Just as incoming funds must equal outgoing funds in a balanced monetary budget, in the total heat budget incoming radiation must equal outgoing radiation. If incoming funds (or deposits) are insufficient, or if outgoing funds (or withdrawals) become too great, serious financial problems arise. In the same way, if less heat were returned to space than is gained, the earth would become hotter, and if more heat were returned than is gained, the earth would become colder. In both cases, the planet would change dramatically.

To follow the deposits and withdrawals in the total heat budget, assume that 100 units of solar energy are incoming to the earth's atmosphere. Refer to figure 6.2 as you read this discussion. Thirty-one of these units are reflected directly back to space from the earth's atmosphere; no heating of the earth or its atmosphere occurs. This leaves 69 units of the 100 units, and 4 of these units are reflected by the earth's surface through the atmosphere to outer space; they take no part in heating the earth or its atmosphere. There are now 65 of the original 100 units left. These 65 units do heat the planet: 47.5 units are absorbed by the earth's surface, and 17.5 units are absorbed by the earth's atmosphere. To balance the budget, 65 units must be lost by **reradiation** of long-wave radiation back to outer space; 5.5 units are reradiated from the earth's surface, and 59.5 units are reradiated from the atmosphere. The budget is balanced; incoming radiation balances outgoing radiation.

Closer inspection shows that although the earth absorbs 47.5 units, it loses only 5.5 units to space, for a gain of 42 heat units, while the atmosphere absorbs 17.5 units and loses 59.5 units to space, for a loss of 42 units. In these circumstances, the atmosphere cools and the earth heats. In order to maintain the near-constant average temperatures of the earth's surface and atmosphere, the 42 excess units gained by the earth must be transferred to the atmosphere to make up for the 42 units the atmosphere has lost. This transfer of heat is accomplished in two ways. First, 29.5 of the units are transferred by evaporation processes, which cool the earth's surface and liberate heat when water vapor condensation occurs in the atmosphere. The remaining 12.5 units are transferred to the atmosphere by conduction and reradiation and are absorbed as heat, raising the temperature of the air. On the world average, the atmosphere is primarily heated from below by heat given off from the earth.

If we consider the heat budget of only a small local portion of the oceans, the following factors must be taken into consideration: total energy absorbed at the sea surface

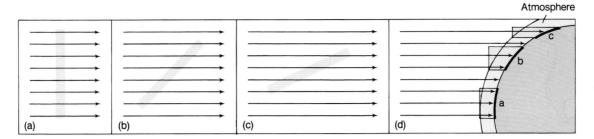

Figure 6.1

Areas of the earth's surface that are equal in size receive different levels of solar radiation as they become more oblique to the sun's rays (see a, b, and c). As latitude increases, the angle between the sun's rays and the earth increases and the solar radiation received on the surface decreases (see d). Note also that the sun's rays must travel an increasing distance through the atmosphere as latitude increases.

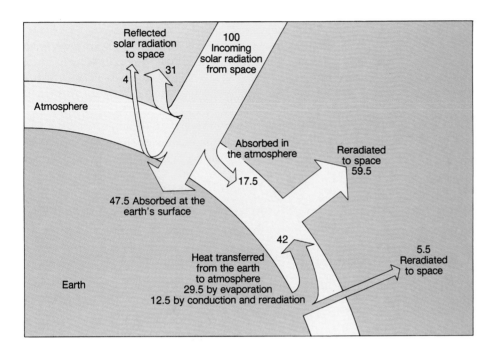

Figure 6.2

The earth's heat budget. Incoming solar energy is balanced by reflected and reradiated energy. The atmosphere's loss of heat is balanced by heat transferred from the earth to the atmosphere by evaporation, conduction, and reradiation.

in that area, loss of energy due to evaporation, transfer of heat into and out of the area by currents, warming or cooling of the overlying atmosphere by heat from the sea surface, and heat reradiated to space from the sea surface. These processes vary with time, showing daily and seasonal variations.

Measurements made over the earth's surface show that more heat over the annual cycle is gained than lost at the equatorial latitudes, while more heat is lost than gained at the higher latitudes (see fig. 6.3). Winds and ocean currents remove the excess heat accumulated in the tropics and release it at high latitudes to maintain the earth's present surface temperature patterns. Figure 6.4 shows summer ocean surface temperatures. North-south deflections in the lines of constant temperature indicate the displacement of surface water by currents carrying warm water from lower to high latitudes and cold water from higher to lower latitudes, redistributing heat energy over the earth's surface.

Annual Cycles of Solar Radiation

The total heat budget and the average distribution of radiant energy with latitude do not include the annual cycle of radiant energy changes related to the seasonal north-south migration of the sun. When these variations are included, the annual cycle of seasonal variation in solar radiation is most pronounced at the middle latitudes, between 40° N and 60° N and between 40° S and 60° S. Here the angle at which the sun's rays strike the earth changes dramatically from summer to winter. At these latitudes, there is a corresponding change in the duration of the daylight hours. The variation is illustrated in figure 6.5. The change in incident radiation, the solar radiation striking the earth's surface, produces seasonal variation in land and sea-surface temperatures due to heat losses or gains. The intensity of solar radiation remains fairly constant at tropic latitudes (between 23 1/2° N and 23 1/2° S) over the year, because the sun's noontime rays are always received at an angle approaching 90°. During the sun's annual migration between 23 1/2° N and 23 1/2° S, its rays cross the intervening latitudes twice. This produces

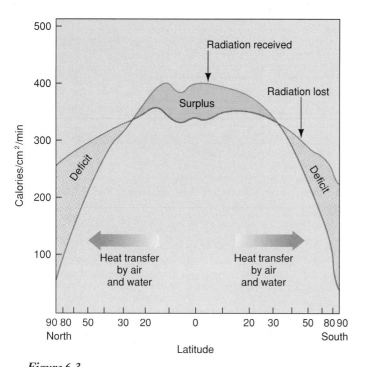

Figure 6.3

Comparison of incoming solar radiation and outgoing long-wave radiation with latitude. A transfer of energy is required to maintain a balance.

Source: NOAA Meteorological Satellite Laboratory.

a small-amplitude, semiannual variation in the intensity of solar radiation. The effect is most evident at the equator and can be seen in figure 6.5. Between 70°N and 90°N and between 70°S and 90°S, the sun's noontime rays always strike the earth at an oblique angle. The long duration of daylight hours in summer produces a high level of incident solar radiation averaged over the day (fig. 6.5). However, the intensity of radiation per unit surface area per minute of daylight is much lower than that found at the lower latitudes (see again fig. 6.1). Refer back to figure 1.3 in chapter 1 to review the relationship of the earth, the sun, and the earth's seasons.

Heat Capacity

Land and ocean areas respond differently to the annual cycle of solar radiation. Land has a low heat capacity; it changes temperature rapidly as it gains and loses heat between day and night or summer and winter. The oceans have a high heat capacity; they absorb and release large amounts of heat at the surface with very little change in temperature.

The annual range of surface temperatures for land and ocean are given in figure 6.6. Note that seasonal land temperature ranges at high latitudes are greater than those for the oceans and that the annual range of ocean surface

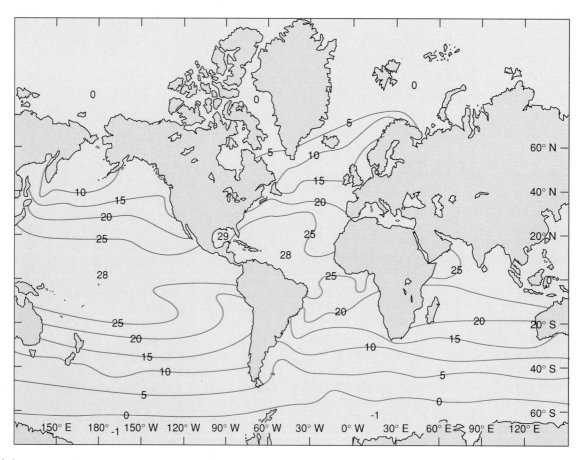

Figure 6.4

Sea-surface temperatures during summer in the Northern Hemisphere, given in degrees Celsius.

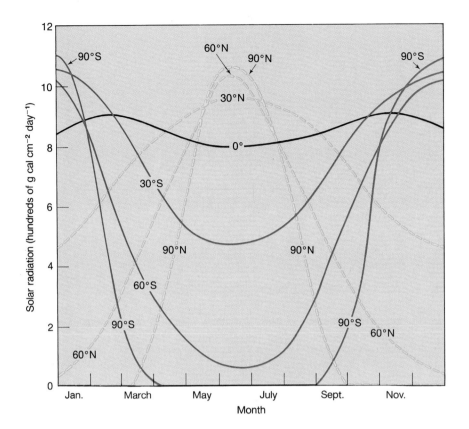

Figure 6.5
Average daily solar radiation values at different latitudes during the year. When it is summer in the Northern Hemisphere it is winter in the Southern Hemisphere; therefore the higher values in the Northern Hemisphere coincide with the lower values in the Southern Hemisphere. Peak values of solar radiation at the higher latitudes occur during summer in each hemisphere as a result of more hours of sunlight.

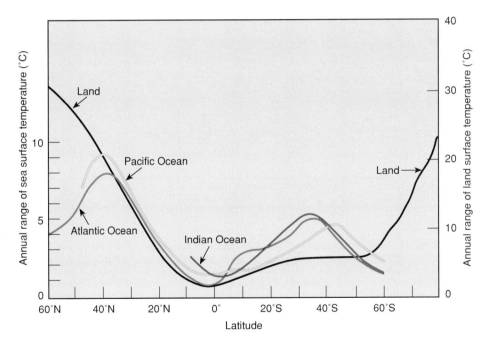

Figure 6.6
The annual range of midocean sea-surface temperatures is considerably less than the annual range of land-surface temperatures at the higher latitudes. The maximum annual range of sea-surface temperatures occurs at the middle latitudes.

temperatures is greatest at the middle latitudes. Because of the unequal distribution of land between the two hemispheres, the summer-to-winter variation in temperature for land at the middle latitudes is much greater in the Northern Hemisphere than it is in the Southern Hemisphere. The lack of land in the Southern Hemisphere means that the oceans, with their high heat capacity, control the earth's annual temperature range at the southern middle latitudes.

Heat that is absorbed at the ocean surface in summer is mixed downward by winds, waves, and currents. In winter, heat is transferred upward toward the cooling surface. The net effect of these processes is the small annual change in midocean surface temperatures: 0° to 2°C in the tropics, 5° to 8°C at the middle latitudes, and 2° to 4°C at the polar latitudes. Global surface temperatures measured by satellite are shown in figure 6.7.

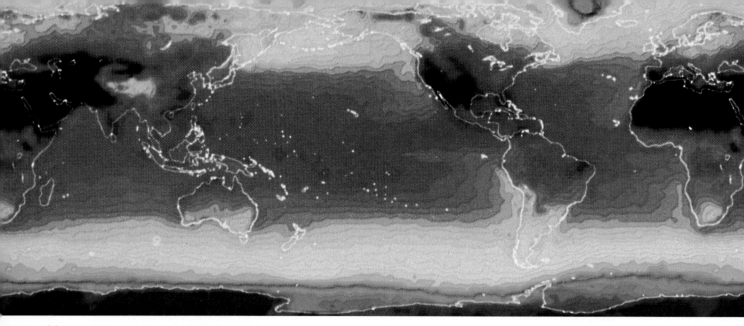

(a)

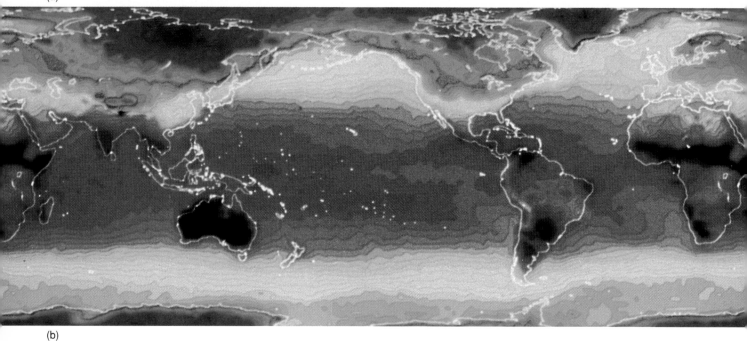

(b)

Figure 6.7

(a) July data shows the Northern Hemisphere landmasses with considerably warmer temperatures, but the ocean waters do not change dramatically from winter to summer. The oceans' warm surface water moves north and south with the change in seasons. North-south currents along the coasts of continents are also visible. (b) Meteorological satellites, such as NOAA's TIROS, carry high-resolution infrared sensors that measure long-wave radiation emitted from the earth's surface and atmosphere. This radiation is related to the earth's surface temperatures. Green and blue indicate temperatures below 0°C warmer temperatures are shown in red and brown. In January, Siberia and Canada show surface temperatures near −30°C; at the same time, latitudes between 30° and 50° south show warm summer temperatures.

| 6.2 | Evaporation and Precipitation Patterns |

The salinity of midocean surface water is controlled by the distribution of the world's evaporation and precipitation zones (see again figs. 5.1 and 5.2). Salinity of ocean water is measured in grams of salt per kilogram of seawater and expressed in parts per thousand (‰). The average ocean salinity is considered to be 35‰. In the tropics, the precipitation is heavy on land and at sea. On land, the tropic rain forest is the result; at sea, the surface water has a low salinity, around 34.5‰. At approximately 30°N and 30°S, the evaporation rates are high. These are the latitudes of the world's land deserts and of increased salinity in the surface waters, around 36.7‰. Farther north and south, from 50°N to 60°N and from 50°S to 60°S, precipitation is again heavy, producing cool but less salty surface water, around 34.0‰, and the heavily forested areas of the Northern Hemisphere. At

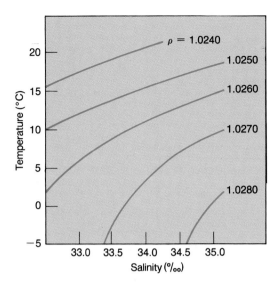

Figure 6.8
The density of seawater, measured in grams per cubic centimeter, is abbreviated as ρ (rho) and varies with temperature and salinity. The pressure factor is not included. Many combinations of salinity and temperature produce the same density. Low densities are at the upper left and high densities at the lower right.

the polar latitudes sea ice is formed in the winter. During the freezing process, the salinity of the water beneath the ice increases, and in the summer the thawing of the ice reduces the surface salinity once more. Climate zones will be discussed further in chapter 7.

<table><tr><td>**6.3**</td><td>**Density Structure and Vertical Circulation**</td></tr></table>

Variations in the surface temperatures and salinities of the oceans combine to control the **density** of the oceans' surface water. Many combinations of salinities and temperatures produce the same density; this is shown in lines of constant density displayed in figure 6.8. As the salinity increases, the density increases; as the temperature increases, the density decreases. Salinity may be increased by evaporation or by the formation of sea ice; it may be decreased by precipitation, by the inflow of river water, by the melting of ice, or by a combination of these factors. Changes in pressure also affect density. As the pressure increases, the density increases. Since pressure plays a minor role in determining the density of surface water, its effects will be ignored in the following discussion.

Surface Processes

Less-dense water remains at the surface (for example, the warm, low-salinity surface water of the equatorial latitudes). Although the surface water at 30°N and 30°S latitudes is warm, it has a higher salinity. Therefore, it is more dense than the warm, low-salinity equatorial water. This 30° latitude surface water sinks below the equatorial surface water; it extends from the surface at 30°N to below the less-dense equatorial layer and back to the surface at 30°S. The combined salinity and temperature in surface waters at

50°–60°N and 50°–60°S produces a water that is denser than either the equatorial or the 30° latitude surface water. The 50°–60° latitude water therefore sinks below the equatorial and 30° surface water and extends from the surface in one hemisphere below the other water types to the surface in the other hemisphere. Winter conditions in the polar regions lower the surface water temperature and, if sea ice forms, increase the surface salinity. The result is a dense surface water that sinks toward the ocean floor at the polar latitudes. These variations in surface water properties and the resulting density changes produce a density layered ocean, for if the density of the seawater is increased at the surface, the water sinks to a level of similar density. This layered system is shown in figure 6.9. The thickness and horizontal extent of each layer is related to the rate at which the water is formed and the size of the surface region over which it is formed.

Changes with Depth

The oceans have a well-mixed surface layer of approximately 100 m (330 feet) and layers of increasing density to a depth of about 1000 m (3300 feet). Below 1000 m the waters of the deep ocean are relatively homogeneous. A region between 100 m and about 1000 m, where density changes rapidly with depth, is known as a **pycnocline** (see fig. 6.10).

Below the 100-m surface layer the temperature decreases rapidly with depth to the 1000-m level. A zone with a rapid change in temperature with depth is called a **thermocline**. Below the thermocline, the temperature is relatively uniform over depth, showing a small decrease to the ocean bottom. A similar situation occurs with salinity. Below the surface water at the middle latitudes, the salinity increases rapidly to about 1000 m; this zone of relatively large change in salinity with depth is called the **halocline**.

Figure 6.9
Water of different densities forms a layered ocean. The density of each layer is determined at the surface by the climate at the latitude at which it is formed. Density values are given in grams per cubic centimeter.

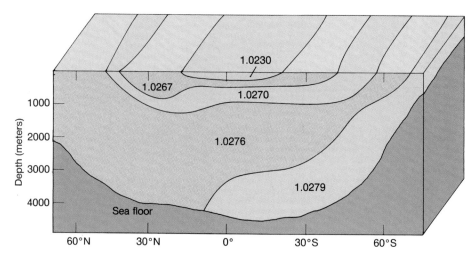

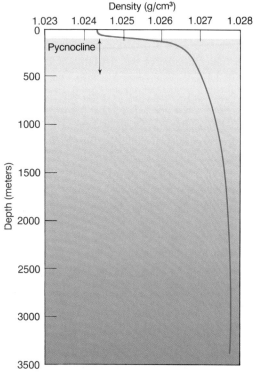

Figure 6.10

Density increases with depth in seawater. The pycnocline is the region in which density changes rapidly with depth. (Based on data from the northeast Pacific Ocean.)

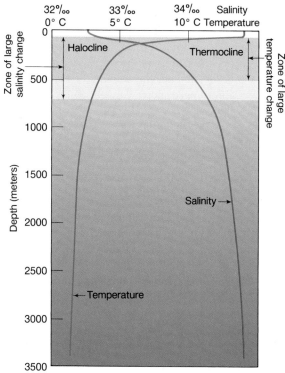

Figure 6.11

Temperature and salinity values change with depth in seawater. Rapid changes in temperature and salinity with depth produce a thermocline and a halocline, respectively. (Based on data from the northeast Pacific Ocean.)

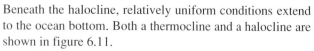

Beneath the halocline, relatively uniform conditions extend to the ocean bottom. Both a thermocline and a halocline are shown in figure 6.11.

If the density of the water increases with depth, the water column from surface to depth is **stable.** If there is more-dense water on top of less-dense water, the water column is **unstable.** An unstable water column cannot persist; the more-dense surface water sinks and the less-dense water at depth rises to replace the surface water. Vertical **overturn** of the water takes place. If the water column has the same density over depth, it has neutral stability and is termed **isopycnal.** A neutrally stable water column is easily

mixed in the vertical by wind, wave action, and currents. If the water temperature is unchanging over depth, the water column is **isothermal;** and if the salinity is constant over depth, it is **isohaline.**

Density-Driven Circulation

Processes that increase the water's density at the surface cause vertical water movement, or **vertical circulation,** which ensures an eventual top-to-bottom exchange of water throughout the oceans. Because the density is normally controlled by changes in temperature and salinity, this circulation is known as **thermohaline circulation.** An excellent

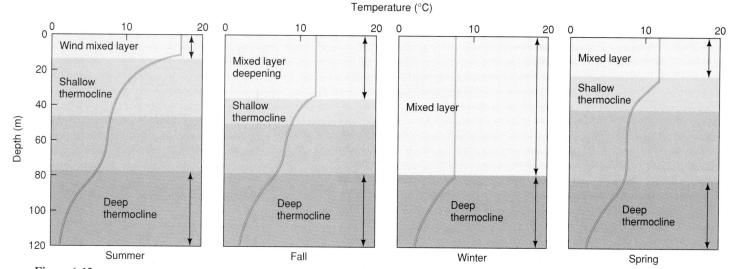

Figure 6.12

The surface layer temperature structure varies over the year. In the absence of strong winds and wave action in the summer, solar heating produces a shallow thermocline. During the fall and winter the surface cooling and storm conditions cause mixing and vertical overturn, which eliminate the shallow thermocline and produce a deep wind-mixed layer. In spring the thermocline re-forms.

example of thermohaline circulation occurs in the Weddell Sea of Antarctica, where the winter cooling and freezing produce dense surface water that sinks down the continental slope to the seafloor. This water descending along the coast of Antarctica is the densest water found in the open oceans.

At the temperate latitudes in the open ocean the surface water's temperature changes with the seasons. This effect is illustrated in figure 6.12. During the summer the surface water warms and the water column is stable, but in the fall the surface water cools, increasing its density, and the water column becomes unstable. The surface water sinks and overturn begins. Winter storms and winter cooling continue the mixing process. The shallow thermocline formed during the previous summer is lost, and the upper portion of the water column becomes isopycnal to greater depths. Spring brings warming, and the thermocline begins to reestablish itself. The water column becomes stable and remains so through the summer.

Seasonal changes in temperature are more important than salinity changes in determining density in the open ocean. For example, in the Atlantic Ocean the surface water at 30°N and 30°S has a high salinity, but it is warm year-round, so it stays at the surface. The water from 50°–60°N in the North Atlantic has a lower salinity, but it is cold, especially during the winter. This cold water sinks and flows below the saltier but warmer surface water.

Close to shore, the salinity of seawater becomes more important than temperature in controlling density. This is particularly true in semi-enclosed bays, sounds, and fjords that receive large amounts of freshwater runoff. Here extremely cold but fresh water is added as the rivers carry precipitation and ice melt to the sea. This water may be only 0° to 1°C, yet its salinity is so low that the water does

not sink when it meets the salt water but remains at the surface as a seaward-moving **freshwater lid.** In polar regions, when the sea ice melts, a layer of fresh water forms at the surface and slowly dilutes the surface water as it mixes.

Sigma-t

Minute changes in density are responsible for relatively large changes in vertical circulation. For this reason, oceanographers routinely measure density to the fifth decimal place, for example, 1.02677 g/cm^3. Because such a number is awkward to handle when performing calculations and processing large quantities of data, a more convenient term is used. Density is converted to a **sigma-t** (σ_t) value. To derive sigma-t from density, we divide the density by 1.000 g/cm^3 (the density of pure water) to remove the units:

$$\frac{1.02677 \text{ g/cm}^3}{1.00000 \text{ g/cm}^3} = 1.02677$$

We then subtract 1 and multiply by 1000:

$$(1.02677 - 1) \times 1000 = 26.77$$

$$\sigma_t = 26.77$$

Pressure is disregarded, and σ_t becomes a factor related only to salinity and temperature, as indicated by the symbols σ for salinity and t for temperature.

6.4 Upwelling and Downwelling

When dense water from the surface sinks and reaches a level at which it is denser than the water above but less dense than the water below, it spreads horizontally as more

Figure 6.13

Upwelling and downwelling are related to horizontal motions of water at the sea surface. Downwelling occurs where surface currents converge, and upwelling occurs where surface currents diverge.

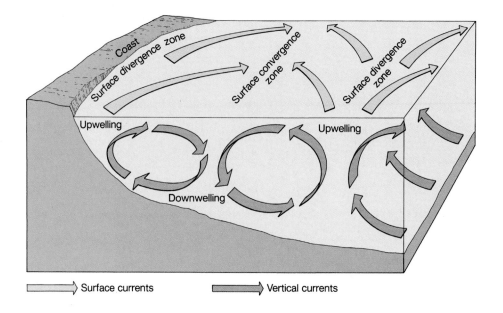

Surface currents   Vertical currents

water descends behind it. At the surface, water moves horizontally into the region where sinking is occurring. The dense water that has descended displaces water upward, completing the cycle. Because water is a fixed quantity in the oceans, it cannot be accumulated or removed at given locations without movement of water between these locations. This concept is called **continuity of flow.** Areas in which water sinks are known as **downwelling zones;** areas of rising waters are **upwelling zones.** Downwelling is a mechanism that transports oxygen-rich surface water to depth, where it is needed for the deep-living animals. Upwelling returns the decay-produced nutrients that have accumulated at depth to the surface waters, where they act as plant fertilizers to promote the production of more oxygen by photosynthesis in the sunlit surface waters.

Upwelling and downwelling are also related to the wind-driven surface currents. When the surface waters are driven together by the wind, a surface **convergence** is formed. Water at a surface convergence sinks, or down-wells. When the wind blows surface waters away from each other or away from a coast, a surface **divergence** occurs. Water from below is upwelled at a divergence (see fig. 6.13). During the slow movement of water from surface to depth and back, the water continually mixes with adjacent layers of water, gradually exchanging chemical and physical properties.

The upwelling and downwelling water moves at rates of 0.1 to 1.5 m (5 ft)/day. Compare this speed to that of oceanic surface currents, which reach speeds of 1.5 m/sec. Horizontal movement at depth due to the thermohaline flow is about 0.01 cm (.004 in)/sec. Water caught in this slow but relentlessly moving cycle may spend 1000 years at depth before it again reaches the surface.

Chemical Tracers

Chemical tracers are substances that occur in small amounts in seawater; they have no effect on water motion but can be used to identify kinds of water as it moves through the ocean system. Transient tracers are also used to identify kinds of water, but these substances change their distribution with time and have usually been added to the oceans by human means. When carefully sampled and analyzed, these transient tracers provide us with the quantitative data used to form computer and mathematical models of water pathways in the oceans.

Atmospheric testing of nuclear weapons in the late 1950s and early 1960s produced radioactive materials that still persist in traces in the oceans. Research carried out by the Geochemical Ocean Sections (GEOSECS) program in 1972–73 showed that water containing radioactive substances from the sea surface had reached the bottom of the North Atlantic, indicating a much faster rate of descent and transport than was previously thought to occur. The 1981 Transit Tracers in the Ocean (TTO) cruises found that small changes had occurred in water properties of the deep North Atlantic water since GEOSECS. This finding was in contrast to earlier research that indicated water properties of the deep ocean remain constant. Current work focuses on **tritium,** a heavy isotope of hydrogen left from these same tests. Since tritium is chemically hydrogen, it is present in the oceans only as part of water molecules and is therefore the perfect tracer for water movement. Tritium is radioactive, making it more easily traceable, and has a half-life of 12.45 years; it decays to an inert isotope of helium. It is possible to compute the time since a water parcel was last at the surface by measuring the tritium-helium isotope concentrations of that water. These age calculations are most accurate for time periods of less than 10 years, and for water movements in shallow areas where there is the least background of naturally occurring helium. Using this method, layers of constant density along which water tends to move and mix have been identified in the North Atlantic.

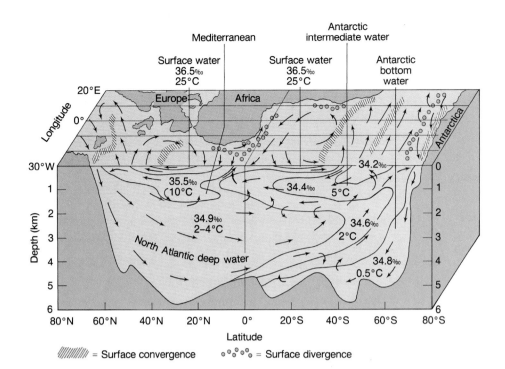

Figure 6.14
The anatomy of the Atlantic Ocean. Surface and subsurface circulation are related here. The surface currents converge to produce downwelling; water sinks to its density level and flows horizontally. Water at depth rises under zones of surface divergence.

6.5   The Layered Oceans

The layered character of each ocean has been determined by taking salinity and temperature measurements over the depth of the water column from many surface positions. Gradually the accumulation of these data allowed oceanographers to identify layers of water that could be traced back to their surface sources. The structure of an ocean is determined by the properties of the layers of water present under the sea surface. Each layer received its characteristic salinity, temperature, and density at the surface. The water's density controls the depth to which the water sinks and forms a layer. The thickness and horizontal extent of each layer is related to the rate at which it is formed at the surface and the size of the surface region over which it is formed. Water that sinks from the surface to spread out at depth eventually rises at another location or slowly mixes with adjacent layers. In all cases, water that sinks to depth displaces an equivalent volume of water upward toward the surface at some other location so that the oceans' vertical circulation may continue.

Structure of the Atlantic Ocean

The properties of the layers of water making up the Atlantic Ocean are shown in figure 6.14. At the surface in the North Atlantic, water from high northern latitudes moves southward, while water from low latitudes moves northward along the coast of North America and then east across the Atlantic. These waters converge in areas of cool temperatures and high precipitation at approximately 50°N to 60°N, at which areas are located the Norwegian Sea and the boundary of the Labrador Current and the Gulf Stream. The resulting mixed water has a salinity of about 34.9‰ and a temperature of 2 to 4°C. This water, known as **North Atlantic deep water**, sinks and moves southward. North Atlantic deep water from the Norwegian Sea moves down the east side of the Atlantic, while water formed at the boundary of the Labrador Current and the Gulf Stream flows along the western side. Above this water at 30°N, a low-density lens of very salty (36.5‰) but very warm (25°C) surface water remains trapped by the circular movement of the major oceanic surface currents. Between this surface water and the North Atlantic deep water lies water of intermediate temperature (10°C) and salinity (35.5‰). This water is a mixture of surface water and the upwelled colder, saltier water from the equatorial regions. It moves northward to reappear at the surface south of the convergence in the North Atlantic.

Near the equator, the upper boundary of the North Atlantic deep water is formed by water produced at the convergence centered about 40°S. This is **Antarctic intermediate water.** Since it is warmer (5°C) and less salty (34.4‰) than the North Atlantic deep water, it is less dense and remains above the more dense and saltier water below. Along the edge of Antarctica very cold (−0.5°C), salty (34.8‰), and dense water is produced at the surface by sea ice formation during the Southern Hemisphere's winter. This is **Antarctic bottom water,** the densest water produced in the world's oceans. After it descends to the ocean floor, it begins to move northward. When it meets North Atlantic deep water it creeps beneath it and continues to move northward along the coast of South America. Antarctic bottom water forms in small quantities and does not accumulate enough thickness to be able to flow over the midocean ridge system and into the basins along the African coast. It is therefore confined to the deep basins on the west side of the South Atlantic and has been found as far north as the equator.

The North Atlantic deep water, meanwhile, trapped between the Antarctic bottom and intermediate waters, rises to the ocean's surface in the area of the 60°S divergence. As it reaches the surface it splits; part moves northward as **South Atlantic surface water** and Antarctic intermediate water; part moves southward toward Antarctica, to be cooled and modified to form Antarctic bottom water. A mixture of North Atlantic deep water and Antarctic bottom water becomes the circumpolar water for the Southern Ocean as it flows around the Antarctica.

Warm (25°C), salty (36.5‰) surface water in the South Atlantic is also caught by the circular current pattern at the surface and is centered about 30°S. Below the southern tips of South America and Africa the water flows eastward, driven by the prevailing westerly winds, which move the water around and around Antarctica.

Because the Atlantic Ocean is a narrow, confined ocean of relatively small volume but great north-south extent, the water types are readily identifiable and their movement can be followed quite easily. Since the bordering nations of the Atlantic have had a long-standing interest in oceanography, the vertical circulation and layering of the Atlantic are both the most studied and the best understood of all the oceans.

Structure of the Pacific Ocean

In the vast Pacific Ocean waters that sink from relatively small areas of surface convergences lose their identity rapidly, making the layers difficult to distinguish. Antarctic bottom water forms in small amounts along the Pacific rim of Antarctica, but it is quickly lost in the great volume of the Pacific Ocean. The deeper water of the South Pacific Ocean is the common mixture of the circumpolar flow. Because the North Pacific is isolated from the Arctic Ocean, only a small amount of water comparable to North Atlantic deep water can be formed. In the extreme western North Pacific, convergence of the southward-flowing cold water from the Bering Sea and the Sea of Okhotsk and the northward-moving water from the equatorial latitudes produces a small volume of water that sinks to mid-depths. There is no dramatic layering as found in the North Atlantic. Warm, salty surface water occurs at subtropic latitudes (30°N and 30°S) in each hemisphere, and Antarctic intermediate water is produced in small quantities, but its influence is small. Deep-water movements in the Pacific are sluggish, and conditions are very uniform below 2000 m (6600 ft).

Structure of the Indian Ocean

Since the Indian Ocean is principally an ocean of the Southern Hemisphere, there is no counterpart of the North Atlantic deep water. Small amounts of Antarctic bottom water are soon lost; the deeper waters tend to be a fairly uniform mixture of Antarctic bottom water and North Atlantic deep water brought into the Indian Ocean by the Antarctic circumpolar current. A counterpart to the Antarctic intermediate water is identifiable, and there is a lens of warm, salty water at the surface.

Mediterranean Sea and Red Sea Water

Two small but specific water types from bordering seas are readily identifiable, one in the North Atlantic and one in the Indian Ocean. The water from the Mediterranean Sea has a temperature of about 13°C and a salinity of 37.3‰ as it leaves the Strait of Gibraltar. This water, mixing with Atlantic Ocean water, creates a water mixture that finds its own density in the North Atlantic at approximately 1000 m (3300 ft) in depth. The influence of Mediterranean water can be traced 2500 km (1500 mi) from the Strait of Gibraltar before it is lost through modification and mixing. In the Indian Ocean very salty (40 to 41‰) water from the Red Sea has been found in a spreading layer at 3000 m (10,000 ft) in depth more than 200 km (124 mi) south of its source.

Comparing the Oceans

Temperature and salinity distributions as a function of depth are shown in figure 6.15 for each of the three oceans. The distributions of Atlantic Ocean salinity and temperature values form patterns with depth (fig. 6.15a and 6.15b) that are clearly identifiable. Salinity values in the Pacific (fig. 6.15c and 6.15d) identify the high-salinity surface water lenses and the mixture of Antarctic bottom water and circumpolar deep water, with a salinity of about 34.7‰ and a temperature of about 1°C. A minor intrusion of low-salinity water in the north is the result of the surface convergence in the far northwest Pacific. Note the large volume of deep water that shows a uniform salinity in this ocean. Temperature values also follow a less definite pattern, emphasizing the uniformity of most of this deep water. The Indian Ocean values in figure 6.15e and 6.15f resemble those for the South Atlantic, without the presence of water that is comparable to North Atlantic deep water.

6.6   T-S Curves

The distribution of salinity and temperature with depth produces distinct patterns in different areas of the world's oceans. If the salinity and temperature values at each observed depth are graphed on a temperature versus salinity diagram, known as a **T-S diagram,** a curve is produced, which is known as a **T-S curve.**

T-S Curves and Water Masses

T-S curves made for large geographic areas of the oceans are similar in shape and fall within narrow zones on a T-S diagram. Figure 6.16 displays specific families of T-S curves which have been given names, and are referred to as **water masses.** Figure 6.17 shows the distribution of these water masses identified by the T-S curves in figure 6.16. Some of the water masses described by these T-S curves have very different characteristics at the surface but are much more uniform with a limited range of salinity and temperature at depth. Where North Atlantic deep water and Antarctic bottom water are formed and sink to depth, the water is not layered, and approximately the same values for

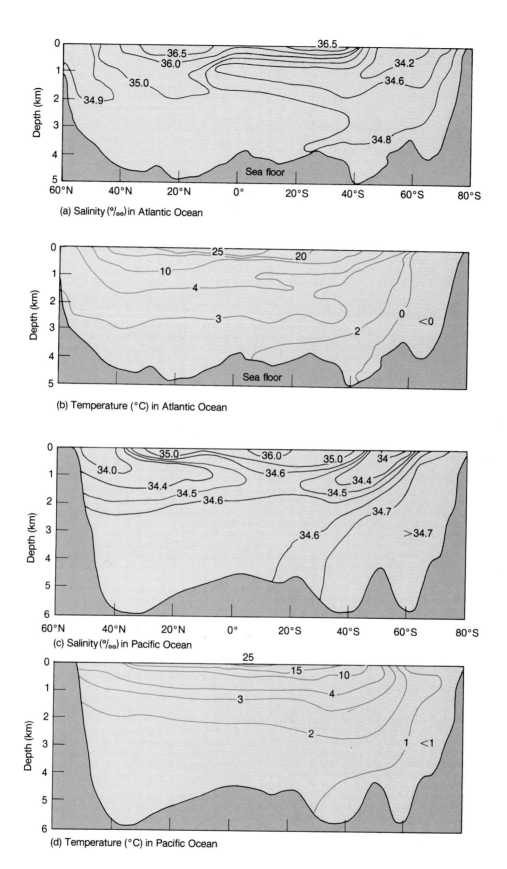

Figure 6.15

Midocean salinity and temperature profiles of the Atlantic, Pacific, and Indian oceans. Temperature and salinity are shown as functions of depth and latitude.

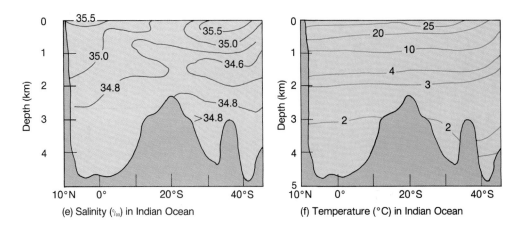

(e) Salinity (‰) in Indian Ocean

(f) Temperature (°C) in Indian Ocean

Figure 6.15

continued

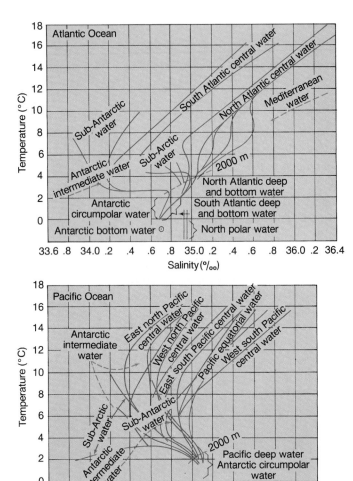

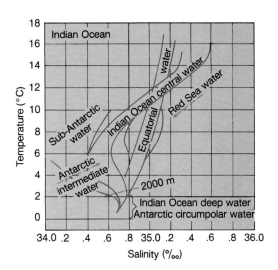

Figure 6.16

The T-S curves that describe the vertical structure of the recognized oceanic water masses.

Sverdrup/Johnson/Fleming, *The Oceans*, © 1942, renewed 1970, p. 741.
Adapted by permission of Prentice-Hall, Inc., Englewood Cliffs, New Jersey.

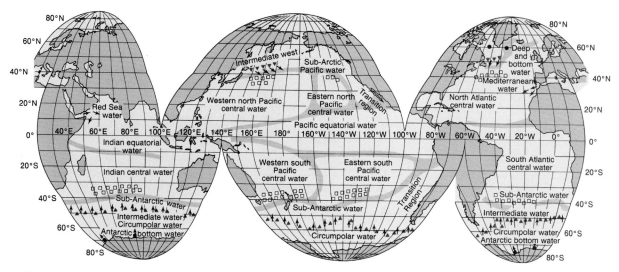

Figure 6.17

The geographic distribution of water masses that have similar vertical salinity and temperature distributions. Open squares mark the region in which the central and shallow waters are formed; triangles indicate the lines along which the intermediate and deep waters sink; solid circles indicate the formation of bottom water.

From Sverdrup/Johnson/Fleming, *The Oceans*, © 1942, renewed 1970, p. 741. Reprinted by permission of Prentice-Hall, Inc., Englewood Cliffs, N.J.

salinity and temperature are found at all depths. A water mass with only one temperature value and one salinity value over its entire extent is known as a **water type.** Antarctic bottom water meets the criteria for a water type, as is shown in figure 6.16.

T-S curves are an oceanographer's diagnostic tools. They are used to find errors in data and to indicate the density stability of the water column at different depths. Remember that water density can also be displayed on a T-S diagram (refer back to fig. 6.8). Oceanographers working with this kind of information become so familiar with these distribution patterns that they are able to inspect a set of temperature-salinity values over a series of depths and know in what area of the world's oceans the measurements were made.

6.7 Measurement Techniques

Determining Water Properties

Collecting water samples for chemical analysis requires specially designed water bottles that are attached to a wire rope or cable known as a **hydrowire** (see fig. 6.18). The spacing between the water bottles and the amount of hydrowire let out is measured by a meter wheel. The water bottles are opened and lowered to the desired depths. A small weight known as a **messenger** is then placed around the hydrowire and released. The messenger slides down the hydrowire at about 200 m (600 ft)/min, and when it strikes the uppermost water bottle, it trips a mechanism that causes the water bottle to close. This action releases another messenger carried on the first water bottle, and it slides down and strikes the next bottle on the wire. This sequence continues, one bottle at a time, until the last bottle has closed. The sequence is shown in figure 6.19.

Figure 6.18

Hanging a water bottle. The cable is measured as it passes over the meter wheel.

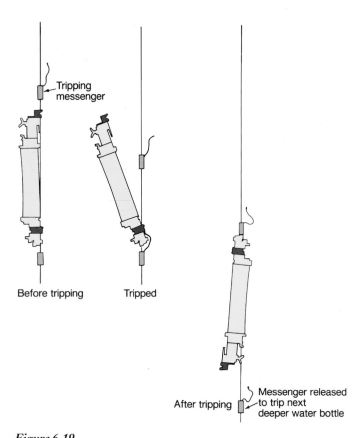

Figure 6.19

The reversing of a messenger-activated water bottle equipped with thermometers, and the release of a second messenger to activate the next deeper bottle.

Figure 6.20

Retrieving a conductivity-temperature-depth sensor, or CTD. The CTD is attached below a rosette of water bottles. Data are relayed to the ship's electronic processing and data center. Water samples are taken when an interesting water structure is found or when water samples are required to calibrate the CTD.

Some water bottles close by turning over, causing their valves to shut; others stay upright but have spring-loaded valves that seal the bottle. Water bottles come in different styles, shapes, and sizes; some are metal and others are plastic. Small-volume samplers of about 1 L (.26 gal) are used if routine chemical analyses are required. If a large water sample is needed the water bottle may be as big as 50 L (13.2 gal). The material that the water bottle is made of becomes important when dealing with trace materials. Metal water bottles are a source of contamination in experiments with trace metals; plastics are a problem if hydrocarbon and organic compounds are being detected.

When all the water bottles have taken their samples, the hydrowire is winched up; the water bottles are removed and placed in a rack on shipboard. The water samples are carefully drained off into storage bottles and are taken for analysis to a chemical laboratory on the ship or on land.

The process of attaching water bottles to a wire, lowering them, tripping them at the desired depth, and then retrieving them is known as taking a **hydrocast.** It is a complicated process and can be a cold, wet, slow, and difficult task at sea under adverse weather conditions. The effort

yields data from specific depths in the sea rather than continuously recorded data from instrument sensors that are lowered through the water column.

Electronic instruments known as conductivity-temperature-depth sensors, or CTDs, (see fig. 6.20), record salinity by directly measuring the electrical conductivity of the seawater. The conductivity and temperature are monitored simultaneously, and the data are returned to the ship as an electronic signal through an insulated wire in the cable that suspends the instrument in the water. On board ship the salinity-depth data signal may be fed directly into the computer, recorded on a chart, or made available as numerical data. Large amounts of data can be acquired in this way, whereas water bottles collect only one sample at each depth for later analysis. Sensors can also be left in place, suspended from a surface vessel or buoy, and the collected data can be stored on magnetic tape in the sensors for retrieval at another time. Although large quantities of data are collected easily and quickly in this way, these instruments do not measure as many water properties as can be measured by the laboratory analysis of collected samples.

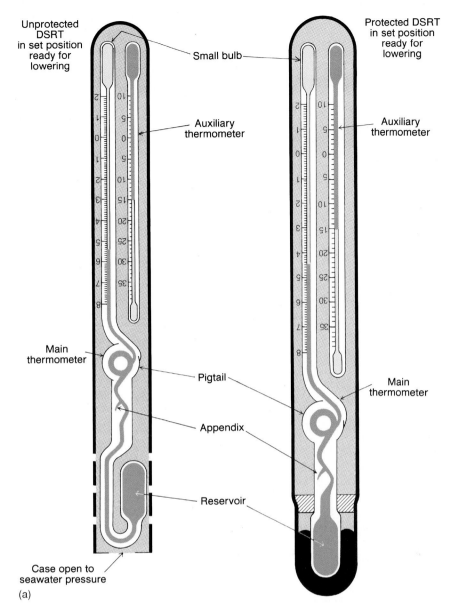

Figure 6.21

(a) Deep-sea reversing thermometers (DSRTs) in set position as they are lowered into the sea.

Temperature Measurements

Water temperatures are obtained by using mercury thermometers. These thermometers are attached in groups of two or three to water-sampling bottles, and together they are lowered into the sea. When the thermometer has come to equilibrium with the temperature at the sample depth, the thermometer must fix this temperature so that the reading will not change as the thermometer passes through layers of different temperatures on its way to the surface.

The **deep-sea reversing thermometer** (DSRT) is a specially designed thermometer that is allowed approximately five minutes to adjust to the water temperature at the desired sampling depth. When a messenger is used to trip the water bottle, the thermometers are reversed (or turned upside down) as the valves of the bottles close. This type of thermometer is specially constructed so that the act of turning it over locks in the reading of the temperature at that depth by isolating the mercury that registers the temperature from the mercury reservoir. When the thermometer is returned to the ship, the seawater temperature indicated by the isolated mercury is read, and any small expansion of this mercury that has occurred on shipboard is corrected by using the difference between the indicated seawater temperature and the reading of an auxiliary thermometer. The auxiliary thermometer gives the shipboard temperature of the thermometer case at the time of reading. Figure 6.21 illustrates the DSRT before and after reversal.

Reversing thermometers are usually used in pairs so that their readings can be compared. Their accuracy is about ±0.01°C. The thermometers are specifically referred

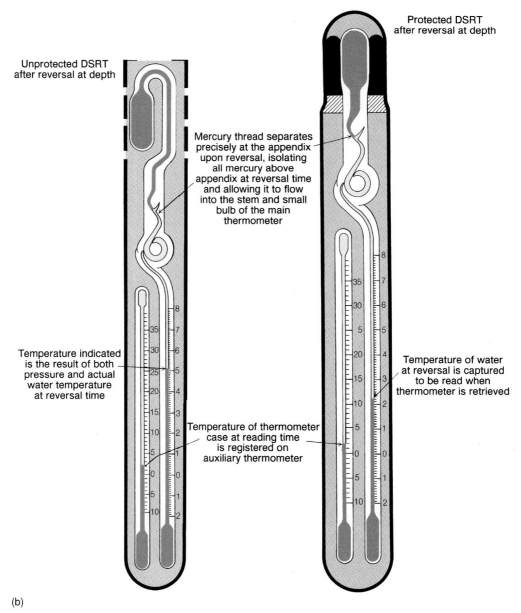

Unprotected DSRT
after reversal at depth

Protected DSRT
after reversal at depth

Mercury thread separates
precisely at the appendix
upon reversal, isolating
all mercury above
appendix at reversal time
and allowing it to flow
into the stem and small
bulb of the main
thermometer

Temperature indicated
is the result of both
pressure and actual
water temperature
at reversal time

Temperature of thermometer
case at reading time
is registered on
auxiliary thermometer

Temperature of water
at reversal is captured
to be read when
thermometer is retrieved

(b)

Figure 6.21 continued

(b) Deep-sea reversing thermometers in reversed position, with the water temperature recorded.

to as protected when they are totally enclosed by a second glass jacket or covering. Other reversing thermometers are unprotected; their outside glass jackets are open, so that the seawater exerts pressure directly on the thermometers. The water pressure pushes more mercury up the thermometer tube, about 1°C for each 100 m (330 ft) of depth, resulting in a higher reading. When an unprotected thermometer is used with a pair of protected thermometers, its elevated reading allows the researchers to calculate the depth at which the temperatures were taken and to verify that the protected thermometers reversed and recorded the temperature at the desired depth as determined by cable length. Today's DSRTs cost about $900 each and must be frequently calibrated to assure their accuracy.

To use CTDs, water bottles, and deep-sea reversing thermometers, the ship must be halted, and readings are taken one sampling effort at a time. Electronic temperature sensors respond very rapidly to temperature changes with an accuracy that is comparable to that of deep-sea reversing thermometers. They can be used to record fluctuations in temperatures over time at a singe location, or they can be lowered through the water, giving a continuous temperature reading with depth. The temperature readings may be recorded internally within the instrument, or they may be sent up the suspending wire to the ship, where they may be fed directly to a computer or used to produce a graph of temperature with depth. The internally recorded data are recovered when the instrument is brought back on board.

New Ways to Measure the Oceans

A new method that is increasing our knowledge of the ocean's structure is known as acoustic tomography. This technique uses sound waves to make three-dimensional images of seawater properties and their fluctuations with time through ocean volumes of 1000 cubic kilometers (158 nmi³) or more. Conventional techniques using ships or instruments moored to the seafloor would be enormously expensive and would require hundreds of instruments deployed by a dozen ships to monitor such large areas of the ocean. Satellites cannot be used for this work because they do not monitor the ocean's interior.

Acoustic tomography uses sound transmitters and sound receivers moored over both area and depth; see Box figure 6.1. This instrument array provides large amounts of data as each sound signal moves along a different path and through a different part of the water column. The amount of data can be increased by using transceivers in place of either transmitters or receivers; transceivers both transmit and receive sound.

Sound speed and refraction are controlled by the salinity, temperature, and pressure of the seawater; refer back to chapter 4. In acoustic tomography a sound pulse from any single transmitter is recorded by every receiver. The distances between the transmitter and the receivers

are known; the time required for the sound pulse to travel between each transmitter and receiver is measured, so the speed with which each sound pulse travels is calculated. The speed of travel is controlled by the properties of the water through which it passed. This process is repeated many times along different pathways. The data is then converted by computer into three-dimensional maps of the ocean's temperature with time in the same way that geologists construct maps of the earth's interior from seismic wave data (see seismic tomography, discussed in chapter 3) or that physicians use CAT scans to form three-dimensional images of the body's internal organs.

In addition the pitch or frequency of a sound pulse is altered by the currents. If a sound pulse of a given frequency is traveling through the water in the same direction that the water is moving, the pitch of the sound pulse increases. If the water motion opposes the sound path the pitch decreases. In this way water movements, including ocean currents, can be detected and mapped.

Acoustic tomography presents the researcher with a versatile system adaptable to many situations. The researcher controls the size of the region investigated and the frequency of the measurements; there is the opportunity to study changes with space and time on both the

large and the small scale. This system allows data to be collected while the experiment is under way, or the instruments could record and store data for future analysis or even transmit their data directly to surface vessels or to laboratories via satellite.

Freely drifting surface buoys with suspended sensors are placed in the ocean to move with the currents. These buoys make measurements of water properties that are then broadcast to passing satellites for relay to land stations. The satellites also determine the changing position of the buoy and relay this information as well. Other instruments are adjusted to drift with the currents at a depth at which their density matches the density of the seawater. These devices gather data and periodically alter their buoyancy to rise to the surface and broadcast their stored information to passing satellites. They then sink again and repeat the process.

The ability to gather information about the oceans is increasing as technology advances. No longer is the study of the oceans limited to data collected from surface vessels using water bottles and reversing thermometers. The toolbox of the oceanographer is increasing; the areas that can be sampled are increasing, and the data is more accurate and more definitive.

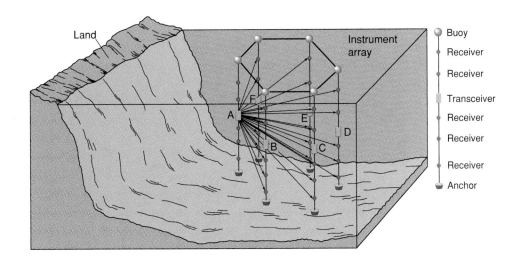

Box figure 6.1

Buoys carrying strings of sound sources and receivers are anchored surrounding a volume of water. Sound pulses from A are picked up by receivers on the other strings, and sound pulses are sent out in turn from the other sound devices (B–F). Changes in the speed of sound describe the properties of the water enclosed by the instrument array.

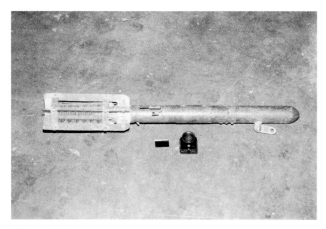

Figure 6.22
The mechanical bathythermograph, or BT, records water temperature with depth on a plated slide. The slide (lower left) is removed from the BT and placed in a reader (lower right).

Figure 6.23
An expendable BT (XBT). The probe (right) is launched from its sleeve (left). The probe sinks at a constant rate and thin copper wire transmits temperature data to the ship.

Electronic thermometers can record five or more readings each second. The reversing thermometer obtains one reading for the hour or so necessary to lower it to the desired depth and then retrieve it.

The mechanical **bathythermograph, or BT,** was the first instrument developed to rapidly measure temperature changes with depth down to 80 m (260 ft) while a ship was under way (see fig. 6.22). This torpedo-shaped device was used by the navy during World War II to detect the effect of temperature on sound waves. The BT traces a temperature versus depth graph on a small glass slide. When the BT is retrieved, the trace is read directly against a grid of temperature versus depth. The mechanical BT has been replaced by the **expendable BT, or XBT** (see fig. 6.23). The XBT is a small bomb-shaped device that is dropped into the water and sinks at a known rate. A wire attached to a recorder on the vessel unrolls from the XBT and transmits the temperature as an electric signal directly to the shipboard recorder. The depth is calculated from the elapsed time. The XBT can be dropped from vessels that are stationary or moving at speeds up to 30 knots. It gives the quick temperature-depth profiles to 500 m (1640 ft) that are needed for sonar work (see chapter 4). The XBT, as indicated by its name, is not recovered.

Although the new electronic instruments increase the precision and speed of the measurements as well as the number of measurements that can be made, they must still be calibrated against control data collected by the old, reliable water bottles and reversing thermometers.

The choice of instruments used to gather data is determined by the cost of the instruments; the time required to take a sample; the need for detail, accuracy, and speed; the personal preference of the researcher; and the conditions under which the sampling is done. On any research cruise, data must be obtained under the conditions present at the

sampling station on arrival. Despite darkness, wind, waves, heat, or cold, data are taken, for there may not be a second chance. In the laboratory, the researcher controls the experiment and varies individual conditions to see the result of each. At sea, the oceanographer has no control over the environment but tries to measure and interpret a system in which all of the conditions change simultaneously and continually. Nature is in control.

6.8 Practical Considerations: Energy Sources

Temperature and salinity variations with depth represent a potential source of energy and an alternative to the consumption of fossil fuels. Since it is solar radiation that is largely responsible for the temperature and salinity distribution, this energy represents an indirect form of solar energy. Because of the transparency and heat capacity of water, large amounts of solar energy can be stored in the ocean. Therefore, the heat can be extracted independent of daily and seasonal changes in the available solar radiation. Several techniques, some more practical than others, are available to extract this energy.

Osmotic Pressure

Power can be produced by exploiting the salt concentration difference between two bodies of water. This technique requires two sources of water with different salt contents (such as river water and seawater) and a semipermeable membrane that permits the water molecules but not salt molecules to move across the membrane from one side to the other. The water flows from the side on which the water is more concentrated (the river water side) to the side on which the water is less concentrated (the saltwater side). If

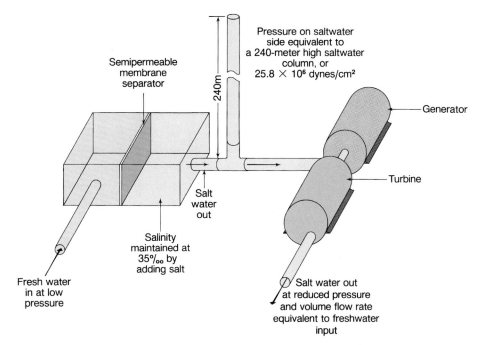

Figure 6.24

The movement of fresh water into salt water through a semipermeable membrane generates pressure on the saltwater side to power a turbine and generator.

the system is to work efficiently, the membrane must have (1) a large surface area to increase the volume of flow and (2) sufficient mechanical strength to withstand the pressure difference between the two sides. Because the movement of water molecules across the membrane into the salt water dilutes the salt water and slows down the process, salt must be added to the saltwater side to maintain the rate of flow.

In a system in which the salt water has the same salt content and temperature as the open ocean and the other water source contains no salt, the fresh water moves through the membrane until the pressure on the saltwater side is equivalent to a column of water 240 m (780 ft) in height. This pressure difference could be used to produce power by passing the water under pressure through a turbine. The pressure that is formed by water moving through a semipermeable membrane is called **osmotic pressure** (see fig. 6.24 and also the discussion at the end of chapter 5).

Ocean Thermal Energy Conversion

An alternative method of energy extraction depends on the difference in temperature between ocean surface water and water at 600–1000-m (2000–3300-ft) depth. This method of power production is known as **OTEC, or ocean thermal energy conversion** (fig. 6.25). There are two different types of OTEC systems: (1) closed cycle, which uses a contained working fluid with a low boiling point, such as ammonia or Freon; and (2) open cycle, which directly converts seawater to steam.

In a closed system (fig. 6.25b), the warm surface water is passed over the evaporator chamber containing the ammonia or Freon, and the ammonia or Freon is vaporized by the

heat derived from the warm seawater. The vapor builds up pressure in a closed system, and this gas under pressure is used to spin a turbine, which generates power. After the pressure has been released, the ammonia or Freon is passed to a condenser, where it is cooled by cold water pumped up from depth. Cooling the ammonia or Freon returns it to its liquid state, and it is pumped as a liquid back to the evaporator to repeat the cycle.

In an open system (fig. 6.25c), large quantities of warm seawater are converted to steam in a low-pressure vacuum chamber, and the steam is used as the working fluid. Since less than 0.5% of the incoming water is turned into steam, large quantities of warm water must be used. The steam passes through a turbine and is condensed by using cold water from depth to cool the condenser. This turns the steam into desalinated water.

OTEC plants using either system can be located onshore, offshore, or on a ship that moves from place to place. However, OTEC requires at least a 20°C difference in temperature between surface and depth to generate useful amounts of energy. This means that these power plants would be located at latitudes between approximately 25°N and 25°S.

The Natural Energy Laboratory of Hawaii at Keahole Point is building a land-based, open-system OTEC plant. This pilot project began producing power in 1992; under 1993 test conditions it produced a net power output of 50 kilowatts with a gross power production of 190 kilowatts. It is the world's only system pumping cold water from depth at this time. The water is pumped through a polyethylene pipe, 1600 m (1 mi) long and 1 m (40 in) in diameter, from

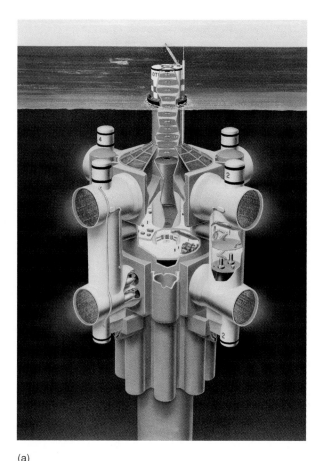

(a)

Figure 6.25

(a) The proposed Lockheed OTEC Spar, designed for generating electrical energy from the ocean temperature gradient. (b) The simplified working system of an OTEC closed-system electrical generator. (c) The simplified working system of an OTEC open-system electrical generator.

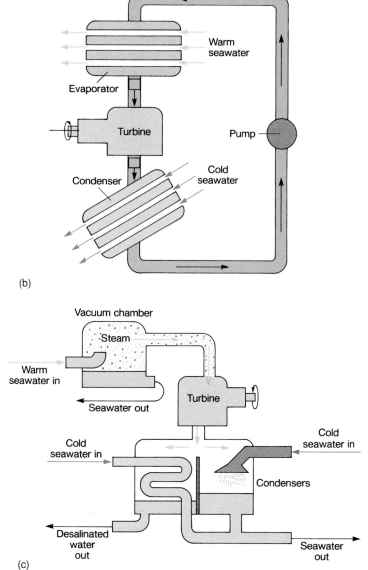

(b)

(c)

850-m depth (2800-ft); the warm water is extracted by a shorter pipe from 25 m (80 ft) below the surface. The cost of the plant is high, $30 million, about ten times the cost of a conventional plant burning coal or oil; however, OTEC fuel is free. Maintenance costs are estimated at about the same as or less than those for a fossil-fuel plant and much less than those for a nuclear plant. The OTEC operation also creates environmental changes that should be considered because heat is removed from the sea and large volumes of cold, nutrient-rich water are pumped to the surface in tropical waters.

Taiwan, facing serious problems in supplying sufficient energy to maintain its growing economy as well as increasing public pressure against nuclear and hydroelectric projects, is exploring a land-based OTEC plant. Along Taiwan's east coast, there is near-shore deep water with a year-round temperature difference of 20°–22° C between surface and depth.

Research on closed-cycle plants is currently focusing on improving heat exchangers and reducing the problems of condenser corrosion and fouling by marine organisms, which settle and grow on the condenser surfaces. The open system is not as fully developed. Problems include the need for larger turbines for use with the low-density steam, large vacuum chambers, and the large flow rates of water needed (2–4 m³/sec [500–1000 gal/sec]). Another difficulty occurs when gases escape from the heated seawater and large quantities of water low in dissolved oxygen are released into the local environment, which might cause problems for the organisms present.

However, there are significant advantages associated with an open system. These include using only seawater as the working fluid so that no contamination of the water is possible; direct contact between heat exchangers and the seawater, which is both more efficient and less expensive;

the ability to use plastic heat exchangers, which cuts down on fouling and corrosion; and, especially, the desalinated water which can be a valuable by-product of the system.

Because of the costs, the only practical competitive use of power generated by OTEC at the present time is on ocean islands, where fuel oil for power generation must be imported. If open systems are used, plants could produce revenue by selling the fresh water, which might make them competitive in areas lacking both power and fresh water, for example the Caribbean. Other uses for the cold water include refrigeration and air-conditioning. Because the cold water is pumped from 1000 m (3300 ft) deep, it contains nutrients that can be used as an artificial upwelling to promote fish and seaweed farming. The French consider the potential for fish-farming, fresh water, and refrigeration to be more important than the net power production of their Tahiti plant. Researchers in Hawaii also believe that fish and seaweed farming are the greatest incentives for commercial development of OTEC plants in their region.

If, in the future, closed-system, floating OTEC installations capable of producing electric power of 100 to 200 megawatts are considered, the large units will require heat exchangers with many square miles of surface and huge pipes (30 m [100 ft] in diameter) to bring very large volumes of cold, deep water up to the near-surface power unit. Constructing, installing, and maintaining a pipe to 1000-m (3300-ft) depth and miles of thin-walled heat exchange coils present considerable problems.

The power produced at these locations must then be transferred to populated areas with their high energy needs. At present, there are no electric cables designed to carry underwater the power produced by large, non-land-based systems. To get around the necessity of such cables, it is considered possible that the electric power produced by a floating OTEC plant might be used to produce hydrogen from seawater or to produce ammonia gas from the atmosphere. These gases could then be transported by tankers for shore conversion to usable power.

Of the two energy-extraction methods explained in this section, OTEC has produced sufficient energy to prove its practical potential, but, as discussed, many problems remain to be overcome before it can produce power at rates that are comparable to those of hydroelectric, fossil fuel, or nuclear power plants. However, it is likely that as the world's ability to produce energy from traditional sources diminishes, the costs associated with an OTEC plant will eventually become acceptable, and the technical and engineering problems will be researched and solved. In this century, though, energy conservation will save more power than the systems described here will be able to produce.

Summary

The intensity of solar radiation over the earth's surface varies with latitude. Incoming radiation and outgoing radiation are equal when averaged over the whole earth. Reflection, reradiation, evaporation, conduction, and absorption keep the heat budget in balance with the incoming solar radiation. In any local area of the oceans there is daily and seasonal variation. Winds and ocean currents move heat from one ocean area to another to maintain the surface temperature patterns.

The oceans gain and lose large quantities of heat, but their temperature changes very little. The heat capacity of the oceans is high compared to that of the land and the atmosphere. Heat absorbed at the surface mixes downward to reduce the temperature change at the surface.

The salinity of surface water reflects the pattern of evaporation and precipitation with latitude.

Cooling, evaporation, and freezing increase the density of the sea-surface water. Heating, precipitation, and ice melt decrease its density. The surface water changes its density with changes in salinity and temperature that are keyed to latitude. The densities at depth are more homogeneous. Thermoclines and haloclines form where the temperature and salt concentrations change rapidly with depth. If the density increases with depth, the water column is stable; unstable water columns overturn and return to a stable distribution. Neutrally stable water columns are mixed vertically by winds and waves.

Vertical circulation that is driven by changes in surface density is known as thermohaline circulation. In the open ocean, temperature is more important than salinity in determining the surface density. Salinity is the more important factor in areas that are close to shore and to land runoff. Sigma-t is a convenient way of representing density.

Water sinks at downwellings and rises at upwellings. A downwelling occurs at the convergence of surface currents and transfers oxygen to depth. Upwellings bring nutrients to the surface and occur at zones of surface current divergence. Chemical tracers are also used.

The oceans are layered systems. The layers (or water types) are identified by specific ranges of temperature and salinity. The water types of the Atlantic Ocean are formed at the surface at different latitudes. They sink and flow northward or southward. The water types of the Pacific Ocean lose their identity in the large volume of this ocean; their movements are sluggish. The water types of the Indian Ocean are less distinct than those of the Atlantic, with no water types that correspond to those formed in the northern latitudes. Mediterranean Sea and Red Sea waters enter the Atlantic and Indian oceans at depth as discrete water types that can be tracked for long distances.

Water masses occupy distinct regions of the oceans and have specific patterns of salinity and temperature from surface to depth. Characteristic values of temperature and salinity for these water masses are identifiable when plotted on T-S diagrams.

Water samples are taken with water bottles that turn over and close at the sampling depth. Analysis of these water samples is done in the laboratory. Direct-reading instruments to determine salinity and temperature in the oceans are also available. The temperature of the water at the depth of a water-bottle sampling station is taken with a deep-sea reversing thermometer. Electronic thermometers and bathythermographs can measure temperature while the vessel is moving.

Temperature and salinity variations in the oceans represent possible sources of energy. Examples are osmotic pressure devices and ocean thermal energy conversion (OTEC).

Key Terms

solar constant	unstable water column
heat budget	overturn
reradiation	isopycnal
density	isothermal
pycnocline	isohaline
thermocline	vertical circulation
halocline	thermohaline circulation
stable water column	freshwater lid
continuity of flow	sigma-t
downwelling zones	water type
upwelling zones	hydrowire
convergence	messenger
divergence	hydrocast
tritium	deep-sea reversing
North Atlantic deep water	thermometer (DSRT)
Antarctic intermediate water	bathythermograph (BT)
Antarctic bottom water	expendable BT (XBT)
South Atlantic surface water	osmotic pressure
T-S diagram	ocean thermal energy
T-S curve	conversion (OTEC)
water mass	

Study Questions

1. Write an equation for the whole earth's heat budget. Include all the factors for incoming and outgoing energy.
2. What natural processes alter the surface salinity of the oceans? How do these processes work at different latitudes?
3. Describe the changes in water density in the upper ocean layer over the annual cycles at tropic, polar, and temperate latitudes. Indicate when periods of stable and unstable conditions exist in the upper water column.
4. Why are water layers more prominent in the Atlantic Ocean than in the Pacific Ocean?
5. Discuss the use of ocean thermal gradients for the production of energy. Would there be environmental side effects?
6. How can heat be transferred from place to place in the oceans? Why is this heat transfer ignored when considering the world's total heat budget?
7. If the upper layers of an ocean area are homogeneous in salinity, explain why the thermocline coincides with the pycnocline.

8. How does Mediterranean Sea water alter the salinity and temperature of the North Atlantic? Use figure 6.14 and figure 6.15a and b. At what depth does this change occur?
9. A water sample taken at 4000 m in the Atlantic Ocean has a salinity of 34.8‰ and a temperature of 3°C. At approximately what latitude was this water last at the surface?
10. The following data were taken from a sampling station located at 79°N, 145°W:

Depth (meters)	Temp. (°C)	Salinity (‰)	Sigma-t
0	1.28	33.29	26.59
50	1.29	33.30	26.59
100	1.36	33.35	26.69
150	1.39	33.55	26.94
200	2.73	33.76	27.01
300	3.07	33.87	27.08
400	3.12	34.03	27.13
500	3.14	34.13	27.21

   a. In what ocean region is this station?
   b. Is the water column stable or unstable?
   c. Does the temperature or the salt content control the density?
   d. How deep is the wind-mixed layer?
   e. At what time of year were these data obtained?
Explain your answers.

11. Why is the surface of the earth not heated equally by the sun's radiation?
12. At the summer solstice, why is the daily average solar radiation at the North Pole greater than that at the equator?
13. Where do the deeper waters in the bottom of the Atlantic Ocean have their origins?
14. Distinguish between the terms in each pair: (a) halocline—thermocline, (b) upwelling—downwelling, (c) water mass—water type, (d) protected thermometer—unprotected thermometer.
15. Why is convective overturn in the sea's upper layers more likely to occur at temperate and high latitudes than at low latitudes?

Suggested Readings

Surface Processes and Ocean Structure

Bryan, K. 1978. The Ocean Heat Balance. *Oceanus* 21 (4): 19–26.
Jenkins, W. J. 1992. Tracers in Oceanography. *Oceanus* 35 (1): 47–55.
Neumann, G., and W. J. Pierson. 1966. *Principles of Physical Oceanography.* Prentice-Hall, Englewood Cliffs, N.J. 545 pp.
Pickard, G. L., and W. J. Emery. 1982. *Descriptive Physical Oceanography, An Introduction.* 4th (SI) ed. Pergamon Press, New York. 249 pp.
Spindel, R. C., and P. F. Worcester. 1990. Ocean Acoustic Tomography. *Scientific American* 263 (4): 94–99.
Von Arx, W. S. 1962. *An Introduction to Physical Oceanography.* Addison-Wesley, Reading, Mass. 422 pp.

Williams, J., J. J. Higginson, and J. P. Rohrbough. 1968. Oceanic Water Masses and Their Circulation. In *Oceanography, Contemporary Readings in Ocean Sciences,* 2d ed., Pirie, R. G., ed. (1977). Oxford Univ., New York. pp. 24–33.

Energy

Brin, A. 1981. *Energy and the Oceans.* Westbury House, Surrey, England. 133 pp.

Bruce, M. 1986. Ocean Energy: Some Perspectives on Economic Viability. In *Ocean Yearbook 6,* Borgese, E. M., and N. Ginsburg, eds. Univ. of Chicago, Chicago. pp. 58–78.

Charlier, R. H. 1983. Water, Energy, and Nonliving Ocean Resources. In *Ocean Yearbook 4,* Borgese, E. M., and N. Ginsburg, eds. Univ. of Chicago, Chicago. pp. 75–120.

Isaacs, J. D., and W. R. Schmitt. 1980. Ocean Energy: Forms and Prospects. *Science* 207 (4428): 265–73.

Lavi, A., ed. 1980. *Ocean Thermal Energy Conversion.* Pergamon Press, New York. 569 pp.

Loupe, D. 1991. The Food Factor. *Sea Frontiers* 37 (2): 22–27.

Penny, T. R., and D. Bharathan. 1987. Power from the Sea. *Scientific American* 256 (1): 86–92. (Review of ocean thermal energy conversion [OTEC].)

Simeons, C. 1980. *Hydro-Power, The Use of Water As an Alternative Source of Energy.* Pergamon, Elmsford, N.Y. 549 pp.

7

The Ocean and the Atmosphere

7.1   The Atmosphere
Structure of the Atmosphere
Composition of Air
Atmospheric Pressure
7.2   Atmospheric Gases of Global Concern
Changing Levels of Carbon Dioxide
The Ozone Problem
7.3   The Role of Sulfur Compounds
7.4   The Atmosphere in Motion

Box: Clouds and Climate

Winds on a Nonrotating Earth
The Effects of Rotation
Wind Bands
7.5   Modifying the Wind Bands
Seasonal Changes
The Monsoon Effect
The Topographic Effect
The Jet Streams
7.6   Hurricanes
7.7   El Niño
**7.8   Practical Considerations: Storm Tides and Storm
       Surges**

Box: A North Atlantic Cool Pool

Summary
Key Terms
Study Questions
Suggested Readings

Aeolus entertained me for a whole month asking me questions all the time about Troy, the Argive fleet, and the return of the Achaeans. I told him exactly how everything had happened, and when I said I must go, and asked him to further me on my way, he made no sort of difficulty, but set about doing so at once. Moreover, he flayed me a prime oxhide to hold the ways of the roaring winds, which he shut up in the hide as in a sack—for Zeus had made him captain over the winds, and he could stir or still each one of them according to his own pleasure. He put the sack in the ship, and bound the mouth so tightly with a silver thread that not even a breath of a side-wind could blow from any quarter. The west wind which was fair for us did he alone let blow as it chose; but it all came to nothing. . . . They (the crew) loosed the sack, whereupon the winds flew howling forth and raised a storm that carried us weeping out to sea and away from our own country.

Homer,
The Odyssey, Book X

Clouds over the coast, Point Reyes, California.

*T*he sun's energy reaches the earth's surface through the atmosphere, a thin shell of mixed gases we call air. The atmosphere and the ocean are in contact over 71% of the earth's surface; their interaction is continuous and dynamic. Processes that occur in the atmosphere are closely related to processes that occur in the oceans, and together they form much of what we call weather and climate. Clouds, winds, storms, rain, and fog are all the result of interplay between the sun's energy, the atmosphere, and the oceans. This complex of interactions provides the earth's average climate and its daily weather, sometimes pleasant and stable, at other times severe and turbulent. Some of these interactions and processes are more predictable than others; some are better understood than others. Understanding the oceans requires an understanding of the atmosphere's influence on them. This chapter presents an overview of these self-adjusting relationships as well as specific examples of their combined effects.

7.1   The Atmosphere

Structure of the Atmosphere

The atmosphere is a nearly homogeneous mixture of gases extending 90 km (54 mi) above the earth. Ninety-nine percent of the mass of atmospheric gases is contained in a layer extending upward 30 km (18 mi), and 90% is within a layer extending only 15 km (9 mi) above the earth's surface. The first layer of atmosphere above the earth is the **troposphere;** here the temperature decreases with altitude, and within 12 km (7 mi) the temperature decreases from a mean earth surface value of 16°C to –60°C. The tropopause marks the minimum temperature zone between the troposphere and the layer above it, the **stratosphere.** In the stratosphere the temperature increases with increasing elevation until the stratopause is reached at 50 km (30 mi). See figure 7.1.

The troposphere is warmed from below by heat reradiated from the earth's surface by conduction, and by the evaporation of water vapor and its condensation in the atmosphere. Refer back to the discussion of the heat budget in chapter 6. Precipitation, evaporation, convective circulation, wind systems, and clouds are all found within the troposphere. **Ozone,** an unstable form of oxygen, occurs in the stratosphere. Each ozone molecule, O_3, is made up of three molecules of oxygen instead of two, as found in the oxygen, O_2, at the earth's surface. Ozone absorbs ultraviolet radiation from sunlight and therefore raises the temperature of the stratosphere. By absorbing ultraviolet radiation, the ozone lowers the incidence of ultraviolet light at the earth's surface, protecting it from the high-intensity ultraviolet radiation that is harmful to living organisms.

Table 7.1

The Composition of Dry Air

Gas	Percentage of mixture by volume
Nitrogen	78.08
Oxygen	20.95
Argon	0.93
Carbon dioxide	0.03
Neon	1.8×10^{-3}
Helium	5×10^{-4}
Krypton	1×10^{-4}
Xenon	1×10^{-5}
Hydrogen	$< 1 \times 10^{-5}$
	~100%

At elevations extending higher than 50 km (30 mi) there is little absorption of solar radiation, so the temperature again decreases with height. This layer is the **mesosphere.** Here the number of molecules of gas per cubic centimeter is reduced by one thousand, and the pressure from the remaining atmosphere is only 1/1000th of the pressure at the earth's surface. The mesosphere extends upward to 90 km (54 mi), and above it is the **thermosphere,** extending out into space. See again figure 7.1

The emphasis of this chapter is on the troposphere; within this layer heat and water move between the earth and the atmosphere, causing the motions that produce the winds, the weather, and the ocean's waves and surface currents. The tropopause is also of interest, because this is the region of the high-altitude winds called jet streams; these winds play an important role in air and water motions at the earth's surface.

Composition of Air

Air is a mixture of gases; see table 7.1 for the gases that make up dry air. Air usually contains about 1.4% water vapor, a small amount when compared to 21% oxygen, but important in controlling the density of air. Water vapor has a density that is less than the density of dry air; when water vapor is present, the total percent of the other gases is reduced, and at a constant temperature and pressure air saturated with water vapor is less dense than dry air. Air becomes less dense when it is warmed, when its **atmospheric pressure** is decreased, and when its water vapor content is increased. Air becomes more dense when it is cooled, when its atmospheric pressure is increased, and when its water vapor content is decreased. Changes in the density of air allow air to move vertically and cause atmospheric convective motion.

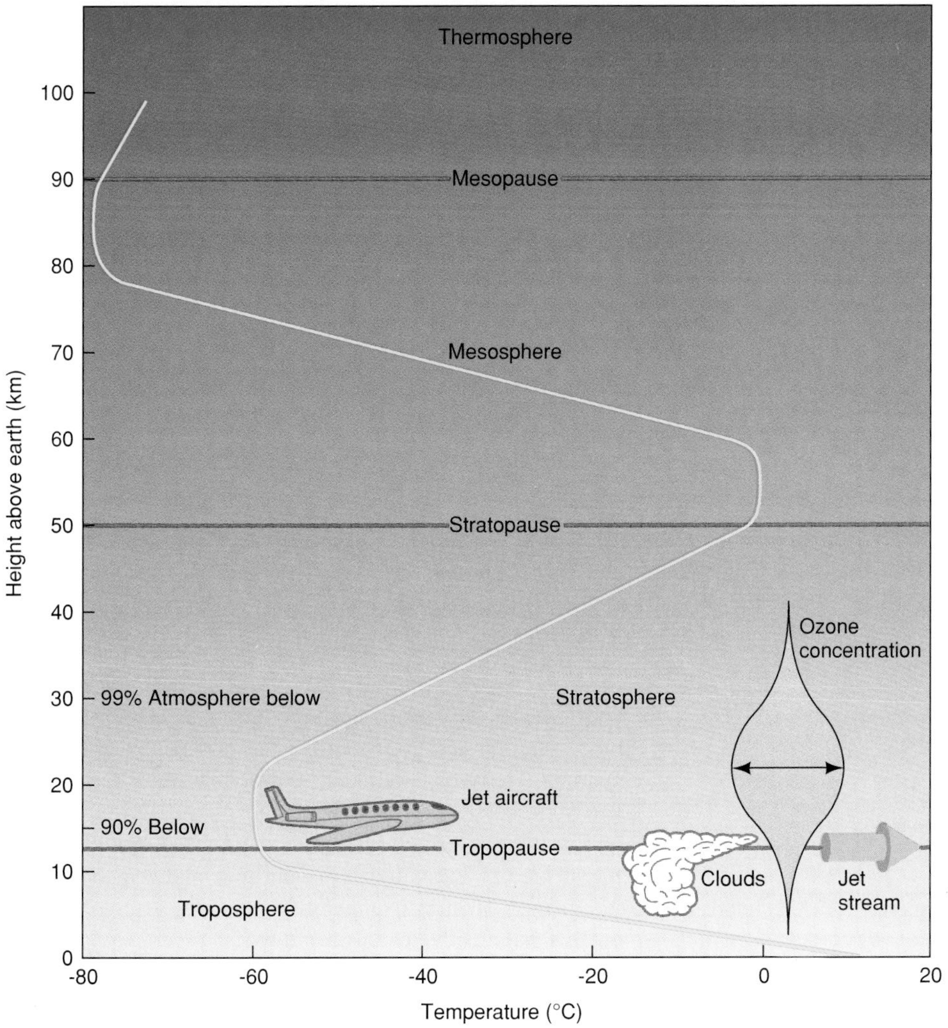

Figure 7.1
The structure of the atmosphere.

Atmospheric Pressure

Atmospheric pressure is the force with which a column of overlying air presses on an area of the earth's surface. The average atmospheric pressure at sea level is 1013.25 millibars (1 bar = 1×10^6 dynes/cm^2), or 14.7 lb/in^2. This standard atmospheric pressure is also equal to the pressure produced by a column of mercury standing 760 mm (29.92 in) high. Barometers may measure atmospheric pressure in millibars or in millimeters or inches of mercury. Pressure can also be recorded in torrs, where 1 torr equals the pressure of a column of mercury 1 millimeter high. Atmospheric pressure distribution is shown on weather charts by lines of constant pressure or **isobars;** see figure 7.2. Where the density of the air is less than average, atmospheric pressure is below average, and a **low-pressure zone** is formed. Regions of air with a density greater than average are known as **high-pressure zones.**

<hr>

7.2   Atmospheric Gases of Global Concern

The atmosphere exerts primary control over the earth's climate, and during the last forty years it has become increasingly evident that there is a change in the balance of the gases that form the earth's atmosphere. Human activities appear to be causing these changes, and there is particular concern about changes in levels of carbon dioxide and ozone.

Changing Levels of Carbon Dioxide

There are three active reservoirs for carbon dioxide (CO_2): the atmosphere, the oceans, and the terrestrial system; in addition there is the inactive reservoir of the earth's crust. The oceans store the largest amount of CO_2, and the atmosphere has the smallest amount; see figure 7.3. The atmosphere is the link between the other reservoirs, and the ocean plays a major part in determining the atmosphere's concentration by physical (mixing and circulation), chemical, and biological means.

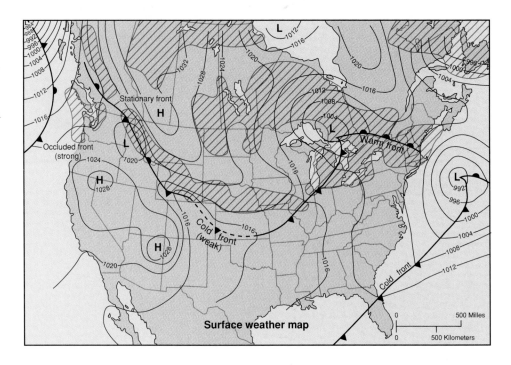

Figure 7.2

Surface weather map. Isobars labeled in millibars. An "H" denotes an area of high pressure and an "L" denotes low pressure. Hatched areas are areas of precipitation. The closer the isobars are together, the stronger the winds.

Source: NOAA; Marine Climate of Washington.

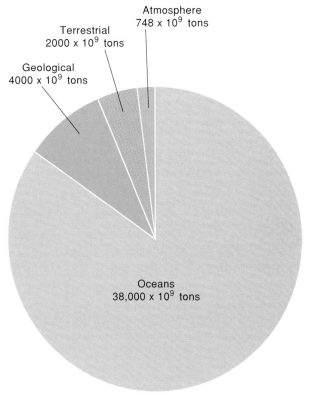

Figure 7.3

World carbon dioxide distribution. Values are in billions of tons.

Atmospheric carbon dioxide (CO_2) is transparent to incoming solar radiation, but at the same time it reduces the rate of outgoing long-wavelength back radiation and warms the earth by what is commonly known as the **greenhouse effect.** In the Northern Hemisphere, the natural cycle of the world's carbon dioxide shows decreasing CO_2 in the spring and summer, when plants increase active photosynthesis

and remove CO_2 in greater amounts than that contributed by respiration and decay. In the fall and winter, the photosynthetic activity is reduced, plants lose their leaves, and decay processes release CO_2; atmospheric carbon dioxide increases. The effects of deforestation and conversion of forest land to agriculture, the burning of fossil fuels, and the growth of human populations has been superimposed on this natural cycle. In preindustrial times, the seasonal cycle was in balance, but current practices have added a huge excess of atmospheric CO_2 to the equation.

Since 1850, the concentration of carbon dioxide in the atmosphere has increased from 280 parts per million (ppm) to 350 ppm. Recently, the average rate of increase has been 1.3 ppm per year. For more than thirty years scientists have been recording a steady increase in the carbon dioxide concentration in the earth's atmosphere, due to the burning of coal, oil, and other fossil fuels (see fig. 7.4). If the increasing trend continues, the concentration of carbon dioxide will double sometime before the middle of the next century. This will warm the earth and alter the earth's average heat budget by reducing the heat lost to space by long-wave radiation.

Based on this trend of increasing CO_2, predictions of global warming of 2°–4°C have been made. A change in temperature of this magnitude in a few decades is comparable to that which has occurred over the last ten thousand years, the time since the last ice age. It is very difficult to predict the extent of the effects brought about by such a change in atmospheric CO_2, because there would also be factors to decrease the warming trend such as increased cloud cover associated with greater evaporation. If such warming does occur, however, it is expected to affect the higher latitudes, causing melting of polar land ice and raising the world sea level by about 1 m (3 ft); changes in local weather could influence agriculture and other human activities.

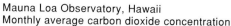

Mauna Loa Observatory, Hawaii
Monthly average carbon dioxide concentration

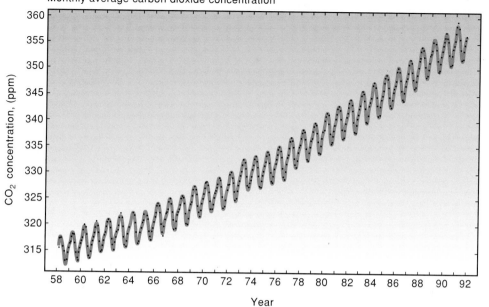

Figure 7.4

Concentration of atmospheric carbon dioxide in parts per million (ppm) of dry air versus time in years observed with a continuously recording non-dispersive infrared gas analyzer at Mauna Loa Observatory, Hawaii. The smooth curve represents a fit of the data to a four harmonic annual cycle which increases linearly with time, and a spline fit of the interannual component of the variation. The dots indicate monthly average concentration.

C. D. Keeling, R. B. Bacastow, A. F. Carter, S. C. Piper, T. P. Whorf, M. Heimann, W. G. Mook, and H. Roeloffzen, "A Three Dimensional Model of Atmospheric CO_2 Transport Based on Observed Winds: Observational Data and Preliminary Analysis," Appendix A, in *Aspects of Climate Variability in the Pacific and the Western Americas*, Geophysical Monograph, American Geophysical Union, vol. 55, 1989 (Nov).

The annual production of CO_2 from fossil fuels and the burning of tropical forests is slightly more than 6 billion tons. The atmospheric CO_2 increase accounts for about half this amount, 3 billion tons. The most recent research on storage of CO_2 in the oceans estimates that about 2 billion tons enters the oceans and ocean sediments; refer back to the discussion of carbonate sediments in chapter 2 and dissolved CO_2 in seawater in chapter 5. At least 1 billion tons is thought to be taken up by the plants, the temperate forests growing back from eighteenth- and nineteenth-century logging and the intense growth of the tropical regions. The exact rates at which this CO_2 is transferred from one reservoir to another are difficult to measure, and the pathways followed by the CO_2 are difficult to trace. During the past decade much has been learned about the global carbon dioxide cycle, but predicting the earth's reactions to increases of greenhouse gases requires further collection and analysis of data.

If global warming were to occur due to increased atmospheric CO_2 and the greenhouse effect, several sea-surface scenarios are possible. In one, a decrease in sea ice could provide more open water for marine plant populations with a corresponding increase of photosynthesis, leading to increased carbon storage in the ocean reservoir. Others are related to the interaction between the clouds, the oceans, and the earth's surface; see the box, Clouds and Climate, in this chapter. Changes in sea-surface temperature could also affect the thermohaline circulation (chapter 6) and the surface wind systems (later in this chapter) that drive the ocean currents. Changes in the ocean current patterns would alter the transfer of heat from low to high latitudes and upset the earth's climate patterns. Also, increases in available CO_2 have the potential to stimulate photosynthesis on both the land and in the oceans with unknown effects.

The Ozone Problem

Depletion of the stratospheric ozone layer that screens the earth from much of the sun's ultraviolet radiation was first reported in 1985 by members of the British Antarctic Survey who reported that significant ozone loss had been occurring over Antarctica since the late 1970s. The Arctic winters are not as cold as those of the Antarctic, and ozone loss over the Arctic is less. In 1991 the Arctic ozone hole extended over parts of Canada and Northern Europe, but in 1992 the cold weather broke in mid-January and a significant depletion of ozone was avoided. It is estimated that the average global loss of ozone since 1978 is about 3%; at polar latitudes the loss is estimated at about 8% per decade.

The most widely accepted theory of ozone destruction is related to the release of chlorine into the atmosphere. Chlorine is commonly released as a component of chlorofluorocarbons (CFCs). CFCs are used as coolants for refrigeration and air-conditioning, as solvents, and in the production of insulating foams. CFCs are distributed throughout the troposphere by the winds and gradually leak into the stratosphere where the ultraviolet light breaks them apart. Gases in the atmosphere react with the chlorine and trap it in inert molecules, but in the presence of stratospheric clouds, which are common during the polar winter, the chlorine gas is freed to react with the ozone. In the presence of sunlight, chlorine and oxygen are formed, and the liberated chlorine is free to attack other ozone molecules.

At ground level ozone is a pollutant and a health hazard; in the stratosphere it absorbs most of the ultraviolet radiation from the sun, protecting life-forms on land and at the sea surface. Increased ultraviolet radiation can affect the growth and reproduction of organisms; it has also been implicated in human skin cancers and cataracts.

Figure 7.5

Ozone hole positions over the Antarctic continent (dark area) for 13 and 29 October 1990. The position of the pack ice (light area) and the mean position of the Antarctic convergence (outer edge of the gray area) are also shown.

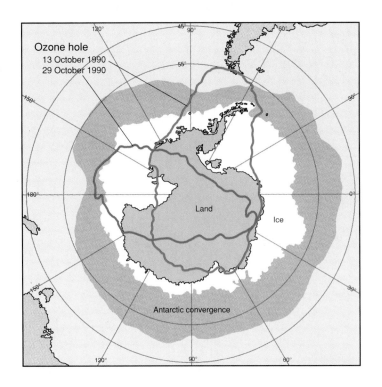

Studies done in 1990 in the Southern Ocean around Antarctica indicated that springtime increases in ultraviolet radiation are affecting the single-celled plants at the sea surface; see figure 7.5. These plants are held close to the surface in the low-density meltwater from sea ice. Higher levels of ultraviolet radiation are estimated to reduce the photosynthesis and reproduction of these organisms by about 6% directly under the ozone hole. When this reduction is adjusted for water and plant life moving in and out of the high-ultraviolet-radiation area, the average reduction of annual plant production is estimated to be 2%–4% of normal. This change may not seem large, but in this area these single-celled plants are at the base of the food chains that support the Antarctic's marine life (see chapters 14 and 15).

7.3 The Role of Sulfur Compounds

Dimethyl sulfide (DMS) is a gas produced by plants at the ocean's surface. It is in part responsible for the characteristic smell of the sea one notices when approaching the coast from the land. It is currently estimated that about 60 million tons of DMS are added to the atmosphere each year. Once in the atmosphere it changes rapidly, and one of its main products is sulfur in the form of sulfate that combines with atmospheric water to form sulfuric acid that is returned to the earth's surface with the rain (figure 7.6). The quantities of sulfuric acid returned to the earth's surface by this process are far below the values associated with the problem of "acid rain." DMS also plays a role in controlling the density of the clouds that form over the ocean; it changes their reflective properties, reducing incoming radiation and decreasing the heating of the ocean's surface (fig. 7.6). If a DMS buildup results in an excess of clouds and sulfur, less

light strikes the sea surface, surface temperatures drop, and the plant production of DMS decreases. If less DMS is produced, there is less cloud cover and less reflection of short-wave radiation. In this way DMS may act as a feedback mechanism or self-regulating thermostat to help control ocean-surface temperatures, another link in the complexity of interactions between the atmosphere and the oceans.

In industrialized areas of the Northern Hemisphere, the burning of fossil fuels creates another source of sulfates. Like the sulfates from DMS, these industrially produced sulfates contribute to acid rain; they form clouds that block solar radiation and cool the earth's surface.

The year 1990 was a record year for high global temperatures until the eruption of the Philippine volcano, Mount Pinatubo. The sulfur-based gases and the particulate materials from this eruption created a globe-circling particulate haze that decreased incoming radiation and caused measurable global cooling. Some researchers predict this will cause a decrease in the world's average temperature; others believe the warming effect of El Niño (discussed later in this chapter) during this period will counteract the cooling and produce a net warming.

7.4 The Atmosphere in Motion

The air moves because at one place less-dense air rises away from the earth, while in another place more-dense air sinks toward the earth. Between these areas the air that flows horizontally along the earth's surface is the wind. This process is shown in figure 7.7; note that there are really two horizontal airflows, or wind levels, moving in opposite directions: one at the earth's surface and one aloft. Air

Sun

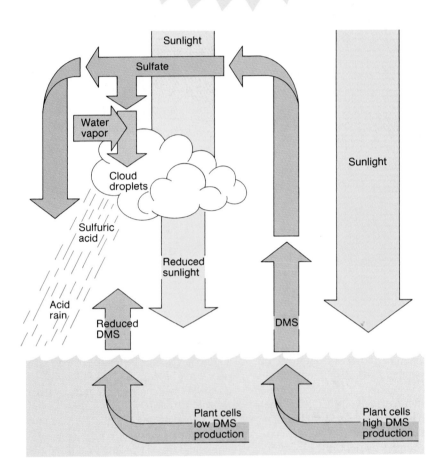

Figure 7.6

A proposed self-supporting thermal control system linking dimethyl sulfide (DMS) to the sea surface and the atmosphere, a cycle among plants, the sun, and clouds.

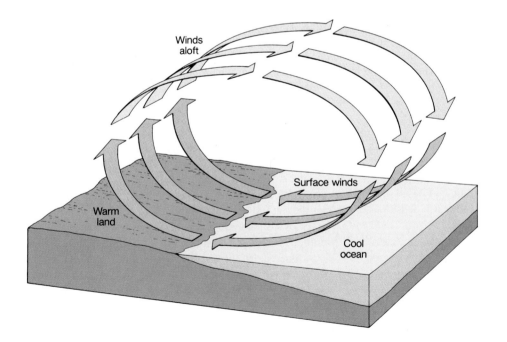

Figure 7.7

A convection cell is formed in the atmosphere when air is warmed at one location and cooled at another.

Clouds and Climate

Clouds are beautiful, always changing, ever moving; we see them white and fluffy, high and wispy, black and threatening. Clouds form from liquid water droplets with temperatures above freezing, supercooled water droplets with temperatures below freezing, and solid particles. In the lower troposphere condensation nuclei, small particles of dust, salt, or other matter, serve as the cores for condensing water vapor droplets (Box 1 figure 1). Clouds of the upper troposphere are composed of crystallized water vapor or ice crystals and snow.

Clouds have a dual role in the atmosphere; they simultaneously heat and cool the earth. Clouds appear white because they reflect shortwave radiation from the sun, and they also absorb some of this incoming shortwave radiation. Both processes reduce incoming radiation to the earth's surface, helping to cool the earth. Clouds intercept longwave radiation from the earth's surface, and because they are generally cooler than the earth, they radiate only part of this longwave radiation to space. This tends to warm the earth and its atmosphere. The net effect is that clouds play a significant part in the earth's heat budget.

Recent research indicates that on a global scale the earth's present cloud cover provides more shielding from incoming radiation than trapping of longwave radiation. At present the result is negative, a total net reduction in radiation to the earth of about 14 to 21 watts per square meter per month. It is estimated that if there were no clouds, average earth surface temperature would warm by about 10°C.

Not only do the clouds affect the earth's climate; they are also affected by

Box 1 figure 1
Storm clouds over western Scotland.

it. This so-called feedback mechanism could impose a new cloud-controlled radiation balance on the earth's climate, if the earth's climate changed. If the present negative radiation balance became less negative, the earth's surface would warm; if cloud feedback produced a more negative radiation balance, the earth's surface would become cooler than it is at present. If the earth warmed due to the greenhouse effect, there could be an increase in evaporation and therefore an increase in low-level water droplet clouds over the ocean, which would absorb and reflect incoming shortwave radiation and help to cool the earth. However, an increase in high-altitude ice clouds would warm the earth because ice clouds pass shortwave radiation and reduce longwave radiation loss.

The formation of clouds is related to the availability of condensation nuclei. If nuclei are sparse, the droplets are fewer and larger, increasing the chance of precipitation and reducing further the amount of nuclei. Many small nuclei produce abundant small droplets, increasing the cloud reflectivity and decreasing incoming radiation. Dimethyl sulfide is a source of condensation nuclei over ocean areas, and its availability is controlled by its own feedback system; refer to section 7.3.

Although clouds are among the most common of atmospheric phenomena, our understanding of cloud formation and distribution and of climate change and feedback mechanisms is presently incomplete. Scientists use computer models to test their ideas, modifying them as new information is acquired. Based on their efforts we hope to be able to predict climate change and better understand the consequences of modifying our atmosphere and its clouds.

---

circulating in this manner forms a convection cell based on vertical air movements due to changes in the air's density. Less-dense air rises, and more-dense air sinks.

Winds on a Nonrotating Earth

The heating and cooling of air and the gains and losses of water vapor in air are related to the unequal distribution of the sun's heat over the earth's surface, the presence or absence of water, and the variation in temperature of the earth's surface materials in response to heating. Imagine an earth with no continents and with no rotation, but heated like the natural earth. On this model covered with uniform layers of atmosphere and water, the wind pattern is very simple. Around the equator the air, warmed from below, rises. Once aloft, the air flows toward the poles where it is cooled from below and sinks to flow back toward the equator. Due to the unequal distribution of the sun's heat over the earth's surface, large amounts of heat and water vapor are transferred to the atmosphere around the equator. This less-dense air rises and as it rises it cools; the water vapor

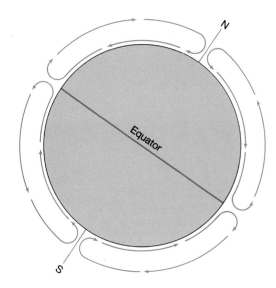

Figure 7.8
Heating at the equator and cooling at the poles produces a single large convection cell in each hemisphere on a nonrotating water-covered earth model.

condenses and rain forms. Equatorial regions are known for their warm, wet climate. The cool, dry air remains aloft and flows toward the poles where it sinks to produce a cold, dry, dense air mass and a zone of high atmospheric pressure. Such air movement is shown in figure 7.8; note the two large convective circulation cells in this figure. In this model, the Northern Hemisphere surface winds blow from north to south, and the upper winds blow from south to north; in the Southern Hemisphere the reverse is true, with the surface winds blowing from south to north and the upper winds blowing from north to south.

It is important to remember that winds are named for the direction *from which they blow*. A north wind blows from north to south, while a south wind blows from south to north. In this model, the Northern Hemisphere surface winds are north winds and the Southern Hemisphere surface winds are south winds.

The Effects of Rotation

Consider next the same model of a water-covered earth, but with rotation added. Gravity holds the earth's atmosphere captive, but the atmosphere is not rigidly attached to the earth's surface. There is little or low friction between the earth's surface and the atmosphere, and the atmosphere moves somewhat independently of the earth's surface. For example, a parcel of air that appears to be stationary above a point on the equator is actually turning with the earth and moving eastward at a speed of about 1700 km (1050 mi)/hr. If a south wind blows this parcel of air due north, it is moved across circles of latitude with progressively smaller circumferences. At these higher latitudes, points on the earth's surface move eastward more slowly. At 60° N, the eastward speed of a point on the earth's surface due to rotation is only half the speed at the equator. For this reason, a parcel of air that was originally moving only northward rel-

ative to the earth at the equator carries with it its equatorial eastward speed, so that it is now moving eastward at a speed that is greater than the eastward speed of the earth's surface at this higher latitude. Therefore the air parcel is displaced to the east, relative to the earth's surface, as it moves from low to high latitudes; in the Northern Hemisphere this deflection is *to the right of the direction of the air motion*. This relationship is illustrated in figure 7.9; follow the arrows at *A*.

If the same situation occurs in the Southern Hemisphere, with a parcel of air moving southward from the equator due to a north wind, the deflection is still to the east relative to the earth's surface. However, this deflection is now *to the left of the direction of the air motion*. See again figure 7.9 and follow the arrows at *B*.

If an air parcel moves southward toward the equator in the Northern Hemisphere, then it moves from a latitude where a position on the earth has a low eastward speed to a latitude where a position on the earth has a higher eastward speed. These positions move to the east, relative to the air, when the air moves from a higher to a lower latitude. This relationship causes the air to fall behind the position on the earth, so that the air is moving westward relative to the earth. This pattern is illustrated in figure 7.10. The deflection is still to the right of the direction of motion in the Northern Hemisphere. A similar pattern in the Southern Hemisphere shows that the deflection is to the left of the direction of motion.

Deflection due to air moving to the east or to the west is shown in figure 7.11. Air moving eastward is moving eastward faster than the earth beneath it, with reference to the earth's axis. The air is affected by a centrifugal force acting outward from the earth's axis of rotation that is stronger than the centrifugal force that is acting at the earth's surface. This small excess force acting on the air is in part acting against the earth's gravity, and in part acting parallel to the earth's surface directed toward lower latitudes. That part of the centrifugal force acting against the earth's gravity is so small that it has very little effect. That part acting parallel to the earth's surface is unopposed and causes a deflection of the moving air to lower latitudes, or to the right of its motion in the Northern Hemisphere. Westward-moving air is moving eastward more slowly, relative to the earth's surface. This air is affected by an outward-acting centrifugal force that is weaker than the centrifugal force that is acting at the earth's surface. Because the centrifugal force acting at the earth's surface is greater than the centrifugal force acting on the air, there is a weak force acting toward the earth's axis of rotation. This force is acting in part in the direction of gravity, and in part toward higher latitudes. Note that the deflection continues to be to the right of the initial wind motion in the Northern Hemisphere. In the Southern Hemisphere, the deflection of eastward-moving air toward lower latitudes and of westward-moving air toward higher latitudes results in a deflection of the air to the left of its original direction of motion. If we were to move from the earth's surface out

Figure 7.9

Because air moving northward from the equator to point *A* carries with it its initial eastward velocity, the air is deflected to the right of the initial wind direction in the Northern Hemisphere. In the Southern Hemisphere, air moving southward to point *B* is deflected to the left.

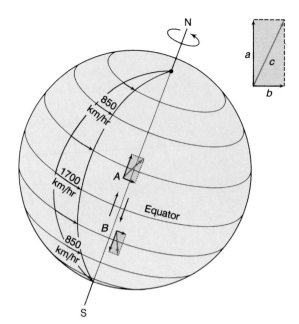

a = Initial north or south wind velocity

b = Initial eastward velocity minus the eastward velocity of the earth at a higher latitude

c = Resultant velocity of the wind

Figure 7.10

Air moving toward the equator passes from a latitude of lower eastward speed to a latitude of higher eastward speed. The result is deflection to the right of the initial wind direction in the Northern Hemisphere and deflection to the left in the Southern Hemisphere. There is no deflection at the equator.

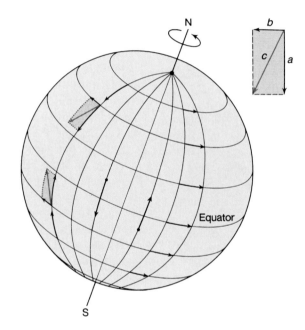

a = Initial north or south wind velocity

b = Initial eastward velocity minus the eastward velocity of the earth at a higher latitude

c = Resultant velocity of the wind

Figure 7.11

The deflection of eastward-moving and westward-moving air. There is no deflection at the equator.

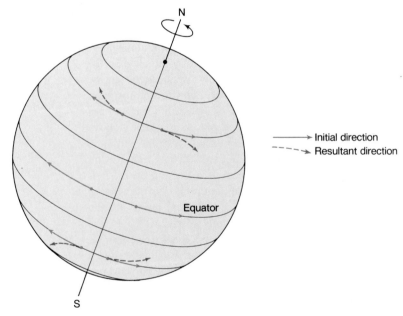

→ Initial direction

--→ Resultant direction

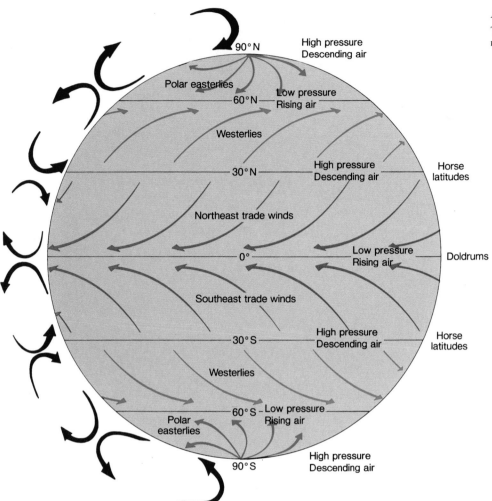

Figure 7.12

The circulation of the earth's atmosphere results in a six-band surface wind system.

into space and were able to watch the motions of earth and atmosphere, we would see the atmosphere's independent motion and also see the earth turning out from under the moving parcels of air. Yet we live on a moving earth's surface that we consider stationary, and we make our measurements of air motion relative to the "stationary" surface. Therefore, from the earth's surface we see the moving air parcels deflected from their paths, deflected to the right in the Northern Hemisphere and to the left in the Southern Hemisphere.

The apparent deflection of the moving air relative to the earth's surface is called the **Coriolis effect,** after Gaspard Gustave de Coriolis (1792–1843), who mathematically solved the problem of deflection in frictionless motion when the motion is referred to a rotating body. Because one of the laws of motion in physics tells us that a body set in motion along a straight line continues to move along that straight line unless it is acted upon by a force, such as a push or a pull, the Coriolis effect is often called the Coriolis force. If the term force is used, remember that it is an apparent force, and that a deflection appears only when the motion of an object that is subject to little or no friction is judged against a rotating frame of reference. The magnitude of the Coriolis force increases with increasing latitude, increases with the speed of the moving air, and is dependent on the rotation rate of the earth on its axis.

Wind Bands

Applying rotation, and therefore the Coriolis effect, to the stationary earth model modifies the winds considerably. Refer to figure 7.12 as you read the following description to understand what happens. The air continues to rise at the equator and flows aloft to the north and the south, but it cannot continue to move northward and southward without being deflected to the right in the Northern Hemisphere and to the left in the Southern Hemisphere. This deflection short-circuits the large, hemispheric, atmospheric convection cells of the stationary-earth model. The deflected air aloft sinks at 30° N and 30° S; it moves along the water-covered surface, either back toward the equator or toward 60° N and 60° S. The air that reaches the poles at the upper level cools and sinks over the poles and moves to lower latitudes, warming and picking up water vapor; at 60° N and 60° S it rises again. The result is three convection cells in each hemisphere for the rotating earth.

Consider the flow of surface air in the three-cell system. Between 0° and 30° N and S the surface winds are deflected relative to the earth, blowing from the north and east in the Northern Hemisphere and from the south and east in the Southern Hemisphere. This deflection creates bands of moving air, known as the **trade winds:** the northeast trade winds are north of the equator, and the southeast

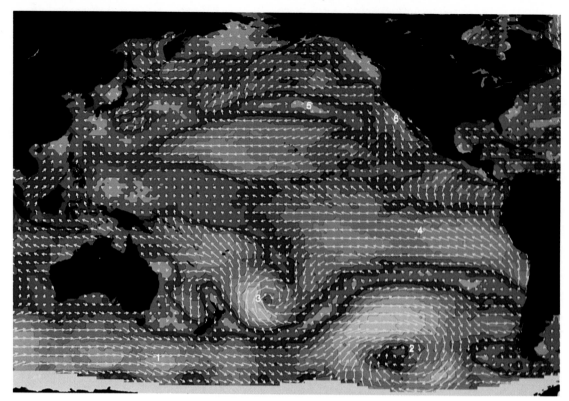

Figure 7.13

Pacific Ocean wind distributions. The SEASAT satellite obtained the first global measurements of ocean wind data during three days in 1978. Arrows point in the direction the winds blow; longer arrows indicate greater wind speed. Light winds (less then 4 m/sec or 9 mph) are colored blue and strong winds (greater than 14 m/sec or 31 mph) are colored yellow. Wind speeds and directions were calculated from radar reflections from the small wind-driven waves that roughened the sea surface. The approximate accuracy is ± 2m/sec and ± 20°. Winds blowing from the west in the Southern Hemisphere (*1*) blow almost continuously around the entire Southern Ocean except near the tip of South America. Intense storms are shown at (*2*) and (*3*). The strong and constant southeast trade winds are shown at (*4*). Summer in the Northern Hemisphere is characterized by a large high-pressure cell in the North Pacific (*5*). Circulation about this cell produces the winds (*6*) along the coast that induce upwelling. Winds and squalls are formed at the boundary between the northeast and southeast trade winds, the doldrums.

trade winds are south of the equator. Between 30° N and 60° N, the deflected surface flow produces winds that blow from the south and west, while between 30° S and 60° S they blow from the north and west. In both hemispheres these winds are called the **westerlies.** Between 60° N and the North Pole, the winds blow from the north and east, while between 60° S and the South Pole, they blow from the south and east. In both cases they are called the **polar easterlies.** The six surface wind bands are shown in figures 7.12 and 7.13.

At 0° and at 60° N and S, moist, low-density air rises; these are areas of low atmospheric pressure, zones of clouds and rain. Zones of high-density descending air at 30° and 90° N and S are areas of high atmospheric pressure, zones of low precipitation and clear skies. Air flows over the earth's surface from regions of high atmospheric pressure to areas of low atmospheric pressure. In the zones of vertical motion, between the wind belts, the surface winds are unsteady. Such areas were troublesome to the early sailors, who depended on steady winds for propulsion. The area of rising air at the equator is known as the **doldrums,** and the high-pressure areas at 30° N and S are known as the **horse latitudes.** In all these areas sailing ships could find themselves becalmed for days. The origin of the word doldrum is obscure, but the horse latitudes are said to have

gotten their name from the stories of ships carrying horses that were thrown overboard when the ships were becalmed and the freshwater supply became too low to support both the sailors and the animals.

7.5 Modifying the Wind Bands

Moving from the rotating, water-covered model of the earth to the real earth requires the consideration of two more factors: (1) seasonal changes in the earth's surface temperature due to solar heating, and (2) the addition of the large continental land blocks. Both ocean and land surfaces remain warm at the equatorial latitudes and cold at the polar latitudes all through the year, but the middle latitudes have seasonal temperature changes. The middle latitudes are warm in summer and cold in winter. Keep in mind that land surface temperatures have a greater seasonal fluctuation than ocean-surface temperatures, because the heat capacity of the ocean water is greater than the heat capacity of land, and because the ocean water has the ability to transfer heat from the surface to depth in summer and from depth to the surface in winter. Land does not transfer heat in this way.

Seasonal Changes

The presence of both land and water in near-equal amounts at the middle latitudes in the Northern Hemisphere produces average seasonal patterns of atmospheric pressure. During the warm summer months, the land is warmer than the ocean. Refer back to figure 6.7. The air over the land is heated from below and rises, creating a low-pressure area, while the air over the sea cools and sinks, producing a high-pressure area over the water. In the summer, low-pressure zones at 60°N and 0° tend to combine over the land, cutting through the high-pressure belt along 30°N latitude. This breaks the high-pressure belt into several high-pressure cells over the oceans during the Northern Hemisphere's summer, rather than maintaining the latitudinal zones of pressure stretching continuously around the earth. In the winter, the reverse is true at the middle latitudes; see again figure 6.7. The land becomes colder than the water, and the air rises over the water and descends over the land, creating a low-pressure zone over the ocean.

Over the land, the polar high-pressure zone spreads toward the high-pressure zone at 30°N, breaking the low-pressure belt centered about 60°N into discrete low-pressure cells that are centered over the warmer ocean water. This middle-latitude, seasonal alternation of high- and low-pressure cells breaks up the latitudinal pressure zones and wind belts that are seen in the water-covered model, to produce the situation shown in figure 7.14.

During the Northern Hemisphere's summer, the air in the high-pressure cells over the central portion of the North Atlantic and North Pacific oceans descends and flows outward toward the continental low-pressure areas. As the descending air moves outward it is deflected to its right, producing winds that spiral in a clockwise direction about the high-pressure cells. On the northern side of these high-pressure cells are the westerlies, and on the southern side are the north easterlies; the eastern side of the cell has northerly winds, and the western side has southerly ones. In the winter the air circulates counterclockwise about the low-pressure cell over the northern oceans, and the prevailing wind directions reverse. The wind directions related to these pressure cells are shown in figure 7.15. Along the Pacific coast of the United States, the northerly winds cool the coastal areas in the summer, and the southerly winds warm them in the winter. The New England coast receives warm, moist air from the low latitudes in the summer, and cold air moves down from the high latitudes in the winter. Although these seasonal changes modify the wind and pressure belts that were developed on the water-covered model, the generalized wind and pressure belts are still identifiable over the earth in the northern latitudes when the atmospheric pressures are averaged over the annual cycle. Rotation of the airflow about high- and low-pressure air cells is reversed in the Southern Hemisphere. At the middle latitudes in the Southern Hemisphere, there is little land to separate the water, and therefore the water temperature predominates. There is little seasonal effect; the atmospheric pressure and wind patterns created by surface temperatures change little over the annual cycle and are very similar to those developed for the water-covered model (see figs. 7.12 and 7.14).

The Monsoon Effect

The differences in temperature between land and water produce large-scale and small-scale effects in coastal areas. In the summer along the west coast of India and in Southeast Asia, the air rises over the hot land, creating a low-pressure air system. The rising air is replaced by warm, moist air carried on the southwest winds from the Indian Ocean. As this onshore airflow rises over the land, it cools and the moisture condenses, producing a steady, heavy rainfall. This is the wet, or summer, **monsoon.** In the winter, a high-pressure cell forms over the land, and northeast winds carry the dry, cool air southward from the Asian mainland and out over the Indian Ocean. This movement produces cool, dry weather over the land, known as the dry, or winter, monsoon. For years, coastal traders in the Indian Ocean who were dependent on sailing craft planned their voyages so that they sailed to their destination on one phase of the monsoon and returned on the next, since the easiest sailing was with the wind.This seasonal reversal in wind pattern is shown in figure 7.16.

The same effect is seen on a smaller local scale along a coastal area or along the shore of a large lake. During the day, the land is warmed faster than the water, and the air rises over the land. The air over the water moves in to replace it, creating an **onshore** breeze. At night, the land cools rapidly, and the water becomes warmer than the land. Air rises over the water, and the air from the land replaces it, this time creating an **offshore** breeze. Such a local diurnal (once-a-day) wind shift is referred to as a land-sea breeze (see fig. 7.17). The onshore breeze reaches its peak in the afternoon, when the temperature difference between the land and the water is at its maximum. The offshore breeze is strongest in the late-night and early-morning hours. Sometimes it is possible to smell the land 20 miles at sea, as the offshore wind carries the odors seaward. This daily wind cycle helps fishing boats that depend only on their sails to leave the harbor early in the morning and return late in the afternoon or early evening. Another example can be seen in San Francisco, California, during the summer. The area east of San Francisco Bay heats up during the day, and the air above it rises. Marine air flows in through the Golden Gate. When the warm marine air from offshore reaches the cold coastal waters, a fog is formed that pours in through the Golden Gate, obscuring first the Golden Gate Bridge and then the city (refer to fig. 4.19). When the air reaches the east side of the bay, it warms, and the fog dissipates. At night the flow is reversed, and the city is swept free of the fog.

The Topographic Effect

Because the continents rise high above the sea surface, they affect the winds in another way. As the winds sweep across the ocean, they reach the land and are forced to rise to continue moving across the land, as shown in figure 7.18. The

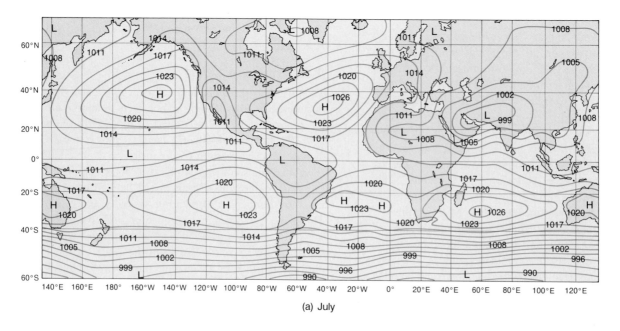

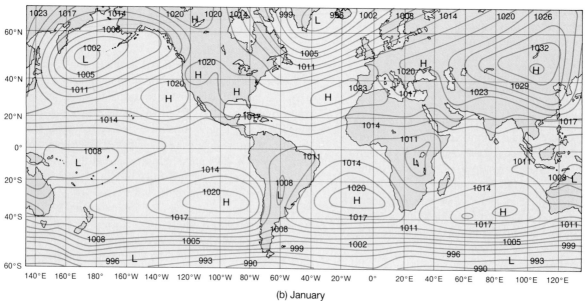

Figure 7.14

Mean sea level pressures expressed in millibars for (a) July and (b) January.

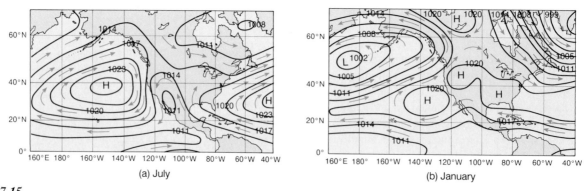

Figure 7.15

(a) Winds flow clockwise about a Northern Hemisphere high-pressure cell in the summer and (b) counterclockwise about a low-pressure cell in the winter.

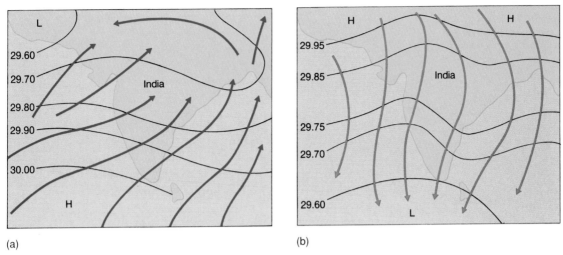

(a)   (b)

Figure 7.16
The seasonal reversal in wind patterns associated with (a) the summer (wet) monsoon and (b) the winter (dry) monsoon. The isolines of pressure are given in inches of mercury.

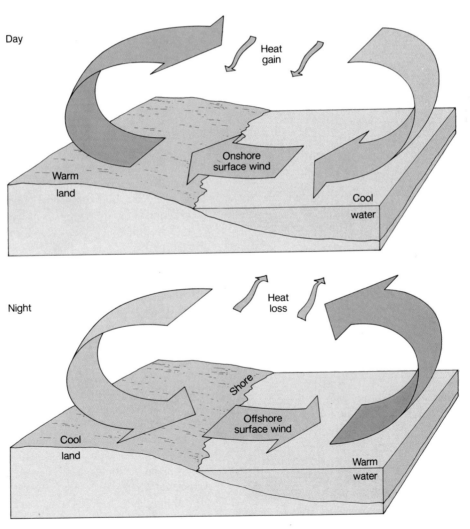

Figure 7.17
Differences between day and night land-sea temperatures produce an onshore breeze during the day and an offshore breeze at night.

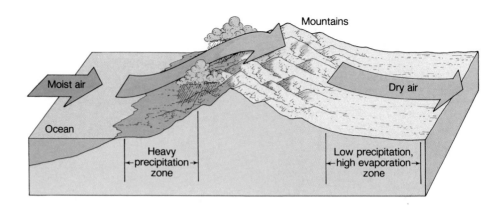

Figure 7.18

Moist air rising over the land loses its moisture on the mountains' windward side. The descending air on the leeward side is dry.

Mountains

Moist air

Ocean

Dry air

Heavy precipitation zone

Low precipitation, high evaporation zone

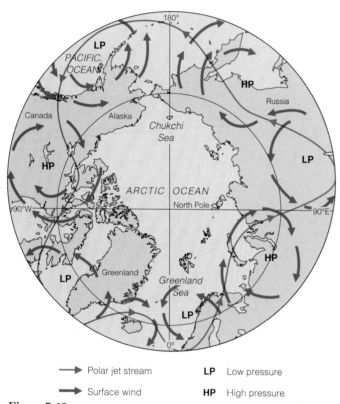

Figure 7.19

The polar jet stream circles the earth in the Northern Hemisphere above the boundary between the polar easterlies and the westerly winds. It is deflected north and south by the alternating air pressure cells of the northern temperate zone.

→ Polar jet stream
→ Surface wind

LP Low pressure
HP High pressure

away, in the rain shadow on the leeward side of the mountains, it is 40 to 50 cm (16–20 in)/yr. The west side of the mountains of Vancouver Island in British Columbia has a high rainfall, while the eastern side of the island is dry and sunny. The adjacent region of channels and islands is noted for its sunshine and scenic cruising. The southeast trade winds on the east coast of South America sweep up and across the lowlands and then rise to cross the Andes Mountains, producing rainfall, large river systems, and lush vegetation on the eastern side of the Andes and a desert on their western slopes. The control of precipitation patterns due to elevation changes is called the **orographic effect.**

The Jet Streams

Centered over zones of sinking and rising air near 60°N and S and 30°N and S are the **jet streams,** high-speed winds of the upper atmosphere. The subtropical jet streams are easterlies; the polar jet streams are westerlies; see figure 7.19. The Northern Hemisphere polar jet stream is used as an example in the following discussion.

The polar jet stream flows rapidly, 65–130 km/hr (40–80 mi/hr), and is displaced more than 2000 km (1250 mi) north and south during changes in the Northern Hemisphere's winter pressure systems. Wind speed and displacement are reduced during the summer. The prevailing westerly winds cause the wave form of the polar jet stream to migrate slowly eastward around the earth. The location and movement of the polar jet stream is controlled by the location and strength of the boundary between subtropical and polar air and the shape of the temperate zone's alternating high- and low-pressure systems. Refer again to figure 7.19.

The position of the polar jet stream and the boundary between subtropical and polar air are important in determining the weather in the temperate zone. The circulation of air around these pressure systems transports subtropical air to higher latitudes and polar air to lower latitudes, an important atmospheric mechanism that moves heat from low to high latitudes. The most important source of heat in this system is the water vapor gained by evaporation in the subtropics and condensed as it cools when displaced to higher latitudes. When the jet stream is displaced from its normal track for long periods, abnormal weather patterns

upward deflection cools the air, causing rain on the windward side of the islands and mountains; on the leeward (or sheltered) side, there is a low-precipitation zone, sometimes called a **rain shadow.** For example, the windward sides of the Hawaiian mountains have high precipitation and lush vegetation, while the leeward sides are much drier and require irrigation. On the west coast of Washington State, the westerlies moving across the North Pacific produce the Olympic Rain Forest as the air rises to clear the Olympic Mountains. On the western side of the Olympic Mountains the rainfall is as much as 5 m (200 in)/yr; 100 km (60 mi)

Figure 7.20
Hurricane Allen in the western Caribbean, photographed by a NOAA satellite in 1980.

result, for example, the long, cold winter of the northeastern U.S. and the heavy rains in California and the Gulf States during the winter of 1991–92.

7.6 Hurricanes

Although the trade winds of the tropics blow steadily, they may develop variations in speed and direction when the winds move over waters of changing temperature. These variations cause moving air to converge and then, as the air oscillates, to diverge; this produces a pressure disturbance known as an **easterly wave.** It is visible as a sharp wrinkle in the isobars over the tropical oceans.

If the sea-surface temperature is above 27° C, the atmospheric pressure decreases; the easterly wave develops into an intense low-pressure cell, and a **tropical depression** is formed. The winds circle this depression and build in strength, and the weather disturbance becomes a tropical storm and then a **hurricane** (fig. 7.20). The strong winds, in excess of 130 km/hr (70 knots), that rotate about the low-pressure system associated with hurricane formation

extract water vapor, and therefore heat, from the sea surface. The large amounts of heat energy liberated from the condensing water vapor fuel the storm winds, raising them to destructive levels, up to 300 km/hr (160 knots). A major hurricane contains energy exceeding that of a large nuclear explosion; fortunately, the energy is released much more slowly. The energy generated by a hurricane is about 3×10^{12} watt-hours per day. This amount is the equivalent to the energy in 1 million tons of TNT. When these storms move over colder water or over land, the hurricane is robbed of its energy source and begins to dissipate. These storms bring not only strong winds but also high precipitation because of the very high rate of condensation associated with the rising warm, moist tropical air.

Hurricanes can form on either side of the equator but not at the equator because the Coriolis effect is necessary to create their spiraling winds. When a storm of this type is formed in the western Pacific Ocean it is called a **typhoon** instead of a hurricane. Areas that give rise to hurricanes and typhoons and their typical storm tracks are shown in figure 7.21.

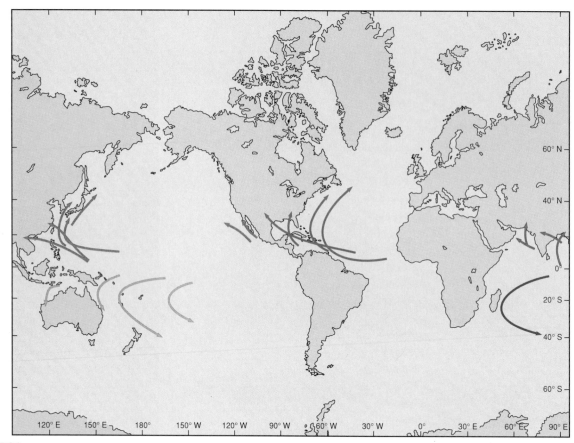

Figure 7.21
Tracks of hurricanes and typhoons. Colored arrows indicate areas in which these storms occur.

7.7   El Niño

On the sheltered or lee side of the tropical landmasses, under the trade winds of the Pacific, the upwelling of deep oceanic water is a nearly constant process. From time to time, this process falters; the trade winds strengthen, and warm tropical surface water accumulates on the west side of the Pacific. This event is followed by the winds losing their driving force; the upwelling lessens, and the mass of warm western Pacific water moves eastward across the ocean to accumulate along the coast of the Americas. This sequence of events is known as **El Niño,** or Christ Child, named for its frequent coincidence with the Christmas season. The increase in coastal surface temperatures usually ends by April, but in some years the large quantities of warm water spread north and south, and surface temperatures remain elevated for more than a year. Severe El Niño conditions occurred in 1953, 1957–58, 1965, 1972–73, 1976–77, 1982–83, 1986–87, and 1991–92. During 1982–83, the surface temperatures off Peru were more than 7°C above normal; the widespread effect of this event found tropical species displaced as far north as the Gulf of Alaska.

A severe El Niño event affects weather systems over large areas of the earth. In 1982–83 the polar jet stream was displaced far southward over the Pacific Ocean, bringing unusually dry conditions to Hawaii and a strong low-pressure system to the Gulf of Alaska that resulted in high winds and high precipitation along the west coast of the United States. Heavy rains occurred in Ecuador, Peru, and Polynesia, while droughts came to Australia, the Sahel of Africa, southern India, and Indonesia. At the same time lower surface temperatures in the North Atlantic made the hurricane season the quietest in over fifty years.

During the 1991–92 event the southern displacement of the polar jet stream over the Pacific brought heavy winter rains to southern California and the U.S. Gulf Coast and a mild, low-precipitation winter to the coastal regions of Oregon and Washington. The jet stream's extreme oscillation, north over the central U.S. and south over the eastern U.S., gave New England and the maritime provinces of Canada extreme cold and heavy snow.

The exact cause of El Niño is still in doubt, but certain processes have been identified with its appearance. One process, in which atmospheric pressure increases on one side of the Pacific, decreases on the other, and then reverses, is known as the Southern Ocean Oscillation. The pressure centers associated with this oscillation lie over Easter Island in the eastern Pacific and Indonesia in the western Pacific. Under normal conditions, there is a high-pressure system over Easter Island and a low-pressure system over Indonesia, the trade winds are strong and constant, and upwelling occurs along the coast of Peru. When the atmospheric pressure system reverses, the southeast trade

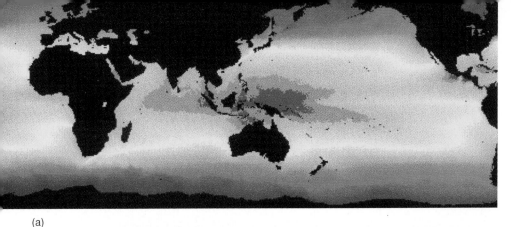

(a)

Figure 7.22

Ocean surface temperature data collected by satellite allows a global view of an El Niño event. Color is keyed to surface temperatures in °C. (a) Annual mean sea-surface temperatures. (b) During an El Niño event the pool of west Pacific warm surface water enlarges and expands across the Pacific and also into the Indian Ocean. Increased sea-surface temperatures modify weather worldwide. (c) The El Niño ceases when the pool of warm surface water dissipates and heat is transported to higher latitudes.

Images processed by Dr. Xiao-Han Yan at University of Delaware.

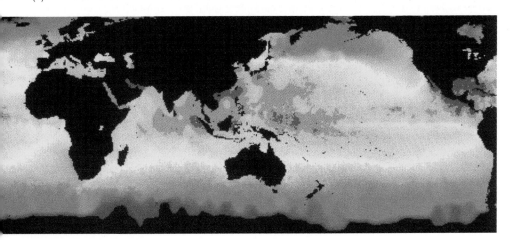

(b)

(c)

winds break down. Westerly winds are formed due to tropical low-pressure areas that develop over accumulated warm water near Indonesia. Over a period of two to three months, warm surface water from the western Pacific surges in a wave form across the Pacific to depress the Peruvian upwelling and raise surface temperatures; see figure 7.22. The elevated surface temperature of the eastern tropical Pacific is due to both the intrusion of warm water and the lack of trade winds to produce upwelled water along the coast. The moving mass of warm surface water carries the overlying low-pressure zone of rising air and precipitation eastward across the Pacific. This movement causes the westerly winds to move eastward along the doldrum belt, helps the eastward movement of the warm water pool, and creates high precipitation in normally dry areas of coastal Peru.

The southward displacement of the intertropical convergence zone, also called the doldrums, is another indication of El Niño. This movement to the south coincides with a lessening of the southeast trade winds and the development of a deep thermocline in the southeast tropical Pacific. This southern shift occurs early in the year, when the southeast trade winds are at a minimum and the sea-surface temperatures are at a maximum. The coming of El Niño is associated with a greater-than-average southward deflection of the doldrum belt.

By midsummer, the El Niño effect lessens; in two to three months the surface water cools as upwelling is reestablished in the eastern tropical Pacific. In November and December another slight warming is often observed; the atmospheric pressure distribution in the Southern Ocean

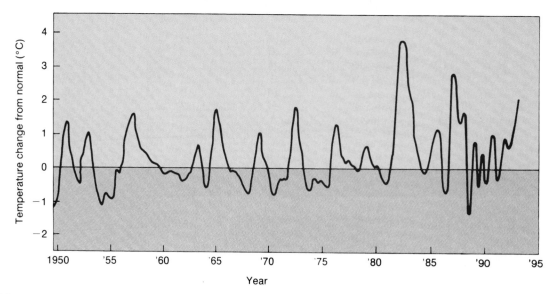

Figure 7.23

Sea-surface temperature changes at Puerto Chicama, Peru, approximately 8°S. El Niño events (above 0 on the chart) and La Niña events (below 0) form an alternating pattern. Normal sea-surface temperature is marked by 0. Note the lack of La Niña events between 1975 and 1987.

Oscillation reverses again, and the trade winds return to their normal state. A severe El Niño usually lasts about fifteen months.

Attempts to forecast El Niño use the change in the Southern Ocean Oscillation and the surge in strength of the trade winds, which appears to precede their decline and the development of the westerlies. In previous years observed events were not always predicted, and in some cases predictions were made of events that did not occur. Computer models successfully predicted the onset of the 1991–92 El Niño event, but the polar jet stream split over the eastern Pacific, and not all the predictions came to pass.

At present two extensive research programs are monitoring sea-surface temperatures in the tropic Pacific to better understand the warm tropical surface water that moves eastward across the Pacific during an El Niño. These are the Tropical Ocean Global Atmosphere (TOGA) and the Coupled Ocean Atmosphere Response Experiment (COARE); they are using research vessels for direct sea-surface measurements and satellites for large-scale observations.

Between El Niño events surface temperatures off Peru may drop below normal; an event of this type is known as **La Niña** or "the girl." See figure 7.23. These colder-than-normal years also produce wide-scale meteorological effects. The trade winds strengthen and surface water temperatures of the eastern tropical Pacific are colder, whereas those to the west are warmer than normal. This helps establish dry conditions over the coastal areas of Peru and Chile while rainfall and flooding increase in India, Burma, and Thailand. The repositioning of cold and warm surface waters in the tropical Pacific also displaces the position of major storm systems and causes waves to form on the jet stream. The drought of 1988 over the central U.S. is thought to have been caused by the northward displacement of the jet stream.

The alternation between these two events has been quite regular for the last hundred years, except for the periods between 1880 and 1900 when La Niña conditions prevailed, and the 1975–88 period of El Niños. Look carefully at figure 7.23 and notice that the 0°C line represents the long-term average sea-surface temperature over the last forty or so years. If this line were redrawn between 0°C on the left and 0.5°C on the right, allowing a half-degree rise in ocean-surface temperature, the regular alternation between El Niño and La Niña events would appear as a more regular pattern. Recent analysis of surface-water temperatures off the southern California coast during this period (1950–1990) indicate an average water temperature increase of 0.8°C. There are at least two possible mechanisms that could lead to such a warming of ocean surface water: (1) the ocean's response to global warming or (2) a reduction in upwelling in the Pacific Ocean associated with changes in North Atlantic circulation (see Box: A North Atlantic Cool Pool).

7.8 Practical Considerations: Storm Tides and Storm Surges

Periods of excessive high water along a coast associated with changes in atmospheric pressure and the wind's action on the sea surface are known as **storm surges** or **storm tides.** These storm surges, combined with normal high-tide conditions, can spell disaster for low-lying coastal areas.

Intense storms at sea, such as hurricanes or typhoons, are centered about intense low-pressure systems in the atmosphere. Under the low-pressure area in the center of the storm, the sea surface rises up into a dome, or hill, while the surface is depressed farther away from the center,

A North Atlantic Cool Pool

Can a change in oceanic and atmospheric circulation cause an abrupt change in climate for a decade or more? The only known instance of such a change occurred after the 1976–77 El Niño when the climate of the Pacific Ocean and the United States underwent a climate change that lasted for nearly ten years. During this period freezing temperatures caused severe damage to Florida agriculture, abnormally high waves along the California coast caused cliff erosion and harbor damage, and the low-pressure system of the Aleutian winter intensified and moved eastward, increasing Alaska's average winter temperatures by 1.5°C. The reasons for this shift are unknown, but they may be associated with the lack of La Niña events between 1975 and 1986.

Another possibility has been detected in the North Atlantic. A large pool of cold, low-salinity surface water appeared in the waters off Greenland, north of Iceland, in 1968. It was about 0.5‰ less salty and 1° to 2°C colder than usual. Within two years it had moved east into the Labrador Sea off eastern Canada; then it crossed the Atlantic and in the mid-1970s moved north into the Norwegian Sea, and it had returned to its place of origin by the early 1980s. Follow the path of this pool of water in Box 2 figure 1. The cause of this pool is not known. It has been suggested that stronger-than-normal winds over the Greenland Sea could have swept excess polar sea ice into warmer water where it melted, forming a pool of lower-salinity cold water. Another theory

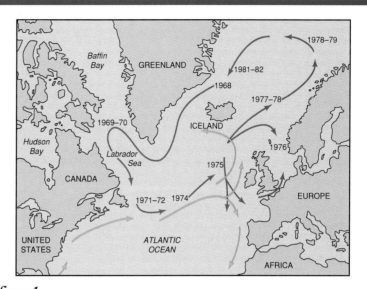

Box 2 figure 1

Purple arrows follow the path of the North Atlantic cool pool. Orange arrows show warm water flow the from the Gulf Stream.

links the ice movement to strong winds blowing all across the Arctic in the late 1960s.

Harsh winters plagued Europe during the late 1960s at the peak of the Atlantic cooling, and the entire Northern Hemisphere had cooler-than-average temperatures for more than ten years. One may conjecture that the presence of this surface pool of cold but low-density water in the North Atlantic slowed the sinking of the surface water that forms most of the deep water of the world's oceans; refer back to chapter 6. This decrease in the sinking of surface water then decreased the Gulf Stream transport of warm surface water into the region,

and this lack of warm water from the south helped to sustain the cool surface pool and the cold land temperatures. Conjecturing further, if the rate of deep-ocean water formation was slowed, then there must have been a complementary decrease in upwelling and a possible relationship between the North Atlantic event and the reduction of La Niñas between 1975 and 1986 due to less-than-normal upwelling elsewhere in the oceans. Again we are looking at an enormously complex series of interactions; they are still far from fully understood, but we are beginning to see the strands that unite the ocean-atmosphere system.

---

where the atmospheric pressure is greater. The atmospheric pressure change between the outside of a hurricane and its center can be as large as 7.5 cm (3 in) of mercury. Since mercury is 13.6 times more dense than water, this pressure change produces a 97-cm (38-in) change in elevation of the water between regions outside the storm and those at the storm's center. In addition, the surface winds spiraling toward the center of the storm add to the sea's elevation under the storm center.

The elevated dome of water travels across the sea under the storm center and raises the water level at the shore when the storm reaches the coast. During the storm the drag of the wind on the sea surface pushes surface water in the approximate direction of the wind. On the side

of the storm where the winds are toward the shore, the water moving toward the shore is piled up against the coast, increasing the height of the sea surface and producing the storm surge. The water will continue to pile against the shore, steepening the slope of the water until the tendency of the water to flow downhill and back to sea is equal to the force of the wind driving the water ashore. If the water along the shore is deep, some of the water moving landward will downwell and return seaward; this decreases the height of the water being driven ashore. Along a shallow shore the landward-moving layer of water extends to the seafloor; there is little downwelling or seaward return flow, and the height of the water along the shore is much greater. A storm surge may sustain a high water level for many hours until the storm winds diminish in the coastal area. Storm surges are often confused with tsunamis and are often incorrectly termed "tidal waves"; tsunamis are great sea waves and are discussed in chapter 9.

Along shallow areas of the East and Gulf coasts of the United States, storm surges have caused considerable damage. In 1900 a storm surge caused "the great Galveston flood." It produced water depths of 4 m (13 ft) over the island of Galveston, Texas, destroying the city and killing 5000 people. Hurricane Camille, in 1969, was one of the strongest storms ever to hit the Gulf Coast. The high water levels caused severe property damage, and several hundred people were killed. In 1989 a storm tide of 5 m (16.5 ft) came ashore with Hurricane Hugo in South Carolina, destroying much property along the coast and in the city of Charleston. Great loss of life was averted by early warnings and mass evacuation of low-lying coastal areas. See figure 7.24. In August 1992 Hurricane Andrew (fig. 7.25) struck Florida with the same intensity that Hurricane Hugo brought to the Carolina coast, but the coastal damage in Florida was much less. Although the winds were high (216 km/hr or 135 mi/hr), the storm surge was only 2.4 m (8 ft). The storm traveled a short distance between the Bahamas and Florida at a fast rate of speed; it did not have enough time to build up energy to produce large waves, and the extremely flat Florida coastline decreased the chance of high-energy waves. In addition some of Andrew's energy was absorbed by offshore reefs and coastal mangrove swamps.

Severe storms over the shallow Bay of Bengal spawned storm surges that took 100,000 lives in 1876 and an estimated 300,000 lives in 1970. Another 10,000 people were killed in 1985, and in 1991 yet another storm surge struck this same area of Bangladesh, killing an estimated 139,000 people. The expanding populations of this area need land for homes and farms, and as fast as new land, barely above sea level, appears in the delta of the Ganges River, the people move seaward. When the storms come there is no way to evacuate the thousands of people in these low-lying areas, and the rising water rapidly covers hundreds of square miles.

A storm over shallow regions of the North Sea in 1953 produced a 3.3-m (11-ft) rise in sea level, which combined with a high tide to break through the protective dike system and flood the Netherlands coast, killing 1800 people. To prevent a repetition of this event, the Dutch are walling off the inlets at the south end of the Rhine delta from the North Sea. They are constructing barrier dams with gates that can be closed when abnormally high water threatens their coastal lowlands. The most complex and costly portion of this barrier became operational in the fall of 1986. The 9.2-km (6-mi)-wide mouth of the Eastern Scheldt estuary in Zeeland was partially blocked by two human-made islands. Between these islands sixty-five concrete piers were set into the seafloor. The piers are equipped with sixty-three steel gates, each weighing

Figure 7.24
Storm tide damage caused by Hurricane Hugo, Pawley's Island, South Carolina, September, 1989.

between 250 and 450 metric tons. These gates are lowered to block an advancing storm surge or extra high-rising tide. The project is scheduled to be completed by the year 2000.

In England, large rotating gates have been installed in the river Thames to protect the lowlands bordering the river. These gates can be raised during a storm tide moving upriver but must be lowered during low tides to prevent flooding by accumulated river water. See figure 7.26.

Storms tides cannot be prevented. In estuary areas of some populated regions the great cost of engineering projects such as those described in England and the Netherlands may be justified, but along many stretches of shore there is no defense. The costs to taxpayers and their governments, for emergency services and cleanup to a populated area historically prone to storm surges and high-water damage, are enormous. Is it wiser to vacate these lands and use them only in nonpermanent ways in order to reduce both the risks and the costs?

Summary

Most clouds and weather occur in the troposphere, where temperature decreases with altitude. In the stratosphere temperature increases with altitude because of the ozone and its ability to absorb ultraviolet radiation. The temperature of the higher mesosphere and thermosphere decreases with height.

The atmosphere is a mixture of gases including water vapor. Atmospheric pressure is the force with which air presses on the earth's surface; high-density air creates high-pressure zones, and low-density air forms low-pressure zones.

Concern that the earth's climate may be changing is related to changes in the balance of the gases in the earth's atmosphere. Increases in CO_2 concentration due to burning and use of fossil fuels are leading to the prediction of global warming because carbon dioxide traps outgoing

Figure 7.25

A satellite image of Hurricane Andrew approaching land south of Miami, August 24, 1992. Maximum sustained winds were 220 km/hour (138 mi/hr).

Figure 7.26

A tidal flood gate system in the Thames River, east of London, England, can be raised to protect upstream lowlands from storm tides.

longwave radiation by the greenhouse effect. There has been a significant depletion of the ozone layer, most probably due to the release of chlorine into the atmosphere. Loss of ozone increases the ultraviolet radiation reaching the earth's surface and is linked to a reduction in single-celled plant life in the oceans surrounding Antarctica.

Dimethyl sulfide produced by plant cells at the sea surface is self-regulated by its role in cloud formation. Sulfur gases from industry and volcanic eruptions are also related to cloud formation and climate change.

The density of air is controlled by air temperature, pressure, and water vapor content. The winds are the horizontal air motion in convection cells produced by heating the atmosphere from below. Winds are named for the direction from which they blow.

Because of the Coriolis effect, winds are deflected to their right in the Northern Hemisphere and to their left in the Southern Hemisphere. This action produces a three-celled wind system in each hemisphere that results in the surface wind bands of the trade winds, the westerlies, and the polar easterlies. Zones of rising air occur at 0° and at 60° N and S; these are low-pressure areas of clouds and rain. Zones of descending air at 30° and 90° N and S are high-pressure areas of clear skies and low precipitation. Surface winds are unsteady and unreliable at the zones of rising and sinking air, producing the doldrums and the horse latitudes. The doldrum belt is displaced north of the earth's geographic equator.

Seasonal atmospheric pressure changes modify these wind bands and cause coastal winds to change direction seasonally in the Northern Hemisphere. Differences in temperature between land and water produce the monsoon effect. The seasonal reversal in wind pattern causes the wet and dry monsoons of the Indian Ocean; a similar daily reversal causes the onshore and offshore winds of any coastal area. Winds from the ocean rising to cross the land also produce heavy rainfall. The jet streams are high-altitude, fast-moving winds found in both hemispheres, when greatly displaced, abnormal weather patterns result. Hurricanes develop from tropical low-pressure systems with high-wind systems; these storms carry enormous amounts of energy.

In some years warm tropical surface water moves eastward across the Pacific and accumulates along the west coast of the Americas, blocking the normal upwelling. This phenomenon is El Niño; it is associated with changes in atmospheric pressure and wind direction. Colder-than-normal surface water temperatures off coastal Peru and associated weather phenomena are known as La Niña; La Niña episodes appear to alternate with El Niño events.

A storm tide or a storm surge is produced by a storm-elevated sea surface when the storm reaches the shore; it is associated with severe coastal flooding and destruction.

Key Terms

troposphere	Coriolis effect
stratosphere	trade winds
ozone	westerlies
mesosphere	polar easterlies
thermosphere	doldrums
atmospheric pressure	horse latitudes
isobar	monsoon
low-pressure zone	onshore
high-pressure zone	offshore
greenhouse effect	rain shadow
dimethyl sulfide	typhoon
orographic effect	El Niño
jet stream	La Niña
easterly wave	storm surge
tropical depression	storm tide
hurricane	

Study Questions

1. How is the troposphere different from the stratosphere?
2. What controls the density of air? Why is moist air less dense than dry air?
3. What is changing the carbon dioxide and ozone concentrations of the atmosphere? Why are scientists concerned about these changes?
4. How do airborne sulfur compounds affect cloud cover and the acid rain problem? What are natural and industrial sources of these compounds?
5. Explain why, on the world average, the earth's atmosphere is unstable. Take into consideration distribution of the earth's surface area with latitude and the variation in heat loss and heat gain with latitude.
6. Why are regions that are noted for their low barometric pressure and rising air famous for their excess precipitation?
7. A frictionless projectile is fired from the North Pole and is aimed along the prime meridian. It takes three hours to reach its landing point, halfway to the equator. Where does it land? (Give latitude and longitude.) If the same projectile is fired from the South Pole under the same circumstances, where does it land? (Give latitude and longitude.)
8. Why does the circulation of the atmosphere depend on its transparency to solar radiation? What would happen to atmospheric circulation if the upper atmosphere absorbed most of the solar radiation?
9. Plot the six major wind belts on a map of the world. Add bands of rising air and descending air, regions of the doldrums, and the horse latitudes.
10. Explain why the northeast and southeast trade winds are steady in strength and direction year-round, while the Northern Hemisphere westerlies alternate from northwest to southwest with summer and winter along the west coast of the United States.
11. Why are the westerlies of the Southern Hemisphere more consistent than the westerlies of the Northern Hemisphere?

12. What are the early signs that alert forecasters to the onset of an El Niño event?

13. In what way does the jet stream influence the transfer of heat from low to high latitudes?

14. How do hurricanes produce storm tides? Why is a storm tide more severe along a coast with a wide, shallow continental shelf than a coast with a narrow continental shelf?

15. Why do the windward sides of the Hawaiian Islands receive more rain than their lee sides?

Suggested Readings

Berner, R. A., and A. C. Lasaga. 1989. Modeling the Geochemical Carbon Cycle. *Scientific American* 260 (3): 74–81.

Canova, P. 1989. The Reclamation of Holland. *Sea Frontiers* 35 (3): 154–64.

Cobb, C. E., Jr. 1993. Bangladesh: When the Water Comes. *National Geographic* 183 (6): 118–34.

Gordon, A. L. 1988. The Southern Ocean and Global Climate. *Oceanus* 31 (2):39–46.

Houghton, R. A., and G. M. Woodwell. 1989. Global Climate Change. *Scientific American* 260 (4): 36–44.

Jensen, P. D., and J. Hovermale. 1990/91. Numerical Air/Sea Environmental Prediction. *Oceanus* 34 (4): 40–49. (Role of data in predicting environmental change.)

Jones, P. D., and T. M. L. Wigley. 1990. Global Warming Trends. *Scientific American* 263 (2): 84–91.

Kretschmer, J. 1990. When the Winds Blow: A Mythology of Gust and Squall. *Sea Frontiers* 36 (3): 40–43.

La Brecque, M. 1990. A Critical Unknown in the Global Equations. *Mosaic* 21 (2): 2–11. (The role of clouds.)

Leetma, A. 1989. The Interplay of El Niño and La Niña. *Oceanus* 32 (2): 30–34.

Olson, D. B. 1990. Monsoons and the Arabian Sea. *Sea Frontiers* 36 (1): 34–41.

Post, W. M., R. Peng, W. R. Emanuel, A. W. King, V. H. Dale, and D. L. De Angelis. 1990. The Global Carbon Cycle. *American Scientist* 78: 310–26.

Takahashi, R., P. P. Tans, and I. Gung. 1992. Balancing the Budget. *Oceanus* 35 (1): 18–28. (Carbon dioxide sources and sinks.)

Toon, O. B., and R. P. Turco. 1991. Polar Stratospheric Clouds and Ozone Depletion. *Scientific American* 264 (6): 68–74.

Webster, P. J. 1981. Monsoons. *Scientific American* 245 (2): 109–18.

Weiner, J. 1989. Glacier Bubbles Are Telling Us What Was in Ice Age Air. *Smithsonian* 20 (2): 78–87.

8

The Currents

8.1 Ocean Surface Currents
The Pacific Ocean Currents
The Atlantic Ocean Currents
The Indian Ocean Currents

8.2 Gyres and Current Flow
The Ekman Spiral and Ekman Transport
Geostrophic Flow
Current Speed
Western Intensification

8.3 Eddies

8.4 Convergence and Divergence
Permanent Zones
Seasonal Zones

8.5 Global Circulation Changes

8.6 Measuring the Currents

Box: *Looking for Carbon: A Worldwide Effort*

8.7 Practical Considerations: Energy from the Currents

Box: *The Great Sneaker Spill*

Summary
Key Terms
Study Questions
Suggested Readings

There is a river in the ocean. In the severest droughts it never fails, and in the mightiest floods it never overflows. Its banks and its bottom are of cold water, while its current is of warm. The Gulf of Mexico is its fountain, and its mouth is in the Arctic Seas. It is the Gulf Stream. There is in the world no other such majestic flow of waters. Its current is more rapid than the Mississippi or the Amazon

Its waters, as far out from the Gulf as the Carolina coasts, are of an indigo blue. They are so distinctly marked, that their line of junction with the common sea-water may be traced by the eye. Often one half of the vessel may be perceived floating in the Gulf Stream water, while the other half is in common water of the sea, so sharp is the line, and such the want of affinity between those waters, and the reluctance, on the part of those of the Gulf Stream to mingle with the common water of the sea.

Mathew Fontaine Maury,
from *The Physical Geography of the Sea*. 1855

A surging wave at the Galápagos Islands.

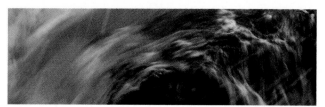

*T*he earth is surrounded by two great oceans: an ocean of air and an ocean of water. Both are in constant motion, driven by the energy of the sun and the gravity of the earth. Their motions are linked; the winds give energy to the sea surface and the currents are the result. The currents carry heat from one location to another, altering the earth's surface temperature patterns and modifying the air above. The interaction between the atmosphere and the ocean is dynamic; as one system drives the other, the driven system acts to alter the properties of the driving system.

In this chapter we explore the formation of the ocean's surface currents. We follow these currents as they flow, merge, and move away from each other. We examine both horizontal and vertical circulation, inspect the coupling of these water motions, and consider the ways in which they are linked to the overall interaction between the atmosphere and the ocean.

8.1 Ocean Surface Currents

When the winds blow over the oceans they set the surface water in motion, driving the large-scale surface currents in nearly constant patterns. The density of water is about one thousand times greater than the density of air, and once in motion, the mass of the moving water is so great that its inertial force keeps it flowing. The currents flow more in response to the average atmospheric circulation than to the daily weather and its short-term changes; however, the major currents do shift slightly in response to long-term seasonal changes in the winds. The currents are further modified by the interactions that occur among the currents, the zones of converging and diverging water, and the landmasses. The major surface currents have been called the rivers of the sea; they have no banks to contain them, but they maintain their average course.

Because the friction coupling between the ocean water and the earth's surface is small, the moving water is deflected by the Coriolis effect in the same way that moving air is deflected. But because water moves more slowly than air, the time required for water to move the same distance as wind is much longer. During this longer time period, the earth rotates farther out from under the water than from under the wind. Therefore, the slower-moving water is deflected to a greater degree than the overlying air. The surface water layer acted upon by the Coriolis effect is deflected to the right of the driving wind direction in the Northern Hemisphere and to the left in the Southern Hemisphere. In the open sea, the deflection of the surface-water layer is at a 45° angle from the wind direction, as shown in figure 8.1. The combined effects of the wind on the surface, the deflection of the water, and the shape and distribution of the landmasses create the large-scale patterns

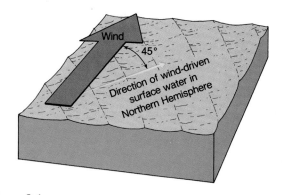

Figure 8.1

A wind-driven surface current moves at an angle of 45° to the direction of the wind; this angle is to the right in the Northern Hemisphere.

of the wind-driven surface currents in each of the oceans. Refer to figure 8.2 to follow the paths of these currents as they are discussed in the next section.

The Pacific Ocean Currents

In the North Pacific Ocean, the northeast trade winds push the water toward the west and northwest; this is the **North Equatorial Current.** The westerlies create the **North Pacific Current,** or **North Pacific Drift,** moving from west to east. Note that the trade winds move the water away from Central and South America and pile it up against Asia, while the westerlies move the water away from Asia and push it against the west coast of North America. The water that accumulates in one area must flow toward areas from which the water has been removed. This movement forms two currents: the **California Current,** moving from north to south along the western coast of North America, and the **Kuroshio Current,** moving from south to north along the east coast of Japan. The Kuroshio and California currents are not completely wind-driven currents; they provide continuity of flow and complete a circular motion centered around 30°N latitude. This circular, clockwise flow of water is called the North Pacific **gyre.** Other major North Pacific currents include the **Oyashio Current,** driven by the polar easterlies, and the **Alaska Current,** fed by water from the North Pacific Current and moving in a counterclockwise gyre in the Gulf of Alaska. There is little exchange of water through the Bering Strait between the North Pacific and the Arctic Ocean; no current exists that is comparable to the Atlantic Ocean's Norwegian Current that moves warm water to very high latitudes.

In the South Pacific Ocean, the southeast trade winds move the water to the left of the wind and westward, forming the **South Equatorial Current.** The westerly winds push the water to the east; at these southern latitudes the surface current so formed can move almost continuously around the earth. This current is the **West Wind Drift.** The tips of South America and Africa act to deflect a portion of this flow northward on the east side of both the South Pacific and South Atlantic oceans. As in the North Pacific, continuity currents form between the South Equatorial Current and the West Wind Drift. The **Peru Current,** or

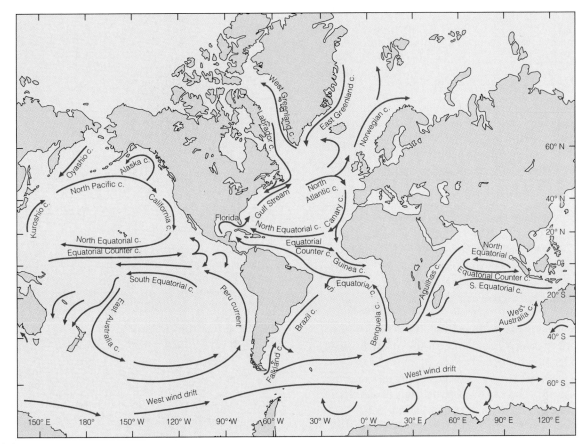

Figure 8.2
The major surface currents of the world's oceans.

Humbolt Current flows from south to north along the coast of South America, while the **East Australia Current** can be seen moving weakly from north to south on the west side of the ocean. These four currents form the counterclockwise South Pacific gyre.

The North Pacific and South Pacific gyres are formed not on either side of the equator (0°) but on either side of 5° N, because the doldrum belt is displaced northward due to the unequal heating of the Northern and Southern Hemispheres. Also between the North and South Equatorial currents and below the doldrums there is a current moving in the opposite direction, from west to east. This is a continuity current known as the **Equatorial Countercurrent,** which helps to return accumulated surface water eastward across the Pacific. Under the South Equatorial Current there is a subsurface current flowing from west to east called the **Cromwell Current.** This cold-water continuity current also returns water accumulated in the western Pacific.

The Atlantic Ocean Currents

The North Atlantic westerly winds move the water eastward as the **North Atlantic Current,** or **North Atlantic Drift.** The northeast trade winds push the water to the west, forming the **North Equatorial Current.** The north-south continuity currents are the **Gulf Stream,** flowing northward along the coast of North America, and the **Canary**

Current, moving to the south on the eastern side of the North Atlantic. The Gulf Stream is fed by the **Florida Current** and the North Equatorial Current. The North Atlantic gyre rotates clockwise. The polar easterlies provide the driving force for the **Labrador** and **East Greenland** currents that balance water flowing into the Arctic Ocean from the **Norwegian Current.**

In the South Atlantic, the westerlies continue the West Wind Drift. The southeast trade winds move the water to the west, but the bulge of Brazil splits the **South Equatorial Current.** Much of this flow is deflected northward over the top of South America, into the Caribbean Sea, and eventually into the Gulf of Mexico, where it exits as the Florida Current, which joins the Gulf Stream. A portion of the South Equatorial Current slides south of the Brazilian bulge along the western side of the South Atlantic to form the **Brazil Current.** The **Benguela Current** moves northward up the African coast. The South Atlantic gyre is complete, and it rotates counterclockwise.

Because much of the South Equatorial Current is deflected across the equator, the Equatorial Countercurrent appears only weakly in the eastern portion of the Mid-Atlantic. The northward movement of South Atlantic water across the equator means that there is a net flow of surface water from the Southern Hemisphere to the Northern Hemisphere. This flow is balanced by a flow of water at depth from the Northern Hemisphere to the Southern

Hemisphere. This deep-water return flow is the North Atlantic deep water, discussed in the previous chapter. Again, the equatorial currents are displaced north of the equator, although not as markedly as in the Pacific Ocean.

The Indian Ocean Currents

The Indian Ocean is mainly a Southern Hemisphere ocean. The southeast trade winds push the water to the west, creating the **South Equatorial Current.** The Southern Hemisphere westerlies still move the water eastward in the West Wind Drift. The gyre is completed by the **West Australia Current** moving northward and the **Agulhas Current** moving southward along the coast of Africa. Since this is a Southern Hemisphere ocean, the currents move to the left of the wind direction, and the gyre rotates counterclockwise. The northeast trade winds in winter drive the **North Equatorial Current** to the west, and the **Equatorial Countercurrent** returns water eastward toward Australia. Again, these equatorial currents are displaced approximately 5°N. With the coming of the wet monsoon season and its west winds, these currents are reduced. The strong seasonal monsoon effect controls the surface flow of the Northern Hemisphere portion of the Indian Ocean. In the summer the winds blow the surface water eastward, and in the winter they blow it westward. This strong seasonal shift is unlike anything found in either the Atlantic or the Pacific oceans.

8.2   Gyres and Current Flow

The Ekman Spiral and Ekman Transport

Wind-driven surface water sets the water immediately below it in motion. But due to low friction coupling in the water, this next deeper layer moves more slowly than the surface layer and is deflected to the right (Northern Hemisphere) or left (Southern Hemisphere) of the surface layer direction. The same is true for the next layer down and the next. The result is a spiral in which each deeper layer moves more slowly and with a greater angle of deflection than the layer above. This current spiral is called the **Ekman spiral,** after the physicist V. Walfrid Ekman, who developed its mathematical relationship. The spiral extends to a depth of approximately 100 to 150 m (330–500 ft), where the much-reduced current will be moving exactly opposite to the surface. Over the depth of the spiral, the average flow of all the water set in motion by the wind, or the net flow (**Ekman transport**), moves 90° to the right or left of the surface wind, depending on the hemisphere in which the motion takes place. This relationship is in contrast to the surface water, which moves at an angle of 45° to the wind direction (see fig. 8.3).

Geostrophic Flow

Wind-driven surface water is deflected toward the center of each current gyre by Ekman transport, producing a surface convergence. This convergence creates a mound of surface water that is elevated to more than 1 m (3 ft) above the

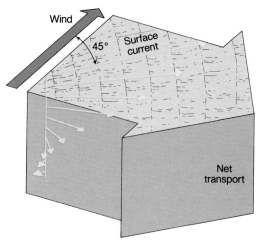

Figure 8.3

Water is set in motion by the wind. The direction and speed of flow change with depth to form the Ekman spiral. This change with depth is a result of the earth's rotation and the inability of water, due to low friction, to transmit a driving force downward with 100% efficiency. The net transport over the wind-driven column is 90° to the right of the wind in the Northern Hemisphere.

equilibrium sea level and also depresses the underlying denser water. The slope of the mound increases until the force of gravity forcing the water downhill and away from the gyre center equals the Coriolis effect acting to deflect the moving water into the raised central mound. At this balance point **geostrophic flow** is said to exist, and no further deflection of the moving water occurs. The currents flow smoothly around the gyre. See figure 8.4 for a diagram of this process. From the subsurface water-density distribution, oceanographers are able to calculate the slope of the sea surface and so calculate the velocity, volume transport, and depth of the currents present in the geostrophic flow around the mound.

     The Sargasso Sea is the classic example of such a situation. It is located in the central North Atlantic Ocean, and its boundaries are the Gulf Stream on the west, the North Atlantic current to the north, the Mid-Atlantic Ridge on the east, and the North Equatorial Current to the south. The circular motion of the current gyre isolates a lens of clear, warm water 1000 m (3000 ft) deep; the water is trapped by the geostrophic flow. The region is famous for the floating mats of *Sargassum,* a brown seaweed, stretching across its surface. The extent of the floating seaweed frightened early sailors, who told stories of ships imprisoned by the weed and sea monsters lurking below the surface. Except for the floating *Sargassum,* with its rich and specialized community of small plants and animals, the clear water is a near biological desert.

Current Speed

The wind-driven open-ocean surface currents move at speeds that are about one one-hundredth of the driving wind speed measured 10 m (30 ft) above the sea surface. The water moves between 0.25 and 1.0 knot, or 0.1 to 0.5 m/s (0.3–1.5 ft/s). Currents flow faster when a large volume

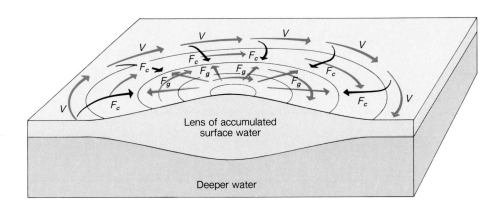

Figure 8.4
Geostrophic flow (*V*) exists around a gyre when F_c, the deflection due to the Coriolis effect, is balanced by F_g, the force due to gravity. The example is of a clockwise gyre in the Northern Hemisphere.

of water is forced to flow through a narrow gap. For example, both the North and South Atlantic equatorial currents flow into the Caribbean Sea, then into the Gulf of Mexico, and finally exit to the North Atlantic as the Florida Current through the narrow gap between Florida and Cuba. The Florida Current's speed may exceed 3 knots, or 1.5 m/sec (5 ft/sec).

Major ocean currents transport very large volumes of water. For example, the Gulf Stream's transport rate to the northeast is 55×10^6 m³/sec (1.9×10^9 ft³/sec); this rate is more than 500 times the flow of the Amazon River. The flow is distributed over the width and depth of the current. When the current expands its cross-sectional area, it slows down; when it decreases its cross-sectional area, it speeds up. Speed of flow then is not always directly related to surface wind speed but can be affected by the depth and width of the current as determined by land barriers, by the presence of another current, or by the rotation of the earth, as explained in the next section.

Western Intensification

In the North Atlantic and North Pacific the currents flowing on the western side of each ocean tend to be much stronger and narrower in cross section than the currents on the eastern side. This phenomenon is known as the **western intensification** of currents. The Gulf Stream and Kuroshio currents are faster and narrower than the Canary and California currents, although both the eastern and western currents transport about the same amount of water in order to preserve continuity of flow. Western intensification of currents traveling from low to high latitudes is related to (1) increases in the Coriolis effect with latitude, (2) the changing strength and direction of the east-west wind field (trade winds and westerlies) with latitude, and (3) the friction between landmasses and ocean water currents. These factors cause a compression of the currents toward the western side of the oceans, where water is moving from lower to higher latitudes. This compression requires that the current speed increase in order to transport the amount of water required in the circulation about the gyre. On the eastern side of the gyre, where currents are moving from higher to lower latitudes, the currents are stretched in the east-west direction. Here the current speed is reduced, but it still transports the required volume of water.

These fast-flowing, western-boundary currents move warm equatorial surface water to higher latitudes. Both the Gulf Stream and the Kuroshio Current bring heat from equatorial latitudes to moderate the climates of Japan and northern Asia (in the case of the Kuroshio) and the British Isles and northern Europe (in the case of the Gulf Stream, via the North Atlantic and Norwegian Currents). Western intensification is obscured in the South Pacific and South Atlantic, because both Africa and South America deflect portions of the West Wind Drift and create strong currents on the eastern sides of these oceans. The deflection of water from the Atlantic's South Equatorial Current to the Northern Hemisphere and the flow of water from the Pacific to the Indian Ocean through the islands of Indonesia also help to prevent the development of strongly flowing currents on the west side of Southern Hemisphere oceans.

8.3 Eddies

When a narrow, fast-moving current moves into the open sea, it displaces the quieter water through which it moves by the force of its flow and captures additional water as it does so. The current oscillates, meanders, and develops waves along its boundary, which break off to form **eddies,** or packets of water moving with a circular motion. Meandering waves hundreds of kilometers long may develop along sharp current boundaries. The large eddies that are cut off take with them energy of motion, which gradually dissipates, due to friction, sometimes far away from the parent current. If the energy of the parent current were not lessened in this way, the current's speed would continue to increase as the winds transfer energy into the water.

As the flow of the Gulf Stream moves away from the coast, it is more likely to develop a meandering path. At times, the western edge of the Gulf Stream develops indentations that are filled by cold water from the Labrador Current side. These indentations pinch off and become eddies, which are displaced to the east and south of the current boundary. The effect is to transfer cold water into warm water. Bulges at the western edge of the Gulf Stream are filled with the warm water from the Sargasso Sea. When these bulges are cut off they drift to the west of the Gulf Stream, into cold water. This process is illustrated in figures 8.5 and 8.6.

Figure 8.5

The western boundary of the Gulf Stream is defined by sharp changes in current velocity and direction. Meanders form at this boundary after the Gulf Stream leaves the U.S. coast at Cape Hatteras. The amplitude of the meanders increases as they move downstream (a and b). Eventually the current flow pinches off the meander (c). The current boundary re-forms, and isolated rotating cells of warm water (W) wander into the cold water, while cells of cold water (C) drift into the warm water (d).

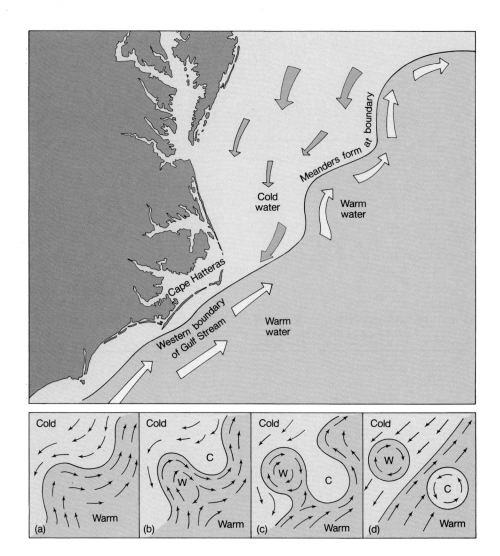

Figure 8.6

A satellite image of the sea surface reveals the warm (orange and yellow) and cold (green and blue) eddies that spin off the Gulf Stream. (Reddish-blue areas at top are the coldest waters.) These eddies may stir the water column right down to the ocean floor, kicking up blizzards of sediment.

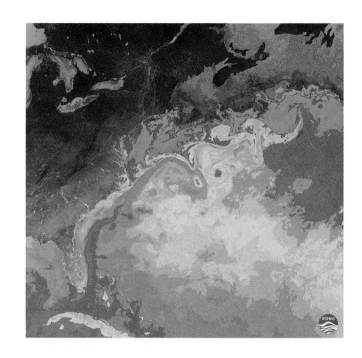

The meandering of strong currents (such as the Gulf Stream) and the formation of large eddies that maintain their physical identity for weeks as they wander about the ocean produce surface flow patterns that differ markedly from the uniform current flows shown on current charts. Current charts show the average current flow, not the daily or weekly variations.

Large and small eddies generated by horizontal flows or currents exist in all parts of the oceans; these eddies are of varying sizes, ranging from tens to several hundred kilometers in diameter. Each eddy contains water with specific chemical and physical properties and maintains its identity as it wanders through the oceans. Eddies may appear at the sea surface or be embedded in waters at any depth (see fig. 8.7).

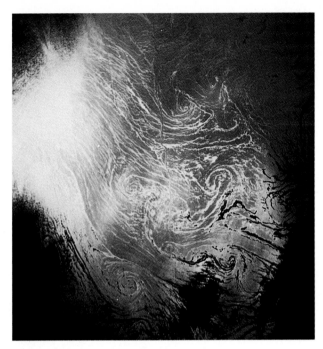

Figure 8.7
Space-shuttle view. Sunlight reflected off the Mediterranean reveals spiral eddies; their effects on climate are being monitored.

Eddies may rotate in a clockwise or counterclockwise direction. These eddies stir the ocean until they gradually dissipate due to fluid friction, losing their chemical and thermal identity and their energy of motion. The water properties of an eddy allow scientists to determine its origin. Small surface eddies found 800 km (500 mi) southeast of Cape Hatteras in the North Atlantic have been found with water properties of the eastern Atlantic near Gibraltar, more than 4000 km (2500 mi) away. These eddies from the Strait of Gibraltar are formed from the salty water of the Mediterranean as it spills into the Atlantic and have been nicknamed "Meddies." Deep-water eddies at Cape Hatteras come from the eastern and western Atlantic, the Caribbean, and Iceland. Researchers estimate that these small eddies are several years old; age determination is based on drift rates, distance from source, and biological consumption of oxygen.

The rotation rate of the large eddies that form at the western boundary of the Gulf Stream is about 0.51 m/sec (1 knot), but because of the water's density, the force of the flow is similar to that generated by a 35-knot wind. The diameter of the eddies may be as much as 325 km (200 mi), and they may reach to the seafloor. At the seafloor the rotation rate is zero; therefore a few meters above the bottom the speed of rotation diminishes very rapidly, and considerable turbulence is generated as the energy of the eddy is dissipated. These eddies are similar in many ways to atmospheric pressure cells, and they are sometimes called abyssal storms. As the eddies wander through the oceans, they stir up bottom sediments, producing ripples and sand waves in their wakes; they also mix the water, creating homogeneous water properties over large areas. Eventually the eddies lose their energy to turbulence and blend into the surrounding water.

Eddies constantly form, migrate, and dissipate at all depths. Eddy motion is superimposed on the mean flow of the oceans. To understand the role of eddies in mixing the oceans we need more data and better tracking of eddy size and position. Satellites are important tools for detecting surface eddies because they can precisely measure temperature, increased elevation, and light reflection of the sea surface; see figure 8.7. Deep-water eddies are monitored by using instruments designed to float at a density layer close to the seafloor. The instruments are caught up in the eddies, moving with them and sending out acoustic signals that are monitored through the Sofar channel. In this way the rate of deep-water eddy formation, the numbers of major eddies, their movements, and their life spans can be observed.

8.4 Convergence and Divergence

When wind-driven surface currents collide with each other or are forced against landmasses, they are said to converge or to produce a surface **convergence.** When surface currents move away from each other or away from a landmass, the currents diverge, and a surface **divergence** is formed. At a convergence, the accumulated surface water sinks; at a divergence, water rises from depth to fill the space left by the diverging flow.

Permanent Zones

There are three major zones of convergence: the **tropical convergence** at the equator, the **subtropical convergences** at approximately 30° to 40°N and S, and the **Arctic** and **Antarctic Convergences** at about 50°N and S. Surface convergence zones are regions of downwelling. These areas are low in nutrients and biological productivity. There are two major divergence zones: the **tropical divergence** and the **Antarctic divergence.** Upwelling associated with divergences delivers nutrients to the surface waters to supply the food chains that support the tropical tuna fisheries and the largely unexploited but richly productive waters of Antarctica. These areas of divergence and convergence are shown in figure 8.8.

In coastal areas where trade winds move the surface waters away from the western side of continents, upwelling occurs nearly continuously throughout the year (for example, off the west coasts of Africa and South America, both of which are very productive and yield large fish catches).

Seasonal Zones

Off the west coast of North America, downwelling and upwelling occur on a seasonal basis as the Northern Hemisphere temperate wind pattern changes from southerly in winter to northerly in summer. The downwelling and upwelling occur because of the change in direction of Ekman transport. Remember that the wind-driven Ekman transport moves at an angle of 90° to the right or left of the wind direction, depending on the hemisphere.

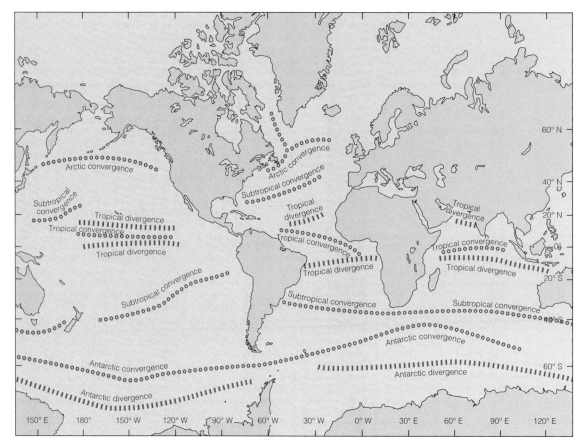

Figure 8.8

The principal zones of surface convergence and divergence associated with the surface currents.

Along this coast, the average wind blows from the north in the summer, and the net movement of the water is to the west, or 90° to the right of the wind. This action results in the offshore movement of the surface water and the upwelling of deeper water along the coast to replace it. The upwelling zone is evident from central California to Vancouver Island. In winter, the winds blow from the south; the wind-driven surface waters move to the east, onshore against the coast; and downwelling occurs. See figure 8.9. The summer upwelling pattern is what produces the band of cold coastal water at San Francisco that helps cause the frequent summer fogs (refer back to chapter 4). Because of the lack of land at the middle latitudes in the Southern Hemisphere, this type of seasonal upwelling is less common.

Convergence and divergence of surface currents result not only in upwellings and downwellings, but also in the mixing of water from different geographic areas. Waters carried by the surface currents converge, share properties due to mixing, and form new water mixtures with specific ranges of temperature and salinity, which then sink to their appropriate density levels. After sinking, the water moves horizontally, blending and sharing its properties with adjacent water, and eventually rises to the surface at a new location. This process is discussed as thermohaline circulation in the section on density-driven circulation in

chapter 6. Thermohaline circulation and wind-driven surface currents are closely related. The connections are so close among formation of water mixtures, upwelling and downwelling, and convergence and divergence, that it is difficult to assign a priority of importance to one process over another. Changes in these flows and events that might cause such changes are discussed in the next section.

8.5 Global Circulation Changes

Over the history of the earth interactions between thermohaline and wind-driven circulation have changed many times. Changes in winds and sea-surface processes occurred during previous glacial and interglacial periods. Farther back in time the changing shapes of the ocean basins and locations of the continents influenced ocean circulation. In 1988 cores obtained near Antarctica by the *JOIDES Resolution* indicated that significant temperature changes have occurred in the ocean depths at these latitudes over the last 45 million years. At present, water temperatures in this region average approximately 2°C, with water temperatures at the 3000-m (9800-ft) depth about 0.5°C colder than water at a nearby 2000-m (6500-ft) site. By analyzing the oxygen isotopes from these cores, researchers

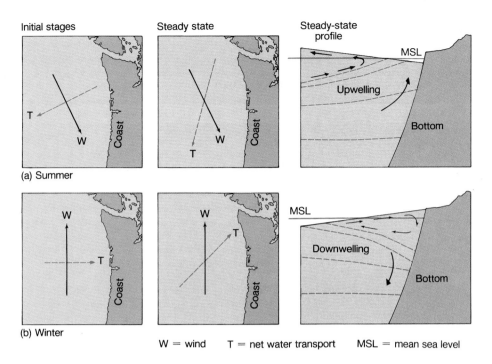

Initial stages   Steady state   Steady-state profile

(a) Summer

(b) Winter

W = wind   T = net water transport   MSL = mean sea level

Figure 8.9

Initially, the Ekman transport is 90° to the right of the wind along the northwest coast of North America. As water is transported (a) away from the shore in summer, or (b) toward the shore in winter, a sea-surface slope is produced. This slope creates a gravitational force that alters the direction of the Ekman transport to produce a geostrophic balance and a new steady-state direction.

From Sverdrup/Johnson/Fleming, *The Oceans,* © 1942, renewed 1970, p. 501. Adapted by permission of Prentice-Hall, Inc., Englewood Cliffs, New Jersey.

have determined that 45 million years ago the average water temperature was 13°C, and it was 2°C warmer at the deeper 3000-m site than at the 2000-m site.

It is thought that at this time the earth was warmer and the sea level was higher, creating extensive shallow seas in subtropical regions. The evaporation rate over such seas could have been very high and could have resulted in dense, high-salinity, warm water which would have sunk to the deep-ocean floor. Thermohaline circulation would have required this water to flow toward the poles, where it would have risen toward the surface and warmed the polar regions. Such a water movement would have been the reverse of today's deep-water circulation system. The sea-level changes required to produce the warm shallow seas could have been caused by the upward displacement of the oceans triggered by a small upward displacement of the seafloor as the midocean ridges were formed. This hypothesis indicates that a small episode can set off or trigger a sequence of events that creates significant large-scale changes in ocean circulation and world climate.

The triggering episode may be nonperiodic, for example, a collision of the earth with an asteroid or a sudden eruption of large-scale vulcanism, or it may be periodic and repeat in a regular cycle. The periodic repositioning of the earth to the sun that causes changes in the distribution of solar energy over the earth's surface is an example of a cyclical phenomenon that is thought to, in part, trigger large-scale climate change. Milutin Milankovitch, a Serbian mathematician, worked out a theory explaining how variations in the earth's orbit alter the amount and distribution of solar radiation over the earth's surface, today called the Milankovitch cycles. These variations include (1) cyclic

changes in the shape of the earth's orbit, from more elliptical to more circular with a period of about 100,000 years; (2) cyclic changes in the tilt of the earth's axis of rotation, varying from 22° to 24.5° every 41,000 years; and most importantly, (3) the precession, or movement of the earth's axis of rotation about a circle, with a period of about 23,000 years. This latter motion gradually causes the Northern Hemisphere's winter to occur at different positions during the earth's orbit about the sun. Today, the winter solstice in the Northern Hemisphere occurs on December 21, when the earth is closest to the sun. In about 12,000 years, the precession of the axis will cause the winter solstice in the Northern Hemisphere to occur when the earth is farthest from the sun.

At present, the earth is in that part of the Milankovitch cycles during which the Northern Hemisphere summers are expected to cool, because when the Northern Hemisphere is tilted toward the sun, the earth is farthest from the sun. Permanent snow fields could be expected to grow as snow accumulates in winter, to be followed by less summer melting due to lower temperatures and increased surface reflectance by the snow. Under these conditions, a gradual increase in permanent land ice in the Northern Hemisphere's high-altitude landmasses is possible and could lead to the onset of a glacial period during the next several thousand years.

This cyclic behavior is a part of the earth's normal dynamic system. Recently, however, humans have begun to change the properties of the earth's surface and atmosphere with increasing rapidity and at a level that is becoming increasingly significant. The greenhouse effect is projected to warm the earth during the time the earth's cooling period

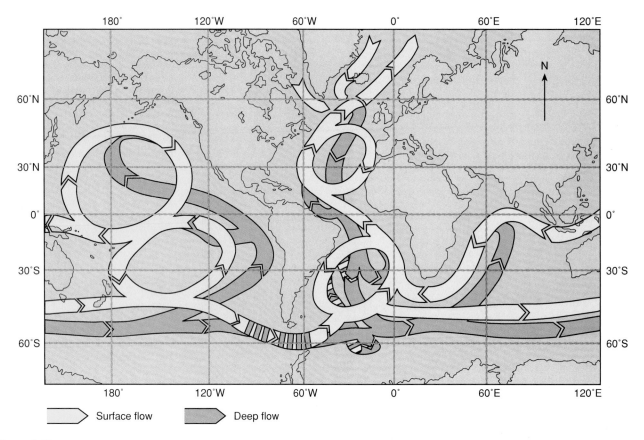

| Surface flow | | Deep flow |

Figure 8.10

The present flow of ocean water within and between oceans circulates at the surface (yellow arrows) and at depth (green arrows). Cold surface water sinks to the seafloor in the North Atlantic Ocean, then flows south to be further cooled by the Antarctic Bottom water formed in the Weddell Sea. This deep water moves eastward around Antarctica, feeding into the surface layer of the Indian Ocean and also into the deep basins of the Pacific Ocean. A return flow of surface water from the Pacific and Indian oceans flows north to replace the surface water in the North Atlantic Ocean.

begins in the Northern Hemisphere. Could this warming action trigger a third set of conditions, inducing large-scale changes in the earth's climatic system?

In the North Atlantic the salinity of the water is higher than in the North Pacific because evaporation exceeds precipitation, while in the Pacific Ocean precipitation exceeds evaporation. The temperature of the North Atlantic water is also warmer at high latitudes, and this surface heat transported from low Atlantic latitudes is very important in warming the air and moderating the climate of northern Europe. Remember that North Atlantic surface waters move northward, lose their heat, cool, mix with Arctic water, and sink, then move at depth to the Southern Hemisphere to join the circumpolar water feeding into the Pacific Ocean. Some of this water slowly migrates northward at depth in the Pacific, rises to the surface, and flows southward to lower latitudes. A flow of surface water from the Pacific to the Atlantic by way of the Indian Ocean completes the flow. This interchange of water between oceans redistributes heat, salt, and dissolved gases (see fig. 8.10).

Is it possible that some small-scale phenomenon could stop this water motion and trigger changes that would have major consequences? By studying the records of past earth surface temperatures laid down in marine sediments,

it can be shown that the temperature of the North Atlantic Ocean and the accumulation and melting of land ice approximately relate to the Milankovitch cycles. The coldest ocean temperatures and the maximum land ice volumes occurred about 6000 years after the minimum solar radiation values of 23,000 years ago. This time lag indicates it takes time for the earth to respond to changes in the solar energy budget.

The warming of the oceans to today's temperatures occurred very rapidly when the earth returned to high solar radiation values 12,000 years ago, again approximately associated with the Milankovitch cycle. Water is a good absorber of solar radiation while ice is a poor absorber and a good reflector, so the land ice melted slowly as the oceans warmed rapidly. Then 11,000 years ago, the surface temperatures of the water in the North Atlantic and air temperatures in northern Europe suddenly decreased again and remained low for about 700 years in a mini-ice age, known as the Younger Dryas event. Large changes (20%) in atmospheric carbon dioxide gas trapped in ice cores of Greenland are associated with this period. This is interpreted to mean that large global atmospheric changes occurred at the same time as the localized cooling of Europe and the North Atlantic. However, this could not have happened unless major changes occurred in the chemical cycles of the earth involving the oceans as part of the system that transports and stores carbon dioxide.

The required connections between the global atmospheric carbon dioxide event and the ocean, and the restriction of decreased temperature to the North Atlantic and northern Europe has led Wallace Broecker of Lamont Doherty Geophysical Observatory to think that something triggered a sudden shutdown of the circulation of the Northern Atlantic; something that prevented the transport of warm surface water northward and interfered with the formation of North Atlantic deep water, which greatly reduced its southward flow at depth. This, of course, would affect oceanic circulation worldwide and alter the oceans' ability to transport and store carbon dioxide, which in turn affects atmospheric carbon dioxide levels.

The trigger proposed by Broecker is the sudden change in the position of meltwater from the receding North American ice sheet entering the Atlantic Ocean, which would have significant effects on the salinity and density of the surface waters of the Atlantic. This would alter the present thermohaline circulation and change the ocean surface temperature as cold, low-density, low-salinity surface water absorbed solar radiation and reduced downward heat and water transfer.

Geological studies support Broecker's ideas. Research shows the majority of the meltwater from the early stages of the receding glaciers in North America converged on the Mississippi River system and drained into the Gulf of Mexico, not directly into the Atlantic. Gradually, as the ice margin receded, meltwater was diverted through the Hudson River drainage system. Further recession of the ice then filled up the Great Lakes and the St. Lawrence River valley, which had been depressed by the weight of the ice. This channel gradually captured most of the meltwater flow and channeled it into the North Atlantic, diluting and cooling the surface water so it no longer supplied heat to the atmosphere that, in turn, had warmed northern Europe.

At the same time, this fresher surface water did not sink and, therefore, the production of North Atlantic deep water was slowed dramatically, which prevented the northward migration of warm equatorial water and affected the circulation in all oceans. This event ceased as rapidly as it started. As the melting of the glaciers slowed, the volume of meltwater was reduced, and surface warming and the thermohaline circulation of the North Atlantic began again.

The lesson to be learned from Broecker's hypothesis and other observations of past changes in land and ocean climate is that, although the earth responds in a cyclic and orderly manner to natural cycles, singular events may trigger sudden departures from normal or expected trends. The El Niño and La Niña events of the 1970s and early 1980s and the low-salinity cold-water pool that occurred in the North Atlantic during approximately the same period were discussed in chapter 7. These are seen by some as mini-versions of the scenario developed by Broecker. In the case of the

greenhouse effect, the warming of the earth to new and higher levels, in opposition to the normal cyclic cooling Milankovitch trend, may trigger a sudden shift to a new energy balance and new climatic conditions. If this were to happen, the earth would be operating outside of its presently understood and assumed rules, and predictive models based on known rules would not function correctly. We would have little understanding of what might happen next.

8.6 Measuring the Currents

Direct measurements of currents fall into two groups: (1) those that follow a parcel of the moving water, and (2) those that measure the speed and direction of the water as it passes a fixed point. Moving waters may be followed with buoys designed to float at predetermined depths. These buoys signal their positions acoustically to the research vessel, allowing those on board to follow their paths and calculate their speed and displacement due to the current. Surface water may be labeled with buoys or with dye that can be photographed from the air. Buoy positions may also be tracked by satellite. A series of pictures or position fixes may be used to calculate the speed and direction of a current from the buoys' drift rates. These buoys can be instrumented to measure other water properties such as temperature (fig. 8.11).

A variation of this technique uses the **drift bottle.** Thousands of sealed bottles, each containing a postcard, are released at a known position. When the bottles are washed ashore the finders are requested to record the time and location of the find and return the card. In this case only the release point, the recovery points, and the elapsed time are known; the actual path of motion is assumed. See Box: The Great Sneaker Spill.

Sensors used to measure current speed and direction are called **current meters.** They consist of a rotor to measure speed and a vane to measure direction of flow (see fig. 8.12). If the current meter is lowered from a stationary vessel, the measurements can be returned to the ship by a cable or can be stored by the meter for reading upon retrieval. If the current meter is attached to an independent, bottom-moored buoy system, the signals can be transmitted to the ship as radio signals or can be stored on tape in the current meter, to be removed when the buoy and current meter are retrieved.

In order to measure a current passing a location, a current meter must not move. Although a vessel can be moored in shallow water so that it does not move, it is very difficult, if not impossible, to moor a ship or surface platform in the open sea so that it will not move and by its motion also move the current meter. The solution is to attach the current meter to a buoy system that is entirely submerged and not affected by winds or waves. See figure 8.13 for a diagram of this taut wire moorage system. The string of anchors, current meters, wire, and floats is preassembled on deck and is launched, surface float first, over

Looking for Carbon: A Worldwide Effort

The Joint Global Ocean Flux Study, or JGOFS, is a multinational effort to understand the oceans' role in the global carbon cycle. This program has two primary goals: (1) to understand the ways that carbon cycling interacts with the atmosphere, the seafloor, and the coastal regions; and (2) to improve our ability to predict the oceans' response to climatic changes related to human activities.

In 1988 JGOFS installed carbon dioxide monitoring stations near Bermuda and Hawaii; three years' data are now being assessed. In 1989 a five-nation study of the North Atlantic used ships, planes, and satellites to follow the path of carbon dioxide. In the Arctic and Antarctic other teams collect data from land bases and ice island installations. In early 1992 ships and planes from the U.S., Canada, Australia, New Zealand, China, and Japan began a series of large-scale surveys of the central equatorial Pacific. All these efforts seek to answer the question of the role of the oceans in removing human-generated carbon dioxide from the atmosphere.

One part of the answer requires understanding the transfer to the seafloor of carbon trapped in organic molecules, the remains of living creatures. Single-celled photosynthetic organisms acquiring carbon from carbon dioxide live in the upper, sunlit waters where they either die or are eaten by other organisms. Their remains aggregate into particles that sink slowly downward, serving as food for the organisms in the lower depths. In the end only about 1% of this material reaches the seafloor.

To determine the rate at which carbon is transferred to the ocean depths and the sediments, specialized sediment traps are deployed from research vessels to catch the particles as they fall through the water. Multiple sets of differently styled traps are attached to surface buoys and set adrift, or a submerged mooring may be used (see Box figure 8.1). The different moorings allow placement of the traps over a wide range of depths. Comparing results from different types of traps allows the efficiency of each type to be gauged; comparisons between traps of the same type ensure statistically significant data.

Box figure 8.1
Deploying particle and sediment traps from the stern of a research vessel.

Sediment trap studies have shown that transfer rates vary from place to place and from depth to depth. There are more particles in the rich coastal waters and upwelling areas; settling varies from season to season and year to year, while waters at different depths can store the organic carbon and its decay product, carbon dioxide, for different lengths of time. It is estimated that from 3 to 6 billion metric tons of carbon are transferred each year to the deep ocean in this way. However, the increasing inputs of carbon dioxide to the atmosphere from human activities do not appear to change the rate or the amount of carbon transferred in the ocean system.

Other studies follow surface water with its dissolved gases as it sinks, travels slowly for long periods of time at great depths, and then rises to exchange its modified gases with the atmosphere at some other location. In the tropics, upwelling of water from moderate depths liberates accumulated CO_2 to the atmosphere at a rate faster than the plants can incorporate it into organic matter, even in upwelling areas that bring nutrients to the surface plants.

In 1993–94 these studies are being supplemented by data from the new SeaWIFS satellite sensor designed to measure chlorophyll from plant production at the sea surface, and future work is planned for the Arabian Sea and the Southern Ocean. Gradually both the global role of carbon dioxide in the oceans and its impact on climate control will be better understood; we must have this understanding if we are to predict the effects of global warming for the entire earth system.

Figure 8.11
Surface buoys (red) carrying instruments
are deployed from the research vessel
Thomas G. Thompson.

(a)

Figure 8.12

An internally recording Aanderaa current meter. The vane orients the
meter to the current while the rotor determines current speed (a). Inside the
protective case (b), the instrument records coded information from its
external sensors. The speed and direction of the current, water temperature,
pressure, and conductivity of the water are recorded on magnetic tape. A
clock controls the frequency at which these data are recorded. The
magnetic tape is retrieved when the current meter is picked up from its taut
wire mooring.

(b)

the stern of the slowly moving ship. As the ship moves
away the float, meters, and cable are stretched out on the
surface. When the vessel reaches the sampling site the
anchor is pushed overboard to pull the entire string down
into the water. The floats and meters are retrieved by grappling from the surface for the ground wire or by sending a
sound signal to a special acoustical link, which detaches the
wire from the anchor. The anchor is discarded, and the
buoyed equipment returns to the surface. The technique is
straightforward, but many problems can occur in launching,

finding, and retrieving instruments from the heaving deck
of a ship at sea. Whenever oceanographers send their
increasingly sophisticated equipment over the side, they
must cross their fingers and hope to see it again.

A new technique for measuring currents does not
need the energy of the moving water to turn the rotor of a
current meter. This technique uses sound echoes and takes
advantage of the change in the pitch or frequency of sound
as it passes through moving water. When sound moves with
the moving water the pitch increases; when it moves

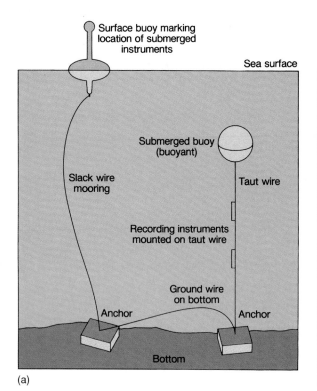

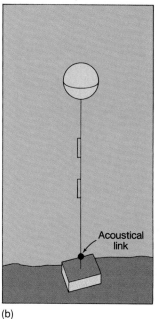

Figure 8.13

Taut wire moorage. (a) Recovery is accomplished by retrieving the surface buoy and hauling in the wire. If the surface buoy is lost, it is possible to grapple for the ground wire. (b) In this system, a sound signal disconnects the anchor, and the equipment floats to the surface.

against the moving water the pitch decreases. This is the **Doppler effect;** the same effect that increases the pitch of the horn or siren of an approaching vehicle and decreases the pitch as the vehicle passes and moves away. To use this method the sound source is held stationary, on a vessel or a buoy system, and two sound beams of precisely known frequency are sent out at right angles. The change in pitch of the echoes provides the speed and direction of the water along each of the two paths; these are used to compute the speed and direction of the resulting current.

8.7 Practical Considerations: Energy from the Currents

The massive oceanic surface currents of the world are untapped reservoirs of energy. Their total energy flux has been estimated at 2.8×10^{14} (280 trillion) watt-hours. Because of their link to winds and surface heating processes, the ocean currents are considered as indirect sources of solar energy. If the total energy of a current was removed by conversion to electric power, that current would cease to exist; but only a small portion of any ocean current's energy can be harnessed, owing to the current's size. Harnessing the energy from these open-ocean currents requires the use of turbine-driven generators anchored in place in the current

stream. Large turbine blades would be driven by the moving water, just as windmill blades are moved by the wind; these blades could be used to turn the generators and to harness the energy of the water flow. (See also the discussion on energy from tidal currents in chapter 10.) Another proposal calls for a barge moored in the current stream with a large cable loop to which parachutes are fastened. The cable would be moved along by the current acting against the open parachutes. When the parachutes reached the end of the loop they would turn the corner and be dragged back against the current while closed. The continuous movement of the cable would be used to turn a generator to produce electricity.

The Florida Current and the Gulf Stream are reasonably swift and continuous currents moving close to shore in areas where there is a demand for power. If ocean currents are developed as energy sources, these currents are among the most likely. But most of the wind-driven oceanic currents generally move too slowly and are found too far from where the power is needed. The impact on other uses of the sea—transport, fishing, recreation—needs to be considered. The cost of constructing, mooring, and maintaining current-driven power-generating devices in the open sea makes them noncompetitive with other sources of power at this time.

The Great Sneaker Spill

In May 1990 a severe storm in the North Pacific caused the container ship, *Hansa Carrier,* en route from Korea to the United States to lose overboard twenty-one deck-cargo containers, each approximately forty feet long (Box figure 1). Among the items lost were 39,466 pairs of NIKE brand athletic shoes, and these began washing up along the beaches of Washington, Oregon, and British Columbia six months to a year later (Box figure 2). Although the shoes had been drifting in the ocean for almost a year, they were wearable after washing and having the barnacles and the oil removed. However, the shoes of a pair had not been tied together for shipping, and pairs did not come ashore together. As beach residents recovered the shoes (some with a retail value of $100 a pair), swap meets were held in coastal communities to match the pairs.

In May of 1991 I was having lunch with my parents, Gene and Paul, and Gene pointed out the news of the beached NIKES. I was intrigued and realized that 78,932 shoes was a very large number of drifting objects compared to the 33,869 drift bottles used in a 1956–59 study of North Pacific currents. I contacted Steve McLeod, an Oregon artist and shoe collector, who had information on locations and dates for some 1600 shoes that had been found between northern California and the Queen Charlotte Islands in British Columbia. I contacted additional beachcombers and constructed a map showing the times and locations where batches of

Box 2 figure 1
After the storm the *Hansa Carrier* docked in Seattle.

Box 2 figure 2
Beached at last.

100 or more shoes had been found (Box figure 3). Next I visited Jim Ingraham at NOAA's National Marine Fisheries Service's offices in Seattle to study his computer model of Pacific Ocean currents and wind systems north of 30° N latitude. Using the spill date (27 May 1990), the spill location (161°W; 48°N), and the dates of the first shoe landings on Vancouver Island and Washington State beaches between Thanksgiving and Christmas 1990, I found that the shoe drift rates agreed with the computer model's predicted currents.

News of Jim's and my interest in the shoe spill reached an Oregon news reporter and was then picked up by the Associated Press resulting in a quick dissemination of the news nationwide. Readers sent letters describing their own shoe finds, and even reports of single shoes were valuable because we found that each shoe had within it a NIKE purchase order number which could be traced to a specific cargo container. We were able to determine from these numbers that only four of the five containers broke open so that only 61,820 shoes were left afloat.

continued

The computer model and previous experiments with satellite tracked drifters showed that there would have been little scattering of the shoes as the ocean currents carried them eastward and approximately 1500 miles from the spill site to shore, but the shoes were found scattered from California to northern British Columbia. The north-south scattering is related to coastal currents that flowed northward in winter carrying the shoes to the Queen Charlotte Islands and southward in spring and summer bringing the shoes to Oregon and California.

I was interested to see where the shoes might have gone if they had been lost on the same date but under different conditions in other years. The computer allowed simulations for May 27 of each year from 1946 to 1991. Box figure 4 shows the wide variation in model-predicted drift routes. If the shoes had been lost in 1951 they would have traveled in the loop of the Alaska Current. If they had been lost in 1982 they would have been carried far to the north during the very strong El Niño of 1982–83, and if lost in 1973 they would have come ashore at the Columbia River.

A Japanese film crew made a video for national broadcast and I made a plea for information concerning any shoes that might arrive in Japan. Because it takes about four and a half years for an object to drift completely around the north Pacific gyre, the shoes should begin to arrive on Japanese beaches during 1993. Perhaps Japanese beachcombers will add another chapter to this story.

Curtis C. Ebbesmeyer, Ph.D.
Evans-Hamilton, Inc.
Seattle, Washington

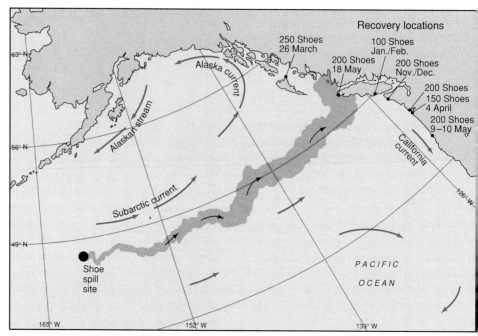

Box 2 figure 3

Site where 80,000 Nike shoes washed overboard on May 27, 1990, and dates and locations where 1300 shoes were discovered by beachcombers (dots at upper right). Drift of the shoes is simulated with a computer model (colored plume).

Box 2 figure 4

Projected drift tracks for the sneakers for the years 1951, 1973, 1982, 1988, and 1990 based on computer modeling of ocean currents and weather.

Figures 3 and 4 from Curtis E. Ebbesmeyer and W. James Ingraham, Jr., "Shoe Spill in the North Pacific" in *American Geophysical Union EOS, Transactions,* Vol. 73, No. 34, August 25, 1992, pp. 361, 365.

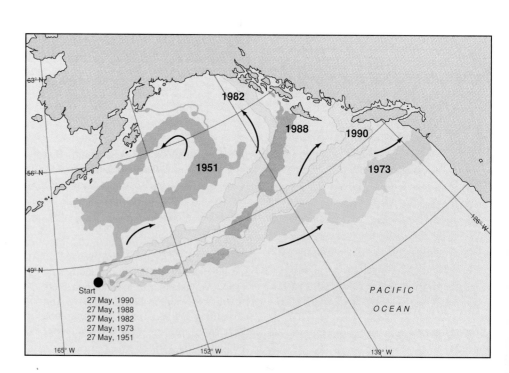

Summary

Winds push the surface water 45° to the right of their direction in the Northern Hemisphere and 45° to the left in the Southern Hemisphere. This action creates the large surface current gyres observed in each ocean. Southern Hemisphere gyres rotate counterclockwise; Northern Hemisphere gyres rotate clockwise. The currents of the northern Indian Ocean change with the seasonal monsoons.

Geostrophic flow is produced when the force of gravity balances the Coriolis effect. The water in the Sargasso Sea is isolated by this circular flow.

A current that is transporting a given volume of water flows faster through a smaller area than through a larger one. The speed of a current is affected by the current's cross-sectional area, by other currents, by westward intensification, and by wind speed. Eddies are formed at the surface when a fast-moving current develops waves along its boundary that break off from the parent current. Eddies occur at all depths, wander long distances, and gradually lose their identity.

Downwelling is produced by converging surface currents, while upwelling is produced by diverging surface currents. Upwelling occurs nearly continuously along the western sides of the continents in the trade wind belts, where water diverges from the coast. Seasonal upwellings and downwellings occur in coastal areas that have changing wind patterns and an alternating coastal flow of water onshore and offshore due to the Ekman transport.

A variety of techniques are available to measure currents: by following the water, by measuring the water's speed and direction as it moves past a fixed point, or by using changes in the pitch of sound.

Deriving energy from the oceanic current flows is not practical at the present time.

Key Terms

gyre	subtropical convergence
Ekman spiral	Arctic convergence
Ekman transport	Antarctic convergence
geostrophic flow	tropical divergence
western intensification	Antarctic divergence
eddy	drift bottle
convergence	current meter
divergence	Doppler effect
tropical convergence	

Study Questions

1. According to the net transport of the Ekman spiral, wind-driven water is directed toward the center of a large oceanic current gyre. Why does the current not flow to the gyre's center, but instead flows in a clockwise circular path about a gyre in the Northern Hemisphere?
2. How is Ekman transport related to coastal upwelling and downwelling?
3. On a map of the world, plot the oceanographic equator, the six major wind belts, the current system of each ocean, and the main areas of convergence and divergence.
4. Why does a flow of water that is constant in volume transport per time increase its speed when it passes through a narrow opening?
5. Explain the way in which winds and currents are named for direction.
6. What are eddies? How are they formed? Where can they be found?
7. Explain why the large Northern Hemisphere midocean gyres tend to flow in a circular path, although the driving force of the winds is to the east on the higher-latitude side of the gyre and to the west on the lower-latitude side.
8. Explain why surface convergences are located at the centers of the large subtropical gyres in both the northern and southern Atlantic and Pacific oceans.
9. Explain why upwellings on the lee side of continents at tropical latitudes function continuously while upwelling on the west side of North America is seasonal.
10. Why is there a net northward flow of surface water across the equator in the Atlantic Ocean but not in the Pacific Ocean?
11. Why would a decrease in the rate of formation of North Atlantic deep water create a decrease in the northward transport of warm surface water by the Gulf Stream?
12. How can the cross section of a current and its average speed be used to calculate the volume of water transport?
13. Why is there a net flow of surface water from the Pacific through the Indian to the Atlantic Ocean?
14. Why is current information obtained from drift bottles limited?
15. How is the Doppler effect used to measure ocean currents?

Suggested Readings

The Water in Motion

Barkley, R. A. 1970. The Kuroshio Current. In *Oceanography, Contemporary Readings in Ocean Sciences,* 2d ed., Pirie, R. G., ed. Oxford Univ., New York. pp. 70–80.

Bowditch, N. 1984 ed. *American Practical Navigator,* Vol. 1. U.S. Defense Mapping Agency Hydrographic Center, Washington, D.C. 1414 pp. (Ocean currents are covered in chapters 31 and 32.)

Broecker, W. S. 1987. The Biggest Chill. *Natural History* 96 (10): 74–80, 82.

Cromie, W. J. 1988. Grappling with Coupled Systems. *Mosaic* 19 (3/4): 12–23.

Dymond, J. 1992. Particles in the Oceans. *Oceanus* 35 (1): 60–67.

Fisher, A. 1988. One Model to Fit All. *Mosaic* 19 (3/4): 52–59.

Huyghe, P. 1990. The Storm Down Below. *Discover* 11 (11): 71–76.

Joyce, T., and P. Wiebe. 1983. Warm-Core Rings of the Gulf Stream. *Oceanus* 26 (2): 34–44.

Kunzig, R. Where the Water Goes. *Discover* 12 (8): 26, 37.

MacLeish, W. H. 1989. The Blue God, Tracking the Mighty Gulf Stream, *Smithsonian* 19 (11): 44–59.

MacLeish, W. H. 1989. Painting A Portrait of the Stream from Miles Above and Below. *Smithsonian* 19 (12): 42–55.

Open University. 1989. *Ocean Circulation.* Pergamon Press, Oxford, England; Open University, Milton Keynes, England. 190 pp.

Richardson, P. L. 1991. SOFAR Floats Give a New View of Ocean Eddies. *Oceanus* 34 (1): 23–31.

Scientific Plan for the World Ocean Circulation Experiment. 1986. World Climate Research Program. WCRP Publication Series 6, WMO/TD No. 122. 83 pp.

Tolmazin, D. 1985. *Elements of Dynamic Oceanography.* Allen and Unwin, Boston. 181 pp.

Whitehead, J. A. 1989. Giant Ocean Cataracts. *Scientific American* 260 (2): 50–57.

Whitworth III, T. 1988. The Antarctic Circumpolar Current. *Oceanus* 31 (2): 53–58.

Wiebe, P. H. 1982. Rings of the Gulf Stream. *Scientific American* 246 (3): 60–70.

Energy

Brin, A. 1981. *Energy and the Oceans.* Westbury House, Surrey, England. 133 pp.

Bruce, M. 1986. Ocean Energy: Some Perspectives on Economic Viability. In *Ocean Yearbook 6,* Borgese, E. M., and N. Ginsburg, eds. Univ. of Chicago, Chicago. pp.58–78.

Charlier, R. H. 1983. Water, Energy, and Nonliving Ocean Resources. In *Ocean Yearbook 4,* Borgese, E. M., and N. Ginsburg, eds. Univ. of Chicago, Chicago. pp. 75–120. (Overview of energy sources.)

Simeons, C. 1980. *Hydro-Power, The Use of Water As an Alternative Source of Energy.* Pergamon, Elmsford, N.Y. 549 pp.

9

The Waves

9.1 **How a Wave Begins**
9.2 **Anatomy of a Wave**
9.3 **Wave Motion**
9.4 **Wave Speed**
9.5 **Deep-Water Waves**
  Storm Centers
  Swell
  Dispersion
  Group Speed
  Wave Interaction
9.6 **Wave Height**
  Episodic Waves
  Wave Energy
  Wave Steepness
9.7 **Shallow-Water Waves**

Box: *Universal Sea State Code*

  Refraction
  Reflection
  Diffraction
  Navigation from Wave Direction
9.8 **The Surf Zone**
  Breakers
  Water Transport
  Energy Release
9.9 **Tsunamis**
9.10 **Internal Waves**
9.11 **Standing Waves**
9.12 **Practical Considerations: Energy from Waves**
Summary
Key Terms
Study Questions
Study Problems
Suggested Readings

None of them knew the color of the sky. Their eyes glanced level, and were fastened upon the waves that swept toward them. These waves were of the hue of slate, save for the tops, which were of foaming white, and all of the men knew the colors of the sea. The horizon narrowed and widened, and dipped and rose, and at all times its edge was jagged with waves that seemed thrust up in points like rocks.

Many a man ought to have a bath-tub larger than the boat which here rode upon the sea. These waves were most wrongfully and barbarously abrupt and tall, and each froth-top was a problem in small boat navigation. . .

A seat in this boat was not unlike a seat upon a bucking bronco, and, by the same token, a bronco is not much smaller. The craft pranced and reared, and plunged like an animal. As each wave came, and she rose for it, she seemed like a horse making at a fence outrageously high. The manner of her scramble over these walls of water is a mystic thing, and, moreover, at the top of them were ordinarily these problems in white water, the foam racing down from the summit of each wave, requiring a new leap, and a leap from the air. Then, after scornfully bumping a crest, she would slide, and race, and splash down a long incline and arrive bobbing and nodding in front of the next menace.

A singular disadvantage of the sea lies in the fact that after successfully surmounting one wave you discover that there is another behind it just as important and just as nervously anxious to do something effective in the way of swamping boats. In a ten-foot dinghy one can get an idea of the resources of the sea in the line of waves that is not probable to the average experience, which is never at sea in a dingey. As each salty wall of water approached, it shut all else from the view of the men in the boat, and it was the final outburst of the ocean, the last effort of the grim water. There was a terrible grace in the move of the waves, and they came in silence, save for the snarling of the crests.

Stephen Crane,
from *The Open Boat*

Waves coming ashore.

*W*e have all seen water waves. This up-and-down motion occurs on the surface of oceans, seas, lakes, and ponds, and we speak of waves, ripples, swells, breakers, white-caps, and surf. Sometimes the waves are related to the wind, or to a passing ship, or even to a stone thrown into the water. Waves may swamp a small boat, or, when large, smash and twist the bow of a supertanker. The storm waves of winter crash against out coasts, eating away the shore and damaging structures, such as docks and breakwaters. Heavy wave action may force commercial vessels to slow their speed and lengthen their sailing time between ports, to the inconvenience and increased cost of the shippers. Special waves associated with seismic disturbances have killed thousands of coastal inhabitants and have severely damaged their cities and towns. Surfers search for the perfect wave, and ancient peoples navigated by the patterns that waves form.

In this chapter we will describe different kinds of waves, investigate their origin and behavior, and study some of their major characteristics. Our study will be somewhat superficial and simplified, because waves are among the most complex of the ocean's phenomena and because no two waves are exactly the same. Books are written about waves; mathematical models have been devised to explain waves; wave tanks are built to study waves. However, there is much we can learn if we follow the sequence of events beginning at sea where a wave is born and ending at last in the spray and surf of a faraway shore.

9.1 How a Wave Begins

Imagine that you are standing on the beach looking out across a perfectly flat, smooth surface of water. In order to produce or generate waves, it is necessary to introduce a disturbing force. You throw a stone into the water, or a large, naturally occurring landslide comes down into the water. A pulse of energy is introduced, and waves are produced; this disturbing force is called a **generating force.** The waves produced by the generating force move or progress away from the point of disturbance. Of course, the magnitude of the disturbance and the resulting wave size are very different in each of these examples.

In the case of the thrown stone, the stone strikes the surface of the water and displaces or pushes aside the water surface. As the stone sinks, the displaced water flows back into the space left behind, and as this water rushes back from all sides, the water at the center is forced upward. The elevated water falls back, causing a depression below the surface which is refilled, starting another cycle. This process sets up a series of waves or oscillations that move outward and away from the point of disturbance. The waves continue to move outward from the point of disturbance until they are dissipated through friction among the water molecules.

Figure 9.1

Wind-generated capillary waves atop larger waves. Capillary wave lengths are measured in centimeters.

In this example the wave-generating force is the stone as it strikes the surface of the water. The force that causes the water to return to its undisturbed surface level is the **restoring force.** If the stone thrown into the water is small, the waves are small, and the restoring force is the **surface tension** of the water surface. (Surface tension, or the elastic quality of the surface due to the cohesive behavior of the water molecules, is discussed in chapter 4.) All very small water waves are affected by surface tension. In the example in which the landslide is the generating force, the much larger waves are pulled back to the undisturbed surface level of the water by the restoring force of gravity. When waves are of sufficiently large size so that the restoring force of the earth's gravity is more important than surface tension, the waves are called **gravity waves.**

The most common generating force for water waves is the moving air, or wind. As the wind blows across a smooth water surface, the friction or drag between the air and the water tends to stretch the surface, resulting in wrinkles; surface tension acts on these wrinkles to restore a smooth surface. The wind and the surface tension create small waves, called **ripples,** or **capillary waves** (see fig. 9.1). Patches of these very small waves are seen forming, moving, and disappearing as they are driven by pulses of wind. These patches darken the surface of the water and move quickly, keeping pace with the gusts of wind; sailors call these fast-moving patches "cat's-paws." These waves die out rapidly due to friction while new ripples form constantly in front of each moving wind gust.

When the wind blows, energy is transferred to the water over large areas, for varying lengths of time, and at different intensities. As waves form, the surface becomes rougher, and it is easier for the wind to grip the roughened water surface and add energy. As the wind energy is increased, waves become larger, and the restoring force changes from surface tension to gravity (see fig. 9.2).

Generating forces include any occurrence that adds energy to the sea surface: wind, landslide, sea-bottom faulting or slipping, moving ships, and thrown objects. The restoring forces are surface tension for capillary waves and ripples, and gravity for larger waves.

Figure 9.2
Wind-generated gravity waves at sea.

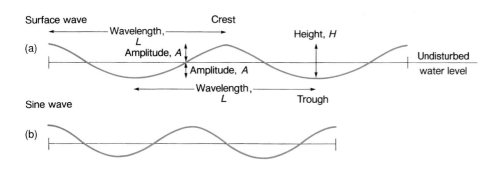

Figure 9.3

The profile of an ideal sea-surface wave (a) differs from the shape of a sine wave (b).

9.2   Anatomy of a Wave

In any discussion of waves, certain terms are used; refer to figure 9.3 during the explanation of these terms. The highest part of the wave that is elevated above the undisturbed sea surface is called the **crest;** the lowest part that is depressed below the surface is called the **trough.** The distance between two successive crests or two successive troughs is the length of the wave, or its **wavelength.** The **wave height** is the vertical distance from the top of the crest to the bottom of the trough. Sometimes the term **amplitude** is used. The amplitude is equal to one-half the wave height, or the distance from either the crest or the trough to the undisturbed water level, or **equilibrium surface.** The oceanographer characterizes a wave not only by its length and height (or amplitude), but also by its **period.** The period is the time required for two successive crests or two successive troughs to pass a point in space. If you are standing on a piling and start a stopwatch as the crest of a wave passes and then stop the stopwatch as the crest of the next wave passes, the time you have measured is the period of the wave.

The dimensions and characteristics of waves vary greatly, but the regularity in the rise and fall of the water's surface and the relationship between wavelength and wave period allow mathematical approximations to be made, which give us insight into the behavior and properties of waves. These calculations are done by relating real waves to simple model waves. Figure 9.3a has been drawn to show that real waves tend to have a trough that is flatter

than the crest; by contrast, figure 9.3b is that of a symmetrical sine wave. This regular wave, which only approximates a water wave in nature, is one of the wave forms used by physical oceanographers and mathematicians to explore and explain wave motion. The relationships presented in sections 9.3 through 9.7 are based on this regular sine-wave form.

9.3   Wave Motion

As a wave form moves across the water surface, particles of water are set in motion by the energy of the wave. Seaward, beyond the surf and breaker zone, where the surface undulates quietly, the water is not moving toward the shore. Such an ocean wave does not represent a flow of water but instead represents a flow of motion or energy from its origin to its eventual dissipation at sea or loss against the land. In order to understand what is happening during the passing of a wave, let us follow the motion of the water particles as the wave moves through them.

As the wave crest approaches, the surface water particles rise and move forward. Immediately under the crest the particles have stopped rising and are moving forward at the speed of the crest. When the crest passes, the particles begin to fall and to slow in their forward motion, reaching a maximum falling speed and a zero forward speed when the midpoint between crest and trough passes. As the trough advances, the particles slow their falling rate and start to move backward, until at the bottom of the trough they have

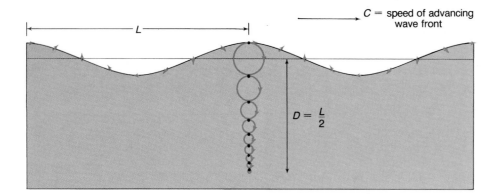

C = speed of advancing
wave front

L

$D = \dfrac{L}{2}$

Figure 9.4

The moving wave form sets the water particles in motion (note the arrows in the diagram). The diameter of a water particle's orbit at the surface is determined by wave height. Below the surface, the diameter decreases and orbital motion ceases at a depth (D) equal to one-half the wavelength.

attained their maximum backward speed and neither rise nor fall. As the remainder of the trough passes, the water particles begin to slow their backward speed and start to rise again, until the midpoint between trough and crest passes. At this point the water particles start their forward motion again as they continue to rise with the advancing crest. This motion (rising, moving forward, falling, reversing direction, and rising again) creates a circular path, or **orbit,** for the water particles. Follow this motion in figure 9.4.

It is the orbital motion of the water particles that causes a floating object to bob, or move up and down, forward and backward, as the waves pass under it. This motion affects a fishing boat, swimmer, sea gull, or any other floating object on the surface seaward of the surf zone. The surface water particles trace an orbit with a diameter equal to the height of the wave. This same type of motion is transferred to the water particles below the surface, but less energy of motion is found at each succeeding depth. The diameter of the orbits becomes smaller and smaller as depth increases. At a depth equal to one-half the wavelength, the orbital motion has decreased to almost zero; notice how the orbit decreases in diameter in figure 9.4. Submarines dive during rough weather for a quiet ride, since the wave motion does not extend far below the surface.

The orbits just described are based on the sine wave (fig. 9.3b), not on real sea waves (fig. 9.3a). The water particles of actual sea waves move in orbits whose forward velocity at the top of the orbit is slightly greater than the reverse velocity at the bottom of the orbit. Therefore, each orbit made by a water particle does move the water slightly forward in the direction the waves travel. This movement is due to the shape of the real waves, whose crests are sharper then their troughs; again see figure 9.3a. This difference means there is a very slow transport of water in the direction of the waves in nature, but this motion is often ignored in calculations based on simple wave models.

9.4   Wave Speed

A wave's speed across the sea surface is related to its wavelength and wave period. The speed of any surface wave (C) is equal to the length of the wave (L) divided by the period (T):

$$\text{Speed} = \frac{\text{length of wave}}{\text{wave period}}$$

or

$$C = \frac{L}{T}.$$

Once a wave has been created, the speed at which the wave moves may change, but *its period remains the same* (period is determined by generating force).

9.5   Deep-Water Waves

"Deep water" has a precise meaning for the oceanographer studying waves. To be a **deep-water wave,** the wave must occur in water that is deeper than one-half the wave's length. Under this condition the orbits of the wave do not reach the seafloor. For example, a 15-m-long wave must occur in water that is deeper than 7½ m in order to be considered a deep-water wave and to behave as the waves described next.

The wavelength (L) of deep-water waves is derived from the wave period (T), and since wavelength (L) is a function of wave period (T), wave speed (C) is also derived from wave period (T). The oceanographer at sea determines the wave period (T) by direct measurement and calculates the wavelength (L) from its relationship with gravity and wave period. In deep water, the wavelength is equal to the earth's acceleration due to gravity (g) divided by 2π times the square of the wave period (T). The value of earth's gravity, g, is 9.81 m/sec^2.

$$L = \frac{g}{2\pi}T^2; \quad L = 1.56 \; m/sec^2 \, T^2.$$

This deep-water wave equation, when combined with the general wave speed equation $C = L/T$, is used to determine wave speed (C) from either wavelength (L) only or wave period (T) only:

$$C = \frac{L}{T} = \frac{g}{2\pi}T \quad \text{or} \quad C^2 = \frac{L^2}{T^2} = \frac{g}{2\pi}L$$

$$C = 1.56 \; m/sec^2 \, T \quad \text{or} \quad C^2 = 1.56 \; m/sec^2 \, L.$$

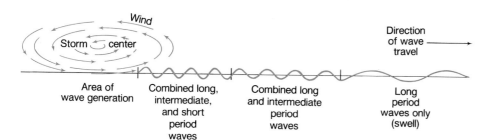

Figure 9.5

Dispersion. The longer waves travel faster than the shorter waves. Waves are shown moving in only one direction in the diagram.

Storm Centers

Most waves observed at sea are **progressive wind waves.** They are generated by the wind, restored by gravity, and progress in a particular direction. These waves are formed in local storm centers or by the steady winds of the trade and westerly wind belts. In an active storm area covering thousands of square kilometers, the winds are not steady but turbulent, varying in strength and direction. Storm area winds flow in a circular pattern about the low-pressure **storm center,** creating waves that move outward and away from the storm in all directions. The storm center may also move across the sea surface following the waves and increasing their heights, supplying energy for a longer time and over a longer distance. You may gain some idea of the size of storms and their direction of travel by viewing the cloud patterns in the satellite photographs that are presented in televised weather forecasts or printed in the newspaper weather sections.

In a storm area, the sea surface appears as a jumble and confusion of waves of all heights, lengths, and periods; there are no regular patterns. Capillary waves ride the backs of small gravity waves, which in turn are superimposed on still higher and longer gravity waves. This turmoil of mixed waves is called a "sea," and sailors use the expression, "There is a sea building," to refer to the growth of these waves under storm conditions. When waves are being generated they are forced to increase in size and speed by the continuing input of energy; these are known as **forced waves.** Due to variations in the winds of the storm area, energy at different intensities is transferred to the sea surface at different rates resulting in waves with a variety of periods and heights. Remember, wave periods are a function of the generating force; the speed at which the wave moves may change, but its period remains the same.

Swell

When the waves move out of the storm center, forced waves become **free waves** moving at speeds due to their periods and wavelengths. Among the waves escaping from a storm center are those with long periods and long wavelengths. These waves have a greater speed than those with short periods and short wavelengths. The faster, longer waves gradually move through and ahead of the shorter, slower waves. They appear as a regular pattern of crests and troughs moving across the sea surface. Once away from the storm these longer-period, uniform waves are called **swell;** they carry considerable energy, which they lose very slowly. Swell can travel for thousands of miles across the ocean. Groups of large, long-period waves created by storms between 40° and 50° S in the Pacific Ocean have been traced across the entire length of that ocean, until they die on the shores and beaches of Alaska.

Dispersion

Waves with long periods and long wavelengths have a greater speed than waves with short periods and short wavelengths. The faster, longer waves gradually move through and ahead of the shorter, slower waves; this process is called **sorting** or **dispersion.** Groups of these faster waves move as **wave trains,** or packets of similar waves with approximately the same period and speed. Because of dispersion, the distribution of observed waves from any single storm changes with time. Near the storm center the waves are not yet sorted, while farther away the faster, longer-period waves are out ahead of the slower, shorter-period waves. This process is shown in figure 9.5. The distribution of the waves from a given storm and the energy associated with particular wave periods change predictably with time, allowing the oceanographer to follow wave trains from a single storm over long distances.

Group Speed

Consider again the waves formed by a stone thrown into the water; the wave group, or train, is seen as a ring of waves moving outward from the point of disturbance. Careful observation shows that waves constantly form on the inside of the train as it moves across the water. As each wave joins the train on the inside of the ring, a wave is lost from the leading edge or outside of the ring and the number of waves remains the same. The outside wave's energy is lost in advancing the wave form into undisturbed water. Therefore, each individual wave in the group moves faster than the leading edge of the wave train, and the wave train moves outward at a speed one-half that of the individual wave. This speed is known as the **group speed,** the speed at which wave energy is transported away from its source under deep-water conditions:

Group speed = ½ wave speed = speed of energy transport;

or

$$V = \frac{C}{2}.$$

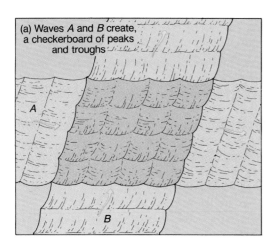

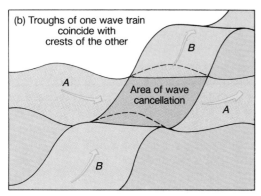

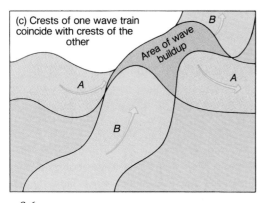

Figure 9.6
Wave trains form interference patterns where they meet. (a) Waves meeting at right angles. (b) Wave crests and wave troughs meet and cancel each other. (c) Wave crests and troughs coincide, building higher waves.

Wave Interaction

Waves that escape a storm and are no longer receiving energy from the storm winds tend to flatten out slightly, and their crests become more rounded. These waves moving across the ocean surface as swell are likely to meet other trains of swell moving out and away from other storm centers. When the two wave trains meet, they pass through each other and continue on. Wave trains may intersect at any angle, and many possible interference patterns may result. If the two wave trains intersect each other sharply as at a right angle, then a checkerboard pattern is formed (fig. 9.6a). If the waves have similar lengths and heights, and if the

crests of one train coincide with the troughs of the other train, the wave trains appear to cancel each other (fig. 9.6b). If the crests or the troughs of the two different wave trains coincide, the profiles are additive and their amplitudes increase (fig. 9.6c). In this way two or more wave trains traveling in the same direction and passing through each other can phase together and suddenly develop large waves unrelated to local storms. If these waves become too high, they may break, lose some of their energy, and create new, smaller waves. If such waves take a vessel by surprise, they can damage it severely.

9.6 Wave Height

The height of actual waves is controlled by the interaction of several factors. The three most important factors are (1) wind speed (how fast the wind is blowing), (2) wind duration (how long the wind blows), and (3) **fetch** (the distance over water that the wind blows in a single direction). The wave height may be limited by any one of these factors. If the wind speed is very low, large waves are not produced, no matter how long the wind blows over an unlimited fetch. If the wind speed is great, but it blows for only a few minutes, no high waves are produced despite unlimited wind strength and fetch. Also, if very strong winds blow for a long period over a very short fetch, no high waves form. When no single one of these three factors is limiting, spectacular wind waves are formed at sea.

Table 9.1 lists the maximum and significant wave heights possible for certain average wind speeds when there is no limiting fetch and no limitation to the wind duration. The significant wave height is defined as the wave height for any storm that is below the higher one-third of the waves and above the lower two-thirds of the waves. Significant wave heights are related to maximum wave heights by calculation and are used to forecast storm waves.

The area of the ocean in the vicinity of 40° to 50°S latitudes is ideal for the production of high waves. Here, in an area noted for high-intensity storms and strong winds of long duration (what sailors have called the "roaring forties and furious fifties"), there are no landmasses to interfere and reduce the fetch length. The westerly winds blow almost continuously around the earth, adding energy to the sea surface for long periods of time and over great distances, resulting in waves that move in the same direction as the wind. Although this area is ideal for the production of high waves, such waves can occur anywhere in the open sea, given the proper storm conditions.

A typical maximum fetch for a local storm over the ocean is approximately 920 km (500 nmi). Keeping in mind that storm winds circulate around a low-pressure disturbance, the winds continue to follow the waves on the side of the storm, along which wave direction is the same as the storm direction. This increases both the fetch and the duration of time over which the wind adds energy to the waves. If the waves move fast enough, their speeds exceed the speed of the moving storm; the waves escape the wind and do not grow larger. Waves 10 to 15 m (33–49 ft) high are not

Table 9.1
The Relationship between Wind Speed and Wave Height

Average wave speed		Significant wave height	Significant wave period	Significant wave speed	Maximum wave height	Minimum fetch[1]	Minimum wave duration[1]
(knots)	(m/sec)	(m)	(sec)	(m/sec)	(m)	(km)	(hours)
10	5.1	1.22	5.5	8.58	2.19	129	11
20	10.2	2.44	7.3	11.39	4.39	240	17
30	15.3	5.79	12.5	19.50	10.43	1017	37
40	20.4	14.33	18.0	28.00	25.79	2590	65
50	25.5	16.77	21.0	32.76	30.19	2775	100

[1]Minimum fetch and minimum wind duration are distances and times required when wind speed is the only limiting factor in wave development.

uncommon under severe storm conditions; the lengths of such waves are typically between 100 and 200 m (330–660 ft). This length is about the same as the length of a modern freighter, and a vessel of this length encounters hazardous sailing conditions, because the ship may become suspended between the crests of two waves and break its back.

Measurements of wave height taken in the North Atlantic over the past twenty-five years by the Institute of Oceanographic Sciences in England show a long-term continuing increase in wave height. Wave height has increased about 25% since 1960. Maximum wave height was 12 m (39 ft) in 1960 and is predicted to reach 18 m (59 ft) in the 1990s if the trend continues. Variations from year to year and season to season are large and there is no way of knowing whether the trend will continue. Weather-ship data from the Atlantic Ocean and the North Sea also show an 11% to 27% increase in wave heights between the early 1960s and 1970s. There is no known reason for this apparent trend in increasing wave height.

Giant waves over 30.5 m (or 100 feet) high are rare. In 1933 the USS *Ramapo,* a Navy tanker, en route from Manila to San Diego, encountered a severe storm, or typhoon. In the course of running the ship downwind to ease the ride, the *Ramapo* was overtaken by waves that, as measured against the ship's superstructure by the officer on watch, were 112 feet (or 34.2 m) high. The period of the waves was measured at 14.8 seconds; the wave speed was calculated at 27 m (90 ft)/sec, and the wavelength at 329 m (1100 ft). Other storm waves in this size category have been reported, but none have been as well documented. It is also probable that ships confronted with such waves do not always survive to report the incidents.

Episodic Waves

Large waves or **episodic waves,** can suddenly appear unrelated to local sea conditions. An episodic wave is an abnormally high wave that occurs due to a combination of intersecting wave trains, changing depths, and currents. We do not know a great deal about these waves, as they do not

exist for long and they can and do swamp ships, often removing any witnesses. They occur most frequently near the edge of the continental shelf, in water about 200 m (660 ft) deep, and in certain geographic areas with particular prevailing wind, wave, and current patterns.

The area where the Agulhas Current sweeps down the east coast of South Africa and meets the storm waves arising in the Southern Ocean is noted for such waves. Storm waves from more than one storm may combine and run into the current and against the continental shelf, producing occasional episodic waves. This area is also one of the world's busiest sea routes, as supertankers carrying oil from the Middle East ride the Agulhas Current on their trip southward to round the Cape of Good Hope en route to Europe and America. The situation is an invitation to disaster, and tankers have been damaged and lost in this area (see fig. 9.7). In the North Atlantic, strong northeasterly gales send large storm waves into the edge of the northward-moving Gulf Stream near the border of the continental shelf, resulting in the formation of large waves. The shallow North Sea also seems to provide suitable conditions for extremely high episodic waves during its severe winter storms.

Researchers studying these waves describe them as having a height equal to a seven- or eight-story building (20–30 m or 70–100 ft) and moving at a speed of 50 knots, with a wavelength approaching a half-mile (0.9 km). If such a wave topples onto a vessel that has dropped its bow into a trough, there is no escape from the thousands of tons of water crashing on its deck. In the North Sea, maximum wave height for an episodic wave is calculated as 33.8 m (111 feet), but 22.9 m (75 feet) is the highest that has been observed. In the Agulhas Current area, researchers searched the storm records of the past 20 years and calculated a possible wave height of about 57.9 m (190 feet).

There are many disappearances of vessels for which episodic waves are now suspected of being the chief cause. It seems that many of the casualties are tankers or bulk carriers of ore, grain, and the like. These vessels are susceptible not only because so many are at sea at any one time, but

Figure 9.7
A giant wave breaking over the bow of the *ESSO Nederland* southbound in the Agulhas Current. The bow of this supertanker is about 25 meters (80 feet) above the water.

also because of their design. Bulk carriers comprise a bow section and an aft (or rear) section that includes the engine and the crew accommodations. These sections are separated by a series of flat-bottomed boxes, or storage tanks, which make up the majority of the vessel's length. Since these vessels are about 300 m (1000 ft) long, they are subject to great wrenching forces if a large portion of the hull is left unsupported or only partially supported while suspended on a wave crest. More traditional and smaller hull forms are stronger, ride more easily, and are less likely to be destroyed by these severe waves.

Wave Energy

A wave's energy is present as **potential energy,** due to the change in elevation of the water surface, and as **kinetic energy,** due to the motion of the water particles in their orbits. The higher the wave, the larger the diameter of the water particle orbit, and the greater the speed of the orbiting particle, therefore the greater the kinetic and potential energy. The energy in a deep-water wave is nearly equally divided between kinetic and potential energy. Wave energy is averaged over one wavelength per unit width (one meter) of crest from sea surface to a depth of $L/2$ and is related to the square of the wave height. This relationship is demonstrated in figure 9.8.

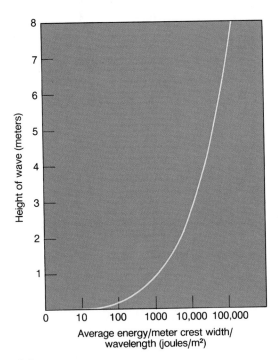

Figure 9.8
Wave energy increases rapidly with the square of the wave height. Average wave energy is calculated per unit width of crest and averaged over the wave length (L) and the depth ($L/2$).

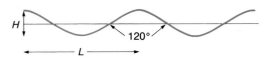

Figure 9.9

Wave steepness. When *H*/*L* approaches 1:7, the wave's crest angle approaches 120° and the wave breaks.

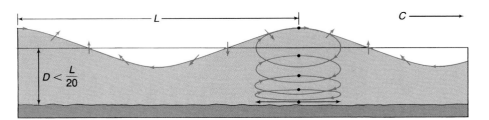

Figure 9.10

Shallow-water wave particles move in elliptic orbits. The orbits flatten with depth due to interference from the seafloor.

Wave Steepness

There is a maximum possible height for any given wavelength. This maximum value is determined by the ratio of the wave's height to the wave's length, and it is the measure of the **steepness** of the wave:

$$\text{Steepness} = \frac{\text{height}}{\text{length}} \quad \text{or} \quad S = \frac{H}{L}.$$

If the ratio of the height to the length exceeds 1:7, the wave becomes too steep and the wave breaks. The angle formed at the wave crest approaches 120°, and the wave becomes unstable. Under these conditions the wave cannot maintain its shape; it collapses and breaks. For example, if the wavelength is 70 meters (230 ft), the wave will break when the wave height reaches 10 meters (33 ft). See figure 9.9.

Small unstable, breaking waves are quite common. When wind speeds reach 8–9 m/sec (16–18 knots), waves known as "whitecaps" can be observed. These waves have short wavelengths (about 1 m), and the wind increases their height rapidly. As each reaches the critical steepness and crest angle, it breaks and is replaced by another wave produced by the rising wind.

Longwaves at sea usually have a height well below their maximum value. There is rarely sufficient wind energy to force them to their maximum height. If a longwave does attain maximum height and breaks in deep water, tons of water are sent crashing to the surface. The energy of the wave is lost in turbulence and in the production of smaller waves. Rather than breaking, under such conditions it is more likely that the top of a large wave will be torn off by the wind and will cascade down the wave face. This action does not completely destroy the wave and is not considered to be the true collapse, or breaking, of such a wave.

In addition, waves break and dissipate if intersecting wave trains pass through each other in proper phase to form a combined wave with sufficient height to exceed the critical steepness. Waves sometimes run into a strong opposing current, forcing the waves to slow down. Remember that the speed of all waves equals the wavelength divided by the period ($C = L/T$) and that a wave's period does not change. If the speed of a wave is reduced by an opposing current, its wavelength must shorten. In such a case, the wave's en-

ergy is confined to a shorter-length wave, so the wave increases its height to satisfy the direct height-energy relationship. If the increase in height exceeds the maximum allowable height to length ratio for the shorter wavelength, the wave breaks. Crossing a sandbar into a harbor or river mouth during an outgoing or falling tide is dangerous because the waves moving against the tidal current steepen and break. Entering a harbor or river should be done at the change of the tide or on the rising tide, when the tidal current moves with the waves, stretching their wavelengths and decreasing their heights.

9.7 Shallow-Water Waves

As a deep-water wave approaches the shore and moves into shallow water, the reduced depth begins to affect the shape of the orbits made by the water particles. The orbits become flattened circles, or ellipses (see fig. 9.10). The wave begins to "feel" the bottom, and the resulting friction and compression of the orbits reduces the forward speed of the wave.

Remember that (1) the speed of all waves is equal to the wavelength divided by period, and (2) the period of a wave does not change. Therefore, when the wave "feels bottom," it slows, and the accompanying reduction in the wavelength and speed results in increased height and steepness as the wave's energy is condensed in a smaller water volume.

When the wave finally enters water with a depth of less than one-twentieth the wavelength ($D < L/20$), the wave becomes a **shallow-water wave.** While the length and speed of a deep-water wave are determined by the wave period, the shallow-water wave's length and speed are controlled only by the water depth. Here the speed and wave length are determined by the square root of the product of the acceleration due to gravity (g) and depth (D):

$$C = \sqrt{gD}; \; C \; m/sec = 3.13 \sqrt{D},$$
$$\text{or } L_m = 3.13 \sqrt{D} \cdot T.$$

When the condition for a shallow-water wave is met, the orbits of the water particles are elliptic; they become flatter with depth until at the seafloor only a back-and-forth oscillatory motion remains (see fig. 9.10). Note that the horizontal

Universal Sea State Code

In 1806, Admiral Sir Francis Beaufort of the British Navy adapted a wind-estimation system from land to sea use. On land, the clues to wind speed included smoke drift, the rustle of leaves, flags flapping, trees swaying, slates blowing from roofs, and the uprooting of trees. Admiral Beaufort used his knowledge of sea waves to relate observations of the sea-surface state to wind speed, and so designed a wind-speed scale known as the Beaufort Scale. The scale runs from 0 to 12, calm to hurricane, with typical wave descriptions for each level of wind speed. The Beaufort Scale was adopted by the U.S. Navy in 1838, and the scale was extended from 0 to 17. At present, a Universal Sea State Code of 0 to 9 based on the Beaufort Scale is in international use for wind speeds and related sea-surface conditions.

Box Table 9.1		
Sea state code	**Description**	**Average wave heights**
SS0	Sea like a mirror; wind less than one knot.	0
SS1	A smooth sea; ripples, no foam; very light winds, 1–3 knots, not felt on face.	0–0.3 m 0–1 ft
SS2	A slight sea; small wavelets; winds light to gentle, 4–6 knots, felt on face; light flags wave.	0.3–0.6 m 1–2 ft
SS3	A moderate sea; large wavelets, crests begin to break; winds gentle to moderate, 7–10 knots; light flags fully extend.	0.6–1.2 m 2–4 ft
SS4	A rough sea; moderate waves, many crests break, whitecaps, some wind-blown spray; winds moderate to strong breeze, 11–27 knots; wind whistles in the rigging.	1.2–2.4 m 4–8 ft
SS5	A very rough sea; waves heap up, forming foam streaks and spindrift; winds moderate to fresh gale, 28–40 knots; wind affects walking.	2.4–4.0 m 8–13 ft
SS6	A high sea; sea begins to roll, forming very definite foam streaks and considerable spray; winds a strong gale, 41–47 knots; loose gear and light canvas may be blown about or ripped.	4.0–6.1 m 13–20 ft
SS7	A very high sea; very high, steep waves with wind-driven overhanging crests; sea surface whitens due to dense coverage with foam; visibility reduced due to wind-blown spray; winds at whole gale force, 48–55 knots.	6.1–9.1 m 20–30 ft
SS8	Mountainous seas; very high-rolling breaking waves; sea surface foam-covered; very poor visibility; winds at storm level, 56–63 knots.	9.1–13.7 m 30–45 ft
SS9	Air filled with foam; sea surface white with spray; winds 64 knots and above.	13.7 m and above 45 ft and above

dimension of the orbit remains unchanged in shallow water. When the water depth is between $L/2$ and $L/20$, the speed of the wave is also slowed. Waves in this depth range are called intermediate waves. There is no simple algebraic equation to determine the speed of these intermediate waves.

Refraction

Waves are refracted, or bent, as they move from deep to shallow water, begin to feel the bottom, and change wavelength and wave speed. When waves from a distant storm center approach the shore, it is likely that they will approach the beach at an angle. One end of the wave crest comes into shallow water and begins to feel bottom, while the other end is in deeper water. The shallow-water end moves more slowly than that portion of the wave in deeper water. The result is that the wave crests bend, or **refract,** and tend to become oriented parallel to the shore. This pattern is shown in figure 9.11. The **wave rays** drawn perpendicular to the crests show the direction of motion of the wave crests. Note that the refraction of water waves is similar to the refraction of light and sound waves, described in chapter 4.

When waves approach an irregular coastline, with headlands jutting into the ocean and bays set back into the land, there is often a submerged ridge seaward from a headland and a depression in front of a bay. When shallow-water waves approach this coastline, the portion of the wave crest over the ridge slows down more than the wave crests on either side. The crests wrap around the headland in the pattern shown in figure 9.12. This refraction pattern results from the waves shortening their wavelengths and focusing their energy on the point of the headland. Since wave energy is proportional to the wave height, the waves gain in height as their wavelength is shortened over the submerged ridge, and more energy is expended on a unit length of shore at the point of the headland than on a unit length of shore elsewhere.

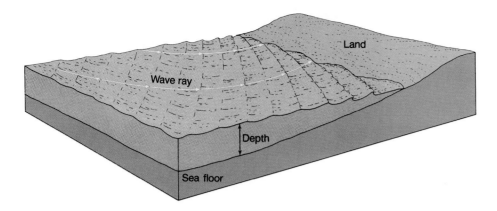

Figure 9.11
Waves moving inshore at an oblique angle to the depth contours are refracted. One end of the wave reaches a depth of $L/2$ or less and slows while the other end of the wave maintains its speed in deeper water. Wave rays drawn perpendicular to the crests show the direction of wave travel and the bending of the wave crests.

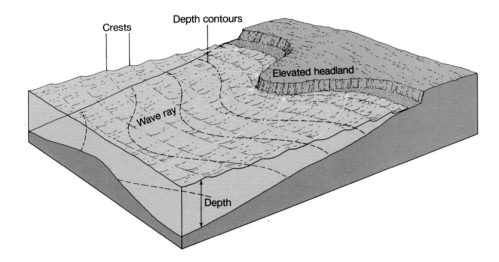

Figure 9.12
Waves refracted over a shallow submerged ridge focus their energy on the headland. The converging wave rays show the wave energy being crowded into a smaller volume of water, increasing the energy per unit length of wave crest as the height of the wave increases.

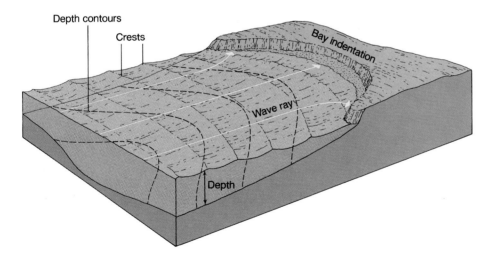

Figure 9.13
Waves refracted by the shallow depths on each side of the bay deliver lower levels of energy inside the bay. The diverging wave rays show the energy being spread over a larger volume of water, decreasing the energy per unit length of wave crest as the wave height decreases.

The central area at the mouth of the bay is usually deeper than the areas to each side, and so the advancing waves slow down more on the sides than in the center. Because the wavelengths remain long in the center and shorten on each side, the wave crests bulge toward the center of the bay. Since the waves in the center of the bay do not shorten as much, they have less height and less energy to expend per length of shoreline. This pattern is shown in figure 9.13. The result is an environment of low wave energy providing sheltered water in the bay. Overall, this unequal distribution of wave energy along the coast re-

sults in a wearing down of the headlands and a filling in of the bays, as the sand and mud settles out in the quieter water. If all coastal materials had the same resistance to wave erosion, this process would lead in time to a straightening of the coastline; but the rocky structure of many headlands resists wave erosion, and therefore the cliffs remain.

Reflection

A straight, smooth, vertical barrier in water deep enough to prevent waves from breaking reflects the waves (fig. 9.14). The barrier may be a cliff, steep beach, breakwater, bulkhead,

Figure 9.14

The wave from a ship's wake is reflected from a concrete wall. The principal wave is moving from lower left to upper right of the photograph. The reflected wave is moving from lower right to upper left. A checkerboard pattern is formed as the principal wave and reflected wave move through each other at approximately right angles.

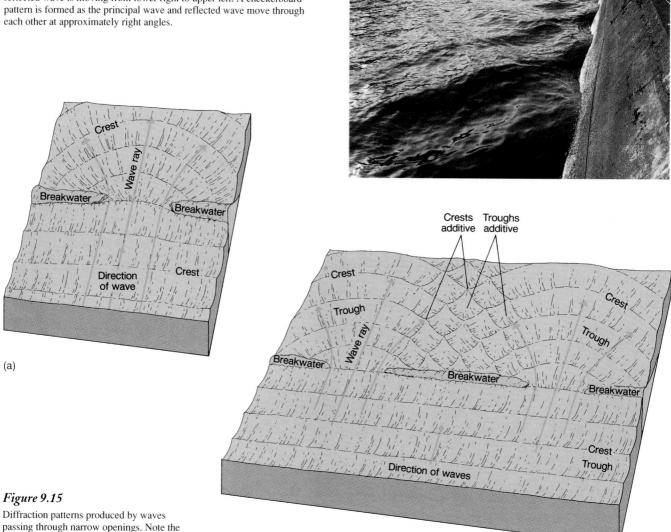

(a)

(b)

Figure 9.15

Diffraction patterns produced by waves passing through narrow openings. Note the interference pattern produced in (b).

or other structure. The reflected waves pass through the incoming waves to produce an interference pattern, and steep, choppy seas often result. If the waves reflect directly back on themselves, the resulting waves appear to stand still, rising and falling in place. If waves are reflected from curved vertical surfaces, the curve may spread the reflected wave rays, dispersing the wave energy, but if the surface curves so that the reflected wave rays converge, the energy is focused. This situation is similar to the reflection of light from curved mirrors. Great care must be taken in designing walls and barriers to protect an area from waves to be sure that reflected waves do not focus their energy and damage another area.

Diffraction

Another phenomenon that is associated with waves as they approach the shore or other obstacles is **diffraction.** Diffraction is caused by the spread of wave energy

sideways to the direction of wave travel. If waves move toward a barrier with a small opening (such as two landmasses separated by a channel or an opening in a breakwater), some wave energy passes through the small opening to the other side. Once through the opening, the wave crests decrease in height, radiating out and away from the gap. A portion of the wave energy is diffracted, transported sideways from its original direction. This effect is shown in figure 9.15a. If more than one gap is open to the waves, the patterns produced by the spreading waves from each opening may intersect and form interference patterns as the waves move through each other (see fig. 9.15b).

If the waves approach a barrier without an opening, diffraction can still occur. Energy will be transported at right angles to the wave crests as the waves pass the end of the barrier (see fig. 9.16). Note that the pattern produced is one-half of the pattern observed when the waves pass through a narrow opening, as in figure 9.15a. Energy is transported behind the sheltered (or lee) side of the barrier.

244   *The Waves*

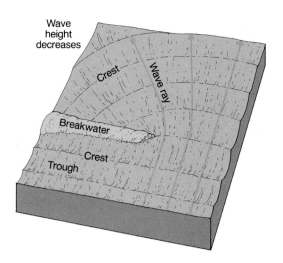

Figure 9.16
Diffraction occurs behind the breakwater.

This effect is an important consideration in planning the construction of breakwaters and other coastal barriers intended to protect vessels in harbors from wave action and possible damage.

Navigation from Wave Direction

In areas of the world where the winds blow steadily and from one direction, as in the trade-wind belts, waves at sea are very regular in their direction of motion, and this regularity allows a vessel to maintain a constant course relative to the waves. Because waves change speed, shape, and height with water depth, and because waves change direction and pattern due to refraction, diffraction, and reflection, it is possible to deduce the presence of shoals, bars, islands, and coasts from the changes in wave patterns.

Careful direct observation of wave patterns, or the sensing of wave patterns through the motions of a small craft, even at night, allows the detection of a change in the angle between waves and wind direction or a change in the wave pattern. Although these changing patterns are subtle, they can be detected many miles from land and can be used to bring a small vessel to shore when no landmarks are visible.

The Polynesians of the past lived with the sea; it was their home and they were acutely sensitive to its changes. They made long voyages in their small canoes using their knowledge of wave patterns. They had no theory to explain these patterns, but they understood their association with winds, shores, and islands. Combining their knowledge of star positions, cloud forms over land and sea, and bird flights with their knowledge of waves, they sailed many hundreds, even thousands, of miles across the open ocean to reach their destinations. No navigational tools were required other than charts constructed from twigs and shells, showing island positions relative to stars, wind direction, and swell. An example of such a chart is found in Prologue figure II. These peoples learned by living with and observing nature in an area where nature cooperated by behaving in regular and predictable ways.

(a)

(b)

Figure 9.17
Breaking waves. A plunger (a) loses energy more quickly than a spiller (b).

9.8   The Surf Zone

The surf zone is the shallow area along the coast in which the waves slow rapidly, steepen, break, and disappear in the turbulence and spray of expended energy. The width of this zone is variable and is related to both the length and height of the arriving waves and the changing depth pattern. Longer, higher waves, which feel the bottom before shorter waves, become unstable and break farther offshore, in deeper water. If shallow depths extend offshore for some distance, the surf zone is wider than it is over a sharply sloping shore.

Breakers

Breakers form in the surf zone because the water-particle motion at depth is affected by the bottom. Orbital motion is slowed and compressed vertically, but the orbit speed of water particles near the crest of the wave is not slowed as much, and these particles move faster toward the shore than the rest of the wave form. This results in the curling of the crest and the eventual breaking of the wave. The two most common types of breakers are **plungers** and **spillers;** these are pictured in figure 9.17.

Figure 9.18
A surfer rides the tube of a large curling wave.

Plunging breakers form on narrow, steep beach slopes. The curling crest outruns the rest of the wave, curves over the air below it, and breaks with a sudden loss of energy and a splash. The more common spilling breaker is found over wider, flatter beaches, where the energy is extracted more gradually as the wave moves over the shallow bottom. This action results in the less dramatic wave form, consisting of turbulent water and bubbles flowing down the collapsing wave face. The spilling breakers last longer than the plungers, because they lose energy more gradually. Therefore, spillers give surfers a longer ride, while plungers give them a more exciting one.

The slow curling-over of the crest observed on some breakers begins at a point on the crest and then moves lengthwise along the wave crest as the wave approaches shore. This movement of the curl along the crest occurs because waves are seldom exactly parallel to a beach. The curl begins at the point on the crest that is in the shallowest water or the point at which the crest height is slightly greater, and it moves along the crest as the wave approaches the beach. The result is the "tube" so sought after by surfers (see fig. 9.18).

If the waves approaching the beach are uniform in length, period, and height, they are the swells from some far-distant storm, which have had time and distance to sort into uniform groups. For example, the long surfing waves of the California beaches in summer begin their lives in the winter storms of the South Pacific and Antarctic oceans. If the waves are of different heights, lengths, and periods and break at varying distances from the beach, then unsorted waves have arrived and are probably the product of a nearby local storm superimposed on the swell.

Water Transport

The small net drift of water in the direction the waves are traveling (refer to the previous discussion on the water particle orbits of real waves) is intensified in the surf zone, as the shoreward motion of the water particles at the crest becomes greater than the return particle motion at the trough. Since the crests usually approach the beach at an angle, the surf zone transport of water flows both toward the beach and along the beach. The result is water accumu-

Figure 9.19
Small rip currents carry turbid water seaward through the surf zone.

lating against the beach and flowing along the beach until it can flow seaward again and return to the area beyond the surf zone. This return flow generally occurs in quieter water with smaller wave heights, for example, in areas with troughs or depressions in the seafloor.

Because regions of seaward return flow may be narrow and some distance apart, the flow in these areas must be swift in order to carry enough water beyond the surf zone to balance the slower but more extensive flow toward the beach. These regions of rapid seaward flow are called **rip currents.** Swimmers who unknowingly venture into a rip current will find themselves carried seaward and unable to swim back to shore against the flow; they must swim parallel to the beach, or across the rip current, and then return to shore. Because of the danger associated with rip currents, swimmers should be on the lookout for indicators of these currents' presence, including (1) turbid water and floating debris moving seaward through the surf zone, (2) areas of reduced wave heights in the surf zone, and (3) depressions in the beach running perpendicular to the shore.

Wave action on the beach stirs up the sand particles and keeps them suspended in the water. The sand is carried along the beach parallel to the shore until the rip current is reached, and the sand is transported seaward. Viewed from a height, such as a high cliff or a low-flying airplane, rip currents are seen as streaks of discolored turbid water extending seaward through the clearer water of the outer surf zone (see fig. 9.19).

Energy Release

Watching the heavy surf pounding a beach from a safe vantage point is an exciting and exhilarating experience; the trick is to determine at what point one is safe. In a narrow surf zone during a period of very large waves, the wave energy must be expended rapidly over a short distance. Under these conditions, the height of the waves and the forward motion of the water particles combine to send the water high up on the beach. The accompanying release of energy is explosive and can result in rocks and debris from the water's edge being hurled high up on the beach by the force of the water. Minot's Lighthouse on the south side of Massachusetts Bay is 30 m (100 ft) high, but it is regularly engulfed in spray. Lonely Tillamook Light off the Oregon coast had to have steel gratings installed to protect the glass that shields the light 40 m (130 ft) above sea level, after the glass had been broken several times by wave-thrown rocks. In every winter storm, waves displace boulders weighing many tons from breakwaters along the world's coasts.

Waves do not always expend their energy on the shore. Some break farther seaward on sandbars such as those associated with river mouths and estuaries. The famous bar at the mouth of the Columbia River is responsible for extremely hazardous conditions for both fishing and commercial vessels, and every winter it produces some casualties. On an ebbing (or falling) tide, the waves approaching the bar against the current stand on end, creating breaking waves up to 20 m (66 ft) high. The Coast Guard uses self-righting or rollover surf-rescue boats to cross the bar when rendering assistance during storm conditions.

The great driftwood logs found stranded on some beaches may be set afloat again at high tide during severe winter storms and can become lethal battering rams when hurled shoreward by the surf. Even with less-severe waves, beach logs lying near the water's edge have a dangerous potential. The unsuspecting vacationer sits or plays on such a log, and when the occasional higher wave rolls it over, the person may become trapped under the log and crushed or drowned by the succeeding waves. Logs in or near the surf represent a great danger for the unwary. Where are you safe? You must be the judge.

9.9 Tsunamis

Sudden movements of the earth's crust may produce **seismic sea waves,** or **tsunamis.** These waves are often incorrectly called tidal waves. Since a seismic sea wave has nothing to do with tides, oceanographers have adopted the Japanese word tsunami to replace the misleading term tidal wave. It has since been pointed out that tsunami means tidal wave in Japanese, but the term is now accepted as a synonym for seismic sea wave.

If a large area, maybe several hundred square kilometers, of the earth's crust below the sea surface is suddenly displaced, it may cause a sudden rise or fall in the sea-surface level above it. In the case of a rise, gravity causes the suddenly elevated water to return to the equilibrium surface level; if a depression is produced, gravity forces cause the surrounding water to flow into it. Both cases result in the production of waves with extremely long wavelengths (100 to 200 km or 60–120 mi) and long periods as well (10–20 minutes). Since the average depth of the oceans is about 4000 m (or 4 km or 13,000 ft), this depth is less than one-twentieth the wavelength of these waves, and tsunamis are shallow-water waves. These seismic waves radiate from the point of the seismic disturbance at a speed determined by the ocean's depth ($C = \sqrt{gD}$) and move across the oceans at about 200 m/sec (400 mph). Since they are shallow-water waves, tsunamis may be refracted, diffracted, or reflected in midocean.

When a tsunami leaves its point of origin, it may have a height of 1 to 2 m (3–6 ft), but this height is distributed over its many-kilometer wavelength. It is not easily seen or felt when superimposed on the other distortions of the sea's surface, and a vessel in the open ocean is in little or no danger if a tsunami passes. The danger occurs only if the vessel has the misfortune to be directly above the area of the original seismic disturbance.

The energy of a tsunami is distributed from the ocean surface to the ocean floor and over the length of the wave. When the path of the wave is blocked by a coast or island, the wave behaves like any other shallow-water wave, and the energy is compressed into a smaller water volume as the depth rapidly decreases. This rapid and sudden increase in energy density causes the wave height to build rapidly, and the loss of energy is also rapid when the wave breaks. A tremendous surge of moving water races up over the land, destroying buildings, docks, and trees; large vessels can be left well up on the beach. The rising and falling of the water level in some bays may not be as destructive as the extreme currents at the entrance to the bay where the large volume of water entering and leaving the bay damages and destroys structures in its way.

The leading edge of the tsunami wave group may be either a crest or a trough. If the initial crustal disturbance was an upward motion, a crest is formed first; if the crustal motion was downward, a trough is formed. If a trough arrives before the first crest, sea level drops rapidly, exposing sea plants and animals. People have drowned after following the receding water to inspect the marine life, for they find, too late, that there is a wall of advancing water that they cannot outrun.

Tsunamis are most likely to occur in ocean basins that are tectonically active. The Pacific Ocean, ringed by crustal faults and volcanic activity, is the birthplace of most tsunamis. They have also appeared in the Caribbean Sea, which is bounded by an active island arc system, and in the Mediterranean Sea as well. The spectacular havoc and destruction caused by these waves is well recorded. In August, 1883, the Indonesian island of Krakatoa erupted and was nearly destroyed in a gigantic steam explosion that hurled several cubic miles of material into the air. A series of tsunamis followed; these waves had unusually long periods of 1 to 2 hours. The town of Merak, 33 miles (53 km) away and on another island, was inundated by waves over

(a)

(b)

(c)

Figure 9.20

Sequential photos of the major wave of a tsunami at Laie Point, Oahu, Hawaii, from an earthquake in the Aleutian Islands, March 9, 1957. The highest wave in Hawaii was 3.6 m (11.8 ft) above sea level.

30.5 m (or 100 feet) high, and a large ship was carried nearly 2 miles (3 km) inland and left stranded 9.2 m (or 30 feet) above sea level. Over 35,000 people died as a result of these enormous waves, and as the waves moved across the oceans, water level recorders were affected as far away as Cape Horn (12,500 km, or 7800 miles) and Panama (18,200 km, or 11,400 miles). The waves were traveling with a speed calculated at about 200 m/sec, or 400 miles per hour.

On April 1, 1946, a disturbance of the crust in the area of the Aleutian Trench off Alaska produced a series of tsunamis. These waves heavily damaged Hilo, Hawaii, killing more than 150 people. Not only the portion of the island facing the oncoming waves was damaged; the waves bent around the island by refraction and were diffracted when passing between islands, producing high waves that struck the lee side of the island as well. The tsunami waves from the 1946 earthquake were apparently highest in the area of the Aleutian Islands, where they destroyed a concrete lighthouse 10 m (33 ft) above sea level at Scotch Cap and killed the crew. A radio mast in the same area mounted 33 m (108 ft) above sea level was also destroyed. In 1957, Hawaii was hit again, with waves higher than the 1946 series; but no lives were lost, due to early warnings and the evacuation of people along the shore (fig. 9.20). In 1964, the Alaska earthquake that severely damaged Anchorage and Alaskan coastal towns produced tsunamis that selectively hit areas on the west coast of Vancouver Island and the northern coast of California.

The frequency with which these waves occur in the Pacific has led to the development of tsunami prediction and warning centers located in Hawaii and Alaska. Whenever an earthquake that might trigger such waves occurs, the time required for a wave to reach population centers is calculated. Areas close to the disturbance report visible wave crests, if any, and monitor tide and water level gauges to detect any rise and fall in sea level. If seismic sea waves have been observed, then populated areas are alerted to a possible tsunami and its predicted arrival time. This warning allows ships to move out of harbors to the open sea and permits inhabitants of low-lying coastal areas to evacuate and move inland.

9.10 Internal Waves

Waves on the ocean surface form at the boundary between two fluids of very different density: water and air. Below the ocean surface, waves form along the density boundaries between water layers of different density just as they do between air and water. These waves are called **internal waves.**

Because the density differences are small, the disturbance needed to begin an internal wave need not be large, and the restoring force is also small. The requirement of small generating forces plus the viscosity of the water results in large-amplitude, slow-moving waves with long wavelengths. Internal waves appear to move like shallow-water waves, with the thickness of the layers playing a role similar to the water depth in surface shallow-water waves.

Internal waves may be caused when the surface layer flows over a more static lower layer in the same way the wind produces surface waves. When the sea surface slopes, as happens in the production of an elevated sea surface under wind stress or a low-pressure storm center, the subsurface boundary may tilt also, but in the opposite direction. The tilting of the subsurface boundary occurs as the waters seek to reestablish a pressure balance at depth. The subsurface boundary oscillates, but the oscillation has a longer period than at the surface. Other possible triggers of internal waves include seismic disturbances, long-term wind changes, and tidal forces.

When the surface layer is relatively shallow, a wave moving along an internal boundary may be high enough to have its crest break the surface. This action creates moving bands of smooth water called **slicks,** each representing a crest position. Slicks are most likely to occur in coastal water, estuaries, or fjords, where the shallow surface layer is diluted by freshwater runoff.

Since internal waves occur below the surface and are usually not visible, instruments that measure the periodic rise and fall of surfaces of constant temperature and salinity record the passing of internal waves. Because they cannot be observed easily and are not as common as surface waves, our knowledge concerning them is limited.

9.11 Standing Waves

Deep-water waves, shallow-water waves, and internal waves are all progressive waves; they have a speed and move in a direction. **Standing waves** do not progress; they are progressive waves reflected back on themselves and appear as an alternation between a trough and a crest at a fixed position. They occur in ocean basins, partly enclosed bays and seas, and estuaries. A standing wave can be demonstrated by slowly lifting one end of a container partially filled with water and then rapidly but gently returning it to a level position. If this is done, the surface alternately rises at one end and falls at the other end of the container. The surface oscillates about a point at the center of the container, the **node,** and the alternations of low and high water at each end are the **antinodes** (see fig. 9.21a). A standing wave is a progressive wave reflected back on itself; the reflection cancels out the forward motions of the initial and reflected waves. If different-sized containers are treated the same way, the period of oscillation increases as the length of the container increases or its depth decreases.

Notice that the single-node standing wave contains one-half of a wave form (fig. 9.21a). The crest is at end A of the container and the trough is at end B. As the wave oscillates, a trough replaces the crest at A, and a crest replaces the trough at B. One wavelength, the distance from crest to crest or from trough to trough, is twice the length of the container. By rapidly tilting the basin back and forth at the correct rate, it is possible to produce a wave with more than one node, shown in figure 9.21b. In the case of two nodes, there is a crest at either end of the container and a trough in the center; this alternates with a trough at each end and the

crest in the center. The two nodes are one-quarter of the basin's length from each end. In this case, note that the wavelength is equal to the basin's length. The oscillation period of the wave with two nodes is one-half that of the wave with a single node.

Standing waves in bays or inlets with an open end behave somewhat differently than standing waves in closed basins. A node is usually located at the entrance to the open-ended bay, so only one-quarter of the wavelength is inside the bay. There is little or no rise and fall of the water surface at the entrance, but a large rise and fall occurs at the closed end of the bay (see fig. 9.22). Multiple nodes may also be present in open-ended basins.

Standing waves that occur in natural basins are called **seiches,** and the oscillation of the surface is called seiching. In natural basins, the length dimension usually greatly exceeds the depth. Therefore, a standing wave of one node in such a basin behaves as a reflecting shallow-water wave, with the wavelength determined by the length of the basin. In water with distinct layers having sharp density boundaries, standing waves may occur along the fluid boundaries as well as at the air-sea boundary. The oscillation of the internal standing waves is slower than the oscillation of the sea surface.

Standing waves may be triggered by tectonic movements that suddenly tilt the basin, causing the water to oscillate at a period defined by the dimensions of the basin. If storm winds create a change in surface level to produce storm surges, the surface may oscillate as a standing wave in the act of returning to its normal level when the wind ceases. The pulsing of a weather disturbance over a lake may also cause periodic water-level changes, reaching a meter or more in height. If the period of the disturbing force is a multiple of the natural period of oscillation of the basin, the height of the standing wave is greatly increased. For example, if a child is riding on a swing, a gentle push timed with each swing period will force the swing higher and higher. The push may be delivered each time the swing passes, every other time, or every third time; all are multiples of the natural period of the swing. In chapter 10, we will learn that repeating tidal forces can produce standing waves in those basins that have natural periods of oscillation approximating the tidal period.

A standing wave in a basin is like a water pendulum. The wave's natural period of oscillation is:

$$T = \left(\frac{1}{n}\right)\left(\frac{L}{\sqrt{gD}}\right),$$

where n is the number of nodes present, D is the depth of water in the basins, g is the earth's acceleration due to gravity, and L is the length of the wave. L equals twice the basin length, l, in a closed basin and four times the basin length in a basin with an open end. This equation is related to the shallow-water wave equation when the number of nodes is equal to 1:

$$\frac{L}{T} = n\sqrt{gD}.$$

Figure 9.21a

(a) A standing wave in a basin oscillating about a single node. The time for one oscillation is the period of the wave, T.

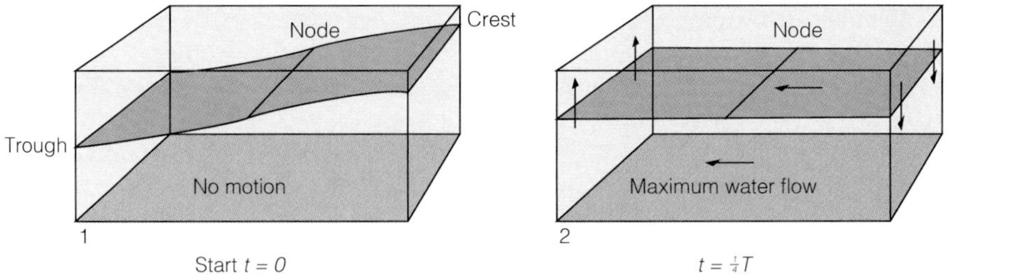

(a)

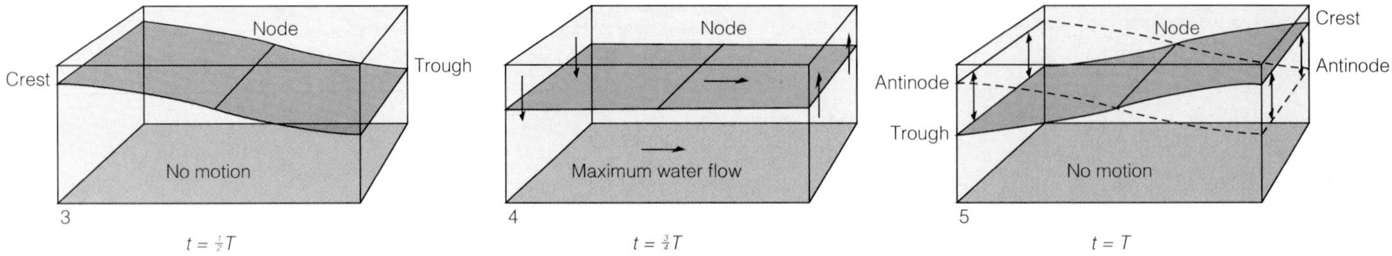

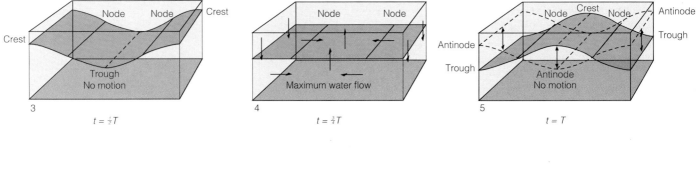

Figure 9.21b

A standing wave oscillating about two nodes. The time for one oscillation is the period of the wave, T.

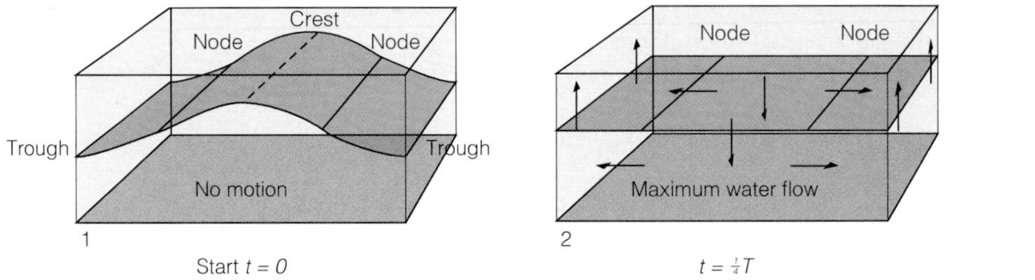

(b)

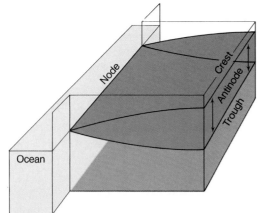

Figure 9.22

A standing wave oscillates about the node located at the opening to a basin. The antinodes produce the rise and fall of water at the closed end of the basin.

A progressive wave directly reflected back on itself produces a standing wave, because the two waves—original and reflected—are moving at the same speed but in opposite directions. The checkerboard interference pattern produced by two matched wave systems approaching each other at an angle also creates standing waves with crests and troughs alternating with each other in fixed positions (refer back to fig. 9.6a).

9.12   Practical Considerations: Energy from Waves

A tremendous amount of energy exists in ocean waves. The power of all waves is estimated at 2.7×10^{12} watts, which is about equal to three thousand times the power-generating capacity of Hoover Dam. Unfortunately for human needs,

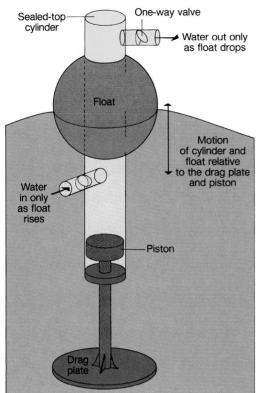

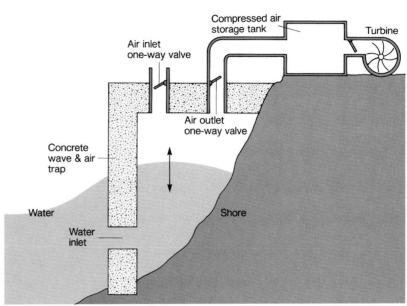

Figure 9.23

The vertical rise and fall of the waves can be used to power a pump.

Figure 9.24

Each rise and fall of the waves pumps pulses of compressed air into a storage tank. A smooth flow of compressed air from the storage tank turns a turbine that generates electricity.

this energy is widely dispersed and is not constant at any given location or time. It is, therefore, difficult to tap this supply in order to produce power, except in small quantities.

Wave energy can be harnessed in three basic ways: (1) using the changing level of the water to lift an object, which can then do useful work because of its potential energy; (2) using the orbital motion of the water particles or the changing tilt of the sea surface to rock an object to and fro; and (3) using the rising water to compress air in a chamber. In each case electric energy may be produced if the wave motion is used either directly or indirectly to turn a generator.

Consider a large buoyant surface float with a hollow cylinder extending down into the sea (see fig. 9.23). Inside the cylinder is a piston, which is connected to a large-diameter disk (a drag plate) at its bottom end. This disk will cause the piston to resist vertical motion, while the up-and-down motion of the surface float will cause the cylinder to move up and down over the piston. If the cylinder is equipped with inlet and outlet pipes with one-way valves, it will take in water as the surface buoy rises on the crest of the wave, causing the cylinder volume to increase, and will squirt water out as the surface buoy drops with the passing of the trough, causing the cylinder volume to decrease. The pumped water can be used to turn a turbine, but since wave energy is distributed over a volume of water this mechanism does not withdraw much of the passing wave's energy. A series of float devices are

being used to power a reverse osmosis desalination plant in Puerto Rico (refer back to chapter 5); modified float systems are being explored in Sweden and Great Britain.

Another system uses a tapered channel at the coast; incoming waves push the water to an increased crest elevation of 2–3 m (6–10 ft), and it spills over the channel edge and down through a turbine. This system is used to generate power by a 75-kilowatt plant on Scotland's Isle of Islay, a 350-kilowatt plant at Toftestalen, Norway, and two 1500-kilowatt plants under construction, one in Java and the other in Australia.

A device called a "sea clam" has been developed by the United Kingdom's Sea Energy Associates. Like a bellows, bags of air are collapsed by wave crests; air is forced through a turbine, and the air flows back into the bags when the wave troughs pass. One-fifth-scale models have been successfully tested in Loch Ness.

Along a wave-exposed coast, air traps can be installed so that the crest of a wave moving into the trap compresses a large volume of air, forcing it through a one-way valve. This compressed air powers a turbine to produce energy. The trough of the wave allows more air to enter the trap, readying it for compression by the following wave crest (see fig. 9.24). This system is used in northern Norway and is planned for Western Ireland.

Wave energy expended against the coasts could also be used to produce power. Shores that are continually pounded by large-amplitude waves are most likely to be developed. Great Britain has a coastline with frequent high-energy waves and an average wave power of about 5.5×10^4 watts (or 55 kilowatts) per meter of coastline. If the wave energy could be completely harnessed along 1000 km (620 mi) of coast, it would generate enough power to supply 50% of Great Britain's present power needs. Along the northern California coast, waves are estimated to expend 23×10^6 kilowatts of power annually; it is thought that 4.6×10^6 kilowatts or 20% could be harvested to generate electrical power. The Pacific Gas & Electric Company, a northern California utility, is considering installing a generating device in a breakwater planned for Fort Bragg, California.

The 1991 estimate for total world energy produced by all operating wave energy systems is less than 500 kilowatts. Hydroelectric power from existing plants costs less than 3 cents per kilowatt-hour. Nuclear power plant costs are about 9 cents per kilowatt-hour; wood-burning plant costs are nearly 8 cents per kilowatt-hour. The cost of wave energy is estimated to be from 8.6 to 20 cents per kilowatt-hour depending on the region, while wind energy cost estimates are between 4.7 and 7.2 cents per kilowatt-hour. When we consider wave energy systems, thoughtful consideration needs to be given to items other than cost. If all the wave energy were extracted from the waves in a coastal area, what effect would this action have on the shore area? If the near-shore areas are covered with wave-energy absorbers 5 to 10 m (15–33 ft) apart, what will the effect be on other ocean uses? Since the individual units collect energy at a slow rate, can they collect enough energy over their projected life span to exceed the energy used to fabricate and maintain them? Your answers to these questions will help you understand that the harvesting of wave energy is not without an effect on the environment, that it may not be either cost- or energy-effective, and that its location may present enormous problems for installation, maintenance, and transport of energy to sites of energy use.

Summary

When the water's surface is disturbed, a wave is formed by the interaction between generating and restoring forces. The wind produces capillary waves, which grow to form gravity waves. The elevated portion of a wave is the crest; the depressed portion is the trough. The wavelength is the distance between two successive crests or troughs. The wave height is the distance between the crest and the trough. Wave period measures the time required for two successive crests or troughs to pass a location. The moving wave form causes water particles to move in orbits. The wave's speed is related to wavelength and period.

Deep-water waves occur in water deeper than one-half the wavelength. Wind waves generated in storm centers are deep-water waves. The period of a wave is a function of its generating force and does not change. Long-period waves move out from the storm center, forming long, regular waves, or swell. The faster waves move through the slower waves and form groups, or trains, of waves. The longer waves are followed by the shorter waves. This process is known as sorting or dispersion. The speed of a group of waves is half the speed of the individual waves in deep water. Swells from different storms cross, cancel, and combine with each other as they move out across the ocean.

Wave height depends on wind speed, wind duration, and fetch. Single large waves unrelated to local conditions are called episodic waves. The energy of a wave is related to its height. When the ratio of the height to the length of a wave, or its steepness, exceeds 1:7, the wave breaks.

Shallow-water waves occur when the depth is less than one-twentieth the wavelength. The speed of a shallow-water wave depends on the depth of the water. As the wave moves toward shore and decreasing depth, it slows, shortens, and increases in height. Waves coming into shore are refracted, reflected, and diffracted. The patterns produced by these processes helped people in ancient times to navigate from island to island.

In the surf zone, breaking waves produce a water movement toward the shore. Breaking waves are classified as plungers or spillers. Water moves along the beach as well as toward it; it is returned seaward through the surf zone by rip currents.

Tsunamis are seismic sea waves. They behave as shallow-water waves, producing severe coastal destruction and flooding.

Internal waves occur between water layers of different densities. Standing waves, or seiches, occur in basins as the sea surface oscillates about a node. Alternate troughs and crests occur at the antinodes.

It is possible to harness the energy of the waves by using either the water-level changes or the changing surface angle associated with them. Difficulties include cost, location, environmental effects, and lack of wave regularity.

Key Terms

generating force	group speed
restoring force	fetch
surface tension	episodic wave
gravity wave	potential energy
ripple	kinetic energy
capillary wave	wave steepness
crest	shallow-water wave
trough	refraction
wavelength	wave ray
wave height	diffraction
amplitude	breaker
equilibrium surface	plunger
wave period	spiller
water particle orbit	rip current
deep-water wave	seismic sea wave
progressive wind wave	tsunami
storm center	internal wave
forced wave	slick
free wave	standing wave
swell	node
dispersion/sorting	antinode
wave train	seiche

Study Questions

1. A surfboard slides downward on the face of a wave. The steepness of the wave face is governed by the decrease of L (wavelength) and the increase of H (wave height) as the wave slows in shallow water. How must the surfer adjust the board in order to stay on the face of the wave as the wave approaches shallow water?

2. Locate a small pond or pool and drop a stone into it. Describe what happens (1) to an individual wave and (2) to the group of waves. Try to determine the group speed and the individual wave speed.

3. Drop two stones into the pond at a short distance from each other. Describe what happens when the wave rings produced by each stone pass through each other. Do the heights of the waves change when they intersect? Do the wave trains pass through each other and continue on?

4. List the forces that act on a smooth-water surface to create a fully developed deep-water wave.

5. If you were ocean-sailing at night in the trade-wind belt, how could you use the waves to keep you on a course of constant direction?

6. Make a sketch of an ideal progressive wind wave in deep water. Label the parts.

7. What happens to a deep-water progressive wave when it moves into shallow water and up a sloping beach?

8. Compare a tsunami and a storm surge (chapter 7). How are they the same? How are they different?

9. Distinguish between (a) sea and swell, (b) wave height and wave steepness, (c) wave height and wave amplitude, (d) plunger and spiller, and (e) node and antinode.

10. What is the effect of sorting (dispersion) on waves moving away from a storm center?

11. How do refraction, reflection, and diffraction affect a wave?

12. How is a standing wave related to a progressive wave?

13. Explain two ways in which wave energy could be harnessed to provide useful power. What are the advantages and disadvantages of each method?

14. If a group of mixed waves is generated in a sudden storm, why does it take more time for the group to pass an island far from the storm center than to pass a nearby island?

15. A depression in the seafloor at right angles to a straight coastline may be the site of a rip current. Why?

Study Problems

1. Using the equations $C = L/T$ and $L = (g/2\pi)T^2$, show that wave speed can be determined from (a) wave period only and (b) wavelength only.

2. What is the period of a wave moving in deep water at 10 m/sec, if its wavelength is 64 m? When it enters shallow water what will happen to the wave's speed, length, period, and height? How high will the wave have to be to break in deep water?

3. A submarine earthquake produces a tsunami in the Gulf of Alaska. How long will it take the tsunami to reach Hawaii, if the average depth of the ocean over which the waves travel is 3.8 km and the distance is 4600 km?

4. Explain wave dispersion. How far from a storm center will waves with periods of twelve seconds, nine seconds, and six seconds have traveled after twelve hours? If the six-second waves arrive at your beach ten hours after the twelve-second waves, how far away is the storm?

5. Fill a rectangular aquarium or dishpan approximately one-third full of water. Measure the water depth (D). Carefully lift one end of the container and set it down rapidly and smoothly. Time the period between successive high waters at one end (T); this is the wave period. The wavelength (L) is twice the length of the container. Show that C (the wave speed) determined from L/T is equal to C determined from $\sqrt{gD}$.

Suggested Readings

Waves and Wave Motion

Bascom, W. 1980. *Waves and Beaches: The Dynamics of the Ocean Survey,* rev. ed. Doubleday, Garden City, N.J. 336 pp.

Bowditch, N. 1984 ed. *American Practical Navigator,* Vol. 1. U.S. Defense Mapping Agency Hydrographic Center, Washington, D.C. 1414 pp. (Waves, breakers, and surf are covered in chapters 33 and 34.)

Dutton, G. 1992. Catch a Wave for Clean Electricity. *The World & I* July: 290–97.

Lissau, S. 1975. Ocean Waves. In *Oceanography, Contemporary Readings in Ocean Sciences,* 2d ed., Pirie, R. G., ed. (1977). Oxford Univ., New York. pp. 88–97.

Lockridge, P. A. 1989. Tsunami: Trouble for Mariners. *Sea Technology* 30 (4): 53–57.

Open University. 1989. *Waves, Tides and Shallow Water Processes.* Pergamon Press, Oxford, England; Open University, Milton Keynes, England. 186 pp.

Energy

Brin, A. 1981. *Energy and the Oceans.* Westbury House, Surrey, England. 133 pp.

Bruce, M. 1986. Ocean Energy: Some Perspectives on Economic Viability. In *Ocean Yearbook 6,* Borgese, E. M., and N. Ginsburg, eds. Univ. of Chicago, Chicago. pp. 58–78.

Changery, M. J., and R. G. Quayle. 1987. Coastal Wave Energy. *Sea Frontiers* 33 (4): 259–62.

Charlier, R. H. 1983. Water, Energy, and Nonliving Ocean Resources. In *Ocean Yearbook 4,* Borgese, E. M., and N. Ginsburg, eds. Univ. of Chicago, Chicago. pp. 75–120. (Overview of energy sources.)

McCormick, M. E. 1986. Ocean Wave Energy Conversion. *Sea Technology* June: 32–34.

Miyazaki, T. 1987. Wave Power Generator Kaimei. *Oceanus* 30 (1): 43–44.

Ross, D. 1979. *Energy from Waves.* Pergamon, Elmsford, N.Y. 121 pp.

Shaw, R. 1982. *Wave Energy: A Design Challenge.* Halsted, N.Y. 202 pp.

Simeons, C. 1980. *Hydro-Power, The Use of Water As an Alternative Source of Energy.* Pergamon, Elmsford, N.Y. 549 pp.

(a)

Going to Sea

(a) The University of Washington's research vessel, *Thomas G. Thompson,* one of the newest vessels in the U.S. oceanographic research fleet, began its research duties in 1992.

Oceanographic research vessels are operated by the U.S. Navy, NOAA, universities, private corporations, and foundations. They are the platforms manned by professional crews that provide accommodations, laboratory space, and the equipment required to handle deep-sea sampling gear. These vessels are often shared by scientific programs that are interested in the same area of the oceans but are directed by different institutions. The university fleet is shared to create the most cost-effective use of all major vessels, reducing ocean travel time between research sites and making maximum research time available on-site. Vessels are scheduled for long periods at sea, so scientists and equipment are flown to and from the nearest port as the scientific mission of the vessel changes. Though ships operate approximately two hundred fifty days at sea per year, a specific research cruise may last only a month.

(b)

(b) The stern of a research vessel getting ready for sea is a bewildering sight. All these scientific supplies and equipment must be properly stowed before the ship can leave the dock.

(c) Safety on shipboard is always a major concern. Here a survival suit is demonstrated.

The deck of a research vessel is the working platform from which all over-the-side sampling gear is launched

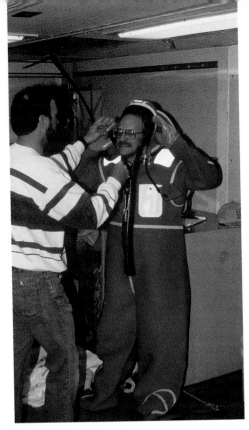

(c)

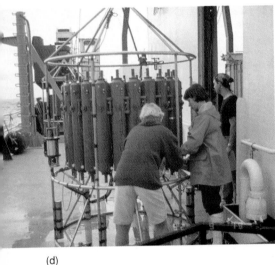

(d)

(e)

and retrieved. Winches, cranes, and A-frames are used to move heavy gear between the deck and the ocean surface. Controlling this equipment and successfully launching and retrieving expensive scientific instruments while the deck rolls and heaves requires close cooperation between the ship's crew operating the vessel and the winches, and the scientists who prepare and use the sampling gear.

(d) Water samples are taken from a ring of sampling bottles that surround a conductivity-temperature-depth (CTD)

recorder. Data from the CTD is transmitted to the ship while the device is underwater; the water samples will be analyzed in the ship's laboratories.

(e) This oceanographer stands in the hydrocage, which juts out from the ship's deck, to maneuver a doppler current meter in preparation for launch. The current meter is moored at sea to monitor currents and is retrieved at a later time to obtain the internally recorded data.

(f) Getting a plankton net ready for its tow. This type of net samples plankton from the sea's surface waters.

(g) This submersible pump, its filtering apparatus, and a 30-liter water bottle are heavy and difficult to handle. It requires a deck crane to remove it from the cable or hydrowire that lowered it into the sea so that it can be secured on deck.

(h) A tangle of gear comes on board. (i) Do you have the end or do I? Sometimes gear is retrieved in a less-than-desirable condition. Patience, a sharp knife, and cable cutters may be

(f)

(h)

(g)

(i)

required before order is restored. The important thing is to save the instrument and its data.

The ship's laboratory spaces are divided into wet labs for handling bulk water samples, geological labs, analytical labs for chemical and biological work, and dry labs with electronic and computer facilities. Plywood, two-by-fours, and steel-angle iron are nailed and bolted together, then attached to deck and overheads to make the required arrangements of tables and benches. Each research group is able to set up its labs to meet its own special requirements. Instruments must be bolted down; storage boxes are tied in place in the lab and on deck. Nothing can be left free to break loose when the ship rolls except the researchers.

(j)

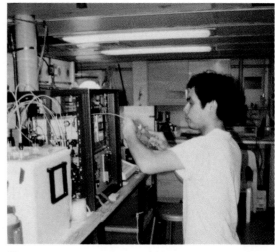

(k)

(l)

(m)

(j) Consulting a seafloor chart in the scientific command center. Here charts are drawn by computer from data as it is received.

(k) Running analyses on water samples.

It costs over $20,000 per sea day to operate a major research vessel, including maintenance, crew costs, fuel, and food, but excluding scientific costs. Because the costs are high and maximum efficiency is required, the vessel works 24 hours each day, taking samples whenever the vessel reaches the required location. Standard shifts or watches are 4 hours on duty and 8 hours off duty over the 24-hour day, and sometimes a schedule of 6 hours on duty and 6 hours off duty is used. Scientists may be on the same sched-ule; they may work 12 hours on duty and 12 hours off duty when research schedules require continuous work, or they may get only short naps during a busy period.

(l) Day and night the work goes on. Here at sunset a sediment sampling device returns from the sea.

(m) Catnapping on deck between watches is pleasant in good weather.

(n)

(o)

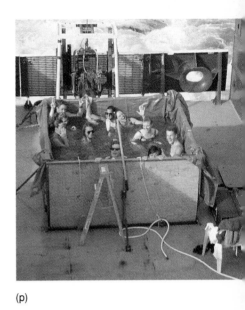

(p)

(q)

(r)

(n) Discussing the day's work in a quiet corner. Analysis of data at sea allows researchers to modify sampling routines and change procedures while their research is in process.

Days at sea are repetitive, and the routine leaves little to distinguish one day from another. Holidays and special occasions are enjoyed and celebrated with gusto.

(o) Fresh fish for dinner brings a change from frozen and canned food. Both crew and scientists often fish during their off-duty hours.

(p) Scientists and students cool off in a research tank.

(q) A first-timer across the equator kneels before King Neptune.

(r) Enjoying a quiet moment on the ship's stern.

10

The Tides

10.1 **Tide Patterns**
10.2 **Tide Levels**
10.3 **Tidal Currents**
10.4 **Equilibrium Tidal Theory**
   The Moon Tide
   The Tidal Day
   The Tide Wave
   The Sun Tide
   Spring Tides and Neap Tides
   Declinational Tides
   Elliptical Orbits
10.5 **Dynamic Tidal Analysis**
   The Tide Wave
   Progressive Tides
   Standing Wave Tides
   Tide Waves in Narrow Basins
10.6 **Tidal Bores**
10.7 **Predicting Tides and Tidal Currents**
   Tide Tables
   Tidal Current Tables
10.8 **Practical Considerations: Energy from Tides**

Box: Undersea Robotic Technology

Summary
Key Terms
Study Questions
Study Problems
Suggested Readings

Spring tides are not the tides of spring as many landsmen suppose. They are the very high and very low tides which occur twice a month, with the new and the full moon, when solar and lunar magnetism pull together to make the circumterrestrial tide wave higher than at other times. The opposite to the spring tide is the neap tide, halfway between these phases of the moon; down East, in Maine, there may be as much as five feet difference in range between springs and neaps.

Spring tides are beloved by all who live by or from the sea. At a spring low, rocky ledges and sandbars which you never see ordinarily are bared; the kinds of seaweed that require air but twice a month appear; sand dollars like tarnished pieces of eight are visible on the bottom. Clam specialists can pick up the big "hen" clams or the quahaugs, and with a stiff wire hook deftly flip out of his long burrow the elusive razor clam. Shore birds—sandpiper, plover and curlew—skitter over the sea-vacated flats, piping softly and gorging themselves on the minor forms of life that cling to this seldom-bared shelf.

Samuel Eliot Morison,
from *Spring Tides*

Low tide along the Oregon coast.

B *est known as the rise and fall of the sea around the edge of the land, the tides are caused by the gravitational attraction between the earth and the sun and between the earth and the moon. Far out at sea tidal changes go unnoticed, but along the shores and beaches the tides govern many of our water-related activities, both commercial and recreational. Early sailors from the Mediterranean Sea, where the daily tidal range is less than 1 m (3 ft), ventured out into the Atlantic and sailed northward to the British Isles; to their amazement, they found a tidal range in excess of 10 m (30 ft). The movement of the tide in and out of bays and harbors has been helpful to sailors* beaching their boats and to food-gatherers searching the shore for edible plants and animals, but it is also recognized as a hazard by navigators and can produce some spectacular effects when rushing through narrow channels.

This chapter surveys tide patterns around the world and explores the tides in two ways: one is a theoretic consideration of the tides on an earth with no land; the other is a study of the natural situation. The chapter also shows you how to use available tide data to predict water level changes and coastal tidal currents.

10.1 Tide Patterns

Measurements of tidal movements around the world show us that the tides behave differently in different places. In some coastal areas there is a regular pattern of one high tide and one low tide each day; this is a **diurnal tide.** In other areas there is a cyclic high water-low water sequence that is

Figure 10.1

The three basic types of tides: (a) a once-daily diurnal tide, (b) a twice-daily semidiurnal tide, and (c) a mixed semidiurnal tide with diurnal inequality.

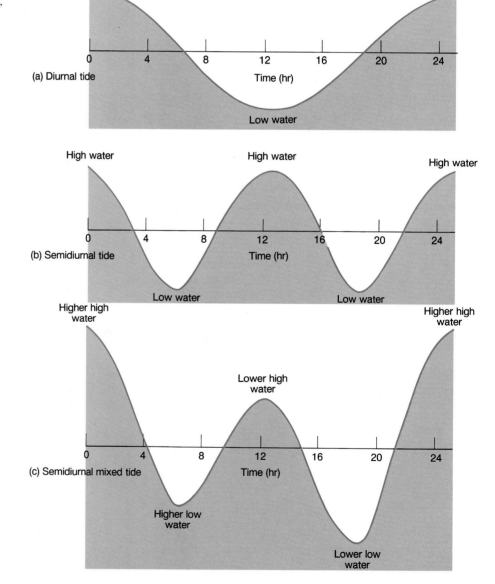

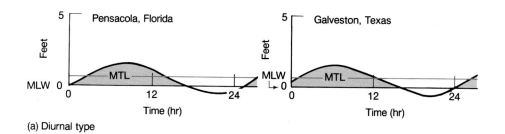

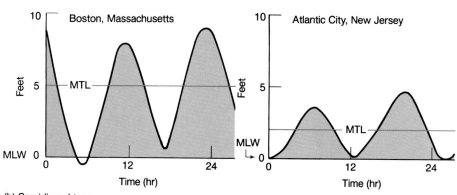

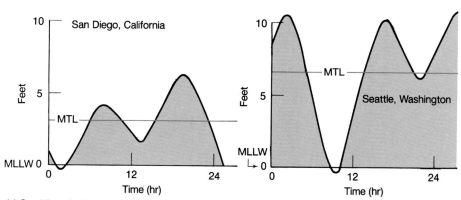

Figure 10.2

Tide types and tidal ranges vary from one coastal area to another. The zero tide level equals mean low water (MLW) or mean lower low water (MLLW), as appropriate. MTL equals mean tide level. All tide curves are for the same date.

repeated twice in one day; this is a **semidiurnal tide.** In a semidiurnal tidal pattern, both high tides reach about the same height and both low tides drop to about the same level. A tide in which the high tides regularly reach different heights and the low tides drop regularly to different levels is called a **semidiurnal mixed tide.** This type of semidiurnal tide has what is called diurnal (or daily) inequality, created by combining diurnal and semidiurnal tide patterns. The tide curves in figure 10.1 show each type of tide. Curves for typical tides at some United States coastal cities are shown in figure 10.2.

<u>10.2</u>   Tide Levels

In a uniform diurnal or semidiurnal tidal system, the greatest height to which the tide rises on any day is known as **high water,** and the lowest point to which it drops is called

low water. In a mixed-tide system, it is necessary to refer to **higher high water** and **lower high water,** as well as **higher low water** and **lower low water** (see fig. 10.1).

Tidal measurements taken over many years are used to calculate the **average** (or **mean**) tide levels. Averaging all water levels over many years gives the local mean tide level. Averages are also calculated for the high-water and low-water levels, as mean high water and mean low water. For mixed tides, mean higher high water, mean lower high water, mean higher low water, and mean lower low water are calculated.

Since the depth of coastal water is important to safe navigation, an average low-water reference level is established; depths are measured from this level for navigational charts. The tide level is added to this charted depth to find the true depth of water under a vessel at any particular time. In areas of uniform diurnal or semidiurnal tide patterns the zero depth reference, or **tidal datum,** is usually equal to

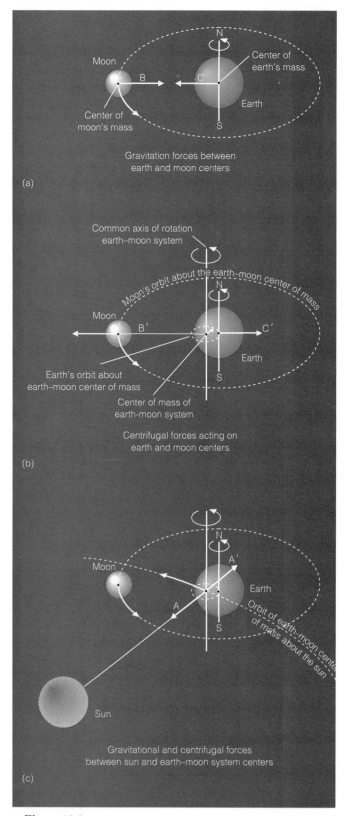

Figure 10.3

Gravitational and centrifugal forces act to keep the earth-moon system in balance.

mean low water. The use of mean low water assures the sailor that the actual depth of the water is, in general, greater than that on the chart. In regions with mixed tides, mean lower low water is used as the tidal datum, for the same reason. Occasionally, the low tide level falls below the mean value used as the tidal datum, producing a **minus tide.** A minus tide can be a hazard to boaters, but it is cherished by clam diggers and students of marine biology, as it exposes shoreline usually covered by the sea.

As the water level along the shore increases in height, the tide is said to be rising or flooding; a rising tide is a **flood tide.** When the water level drops, the tide is falling or ebbing; a falling tide is an **ebb tide.**

10.3 Tidal Currents

Currents are associated with the rising and falling of the tide in coastal waters. These **tidal currents** may be extremely swift and dangerous as they move the water into a region on the flood tide and remove it on the ebb. When the tide turns or changes from an ebb to a flood or vice versa, there is a period of **slack water,** during which the tidal currents slow and then reverse. Slack water may be the only time that a vessel can safely navigate a narrow channel with swiftly moving tidal currents, sometimes in excess of 5 m/sec (10 knots) (see forward to table 10.4 in the Tidal Current Tables section in this chapter). The relationship of tidal currents to standing wave tides, progressive tides, and tidal current prediction is discussed in later sections of this chapter.

10.4 Equilibrium Tidal Theory

Oceanographers analyze tides in two ways. The tides are studied as mathematically ideal wave forms behaving uniformly in response to the laws of physics. This method of study is called **equilibrium tidal theory.** It is based on an earth covered with a uniform layer of water, in order to simplify the study of relationships between the oceans and the tide-rising bodies, the moon and the sun. The tides are also studied as they occur naturally; this method is called **dynamic tidal analysis.** It studies the oceans' tides as they occur, modified by the landmasses, the geometry of the ocean basins, and the earth's rotation.

The effect of the sun's and moon's gravity and the rotation of the earth on tides is most easily explained by studying equilibrium tides. In this discussion the earth and the moon act as a single unit, the earth-moon system that orbits the sun; refer to figure 10.3. The moon orbits the earth, held by the earth's gravitational force acting on the moon (B). There is also a force acting to pull the moon away from the earth and send it spinning out into space (B′); this is considered a **centrifugal force** in our discussion. Forces B and B′ must be equal and opposite to keep the moon in its orbit. Likewise, the moon's gravitational force acting on the earth (C) must be balanced by the centrifugal force (C′). B′ and C′ are caused by the earth-moon system rotating about an axis at the center of the system's

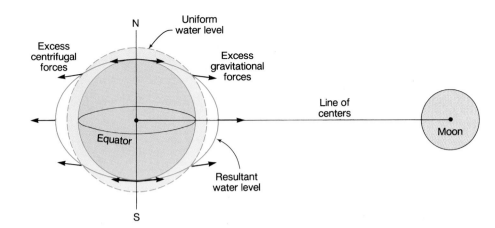

Figure 10.4

Distribution of tide-raising forces on the earth. Excess lunar gravitational and centrifugal forces distort the earth model's water envelope to produce bulges and depressions.

mass (fig. 10.3b), which is a point 4640 km (2880 mi) from the earth's center along a line between the earth and the moon. The earth-moon system is held in orbit about the sun by the sun's gravitational attraction (A). A centrifugal force again acts to pull the earth-moon system away from the sun (A′). To remain in this orbit, the earth-moon system requires that the gravitational forces equal the centrifugal forces (fig. 10.3c).

Sir Isaac Newton's universal law of gravitation tells us that the force of attraction between any two bodies is proportional to the product of the two masses divided by the square of the distance between the centers of the masses:

$$F = G \left(\frac{m_1 m_2}{R^2} \right),$$

where

G = universal gravitational constant,
  6.67×10^8 cm^3/g/sec^2;
m_1 = mass of body 1 in grams;
m_2 = mass of body 2 in grams; and
R = distance between centers of masses in
  centimeters.

If the distribution of gravitational and centrifugal forces is calculated for each gram of material at the earth's surface and also at its center, the forces at the earth's surface are found to be different from those at its center because of the change in value of the distance between a particle and the tide-raising body and also because of the distance between a particle and the earth-moon system's axis of rotation. The earth-moon-sun system is in balance at the earth's center (gravitational forces balanced by centrifugal forces), but these forces are not in balance between points on the earth's surface. At each point, the force difference per unit mass is found to be proportional to $G(M/R^3)$, where G is the gravitational constant, M is the mass of the sun or the moon, and R is the distance between the earth and the sun or between the earth and the moon.

In the earth-moon-sun system, the mass of the sun is very great, but the sun is very far away. By contrast, the moon is small but it is close to the earth. When the distribution of these forces is calculated for each water particle at the earth's surface, the moon has a greater attractive effect on the water particles than the sun. In the following discussion the effects of each tide-raising body will be considered separately.

The Moon Tide

The water particles on the side of the earth facing the moon are closest to the moon and are acted on by a larger moon gravitational force than is present at the earth's center. Because the water covering is liquid and deformable, this force moves the water particles toward a point directly under the moon. This movement produces a bulge in the water covering. At the same time, the centrifugal force of the earth-moon system acting on the water particles at the earth's surface opposite the moon is larger than that present at the earth's center, creating a bulge. This centrifugal force is equal to the magnitude of gravity's tide-raising force and proportional to $-G(M/R^3)$. The minus sign indicates that this centrifugal force is acting opposite to the moon's tide-producing gravity force. If we place the moon opposite the earth's equator and then stop the earth model and its moon in space and time, we see the bulges in the water covering, as shown in figure 10.4. Remember that our model earth initially had a water covering of uniform depth; therefore, as the two bulges are created, an area of low-water level is formed between the bulges. We now have a water covering with two bulges and two intervening depressions (or two crests and two troughs, or two high-tide levels and two low-tide levels) distributed around the equator.

The earth makes one rotation in about twenty-four hours, and the bulges (or crests) in the water covering tend to stay under the tide-producing body as the earth turns. As a result of this movement, a point on earth that is initially at a crest (or high tide) passes to a trough (or low tide), to another high tide, to another low tide, and back to the original high tide as the earth completes one revolution. You can follow this process in figure 10.5. The effect created by the motion of the earth as it turns to the east within its water covering when observed from space can also be interpreted as the wave form of the water covering moving westward when observed relative to a fixed position on the earth.

Figure 10.5

The change in water level at point *A* during one earth rotation through the distorted water envelope. (Refer to fig. 10.4.) Fractions indicate portions of a revolution.

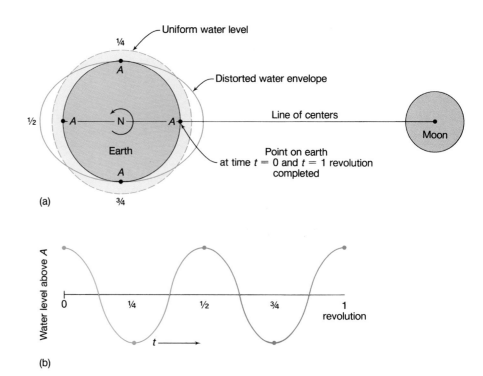

Figure 10.6

Point *A* requires twenty-four hours to complete one earth rotation. During this time the moon moves 12° east along its orbit, carrying with it the tide crest. To move from *A* to *A′* requires an additional fifty minutes to complete a tidal day.

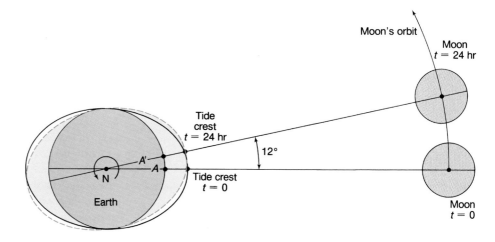

The Tidal Day

While the earth turns upon its axis, the moon is moving in the same direction along its orbit about the earth. After twenty-four hours the earth point that began directly under the moon is no longer directly under the moon. The earth must turn for an additional fifty minutes, about 12°, to bring the starting point on earth back in line with the moon. Therefore a **tidal day** is not twenty-four hours long, but twenty-four hours and fifty minutes. This difference also explains why corresponding tides arrive at any location about one hour later each day. This relationship is shown in figure 10.6.

The Tide Wave

The tides produced in this example are semidiurnal with two highs and two lows each day. The tidal distortion of the model's water covering produces a wave form known as the **tide wave.** The crest of the tide wave is the high-water level, and the trough of the tide wave is the low-water level. The wavelength in this example is half the circumference of the earth, and the tide wave's period is about twelve hours and twenty-five minutes.

The Sun Tide

Although the moon plays the greater role in the tide-producing process, the sun produces its own tide wave. Despite the sun's large mass, it is so far away from the earth that its tide-raising force is only 46% that of the moon. The time required for the earth to make a revolution with respect to the sun is on the average twenty-four hours, and not twenty-four hours and fifty minutes, as in the case of the earth-moon system. For this reason, the tide wave produced by the moon is not only of greater magnitude than that produced by the sun, but it continually moves eastward relative to the tide wave produced by the sun. Because the

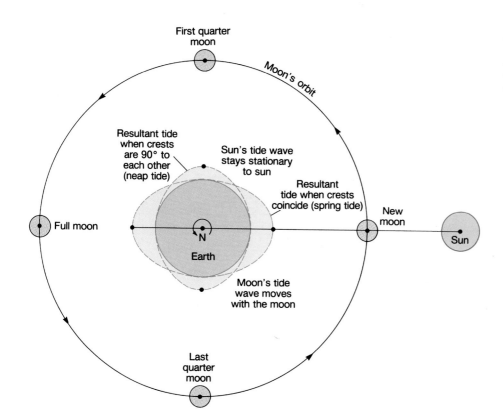

Figure 10.7

Spring tides result from the alignment of the
earth, sun, and moon during the full moon and
the new moon. During the moon's first and last
quarters, neap tides are produced.

tidal forces of the moon are greater than those of the sun,
the tidal period of the moon is more important; and the tidal
day is considered to be twenty-four hours and fifty minutes.

Spring Tides and Neap Tides

The moon's orbit requires 29½ days relative to the earth.
During this period the sun, the earth, and the moon move in
and out of phase with each other. At the new moon, the
moon and sun are on the same side of the earth, so that the
high tides, or bulges, produced independently by each coincide (see fig. 10.7). Since the water level is the result of
adding the two wave forms together, tides of maximum
height and depression, or tides with the greatest **range**
between high water and low water, are produced. These
tides are known as **spring tides.** The vertical displacement
or amplitude of the tide is one-half the range—the distance
above or below mean sea level.

In a week's time, the moon is in its first quarter; it
has moved along its orbit (about 12° per day) and is located
approximately at right angles (or 90°) to the line of centers
of the earth and sun. The crest or bulge of the moon tide is
at right angles to the tide wave created by the sun; the
crests of the moon tide will coincide with the troughs of the
sun tide, and the same will be true of the sun's tide crests
and the moon's tide troughs (see fig. 10.7). The crests and
troughs tend to cancel each other out, and the range
between high water and low water is small, producing low-
amplitude **neap tides.**

At the end of another week the moon is full, and the
sun, moon, and earth are again lined up, producing crests
that coincide and tides with the greatest range between high

and low waters, or spring tides. These spring tides are followed by another period of neap tides, produced by the
moon in its last quarter when it again stands at right angles
to the sun (see fig. 10.7). The tides follow a four-week
cycle of changing tidal amplitude, with spring tides occurring every two weeks, and a period of neap tides occurring
in between. This progression can be seen in the portions of
the tide records reproduced in figure 10.8. The effect
occurs each lunar month and is the result of the moon's tide
wave moving around the earth relative to the sun's tide
wave.

Declinational Tides

If the moon stands north or south of the earth's equator, one
bulge or high water is in the Northern Hemisphere and the
other is in the Southern Hemisphere (see fig. 10.9). Under
these conditions, a point at the middle latitudes on the
earth's surface passes through only one crest or high tide
and one trough or low tide each tidal day. A diurnal tide,
often called a **declinational tide,** is formed, because the
moon is said to have declination when it stands above or
below the equator.

Declinational (or diurnal) tides are influenced by both
the moon and the sun. The sun stands above 23½°N at the
summer solstice and above 23½°S at the winter solstice.
This variation causes the bulge created by the sun to oscillate north and south of the equator in a regular fashion each
year and tends to create more diurnal sun tides during the
winter and summer seasons of the year. The moon's declination varies between 28½°N and 28½°S with reference to
the earth. The moon's orbit is inclined 5° to the earth-sun

Figure 10.8

Spring and neap tides alternate during the tides' monthly cycle. MHWS is the mean high water spring tides; MLWS is mean low water spring tides. (a) A semidiurnal tide from Port Adelaide, Australia. (b) A diurnal tide from Pakhoi, China.

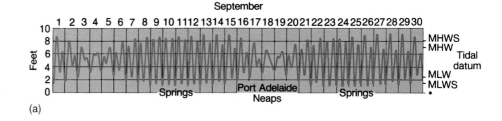

(a)

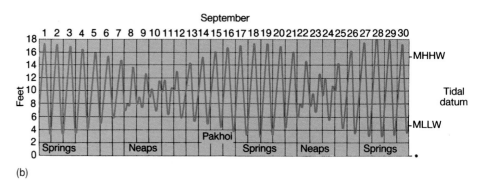

(b)

orbit, and it takes 18.6 years for the moon to complete its cycle of maximum declination. Occasionally, the sun's and the moon's declinations coincide; when this happens, both tide waves become more diurnal.

Elliptical Orbits

The moon does not move about the earth in a perfectly circular orbit, nor does the earth circle the sun at a constant distance. These orbits are elliptical, and therefore there are times during an orbit when the earth is closer to a certain tide-raising body. During the Northern Hemisphere's winter, the earth is closest to the sun, and therefore the sun plays a greater role as a tide producer in winter than in summer.

10.5 Dynamic Tidal Analysis

Equilibrium tidal theory helps us understand the distribution of wave-level changes and tide-raising forces, but it does not explain the tides as observed on the real earth. Return to figure 10.2 and notice the variety of tidal ranges and tidal periods that appear at different locations on the same date. Refer also to figure 10.8, showing different tides at different places during the same time period. The sun-moon-earth system for all locations in both figures is the same, but the equilibrium tidal theory does not explain natural tides at any particular location. Investigating the actual tides requires the dynamic approach, a mathematical study of tide waves as they occur.

The Tide Wave

Although the calculated tide-raising forces are the same, the behavior of the natural tide wave varies considerably from the tide wave of the water-covered model used to explain equilibrium theory. Because the continents separate the oceans, the tide wave is discontinuous; the wave starts at the shore, moves across the ocean, and stops at the next shore. Only in the Southern Ocean around Antarctica do the tide waves move continuously around the earth. A tide wave has a long wavelength compared to the depth of the oceans; therefore it behaves like a shallow-water wave, with its speed controlled by the depth of the water. Because the wave is contained within the ocean basins, it can oscillate in the basin as a standing wave, and it is also reflected from the edge of the continents, refracted by the change in water depth, and diffracted as it passes through gaps between continents. In addition the persistence of the tidal motion and the scale on which it occurs are so great that the Coriolis effect plays a role in the water's movement. All these factors together produce the earth's real tides, and because their interactions are complex, it is not possible to understand them together until each is first considered separately.

The tide wave's speed as a **free wave** moving across the water's surface is determined by the depth of the water. The tide wave as a free wave moves at about 200 m/sec (or 400 miles/hour), but at the equator the earth moves eastward under the tide wave at 463 m/sec (1044 mph), more than twice the speed at which the tide can travel freely as a shallow-water wave. Therefore, the tide moves as a **forced wave** that is the result of the moon's attractive force and the earth's rotation. Since the earth turns eastward faster than the tide wave moves freely westward, friction displaces the tide crest to the east of its expected position under the moon. This eastward displacement continues until the friction force is balanced by a portion of the moon's attractive force. These two forces, when balanced, hold the tide crest in a position to the east of the moon rather than directly under it. This process is illustrated in figure 10.10.

Above 60° N or 60° S, the distance around a latitude circle is less than one-half the distance around the equator. Here, the free propagation speed of the tide wave equals the speed at which the seafloor moves under the wave form. Under these conditions, the crest of the tide stays aligned

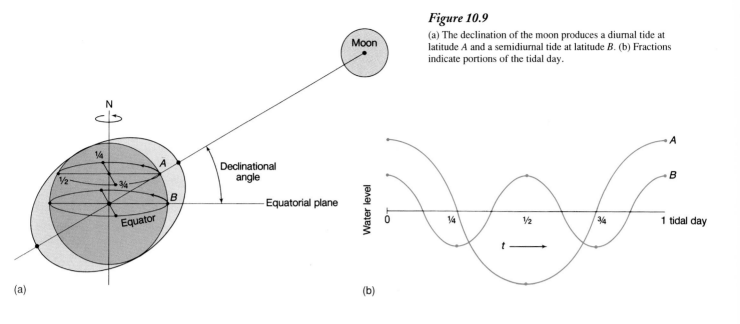

Figure 10.9

(a) The declination of the moon produces a diurnal tide at latitude *A* and a semidiurnal tide at latitude *B*. (b) Fractions indicate portions of the tidal day.

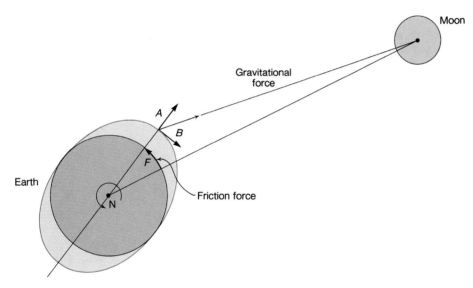

Figure 10.10

The crest of the tide wave is displaced eastward until the *B* component of the gravitational force balances the friction (*F*) between the earth and the tide wave. Component *A* of the gravitational force is the tide-raising force. Component *B* causes the tide wave to move as a forced wave.

with the moon, and less friction is generated between the rotating earth and the moving wave form. Friction between the moving tide wave and the turning earth also acts to slow the rotation rate of the earth, adding about 1½ milliseconds per 100 years to the length of a day.

Progressive Tides

In a large ocean basin the tide wave moving across the sea surface like a shallow-water wave is a **progressive tide.** Examples of progressive tides are found in the western North Pacific, the eastern South Pacific, and the South Atlantic oceans. **Cotidal lines** are drawn on charts to mark the location of the tide crest at set time intervals, generally one hour apart. The cotidal lines for the world's ocean tides are shown in figure 10.11.

Because the tide wave is a shallow-water wave, the water particles move in elliptic orbits, and the movement extends to the seafloor. The horizontal component of the motion greatly exceeds the vertical motion. Because the time in which the water particles move in one direction is so long (one-half the tide period), the Coriolis effect

becomes important. In the Northern Hemisphere the water particles are deflected to the right, and in the Southern Hemisphere they are deflected to the left. This deflection causes a clockwise rotation of the water in the Northern Hemisphere and a counterclockwise rotation in the Southern Hemisphere. This circular (or rotary) movement is the oceanic tidal current described at the end of the next section.

Standing Wave Tides

In some ocean basins or parts of ocean basins, the tide wave is reflected from the edge of the continents, and a **standing wave tide** is produced; refer back to chapter 9. Remember that if a container of water is tipped so that the water level is high at one end and low at the other, the water flows to the low end, raising it as the water level at the high end drops. This movement produces a wave having a wavelength that is twice the length of the container, with antinodes at the ends of the basin and a node at the basin's center. This same process occurs in ocean basins, but with some important modifications.

Figure 10.11

Cotidal lines for the world's oceans. The high tide crest occurs at the same time along each cotidal line. Positions of the high tide are indicated for each hour over a twelve-hour period for semidiurnal tides.

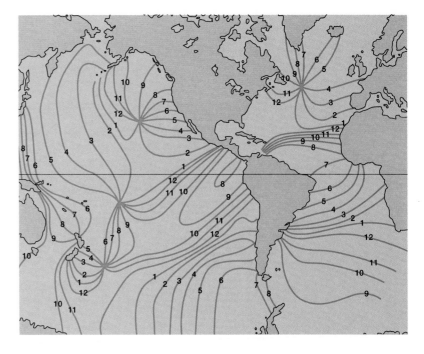

Figure 10.12

Corange lines for the world's oceans. Corange lines connect positions with the same spring tidal range. Open-ocean tidal ranges are less well known than nearshore ranges, where tide level recorders are in common use.

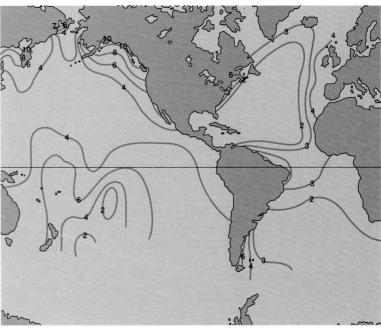

For the water to flow from the high-water side to the low-water side of an ocean basin requires a long time, and the Coriolis effect must be included. The moving water, deflected to the right in the Northern Hemisphere, does not reach the low-tide end but instead is deflected to a position to the right of the high-tide position. This movement causes the tide crest to be displaced counterclockwise around the basin in which it oscillates, but the tidal current rotates clockwise, because the current is deflected to the right in the Northern Hemisphere. Follow this rotation in figure 10.13. In the Southern Hemisphere these directions are reversed.

In the **rotary standing tide wave,** the node becomes reduced to a central point, while the tide crests (shown as cotidal lines) progress around the edges of the basin. (See fig. 10.11 for a demonstration of this pattern in the northeast and southwest Pacific and the north Atlantic.) The central point, or node, for a rotary tide is called the **amphidromic point.** The distance between low- and high-water levels, or the tidal range, for a rotary tide is shown on a chart by a series of lines decreasing in value as they approach the amphidromic point. The lines of equal tidal range are called **corange lines** (see fig. 10.12). Near the amphidromic point the tidal range is small; the farther from

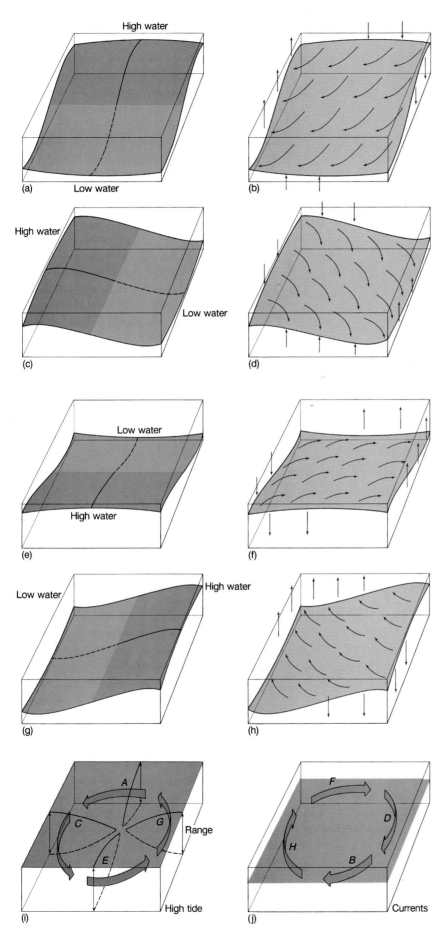

(a) High water

Low water

(b)

(c) High water

Low water

(d)

(e) Low water

High water

(f)

(g) Low water

High water

(h)

(i) Range

High tide

(j) Currents

Figure 10.13

Rotary standing tide waves. As water at the high-water side of a basin (shaded areas) begins to flow to the low-water side, it is deflected to its right in the Northern Hemisphere (a and c). This deflection causes the tidal current to move in the direction shown in (b). The tide crest (a, c, e, and g) continues to move counterclockwise, while the tidal current (b, d, f, and h) moves clockwise. The movements are summarized in (i) and (j).

Table 10.1 Dimensions of Closed Ocean Basins with Natural Periods Equaling Tidal Periods		
Tidal period	**Depth (m)**	**Length or width (km)**
Semidiurnal	4000	4428
12.42 hr	3000	3835
44,712 sec	2000	3131
	1000	2214
	500	1566
	100	700
	50	495
Diurnal	4000	8853
24.83 hr	3000	7667
89,388 sec	2000	6260
	1000	4427
	500	3130
	100	1399
	50	989

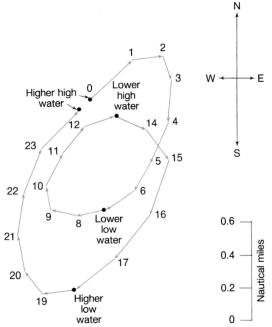

Figure 10.14

Ocean tidal currents are rotary currents. The arrows trace the path followed by water particles in a tide wave during a mixed tidal cycle. Two unequal tidal cycles are shown. Numbers indicate consecutive hours in each tidal stage. The Coriolis effect deflects the horizontal component of the water particles' orbital motion, causing them to move in a circular path.

the amphidromic point, the greater the range. Since the amphidromic point is located near the center of an ocean basin, many midocean areas have small tidal ranges, while the shores of the landmasses forming the sides of the basins have larger tidal ranges. The value and position of corange lines in midocean are not as well known as they are near shore.

The flow of water from the high-water side to the low-water side of a standing tide wave produces a rotating tidal current, as shown in figure 10.13. A rotating tidal current is also produced by the orbital motion of water particles in a progressive tide wave (review section 10.3 on tidal currents). If the tide wave is diurnal, the water particles travel in one complete circle in a tidal day. A semidiurnal tide causes two circles, and a mixed tide produces two circles of unequal size. An example of a rotating mixed tidal current recorded at the Columbia River lightship in the North Pacific is presented in figure 10.14.

Rotary standing tides occur in basins in which the natural period of the basin approximates the tidal period. Table 10.1 relates basin depths and lengths or widths that produce equal tidal periods. Remember from chapter 9 that the natural period of oscillation of a wave in a basin is

$$T = \left(\frac{1}{n}\right)\left(\frac{L}{\sqrt{gD}}\right),$$

where L, the wavelength, is twice the basin length, l, in closed basins and four times l in open basins. A comparison of the values given in table 10.1 shows that deep-ocean basins must have great length to accommodate standing waves with tidal periods, while shallow basins may be much shorter. Most tides are semidiurnal, but dimensions of some basins cause the basin to resonate with a diurnal tidal period rather than a semidiurnal period. See the tide curves in figure 10.2 for Pensacola, Florida, and Galveston, Texas. In an open-ended tidal basin with a mixed tide, the

semidiurnal portion of the tide may cause the basin to resonate with a node at the entrance to the basin and another node within the basin. The diurnal portion of the tide may have only a single node at the basin entrance. The result is a diurnal water-level pattern at the second node of the semidiurnal tide and a mixed pattern in all other parts of the basin. The harbor of Victoria, British Columbia, is located near the semidiurnal node so that it registers a diurnal tidal pattern in an inlet system of mixed tides.

Ocean tides are the result of combining progressive and standing wave tides with diurnal and semidiurnal characteristics. Different kinds of tides interact with each other along their boundaries, and the results are exceedingly complex.

Tide Waves in Narrow Basins

Unlike ocean basins, coastal bays and channels are often long and narrow with a length that is considerably greater than their width. These narrow basins have an open end toward the sea, so that the reflection of the tide wave occurs only at the head of the basin. In order to resonate with a tidal period, the length dimension need only be one-half the length cited for the closed basins in table 10.1. In this type of open basin, the node is at the entrance and only one-fourth of the tide wave's length is present; an antinode is at the head of the bay. This is illustrated in chapter 9, figure 9.22. If the open basin is very narrow, oscillation occurs only along the length of the basin; there is no rotary motion because the basin is too narrow. For example, the Bay of Fundy in northeast Canada has a tidal range near the entrance node of about 2 m (6.6 ft), while the range at

(a)

(b)

Figure 10.15

Low tide (a) and high tide (b) at Deer Island in the Bay of Fundy. The tidal range at the head of the bay exceeds 10 meters.

Figure 10.16

The fast-rising tide in the Bay of Fundy produces a tidal bore that sweeps across the shallows.

the head of the bay is 10.7 m (35 ft). This particular bay has a natural oscillation period that is so well matched to the tidal period that every tidal impulse at its entrance creates a large oscillation at the head of the bay (see fig. 10.15). In another bay along another coast, the shape of the basin may be such that it decreases rather than increases the tide's range. Every naturally occurring basin is unique in this regard.

10.6 Tidal Bores

In some areas of the world, large-amplitude tides cause large and rapid changes in water volume along shallow coasts, bays, or river mouths. Under these conditions, the tide wave is forced to move toward the land at a speed greater than the shallow-water wave speed determined by

the depth of the water or the speed of the opposing river flow. When the forced tide wave breaks, it forms a spilling wave front that moves into the shallow water or up into the river. This wave front appears as a wall of turbulent water called a **tidal bore** and produces an abrupt change in water levels as it passes. There may be a single bore formed, or a series of bores may be produced. The bores are usually less than a meter in height but can be as much as 8 m (26 ft) high, as in the case of spring tides on the Qiantang River of China. The Amazon, Trent, and Seine rivers have bores. Fast-rising tides also send bores across the sand flats surrounding Mont Saint Michel in France. The bore in the Bay of Fundy in Canada (see fig. 10.16) has been reduced by the construction of a causeway. Towns in areas having tidal bores often post warnings; their turbulence can be a severe hazard because they suddenly flood areas that have been open stretches of beach only minutes before.

10.7 Predicting Tides and Tidal Currents

Because of all the natural combinations of progressive and standing tides and the factors that affect them, it is not possible to predict the earth's tides from knowledge of the tide-raising bodies alone; equilibrium tidal theory is not adequate for the task. Accurate, dependable daily tidal predictions are made by combining actual local measurements with astronomical data.

Water-level recorders are installed at coastal sites, and the rise and fall of the tides are measured over a period of years. Primary tide stations make these water-level measurements for at least 19 years, to allow for the 18.6-year declinational period of the moon. From these data, mean tide levels are calculated. Oceanographers use a technique

Table 10.2

Times and Heights of High and Low Waters, Aberdeen, Washington, August 1993

Day	Time (h m)[1]	Height ft	Height cm	Day	Time (h m)[1]	Height ft	Height cm	Day	Time (h m)[1]	Height ft	Height cm
1 Su	0627	−1.2	−37	11 W	0125	1.4	43	21 Sa	0329	9.3	283
	1301	8.2	250		0726	6.4	195		0947	0.2	6
	1836	2.3	70		1310	3.5	107		1550	10.7	326
2 M	0027	9.9	302		1914	9.0	274		2225	−0.5	−15
	0709	−1.2	−37	12 Th	0229	1.1	34	22 Su	0424	8.6	262
	1339	8.5	259		0836	6.4	195		1033	1.0	30
	1922	2.0	61		1423	3.8	116		1636	10.5	320
3 Tu	0111	9.7	296		2014	9.1	277		2320	−0.4	−12
	0750	−1.1	−34	13 F	0330	0.6	18	23 M	0525	7.9	241
	1412	8.7	265		0943	6.7	204		1122	1.9	58
	2006	1.8	55		1533	3.7	113		1728	10.2	311
4 W	0150	9.4	287		2114	9.4	287	24 Tu	0016	−0.1	−3
	0827	−0.7	−21	14 Sa	0429	0.0	0		0631	7.3	223
	1440	8.9	271		1044	7.3	223		1218	2.7	82
	2045	1.6	49		1633	3.3	101		1828	9.9	302
5 Th	0229	9.0	274		2215	9.8	299	25 W	0118	0.2	6
	0904	−0.3	−9	15 Su	0521	−0.6	−18		0744	7.0	213
	1509	9.0	274		1136	7.9	241		1320	3.3	101
	2126	1.5	46		1729	2.6	79		1932	9.6	293
6 F	0308	8.6	262		2310	10.1	308	26 Th	0224	0.3	9
	0939	0.3	9	16 M	0611	−1.1	−34		0900	7.0	213
	1541	9.0	274		1222	8.6	262		1429	3.5	107
	2206	1.5	46		1822	1.9	58		2038	9.4	287
7 Sa	0347	8.1	247	17 Tu	0003	10.4	317	27 F	0328	0.3	9
	1015	1.0	30		0656	−1.4	−43		1009	7.3	223
	1612	9.1	277		1304	9.3	283		1538	3.3	101
	2248	1.5	46		1912	1.1	34		2140	9.4	287
8 Su	0430	7.6	232	18 W	0053	10.5	320	28 Sa	0428	0.1	3
	1050	1.7	52		0739	−1.4	−43		1105	7.8	238
	1648	9.1	277		1343	9.9	302		1638	2.9	88
	2334	1.5	46		1958	0.4	12		2239	9.4	287
9 M	0520	7.1	216	19 Th	0144	10.3	314	29 Su	0518	−0.2	−6
	1129	2.4	73		0821	−1.1	−34		1153	8.3	253
	1728	9.0	274		1425	10.3	314		1732	2.4	73
10 Tu	0026	1.5	46		2047	−0.2	−6		2331	9.5	290
	0619	6.7	204	20 F	0235	9.9	302	30 M	0601	−0.3	−9
	1211	3.0	91		0904	−0.6	−18		1229	8.7	265
	1817	9.0	274		1507	10.6	323		1819	1.8	55
					2136	−0.5	−15	31 Tu	0016	9.5	290
									0643	−0.3	−9
									1301	9.0	274
									1902	1.4	43

Time meridian 120° W. 0000 is midnight. 1200 is noon.

Heights are referred to mean lower low water which is the chart datum of soundings.

[1](h m) = hours, minutes.

Source: U.S. Department of Commerce, National Oceanic and Atmospheric Administration, National Ocean Survey, 1993.

called **harmonic analysis** to separate the tide record into components with magnitudes and periods that match the tide-raising forces of the sun and the moon. They are then able to isolate the effect of the local geography known as the **local effect.** To predict tides for any location the local effect is combined with the astronomer's predicted data. Complex and cumbersome mechanical computers or tide machines were once used to predict the tides, but today computers quickly and easily recombine the data and predict the time, date, and elevation of each high-water and low-water level.

Tide Tables

The National Oceanic and Atmospheric Administration (NOAA) of the U.S. Department of Commerce has the responsibility for determining and publishing an annual

	Place	Position Lat. N. °	'	Long. W. °	'	Time High water h	m	Low water h	m	Differences Height High water ft	Low water ft	Mean ft	Ranges Diurnal ft	Mean tide level ft
	Washington Coast					(Based on Aberdeen tides)[1]								
919	Long Beach 46	21		124	03	–1	05	–0	59	*0.80	*0.80	6.2	8.1	4.4
	Willapa Bay and River													
921	Willapa Bay entrance 46	43		124	04	–0	37	–0	43	*0.80	*0.80	6.2	8.1	4.4
923	Nahcotta, Willapa Bay 46	30		124	01	+0	20	+0	16	0.0	–0.1	8.0	10.2	5.4
925	Tarlatt Slough, Willapa Bay 46	22		124	00	+0	21	+0	54	*0.91	*0.47	7.9	9.4	4.6
927	Bay Center, Palix River 46	38		123	57	–0	09	+0	04	–1.3	–0.2	6.8	8.9	4.7
929	Toke Point, Willapa Bay..........46	42		123	58	–0	36	–0	19	–1.6	–0.2	6.5	8.5	4.5
931	South Bend, Willapa River.......46	40		123	48	–0	04	–0	08	–0.3	–0.2	7.8	9.8	5.2
933	Raymond, Willapa River 46	41		123	45	+0	02	–0	03	–0.2	–0.1	7.8	9.9	5.3
935	Grayland.........................46	49		124	06	–1	05	–0	59	*0.80	*0.80	6.2	8.1	4.4
937	Westport (ocean) 46	53		124	07	–1	05	–0	56	*0.84	*0.84	6.4	8.5	4.6
	Grays Harbor													
939	Point Chehalis 46	55		124	07	–0	32	–0	43	–1.1	–0.1	6.9	9.0	4.8
941	Bay City 46	52		124	04	–0	15	–0	32	–0.9	–0.1	7.1	9.2	4.9
943	Markham.................................46	54		124	00	–0	19	–0	12	–0.9	–0.2	7.2	9.2	4.9
945	North Channel.........................46	58		123	57	–0	06	–0	01	–0.4	–0.1	7.6	9.7	5.2
947	ABERDEEN 46	58		123	51	Daily predictions						7.9	10.1	5.4
949	Montesano, Chehalis River 46	58		123	36	+1	21	+1	48	*0.80	*0.53	6.7	8.1	4.1
951	Pacific Beach 47	13		124	12	–1	02	–0	59	*0.85	*0.85	6.5	8.6	4.6
953	Point Grenville 47	18		124	16	–1	02	–0	59	*0.85	*0.85	6.5	8.6	4.6
955	Destruction Island 47	40		124	29	–1	01	–1	03	*0.87	*0.87	6.6	8.7	4.7
957	La Push, Quillayute River 47	55		124	38	–1	00	–0	47	*0.84	*0.84	6.5	8.5	4.6
959	Cape Alava (Flattery Rocks)........48	10		124	44	–0	53	–0	39	*0.81	*0.81	6.0	8.2	4.4

predicted tide table for North and South America, Alaska, Hawaii, and the coast of Asia. These tables give the dates, times, and water levels for high and low water at primary tide stations (see table 10.2). There are 196 primary tide stations in these volumes, but many more localities require accurate tide predictions. The data for these other stations are determined by consulting a list of 6000 auxiliary stations and applying corrections for time and height to the appropriate primary station data (see table 10.3). The pocket tide tables provided by sporting goods stores and marine suppliers contain local information selected from the NOAA tables for the area's boaters, beachcombers, and fishing enthusiasts.

Tidal Current Tables

Tidal currents in the open ocean have been explained as rotary currents formed by the passing tide wave form and the deflection of water particles due to the Coriolis effect. Tidal currents in the deep sea are of scientific interest to oceanographers concerned with removing this circular motion from their data to obtain the net flows of the major ocean currents. Tidal currents in harbors and coastal waters are of major interest to commercial vessels and pleasure boaters, since these currents can be very strong and must be taken into account by those wishing to navigate in such waters.

Like the tides, the tidal currents are first measured at selected primary locations in important inland waterways. These current data are studied to determine how the speed and direction of the tidal current are related to the predicted tide level changes. As before, the local effect is determined and is used to predict future tidal currents based on the tide tables.

Table 10.4

Tidal Currents for August 1993, Seymour Narrows, B.C., Canada

Day	Slack water time (h m)	Maximum current Time (h m)	Maximum current Vel. knots	Day	Slack water time (h m)	Maximum current Time (h m)	Maximum current Vel. knots	Day	Slack water time (h m)	Maximum current Time (h m)	Maximum current Vel. knots
1 Su		0100	7.7F	7 Sa	0210	0505	8.3F	13 F	0020	0410	7.2E
	0345	0710	11.5E		0820	1120	7.7E		0720	1045	9.1F
	1005	1325	13.4F		1410	1710	7.8F		1425	1730	6.8E
	1655	2000	11.9E		2015	2335	9.3E		2040	2300	4.3F
	2305										
2 M		0145	8.5F	8 Su	0250	0550	7.7F	14 Sa	0130	0515	8.3E
	0440	0755	11.7E		0905	1205	6.4E		0815	1135	10.5F
	1050	1405	13.2F		1455	1750	6.3F		1510	1820	8.4E
	1730	2040	12.2E		2045				2125	2355	5.8F
	2345										
3 Tu		0230	9.0F	9 M		0015	8.3E	15 Su	0230	0605	9.6E
	0525	0840	11.5E		0335	0635	7.2F		0905	1220	11.8F
	1135	1445	12.7F		1005	1255	5.3E		1555	1900	10.0E
	1805	2115	12.2E		1550	1835	4.8F		2205		
					2125						
4 W	0025	0310	9.2F	10 Tu		0100	7.4E	16 M		0040	7.4F
	0610	0920	10.9E		0420	0735	7.0F		0325	0655	10.9E
	1215	1520	11.8F		1110	1405	4.5E		0950	1305	12.9F
	1840	2150	11.8E		1700	1935	3.7F		1630	1940	11.4E
					2210				2245		
5 Th	0100	0345	9.1F	11 W		0155	6.8E	17 Tu		0125	9.0F
	0650	1000	10.0E		0520	0840	7.2F		0415	0740	12.0E
	1250	1555	10.7F		1225	1520	4.5E		1035	1345	13.6F
	1910	2225	11.1E		1825	2045	3.1F		1710	2015	12.5E
					2310				2320		
6 F	0135	0425	8.8F	12 Th		0305	6.7E	18 W		0205	10.5F
	0735	1040	8.9E		0620	0945	7.9F		0505	0825	12.7E
	1330	1630	9.3F		1330	1635	5.4E		1120	1425	13.7F
	1945	2300	10.3E		1940	2200	3.4F		1745	2055	13.3E

Source: U.S. Department of Commerce, National Oceanic and Atmospheric Administration, National Ocean Survey, 1993.

NOAA publishes tidal current data in a format similar to the tide tables (see table 10.4). The times of slack water, maximum flood currents and maximum ebb currents; the speed of the currents in knots; and the direction of flow for ebb and flood currents are given. Auxiliary tidal current stations are listed and are keyed to the primary stations with correction factors to determine current speed, time, and direction at the secondary stations (see table 10.5). This information allows the master of a vessel to decide at what time to arrive at a particular channel in order to find the current flowing in the right direction or how long to wait for slack water before choosing to proceed through a particularly swift and turbulent passage.

10.8 Practical Considerations: Energy from Tides

The possibility of obtaining energy from the tides exists in coastal areas with large tidal ranges or in narrow channels with swift tidal currents. There are two systems for extracting energy from the rise and fall of the tides. Both require building a dam across a bay or estuary so that seawater can be held in the bay at high tide. When the tide ebbs, a difference in water-level height is produced between the water behind the dam and the ebbing tide. When the elevation difference becomes sufficient, the seawater behind the dam is released through turbines to produce electric power. The reservoir behind the dam is refilled on the next rising tide by opening gates in the dam. This single-action system

	Slack water time	Maximum current Time	Vel.		Slack water time	Maximum current Time	Vel.		Slack water time	Maximum current Time	Vel.
Day				Day				Day			
	(h m)	(h m)	knots		(h m)	(h m)	knots		(h m)	(h m)	knots
19 Th	0000	0250	11.6F	24 Tu		0035	10.6E	29 Su	0245	0610	10.1E
	0555	0905	12.9E		0350	0705	10.6F		0905	1220	11.5F
	1205	1505	13.3F		1035	1335	7.8E		1545	1850	10.9E
	1820	2135	13.6E		1640	1920	5.9F		2200		
					2205						
20 F	0040	0335	12.3F	25 W		0135	9.4E	30 M		0045	8.4F
	0640	0950	12.5E		0450	0815	10.0F		0340	0655	10.9E
	1250	1550	12.4F		1150	1450	7.3E		0955	1305	11.8F
	1855	2215	13.4E		1800	2035	5.1F		1625	1935	11.6E
					2310				2235		
21 Sa	0120	0420	12.4F	26 Th		0250	8.7E	31 Tu		0125	9.4F
	0735	1040	11.6E		0600	0925	9.9F		0430	0740	11.3E
	1340	1635	11.0F		1300	1605	7.6E		1035	1340	11.7F
	1935	2255	12.8E		1920	2150	5.2F		1700	2010	12.0E
									2315		
22 Su	0205	0510	12.0F	27 F	0030	0405	8.7E				
	0830	1130	10.4E		0710	1030	10.4F				
	1430	1720	9.3F		1405	1710	8.6E				
	2020	2340	11.8E		2020	2300	6.1F				
23 M	0255	0605	11.4F	28 Sa	0145	0510	9.3E				
	0930	1225	8.9E		0810	1130	11.0F				
	1530	1815	7.5F		1500	1805	9.8E				
	2105				2115	2355	7.3F				

produces power only during a portion of each ebb tide (see fig. 10.17a). A tidal range of about 7 m (23 ft) is required to produce power using this system.

This same arrangement can be used as a double-action system. In this system, power is produced on the ebbing tide. At the end of this power cycle, when the level behind the dam is not sufficient to produce power, the gates are opened and the last of the water in the reservoir is spilled out to sea. The gates are then closed and some of the remaining water in the bay is pumped seaward to further decrease the water level. Although the pumps consume power, this expenditure of energy is worthwhile if the pumping period occurs when power demands are low and if the additional increase in height between water levels allows more power to be produced on the next cycle.

The gates are kept closed and the tide rises on the seaward side of the dam while the reservoir level remains low. When the difference in height is great enough, the incoming water is released through the turbines into the reservoir to produce power. At the end of the rising tide the dam is sealed, and water is pumped into the reservoir to raise the reservoir to its maximum level; when the tide drops on the seaward side, the cycle is repeated. This system requires very specialized turbine systems because the water moves through the turbines in two directions, and the turbines must be able to generate power on both the flood and ebb tides. A sketch of the power cycle of such a double-action system is presented in figure 10.17b.

This system appears to be a simple and cost-effective method for producing electric power, but there are few places in the world where the tidal range is sufficient and

Table 10.5

Current Differences and Other Constants for Georgia Strait Region Based on Seymour Narrows, B.C., Canada

Tide current station no.	Place	Position				Time differences					
						Slack water before flood		Flood		Slack water before ebb	
		Lat. N		Long. W							
		°	′	°	′	h	m	h	m	h	m
	Passages north of Vancouver Island										
	Time meridian, 120°W					*Based on Seymour Narrows*					
1980	Surge Narrows, Okisollo Channel 50		14	125	10	−0	45	−0	45	−0	45
1985	Hole-in-the-Wall, Okisollo Channel.................... 50		18	125	14	−0	55	−0	55	−0	55
1990	Rapids, near Pulton Bay, Okisollo Channel......... 50		19	125	16	−0	50	−0	55	−0	55
1995	Arran Rapids, north of Stuart Island 50		25	125	09	−0	45	−0	45	−0	45
2000	Yuculta Rapids, southwest of Stuart Island 50		21	125	09	−0	40	−0	40	−0	40
2010	Godwin Point, Cordero Island 50		28	125	25	−0	55	−0	55	−0	55
2015	Shell Point, Blind Channel.................................. 50		26	125	31	−1	10	−1	10	−1	10
2020	Green Point Rapids, Cordero Channel................. 50		25	125	32	−1	30	−1	30	−1	30
2025	Whirlpool Rapids, Wellbore Channel................. 50		27	125	47	−1	50	−1	50	−1	50
2030	Shaw Point, Sunderland Channel......................... 50		28	125	56	−1	05	−1	05	−1	05
2035	Root Point, Chatham Channel.............................. 50		35	126	12	−1	05	−1	05	−1	05
2040	Littleton Point, Chatham Channel....................... 50		37	126	17	−1	05	−1	05	−1	05
2045	Ripple Bluff, Knight Inlet 50		38	126	31	−1	15	−1	15	−1	15
2050	Owl Island, main entrance to Knight Inlet........... 50		38	126	41	−1	20	−1	20	−1	20

[1]Current values at Seymour Narrows are multiplied by these ratios to obtain currents at the above secondary tidal current stations.
Source: U.S. Department of Commerce, National Oceanic and Atmospheric Administration, National Ocean Survey, 1993.

Figure 10.17

(a) A single-action tidal power system. Power is generated on the ebb tide. (b) A double-action tidal power system. Power is generated on the ebb and flood tides.

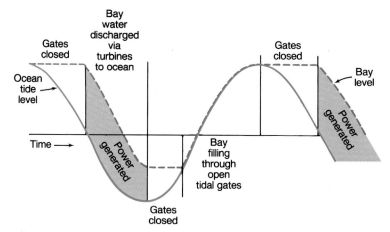

(a) Single-action power cycle; ebb only

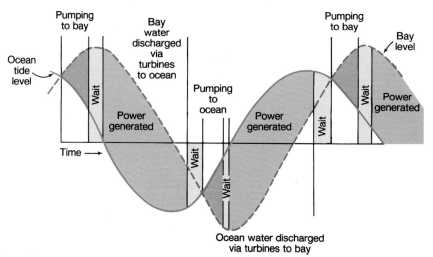

(b) Double-action power cycle; ebb and flood

Ebb		Speed Ratios[1]		Average Speeds and direction			
		Flood	Ebb	Maximum flood		Maximum ebb	
h	m			knots	deg.	knots	deg.
–0	45	0.7	0.7	7.0	140	7.0	320
–0	55	0.8	0.8	7.5	060	7.5	240
–0	55	0.7	0.7	6.5	072	6.5	252
–0	45	0.7	0.7	7.0	065	7.0	245
–0	40	0.5	0.5	5.0	145	5.0	325
–0	55	0.2	0.2	2.2	050	2.2	230
–1	10	0.5	0.5	5.0	170	5.0	350
–1	30	0.5	0.5	5.0	145	5.0	325
–1	50	0.6	0.6	6.0	185	6.0	005
–1	05	0.2	0.2	1.5	060	1.5	240
–1	05	0.6	0.6	5.5	110	5.5	290
–1	05	0.4	0.4	3.5	130	3.5	310
–1	15	0.3	0.3	2.5	105	2.5	285
–1	20	0.3	0.3	2.5	120	2.5	300

Figure 10.18
The Annapolis River tidal power project is the first tidal power plant in North America.

where natural bays or estuaries can be dammed at their entrances at reasonable cost and effort. Moreover, the appropriate tides and bays are not necessarily located near population centers that need the power. Installation and power-distribution costs in addition to periodic low power production because of the changing tidal amplitude over the tide's monthly cycle make this type of power expensive in comparison to other sources.

There is a commercial tidal power installation on the La Rance River Estuary in France that produces 5.4×10^{10} watt-hours per year. Present global energy demands could be satisfied by 250,000 plants of this capacity, but there are only about 255 sites that have been identified around the world with the potential for tidal energy development.

Tidal power has been under consideration for the Bay of Fundy since the 1930s. Canadian and American interest was casual due to the expensive nature of the project, until the rising cost of fossil fuels focused interest on alternative energy sources. Nova Scotia is currently actively pursuing the development of sites in the Minas and Cumberland basins at the head of the Bay of Fundy. The Minas Basin site would have a capacity of 5300 megawatts and the Cumberland Basin site would produce 1400 megawatts. Ultimately, the development of one or both of these sites will depend upon obtaining financing for the multimillion-dollar projects, securing markets in the northeastern United States for excess power, coordinating intermittent tidal energy with other, more conventional methods of producing electrical power, and mitigating potential environmental effects of the project. On the United States side of the bay, Cobscook Bay and Half-Moon Cove in Maine are being considered as potential power-generating sites. The U.S. Army Corps of Engineers is interested in the Cobscook Bay site, and the Passamaquoddy Indian Tribe is proposing a 12-megawatt development at Half-Moon Cove.

The Province of Nova Scotia commissioned a power station in the tidal estuary of the Annapolis River in 1984 (fig. 10.18). The world's largest straight-flow/rim type turbine generator has been used to produce power in a single-basin/single-effect scheme. Tidal ranges at the Annapolis site vary from 8.7 m (29 ft) during spring tides to 4.4 m (14 ft) during neap tides. The unit generates up to 20 megawatts of power, from the head of water developed between the upstream basin and the sea level downstream at low tide. Initially, this project was intended as a pilot project to demonstrate the feasibility of a large-scale straight-flow turbine in a tidal setting. The station has now been added to the Province's principal electrical utility's hydro-generating system. Annual production from the unit is between 30,000 and 40,000 megawatt hours. Power availability has been in excess of 95%.

Although tidal power does not release pollutants, it is not without environmental consequences. The dams isolate the bay from the rivers and estuaries with which it was previously connected. At present, the natural period of oscillation of the Bay of Fundy is about thirty minutes longer than the tidal period; these periods are sufficiently alike to resonate with the tides. Damming the bay will shorten the period of oscillation and increase the resonance in the bay. It is estimated that this will increase tidal ranges by 0.5 m (20 in) and will increase the tidal currents by 5% along the coast of Maine. Coastal residents fear increased erosion and changes affecting the shellfish populations.

The damming of a bay or estuary interferes with ship travel and with port facilities. The dams are barriers to migratory species and alter the circulation patterns of the isolated basin, as well as forming traps for deep water and slowing or eliminating its exchange with the outside seawater.

The deep-ocean floor is at or near the operating limits of manned submersibles, and sampling programs for this area of the oceans are difficult and expensive. To decrease the risks and costs associated with deep-sea observations, a variety of remotely operated devices and vehicles have been developed to act as eyes, samplers, and manipulators for oceanographers in submersibles and surface vessels.

Some of these are true robots, internally programmed to perform specific tasks and to modify these tasks in response to sensors monitoring their environment; most, however, perform their tasks directed by computer-aided operators. The latter are known as remotely operated vehicles, or ROVs, and are under the continuous control of their operators who respond to the ROVs' observations. These ROVs are tethered to a surface vessel, a manned submersible, or an intermediate, below-the-surface-system that isolates the ROV from ship motion. The tether is the umbilical cord between ROV and operator; it supplies power and transmits information. ROVs use video, digital, and still cameras accompanied by lights to explore their environment; they are equipped with mechanical hands to manipulate objects of interest and, depending on their mission, other sensors such as sonar, and temperature and salinity monitors. The operator flies the ROV over the seafloor, and the ROV sends data to the operator and receives directions to change position, to manipulate or retrieve objects, or to use cameras and other sensors. Full-ocean-depth vehicles required to perform a variety of tasks over long periods weigh several tons; small, limited-use, shallow-water vehicles may weigh less than 45 kg (100 lb). These smaller robotic devices are used to service, inspect, and aid in construction of offshore oil and gas facilities. They are also used to inspect bridge piers, dam footings, and cable installations.

ROV *Jason* is a tethered vehicle operated by the Woods Hole Oceanographic Institution. *Jason* weighs about 1350 kg (3000 lb), can operate to a depth of 6000 m (20,000 ft), and provides simultaneous use of up to four-color video channels, sonar, electronic cameras, and manipulators. A smaller ROV, *Jason Jr.* or *JJ,* operated from the submersible *Alvin,* was used to inspect the wreck of the *Titanic* and photograph its interior spaces (see Box figure 10.1 and Prologue box, Marine Archaeology).

Other vehicles are untethered and move freely with no restrictions. Most carry their own power supply and carry on their activities according to their internal program, but some, for example, the Navy's Advanced Underwater Search System (AUSS) and the French *Epaulard,* may be supervised by acoustical communication signals sent by a surface vessel.

Another class of robotic devices is designed to monitor an area of the seafloor over extended time periods. Recording cameras or video systems and other sensors are placed on the seafloor to take measurements on a preset schedule; refer back to figure 2.15. When their operation is complete, a surface vessel signals the robot to discard its anchor and rise to the surface. The Autonomous Benthic Explorer (ABE) now being developed at WHOI is designed to survey deep-sea hydrothermal vent areas for up to one year. Small preprogrammed torpedoes have been developed to make measurements and record data under the sea ice; after their runs the torpedoes return to the research vessel or to the hole in the ice through which they were released. See Box figure 10.2.

Long-range ROVs are being planned in the United Kingdom and Japan. The U.K. has plans under way for a Deep Ocean Geological and Geophysical Instrumented Explorer (DOGGIE) with a working range of 900 km (500 mi) and a Deep Ocean Long Path Hydrographic Instrumentation (DOLPHIN) robot with a range of several thousand miles. The Japanese are planning R1, an untethered vehicle, to survey the Mid-Atlantic Ridge.

Autonomous ROVs may someday become underwater observation platforms for data collection similar to satellites that are presently collecting earth data. As design and development continue, undersea robotic research vehicles will become more complex and more sophisticated, but at the same time, the technology is being improved and simplified, and smaller, less-expensive ROVs will be more available to industry and possibly even to schools and recreationists.

Box figure 10.1

The camera-equipped ROV, *Jason Jr.,* peers into a cabin of the luxury liner *Titanic,* 3800 m (12,500 ft) below the surface. A tether connecting *Jason Jr.* to the submersible *Alvin* is seen at left.

(a)

(b)

Box figure 10.2

Programmed torpedoes explore waters under the sea ice. (a) Programming the torpedo with speed and direction. (b) Retrieving the torpedo beneath the Arctic Ocean ice.

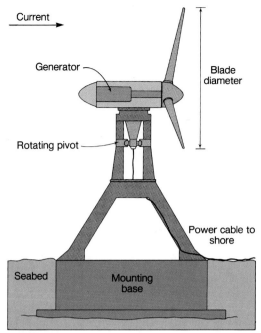

Figure 10.19

A water mill designed to utilize tidal current energy. Power production is related to blade size and current speed.

Table 10.6 Estimated Power Generation (in kw) for Water Mills at Various Current Velocities			
Blade diameter (m)	**Current speed**		
	5 knots 2.5 m/sec	*4.5 knots 2.5 m/sec*	*3.5 knots 1.75 m/sec*
2	8.5	5.5	3.7
5	53	35	23
10	210	140	90
15	480	310	205
20	850	550	370
30	1910	1250	820

Swift tidal currents in inshore channels represent another possible energy source. Flowing water has been used for several centuries to turn the equivalent of windmills or water wheels for limited power. Since the tidal currents reverse with the tide, these "water mills" must be installed to operate with the current flowing in either direction (see fig. 10.19).

Power generated by windmills is dependent on the density of the air, the blade diameter, and the cube of the wind speed. Water mills with a similar design are dependent on the density of the water, the blade diameter, and the cube of the current speed (see table 10.6). A windmill in a 20-knot wind will produce about the same power as a water mill with blades of the same diameter in a 2-knot current, because the density of water is about 1000 times the density of air:

$$0.001 \text{ g/cm}^3 \times (20\text{Kts})^3 = 1\text{g/cm}^3 \times (2\text{Kts})^3$$
$$\text{air density} \times (\text{speed})^3 = \text{water density} \times (\text{speed})^3.$$

A 2×10^6 watt wind-generating unit installed on land has a blade diameter of nearly 210 feet mounted on a 120-foot tower. Installation costs are around $1.6 million. The cost of installing and maintaining a similar-sized unit in a tidal channel submerged in seawater would be much greater, and such a unit would still suffer from periodic loss of power production during slack water. These enormous units could be navigation hazards, and channels of sufficient depth with currents of the required speed are few and are not easily accessible. Although the installation of large water mills has been contemplated, economic and practical realities have prevented the construction of even one test unit to date. There is, however, an effort once again to use small water mills to produce limited power to satisfy local needs.

Summary

Diurnal tides have one high tide and one low tide each tidal day; semidiurnal tides have two high tides and two low tides. A mixed tide has two high tides and two low tides, but the high tides reach different heights and the low tides drop to different levels. For diurnal and semidiurnal tides, the greatest height reached by the water is high water, and the lowest point is low water. Mixed tides have higher high water, lower high water, higher low water, and lower low water. The zero depth on charts is referenced to mean low water or mean lower low water; low tides falling below these levels are minus tides. Rising tides are flood tides; falling tides are ebb tides.

Equilibrium tidal theory is used to explain the tides as a balance between gravitational and centrifugal forces. An equatorial tide is a semidiurnal tide, or a tide wave with two crests and two troughs. Because the moon moves along its orbit as the earth rotates on its axis, a tidal day is twenty-four hours and fifty minutes long. The period of the semidiurnal tide is therefore twelve hours and twenty-five minutes.

The sun's effect is less than half that of the moon's, and the tidal day with respect to the sun is twenty-four hours. Because the tidal force of the moon is greater than that of the sun, the tidal day is still considered to be twenty-four hours and fifty minutes.

Spring tides have the greatest range between high and low water; they occur at the new and full moons, when the earth, sun, and moon are in line. Neap tides have the least tidal range; they occur at the moon's first and last quarters, when the moon is at right angles to the sun.

When the moon or sun stands above or below the equator, its tides become more diurnal. Diurnal tides are often called declinational tides. The elliptic orbits of the earth and moon also influence the tide.

The dynamic approach to tides investigates the actual tides as they occur in the ocean basins. The tide wave is discontinuous, except in the Southern Ocean. It is a shallow-water wave, which oscillates in some ocean basins as a standing wave, and its motions persist long enough to be acted on by the Coriolis effect. The tide wave is reflected, refracted, and diffracted along its route. Because a point on the earth moves eastward faster at the equator than the tide wave progresses westward as a free wave, the tide wave moves as a forced wave, and its crest is displaced to the east of the tide-raising body. Above 60°N and 60°S latitudes, the crest is more nearly in line with the moon, as the speeds of earth and the tide wave match more closely.

In large ocean basins, the tide wave can move as a progressive wave. The Coriolis effect causes a rotary tidal current. Standing wave tides can also form in ocean basins; they rotate around an amphidromic point as they oscillate. Cotidal lines mark the progression of the tide crest, and regions of equal tidal range are identified by corange lines. The tidal range increases with the distance from an amphidromic point. Standing wave tides in narrow open-ended basins oscillate about a node at the entrance to the bay or basin; the antinode is at the head of the basin, as in the Bay of Fundy. There is no rotary motion when a bay is very narrow.

A rapidly moving tidal bore is caused by a large-amplitude tide wave moving into a shallow bay or river.

Tidal heights and currents are predicted from astronomic data and actual local measurements. NOAA determines and publishes annual tide and tidal current tables.

Single- and double-action dam and turbine systems extract energy from the tides. Tidal power plants are in use in France, the Soviet Union, and Canada. Few places have large enough tidal ranges and suitable locations for tidal dams. Tidal power has environmental drawbacks as well as high developmental costs. Tidal currents are another energy-producing possibility, but installation and service costs are considered high.

Key Terms

diurnal tide	tidal current
semidiurnal tide	slack water
semidiurnal mixed tide	equilibrium tidal theory
high water	dynamic tidal analysis
low water	centrifugal force
higher high water	tidal day
lower high water	tide wave
higher low water	range
lower low water	spring tide
mean tide/average tide	neap tide
tidal datum	declinational tide
minus tide	free wave
flood tide	forced wave
ebb tide	progressive tide

cotidal line	corange line
standing wave tide	tidal bore
rotary standing tide wave	harmonic analysis
amphidromic point	local effect

Study Questions

1. Distinguish between the terms in each pair:
   a. Diurnal tide; semidiurnal tide.
   b. Tidal day; tidal period.
   c. Spring tide; neap tide.
   d. Flood tide; ebb tide.
   e. Cotidal lines; corange lines.
2. What is the path of a water particle in the tidal current shown in figure 10.14 if a 1-knot current flowing south is also present?
3. Why is it more efficient to generate power by means of a tidal dam than to erect water mills in tidal currents?
4. Why are standing wave tides produced in small coastal basins as well as in large ocean basins? Use table 10.1.
5. Explain why it is necessary to have both a large tidal range and a relatively large volume of water behind a tidal dam in order to generate electric power from the rise and fall of the tides.
6. Sketch each of the three different tidal patterns during a spring tide and a neap tide. Label the tide levels of each spring tide.
7. Explain why a tide is a wave.
8. Explain the relationship between the tides and tidal currents.
9. Why does the tide act as a shallow-water forced wave in the ocean basins?
10. How do progressive tides and standing wave tides differ?
11. How is a tidal bore produced?
12. What would happen if there was no force counteracting the sun's gravitional force on the center of mass of the earth-moon system?
13. Why are there higher high tides during the winter at midlatitudes in the Northern Hemisphere?
14. Why must the tides be measured for approximately 19 years at a location before the data can be used in tide forecasting?
15. Why are accurate satellite sea-surface elevation measurements over the entire earth so important to oceanography?

Study Problems

1. How many days will pass before a high tide reoccurs at the same clock time?
2. Using the information in table 10.2 plot the Aberdeen, Wash., tide curves for August 2 and 10. Which is a spring tide? Which is a neap tide? What type of tide is this? Label the water levels on each curve.
3. Choose the best date and time for clam digging during early-morning hours in the month of August at Aberdeen, Wash. Refer to table 10.2.
4. Use tables 10.2 and 10.3 to correct the Aberdeen tides to tides at Point Grenville for August 20.
5. At what time on August 17 do you arrive at Seymour Narrows to navigate the narrows at slack water between breakfast and dinnertime? See table 10.4.

Suggested Readings

Theory and Dynamics

Bowditch, N. 1984 ed. *American Practical Navigator,* vol. 1. U.S. Defense Mapping Agency Hydrographic Center, Washington, D.C. 1414 pp. (Chapters 12 and 31 are on tide prediction, tides, and tidal currents.)

Darwin, G. H. 1898. *The Tides and Kindred Phenomena in the Solar System.* Freeman, San Francisco (1962). 378 pp.

Fisher, A. 1989. The Model Makers. *Oceanus* 32 (2): 16–21.

Lynch, D. K. 1982. Tidal Bores. *Scientific American* 247 (4): 146–56.

National Oceanic and Atmospheric Administration. *National Ocean Survey—Tide and Tidal Current Predictions.* (Published annually for specific geographic areas.)

Open University. 1989. *Waves, Tides and Shallow Water Processes.* Pergamon Press, Oxford, England; Open University, Milton Keynes, England. 186 pp.

Redfield, A. C. 1980. *Introduction to Tides.* Marine Science International, Woods Hole, Mass. 108 pp.

Yoerger, D. R. 1991. Robotic Undersea Technology. *Oceanus* 34 (1): 32–37.

Energy

Booda, L. 1985. River Rance Tidal Power Plant Nears Twenty Years in Operation. *Sea Technology.* September: 22–26.

Brin, A. 1981. *Energy and the Oceans.* Westbury House, Surrey, England. 133 pp.

Bruce, M. 1986. Ocean Energy: Some Perspectives on Economic Viability. In *Ocean Yearbook 6,* Borgese, E. M., and N. Ginsburg, eds. Univ. of Chicago–Chicago. pp. 58–78.

Charlier, R. H. 1983. Water, Energy, and Nonliving Ocean Resources. In *Ocean Yearbook 4,* Borgese, E. M., and N. Ginsburg, eds. Univ. of Chicago–Chicago. pp. 75–120. (Overview of energy sources.)

Greenberg, D. A. 1987. Modeling Tidal Power. *Scientific American* 257 (5): 128–31.

Holloway, T. 1989. Eling Tide Mill. *Sea Frontiers* 35 (2): 114–19.

Simeons, C. 1980. *Hydro-Power, The Use of Water As an Alternative Source of Energy.* Pergamon, Elmsford, N.Y. 549 pp.

11

Coasts, Shores, and Beaches

11.1 **Major Zones**
11.2 **Types of Coasts**
    Primary Coasts
    Secondary Coasts

    Box: Rising Sea Level

11.3 **Anatomy of a Beach**
11.4 **Beach Types**
11.5 **Beach Dynamics**
    Natural Processes
    Coastal Circulation
    Artificial Processes
11.6 **Practical Considerations: Case Histories**
      of Two Harbors
    The Santa Barbara Story
    The History of Ediz Hook
Summary
Key Terms
Study Questions
Suggested Readings

The seashore is a sort of neutral ground, a most advantageous point from which to contemplate this world. It is even a trivial place. The waves forever rolling to the land are too far-traveled and untamable to be familiar. Creeping along the endless beach amid the sun-squawl and the foam, it occurs to us that we, too, are the product of sea-slime.

It is a wild, rank place, and there is no flattery in it. Strewn with crabs, horseshoes, and razor-clams, and whatever the sea casts up—a vast morgue, where famished dogs may range in packs, and crows come daily to glean the pittance which the tide leaves them. The carcasses of men and beasts together lie stately up upon its shelf, rotting and bleaching in the sun and waves, and each tide turns them in their beds, and tucks fresh sand under them. There is naked Nature—inhumanly sincere, wasting no thought on man, nibbling at the cliffy shore where gulls wheel amid the spray.

Henry David Thoreau,
from *Cape Cod*

Land's End, Cornwall, England.

*T*o most people, the most familiar areas of the oceans are *the shores and beaches. We visit the coast to see the ocean, to play along the beach, to enjoy the constant interaction between moving water and what appears to be the stable land. But even the casual visitor senses changes in this area: on any visit the tide may be high or low, the logs and drift may have changed position since the last visit, and the dunes and sandbars may have shifted since the previous summer. A visit to a beach farther along the coast or along a different ocean presents a different picture. The sand is a different color or there is no sand at all; the waves break higher or lower, closer or farther away; the slope of the beach is steeper or flatter; and so on.*

Coasts and beaches are dynamic, not static, and no two regions are exactly the same. People play an important role along coasts and beaches because people and their structures change these areas more than they realize. In this chapter we study the types of coasts and beaches and the natural processes that create and maintain them. We learn how humans affect these areas, often destroy them, and sometimes try to save them.

11.1   Major Zones

The **coasts** of the world's continents are the areas where the land meets the sea. The terms coast, coastal area, and coastal zone are used descriptively to designate land areas that are associated with the sea and may include areas of cliffs, dunes, beaches, and even hills and plains. Some examples of the various kinds of coasts are presented in figures 11.1 and 11.2. The width of the coast, or the distance to which the coast extends inland, varies and is determined by local geography, climate, vegetation, even social customs and culture. However, the coast is most generally described as the land area that is or has been affected by marine processes such as tides, winds, and waves, even though the direct effect of these processes may be felt only under extreme storm conditions. The seaward limit of the coast usually coincides with the beginning of the beach or shore, but sometimes includes nearby offshore islands. The term **coastal zone** includes the open coast as well as the bays and estuaries found in coastal indentations; it incorporates both land and water areas.

Coastal areas are regions of change in which the sea acts to alter the shape and configuration of the land. Sometimes these changes are extreme and occur rapidly (for instance, the damage caused by a hurricane occurs in a period of hours). Sometimes the changes are subtle and so slow that they are not perceived by people during their lifetimes,

Figure 11.1

(a) The Oregon coast is famed for its bold rocky headlands and its intervening pocket beaches. (b) These chalky cliffs along the Dorset coast of England were formed from the remains of foraminifera, principally *Globigerina*. (c) Gently sloping pastureland runs down to the sea along the eroded coast of western Scotland.

(a)

(b)

(c)

(a)

(b)

(c)

Figure 11.2

(a) Bayou La Loutre at Ysclosky, Louisiana. The bayou is an old stream course of the Mississippi River. The salt content of the water changes as the wind drives fresh water from the river or salt water from the coast through the delta. (b) A backwater tidal slough near the mouth of the Columbia River. Marshes and mudflats of these sloughs are highly productive nurseries for many marine species. (c) Glacial bays and fjords produce a rugged topography along the coast of Glacier Bay in southeastern Alaska.

but when considered over long periods of the earth's history these slow changes are seen to be impressive (for example, the formation of the delta of the Mississippi River or the gradual erosion of Cape Hatteras, North Carolina). In general, coasts that are composed of soft, unconsolidated materials, such as sand, change more rapidly than coasts composed of rock.

The **shore** is that region from the outer limit of wave action on the bottom, seaward of the lowest tide level to the limit of the waves' direct influence on the land. This limit may be marked by a cliff or an elevation of the land above which the sea waves cannot break. Such features act as barriers to the wave-tossed drift of logs, seaweeds, and human debris. The **beach** is an accumulation of sediment (sand or gravel) that occupies a portion of the shore. The beach is not static, but is moving and dynamic, because the beach sediments are constantly being moved seaward, landward, and along the shore by nearshore wave and current action.

Between the high-tide mark and the upper limit of the shore there may be dunes or grass flats spotted with occasional drift logs left behind by an exceptionally high tide or severe storm (see fig. 11.3).

11.2 Types of Coasts

In this section the different types of coasts, the processes that formed each, and their special characteristics are emphasized. The coastal categories used in this discussion follow the system of the late Francis P. Shepard of the Scripps Institution of Oceanography.

All coastal areas belong in one of two major categories: (1) coasts that owe their character and appearance to processes that occur at the land-air boundary and (2) coasts that owe their character and appearance to processes that

Figure 11.3
Driftwood accumulates at the high-tide line in areas where timber is plentiful.

are primarily of marine origin. Classification depends on the large-scale features created by tectonic, depositional, erosional, volcanic, or biological processes.

In the first category are coasts that have been formed by (1) erosion of the land by running surface water, wind, or land ice, followed by a sinking of the land or a rise in sea level; (2) deposits of sediments carried by rivers, glaciers, or the wind; (3) volcanic activity, including lava flows; and (4) uplift and subsidence of the land by earthquakes and associated crustal movements. Coasts formed by these processes are **primary coasts,** since there has not yet been time for the sea to substantially alter or modify the appearance given to them by nonmarine processes. The second category includes coasts formed by (1) erosion due to waves, currents, or the dissolving action of the seawater; (2) deposition of sediments by waves, tides, and currents; and (3) alteration by marine plants and animals. These coasts are **secondary coasts;** their character, even though it may have been originally land-derived, is now distinctly a result of the sea and its processes.

Since a relatively young coast may be rapidly modified by the sea, while a coast that is old in time may retain its land-derived characteristics, the terms primary and secondary may be confusing. Keep in mind that absolute age is not really important, because the classification is based only on whether the characteristics are derived from the land or from the sea, and it is possible to find both types of features along the same coast.

Primary Coasts

Coasts formed by erosion at the land-air boundary followed by a sinking of the land or a rise in sea level include those that were covered by glaciers during the ice ages. During these glacial periods, sea level was lower than it is at present, because much of the water was held as ice on the continents,

and the glaciers moved slowly across the land, scouring out valleys as they inched along to the sea. The weight of the ice caused the land to subside, and when the ice began to melt, in some cases sea level rose faster than land could rebound upward from its depressed state. In other cases, the glacial troughs were scoured below sea level and filled with seawater as the ice receded. Both cases resulted in the formation of the fjords of Norway, Greenland, New Zealand, Chile, and southeastern Alaska. Fjords are long, deep, narrow channels, U-shaped in cross section (see fig. 11.4). At their sea end there is often a collection of debris left by the glacier where it met the sea, and this debris forms a lip that creates a shallow entrance, or **sill.** In some areas, land that was heavily covered by ice during the last ice age is still slowly rising; in Scandinavia the rate of rise is about 1 to 5 cm (0.4–2 in)/year. Along other coasts, tectonic forces have caused land uplift, for example, the west coasts of North and South America. Along these coasts, old wave-cut terraces have been elevated above sea level.

When a glacier or ice sheet ceases its forward motion and retreats, it leaves a mound of rubble at the point of its greatest extension. If the glacier has reached the edge of a continent, this material becomes a part of the coastal area. Such deposits are called **moraines.** Long Island, off the New York and Connecticut coasts, is a moraine; it acts as a protective barrier to the continental coast and modifies this area of the coastline.

When sea level was lowered during the ice ages, rivers flowed not only across the land but also over the exposed shore to the sea. This process cut typical V-shaped river and stream channels in these areas, and in many cases the channels had numerous side branches formed by feeder streams. As the sea level rose, these channels were filled with seawater, producing coastal areas such as Chesapeake Bay and Delaware Bay on the East Coast of the United States. A coastal feature of this type is called a **drowned river valley** or **ria coast,** as shown in figure 11.5.

Rivers carrying extremely heavy sediment loads build **deltas** at the edge of the sea. The delta, or deposit of river-borne sediment left at the river mouth, produces a flat, fertile coastal area. Good examples of these deposits are found at the mouths of the Mississippi, Ganges, Nile, and Amazon rivers. A similar type of coast is produced when eroded material is carried down from the hills by surface runoff and many small rivers join together to form an **alluvial plain.** The eastern seacoast of the United States south of Cape Hatteras has been produced in this manner.

It is estimated that during each second the rivers of the world carry 530 tons of sediment to the sea. This rate of removal is equal to the erosion of a layer 6 cm (2.4 in) thick from all land above sea level every 1000 years. Much of this sediment helps to form and maintain the world's beaches, and some of it finally finds its way to the deep-ocean floor. All of it passes through the coastal zone and

Figure 11.4 (facing page)
The narrow channel of a fjord, Milford Sound, New Zealand.

takes part in coastal processes. Refer back to chapter 2 for sources of terrigenous material. A **dune coast** (fig. 11.6) is a wind-modified depositional coast. In Africa, the Western Sahara is gradually growing westward toward the Atlantic Ocean as the prevailing winds move sand from the inland desert areas to the coast. Along other dune coasts the winds move the sands inland to form dunes, and elsewhere dunes driven by the wind migrate along the shore.

The Hawaiian Islands are the tops of large volcanoes rising from the seafloor and have excellent examples of coasts formed by volcanic activity. Lava flows extend to the sea, forming black sand beaches; these are **lava coasts.** Volcanic explosions just beneath the sea surface at the water's edge formed craters that became concave bays when they lost their rims on the seaward side and formed **cratered coasts** (see fig. 11.7).

When tectonic activity results in faulting and displacement of the earth's crust, the coast is changed in characteristic ways. The California San Andreas Fault system lies along a boundary where crustal plates are moving parallel to each other along a transform fault. (Plate movements and transform faults are described in chapter 3.) Over long periods of time, faults formed in this area filled with seawater. The Gulf of California, also known as the Sea of Cortez, located between Baja California and the mainland of Mexico, is found at the southern end of the fault system. At the northern end lie San Francisco and Tomales Bay, where the fault runs out into the Pacific Ocean. Tomales Bay is a particularly good example of **fault**

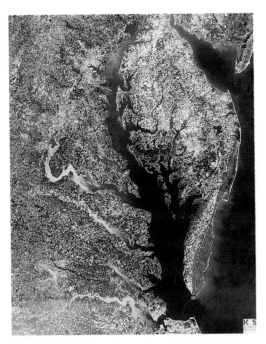

Figure 11.5
Delaware Bay (upper right) and Chesapeake Bay (center) are examples of drowned river valleys.

Figure 11.6
Coastal sand dunes near Florence, Oregon.

Figure 11.7
A volcanic crater coast. Hanauma Bay, Hawaii, is a crater that has lost the seaward portion of its rim. Today it is a park and marine preserve.

Figure 11.8
The San Andreas Fault runs along the California coast to the west of San Francisco, separating Point Reyes from the mainland and producing Bolinas and Tomales bays, seen as a dark line at the upper left.

bay. See the satellite image of this fault system in figure 11.8. In other parts of the world, the Red Sea and the Scottish coast provide examples of **fault coasts.**

Secondary Coasts

Secondary coasts owe their present-day appearance to marine processes. As waves batter against the coast, they constantly erode and grind away the shore; rocks and cliffs are undercut by the wave action and fall into the sea, where they are ground into sand. A coastal area formed of uniform material may have originally been irregular in shape, composed of headlands and bays, but as the waves approach the shore they concentrate their energy on the headlands, wearing them down more rapidly than the straighter shore and cove areas. With time, this action tends to produce a more regular coastline; examples are found in southern California, southern Australia, and England's cliffs of Dover. However, if the original coastline is made up of materials that vary greatly in composition and resistance to wave erosion, then the result will be an increasingly irregular coastline. In this case, headlands of hard rock jut into the sea and are separated by sandy coves formed by the erosion of softer materials. The headlands erode irregularly as the shore is cut away, and the result is the formation of small rock islands and tall slender pinnacles of resistant rock known as **sea stacks** (see fig. 11.9). Headlands still connected to the mainland may be undercut to produce sea caves or even cut through to produce arches or, if the opening is small, a window.

In some cases, eroded materials are carried seaward by the waves and currents to areas just off the coast. If sufficient sandy material is deposited in offshore shallows paralleling the beach, **bars** are produced. If still greater amounts of material are added, the bars may grow until they break the surface, forming **barrier islands.** Barrier island formation may also be related to periods of glaciation, when sea level drops to expose offshore bars or when

Figure 11.9
Sea stacks are common features along the coasts of northern California, Oregon, and Washington, occurring in a wide variety of shapes and sizes.

sea level rises to inundate low-lying coastal areas. Features of continental margins that help to form the continental shelves by trapping sediments are discussed in chapter 2; these features are a part of the complex process of barrier island formation. Once a barrier island is formed, plants begin to grow and the vegetation helps to stabilize the sand and increase the island's elevation by trapping sediments and accumulating organic matter. Attempts to halt storm-caused erosion by building breakwaters and beach-holding devices have not always been successful and, although well intentioned, have often resulted in aggravating the loss of material from barrier islands rather than halting it.

A line of these islands along a coast protects the continental coastline from the waves; such barrier islands are found all along the Gulf and East coasts of the United States (fig. 11.10). Between the islands and the mainland lies a protected water route known as the Intercoastal Waterway, which is ideal for small-boat navigation. Although these barrier islands do protect the mainland from severe storms, the islands themselves sustain the damage that the continental coast is spared. This situation is quite as nature intended, but people who live on these islands look to the government to protect their property from these natural occurrences. During 1989's Hurricane Hugo, Folly Island, along the coast of South Carolina, suffered extensive storm damage. Eighty-six of 290 oceanfront structures were more than 50% destroyed, and fifty were damaged beyond repair. Rebuilding requires raising the elevation of the structures and in some cases relocating them landward. The U.S. National Flood Insurance Program issued nearly $3 million in damage payments to Folly Island property

owners. Although 1992's Hurricane Andrew did extreme and extensive inland damage, less property damage was reported along the Florida coast.

Sand spits and **hooks** are bars connected to the shore at one end. They are the result of similar sea forces (see figs. 11.11 and 11.26). They may grow, shift position, wash away in a storm, or rebuild under more moderate conditions. The area between the mainland and these spits and hooks is protected from turbulence and is therefore often the site of beach flats formed of sand or mud. If a spit grows sufficiently to close off the mouth of an inlet, a shallow lagoon is formed. Water percolates through the gravel and sand of the spit, and inside the lagoon the water rises and falls with the tides.

In some parts of the world, **reef coasts** are the result of the activities of sea organisms. In tropic areas, corals grow in the shallow, warm waters surrounding a landmass, and the small animals gradually build a fringing reef, which is attached directly to the landmass. In other places, a lagoon of quiet water may lie between the barrier reef and the land, and, in a few cases, the coral encircles a submerged island to form an atoll. The formation of these reef types is discussed in chapter 2.

The Great Barrier Reef of Australia, stretching along its northeast coast toward New Guinea, is the largest and most famous of the world's coral reefs. Coral atolls include the Pacific islands of Tarawa, Kwajalein, Eniwetok, and Bikini. The reef-encircled islands of Iwo Jima and Okinawa became familiar as the sites of major battles in the southwest Pacific during World War II.

Figure 11.10

Sea Island, Georgia, is a barrier island that has been extensively developed. The shallow water between the island and the coast is visible on the left.

Figure 11.11

A spit has formed across the entrance to Sequim Bay, Washington.

Rising Sea Level

Sea level is once again on the rise around the coasts of the world. Over the past century, sea level has risen about 15 cm (6 in), but recently the rate of rise has increased. It is thought that sea level may gain another 20 cm (8 in) in the next fifty years, a rate that is about three times greater than that experienced over the last century. By the year 2100 sea level is expected to be 31 to 110 cm (1–3.5 ft) higher than at present; the best estimate is an increase of 66 cm (2 ft). If such an increase occurs, the edge of the shore will move inland from hundreds to thousands of meters along low-lying coasts, causing great damage to homes, towns, ports, farms, and industry as well as natural features. As sea level rates increase they may exceed the upward growth rate of tropical coral reefs, about 1 cm (2.5 in) per year, jeopardizing their existence. Sea level rates greater than 5 mm (0.2 in)/year result in the loss of river delta wetlands, and any rise in sea level accelerates the erosion of coastal areas.

Some believe coastal planners should act now to minimize the damage and loss which will be caused by such a rise. Highly populated shore areas such as Long Island, New York; Atlantic City, New Jersey; Miami Beach, Florida; Ocean City, Maryland; and Galveston, Texas, may be heavily damaged.

Many factors change sea level and alter the boundary where the land meets the sea. Globally, the most significant sea level changes are governed by the amount of water stored in the earth's reservoirs and changes in the size of the ocean basins. The transfer of water from the ocean reservoir to land as ice during glacial periods and back to the ocean

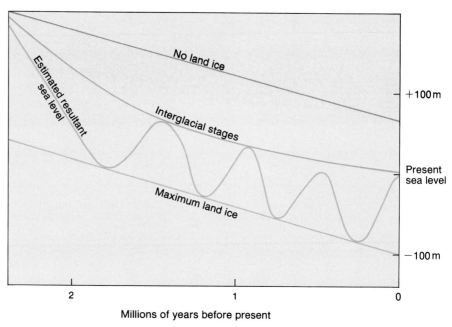

Box figure 11.1

Estimated sea-level changes during the Quaternary. Crustal movements that modify the elevation and shape of ocean basins produce sea level changes along the "No land ice" curve. Sea levels between glacial periods are represented by the "Interglacial stages" curve. Sea level during periods of maximum ice accumulation on land are estimated by the "Maximum land ice" curve. The sea level resulting from these processes is shown as a periodic rise and fall.

during interglacial periods can account for a 150- to 200-m (656-ft) change in sea level. See Box figure 11.1. As tectonic forces rearrange the continents, the shape of the ocean basins changes and sea level tends to fall as landmasses combine and to rise as landmasses disperse. The more minor effect of sediment accumulation in ocean basins displaces seawater upward as well.

Continental landmasses rise or sink over large areas or very local areas. As landmasses are compressed they may first rise and then slowly subside, seeking to reestablish isostatic balance. Heat from the earth's interior trapped under landmasses may increase the land's buoyancy and raise the land. The weight of ice may depress a land area, which then rebounds as the ice melts. Soils eroded from the land and deposited in the coastal zone may first raise the elevation of the land, followed by a reduction in elevation as weight loading causes subsidence and the soils compact. Human activities may also create immediate changes as oil, gas, or

Other marine animals form reef-like structures as their shells are deposited layer on layer, gradually building up a mass of hard material. There are large reef deposits of oyster shells in the Gulf of Mexico off the coasts of Louisiana and Texas. These reefs are so large that the shells are harvested commercially for lime production. Along the east coast of Florida, large populations of shell-bearing animals have contributed their shells directly to the shore as the naturally occurring small shell fragments that form the sand on the beaches.

Plants as well as animals may modify a coastal area. Along low-lying coasts in warm climates, mangrove trees grow in the shallow water. Their great roots form a nearly impenetrable tangle, providing shelter for a unique community of other plants and animals (fig. 11.12). Coastal mangrove swamps are found along the Florida coast, northern Australia, the Bay of Bengal, and in the West Indies. In more temperate climates, low-lying protected coasts with areas of sand and mud are often thickly covered with grasses, forming another type of plant-maintained environment water are extracted from coastal subterranean reservoirs, leaving spaces that are then filled as the overlying earth settles.

There are also changes in sea level driven by the expansion or contraction of water as it changes temperature. If the oceans of the world increased their average temperature by 1°C, it is estimated that sea level would rise by about 60 cm (24 in). In temperate latitudes, an annual sea-surface change of 10°C over a depth of 10 m (33 ft) can cause a seasonal rise and fall in sea level of about 15.5 mm (0.6 in).

In the tropics, sea temperatures are nearly constant over the annual cycle, so such thermal expansion changes are small, but the seasonal oscillation of the meteorological equator (the intertropical convergence zone) between approximately 0° and 11°N causes sea level to vary in response to atmospheric pressure changes. In the North Atlantic and North Pacific the annual seasonal shift from high pressure to low pressure over the oceans causes a related change in sea level, rising under low atmospheric pressure and falling under high atmospheric pressure. A 1-millibar pressure change modifies sea level by about 10 mm (0.4 in).

Sea level is constantly changing, and the global average change may not be reflected in the change observed at any particular location because of the sum total of the many and varied factors that affect the change.

The increasing rate of sea-level rise has been partly attributed to the greenhouse effect, discussed in chapter 7. The 25% increase in the amount of atmospheric CO_2 since 1900 has increased the surface temperature of the oceans by about 0.5°C and warmed the air over the land by about the same amount. If this trend continues, and if the amount of CO_2 in the atmosphere doubles during the next 100 years, the ocean surface will warm an additional 2–4°C. The greatest warming will occur at the higher latitudes, where the continental ice masses will melt and increase the volume of water in the oceans (see chapter 1, Reservoirs and Residence Time). As the oceans warm, the water will expand, introducing another factor that will affect the sea level.

The rate at which sea level will rise is not clear. The warming process will create changes in cloud cover and precipitation patterns; these will, in turn, change the distribution of incoming solar radiation, which will affect not only ice melt but also wind and storm systems. If the storm patterns that drive the ocean waves change, the rate and location of shore erosion will also change.

Long-term climate cycles, including cold and severe storms alternating with milder climatic periods, play additional roles in water-level changes and coastal erosion. In addition, rivers modified by dams and dikes cut off the delivery of sediments to the beaches, and the construction of seawalls, groins, bulkheads, and breakwaters alters the current and wave patterns along the coasts. Again, these are changes that influence coastal erosion rates as the sea level rises. All together, these events and processes weave a complex unity from which it is extremely difficult to isolate any single item as causing the changes that are being forecast. Sea-level measurements are relative measurements because they are made against the land's motion. To make more accurate measurements of sea-level rise scientists are looking to the global satellite network and altimeters that measure the distance between the satellite and the earth's surface (see chapter 1 Box: Satellite Oceanography). To determine exact satellite position and to detect small absolute sea-level changes, scientists are hoping to establish a stable framework using quasars (distant space bodies) as unchanging reference points.

Whatever the predictions, it is unlikely that people will give up their dreams of a summer cabin on the shore or a home on the beach. However, people are beginning to become more sensitive to the encroaching waters. The 1982 Coastal Barrier Resources Act prohibits federal subsidies for roads, bridges, piers, and flood insurance coverage for homes built after 1983 in certain designated undeveloped barrier island areas. Western Galveston Island in Texas prohibits new construction and reconstruction of storm-damaged property in some shore areas. States from Maine to North Carolina to Washington have regulated construction of seawalls, jetties, groins, and other permanent structures in the coastal zone. While such regulations do point to an increased awareness of sea-level and coastal erosion problems, it may be that human efforts can only be considered too little and too late when confronting these combined natural forces.

(fig. 11.13). These **salt marshes** may extend inland a considerable distance if the land is flat enough to permit periodic tidal flooding. Such marshes also form around protected bays and coves that have large changes in tide level. Salt marshes are extremely productive in terms of organic matter and extend the shoreline seaward by trapping sediments. These marshes are rapidly disappearing under the pressures of human habitation and the demand for flat industrial sites, harbor space, and recreational property adjacent to the water.

11.3   Anatomy of a Beach

The coast, regardless of the distance it extends inland, is commonly considered to end on the seaward side at the shore. The shore has been defined as the region from just seaward of the lowest tide level to the upper limit of the waves' direct influence on land. The beach is considered to include the sand and sediment lying along the shore, together with the sediments carried along the shore by the nearshore waves and currents. Let us first look only at the beach, along which we may walk or on which we may sunbathe and relax.

Figure 11.12
Mangrove trees growing along the shore of Guam.

Many studies have been made of beach areas around the world, and from these studies there has developed a specific terminology applicable to beaches and beach features. This standardization of terms prevents confusion and allows each area to be described as completely as possible. A beach profile represents a cut through a beach perpendicular to the coast and in this case (fig. 11.14) shows the major features that might appear on any beach. Use figure 11.14 while reading this section, remembering that a given beach may not have all these features, for each beach is unique.

The **backshore** is the dry region of beach that is submerged only during the highest tides and the severest storms; the **foreshore** extends out past the low-tide level, and the **offshore** includes the shallow-water areas seaward of the low-tide level out to the limit of wave action on the bottom.

Terraces appear on beaches in the area where foreshore and backshore meet. These terraces are **berms** formed by wave-deposited material; they have relatively flat, terracelike tops. Berms are recognizable by their slope or rise in elevation. Some berms have a slight ridge crest that runs parallel to the beach; if there is a ridge, it is called

Figure 11.13
A coastal saltwater marsh, Gaspé, Québec, Canada.

Figure 11.14

A typical beach profile with associated features.

(a)

(b)

Figure 11.15

(a) Berms on a gravel beach. The winter storm berm with drift logs is located high on the beach. The summer berm is between the winter berm and the beach face in the foreground. (b) A wave-cut scarp on a gravel beach.

the **berm crest.** When two berms are found on a beach, as in figures 11.14 and 11.15, the berm higher up the beach, or closer to the coastline, is the **winter berm.** The winter berm is formed during severe winter storms when the waves reach farthest up the beach and pile up material along the backshore. The seaward berm, the berm closer to the water, is the **summer berm** formed by less-severe waves that do not reach as far up the beach. Once the winter berm is formed, it is not disturbed by the less-intense storms during the year. The waves that produce a winter berm erase the summer berm, but after the storm season is over, a new summer berm is formed by gentler wave action.

Between the berm and the water level there may be a wave-cut **scarp** at the high-water level. The scarp is an abrupt change in the beach slope that is caused by the cutting action of waves at normal high tide; see figure 11.15b. Berms and scarps do not form between the normal high- and low-tide levels because of the wave action and the continual rise and fall of the water. This relatively featureless area is known as the foreshore; the portion of the foreshore below the low-tide level is often quite flat, forming a **low-tide terrace.** The upper portion of the foreshore has a steeper slope and is known as the **beach face.**

Figure 11.16

A series of pocket beaches cut an elevated marine terrace along the California coast.

Seaward of the low-tide level, in the offshore region, there may be **troughs** and bars that run parallel to the beach. These structures change seasonally as beach sediments move seaward during periods of large waves (winter storms), to enlarge existing bars, and then shoreward during periods of small waves (summer), to diminish the same bars. When a bar accumulates enough sediment to break the surface and is stabilized by vegetation, it becomes an island of the type described for a barrier island coast.

Beaches do not occur along all shores. Along rocky cliffs, no beach area may be exposed between low- and high-tide levels, but there may be small pocket beaches, each separated from the next by rocky headlands or cliffs (see fig. 11.16).

11.4 Beach Types

Beaches are also described in terms of (1) shape and structure, (2) composition of beach materials, (3) size of beach materials, and (4) color. In the first category, beaches are described as wide or narrow, steep or flat, and long or discontinuous (pocket beaches). A beach area that extends outward from the main beach and turns and parallels the shore is called a spit (refer back to fig. 11.11). Spits frequently change their shape in response to waves, currents, and storms. In a wave-protected environment, a spit may extend out at an angle from the shore to an offshore island. If this spit builds until it connects the rock or island to the shore, it forms what is called a **tombolo.** A spit extending offshore in a wide sweeping arc bending in the direction of the prevailing current is called a hook. The sediment moves around the end of a hook and is deposited in the quiet water behind the point, producing a broad point on its end. See figure 11.17 for an example of a tombolo and figure 11.26 for an example of a hook.

Materials that form beaches include shell, coral, rock, lava, and shingle. Shingles are flat, circular, smooth stones; they are formed when the beach slope, wave action, and stone size and structure combine to cause the stones to slide back and forth with the water movement. When the stones roll instead of slide, the water action produces round stones, or cobbles. The terms sand, gravel, pebble, cobble, and boulder are used to describe the size of beach particles (see table 2.3 in chapter 2).

Both the composition of the beach material and its size are related to the source of the material and the forces acting on the beach. Land materials are brought to the coast by the rivers or are derived from the local cliffs by wave erosion. The sands of many beaches are the ground-up, eroded products of land materials rich in minerals like quartz and feldspar. Small-sized particles such as sand, mud, and clay are easily transported and redistributed by waves and currents; larger-sized rocks are usually found in the area close to their source, because they are too large to be moved any distance. The finer particles are carried away by the currents, and the larger rocks are left scattered on the beach. In general, a beach littered with large rocks is an eroded beach, and the remaining large rocks are called **lag deposits.** They may accumulate in sufficient quantity to protect a beach from further wave and water erosion; this type of beach is known as an **armored beach** (see fig. 11.18).

Figure 11.17

A spit connected to an offshore island forms a tombolo.

Figure 11.18

An armored beach. Lag deposits of large rocks are left on an eroded beach.

Some beach materials come to the beach from off-shore areas. Coral and shell particles broken by the pounding action of the waves are carried to the beaches by the moving water. A beach covered with uniform small particles of sand or mud is or has been a depositional beach.

Some of the world's beaches have distinctive colors. In Hawaii, there are white sand beaches derived from coral, and black sand beaches derived from lava. Green sands can be found in areas where a specific mineral (like olivine or glauconite) is available in large enough quantities, and pink sands occur in regions with sufficient shell material.

11.5 Beach Dynamics

A sand beach exists because there is a balance between the supply and the removal of the material that forms it. A sand beach that does not appear to change with time is not necessarily static (no new material supplied, no old material removed), but is more likely to represent a **dynamic equilibrium,** the supply of material equaling the removal of material. The beach is continually changing, but it appears unchanged and remains in balance.

(a)

(b)

Figure 11.19

Seasonal changes of La Jolla Beach: (a) winter and (b) summer.

Natural Processes

The gentler waves of summer move sand shoreward and deposit it, where it remains until the winter season. The large storm waves of winter remove the sand from the beach and transport it back offshore, to the sandbar. These processes occur alternately, winter and summer, leaving the beach rocky and bare each winter and covered with sand during the summer months (see fig. 11.19). These seasonal changes are fluctuations about the equilibrium state of the beach over a yearly cycle. Other changes also appear in a cyclic fashion, such as the arrival of more sand during periods of late winter and spring river flooding. If a beach receives as much material as it loses, the beach is in equilibrium and does not appear to change from year to year.

Single violent events may also alter a beach. A great storm may arrive from such a direction that the usual beach current is reversed. A landslide may pile rubble across the beach and some distance into the water, interfering with the transport of sand along the beach. In either case, the beach changes due to changes in rates of sediment supply, transport, and removal.

Waves moving toward a beach produce a current in the surf zone that moves water onshore and along the beach. This landward motion of water is called the **onshore current,** and this current transports suspended sediment toward the shore in what is called **onshore transport.** Waves usually do not approach a shore with their crests completely parallel to the beach but strike the shore at a slight angle. This pattern sets up a surf zone **longshore current** that moves down the beach. Both onshore and longshore flow in the surf zone are illustrated in figure 11.20. The turbulence that occurs as the waves break in the surf zone tumbles the beach material into suspension in the

water. The wave-produced longshore current displaces this sediment down the shore in the surf zone, producing a **longshore transport.** Along both coasts of the North American continent, the predominant longshore transport steadily moves the sediments in a southerly direction.

Each breaking wave's oblique uprush of water, or **swash,** along the beach face moves the sand particles diagonally up and along the beach in the direction of the longshore current. The backwash from the receding wave moves downslope toward the surf zone, but the backwash is weaker than the swash because much of the receding water percolates down into the sand. The combined action of the swash and backwash moves particles in a zigzag or sawtooth path along the swash zone of the beach as part of the longshore transport.

The path over which beach sediments travel as they are transported from their source to their area of deposition is called a **drift sector.** Beaches within the drift sector usually remain in dynamic equilibrium and change little in appearance. Although there is often a great flow of materials along the central portion of a drift sector, the supply usually equals the amount removed in this area. At either end of the drift sector, the beaches change with time; the beaches at the sediment source end of a drift sector are eroded, and those at the other end are depositional. The depositional beaches are growing, or **accreting.** These processes are shown in figure 11.21. The wave-produced longshore current in the surf zone provides the transport mechanism in a drift sector, but tidal currents and coastal currents associated with large-scale oceanic circulation also affect the sediment transport process in some places.

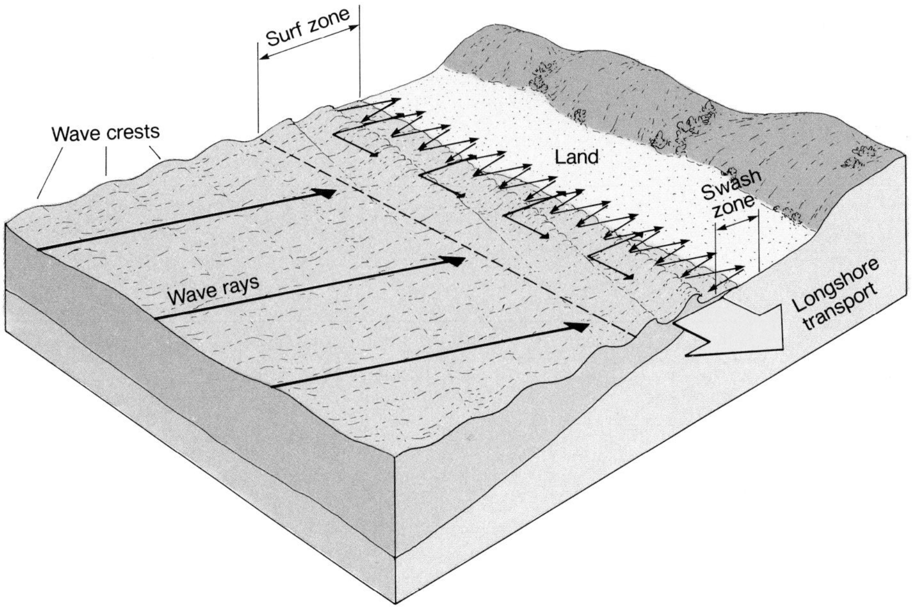

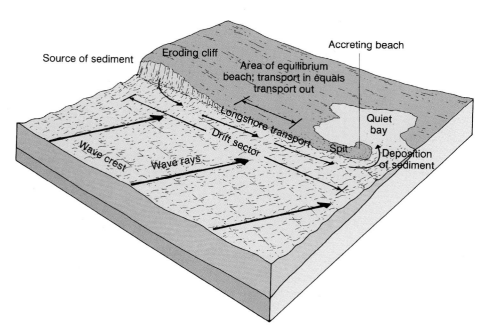

Figure 11.20

Waves in the surf zone produce a longshore current that transports sediments down the beach. Arrows indicate onshore and longshore water movement.

Surf zone

Wave crests

Land

Swash zone

Wave rays

Longshore transport

Figure 11.21

A drift sector extends down the coast from the sediment source to the area of deposit.

Accreting beach

Source of sediment

Eroding cliff

Area of equilibrium beach; transport in equals transport out

Longshore transport

Quiet bay

Drift sector

Spit

Wave crest

Wave rays

Deposition of sediment

Coastal Circulation

Onshore transport accumulates water as well as sand along a beach. This water must flow back out to sea and will do so one way or another. A headland may deflect the longshore current seaward, or the water flowing along the beach can return to the area beyond the surf zone in quieter water, such as areas over troughs or depressions in the seafloor. Regions of seaward return are frequently narrow and fast-moving. These are the rip currents that carry sediment through the surf zone (see fig. 11.22); rip currents are also discussed in chapter 9 (see fig. 9.19).

Just seaward of the surf, the rip current dissipates into an eddy. Some of the sediment is deposited here in quieter deeper water and is lost from the downbeach flow.

However, some is returned to shore with the onshore transport on either side of the rip current. This cycle of partial return of sediments to the beach, transport along the beach, transport back to sea in another rip current, and return again to the beach is a part of a drift sector.

A series of these drift sectors, when linked together along a stretch of coast, forms a **coastal circulation cell.** In a major coastal circulation cell, the seaward transport of sediment results in deposits far enough offshore to prevent any recycling. At the end point of a coastal circulation cell, the longshore current is deflected away from the beach and the sand is deposited in deep water offshore. This is often caused by the sand moving down a submarine canyon to the adjacent ocean basin floor. This removes the sand from

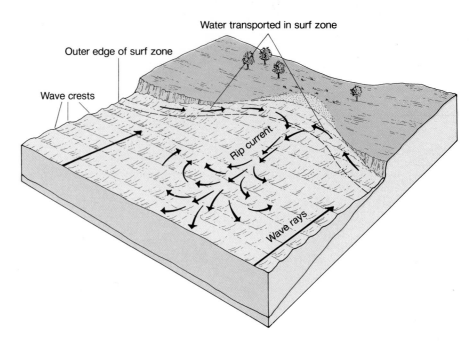

Figure 11.22

Rip currents form along a beach in areas of low surf and reduced onshore flow.

Water transported in surf zone

Outer edge of surf zone

Wave crests

Rip current

Wave rays

Figure 11.23

Major coastal sediment circulation cells along the California coast. Each cell starts with a sediment source and ends where beach material is transported into a submarine canyon.

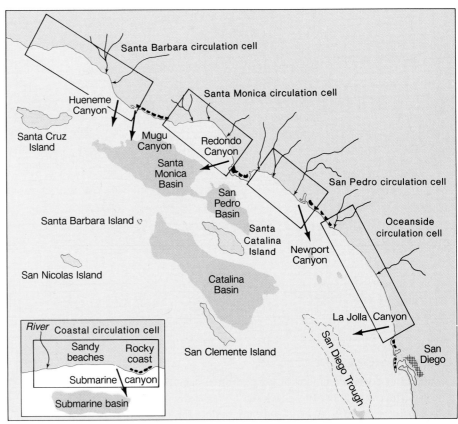

Santa Barbara circulation cell

Santa Monica circulation cell

Hueneme Canyon

Santa Cruz Island

Mugu Canyon

Redondo Canyon

San Pedro circulation cell

Santa Monica Basin

San Pedro Basin

Santa Barbara Island

Oceanside circulation cell

Santa Catalina Island

San Nicolas Island

Newport Canyon

Catalina Basin

La Jolla Canyon

San Clemente Island

San Diego Trough

San Diego

River Coastal circulation cell

Sandy beaches

Rocky coast

Submarine canyon

Submarine basin

its journey along the coast, and sand beaches disappear below this point until a new source contributes sand to form the next major coastal circulation cell.

South of Point Conception in southern California, oceanographers recognize four distinct coastal circulation cells (fig. 11.23). Each cell begins and ends in a region of rocky headlands, where beaches are sparse and submarine canyons are found offshore. Beaches become wider as river sources contribute their sand to the longshore current, and

at the end of each cell, the current deposits the sediment in a submarine canyon, where it cascades to the ocean basin floor. Beaches just south of the canyon are sparse, and a new cell begins. Refer back to chapter 2 for a discussion of submarine canyons.

Estimates of the rate at which sediments are moved along a beach are made by observing the rate at which sand is deposited, or accreted, on the upstream side of an obstruction, or by observing the rate at which sand spits

Figure 11.24
The entrance to Grays Harbor, Washington, is protected from southerly swells by a long jetty extending into the ocean. Groins built perpendicular to the land keep the tidal currents from eroding the point. Note the buildup of sand in the angles formed by the groins and the shore.

migrate. Transport of sand along a section of coast varies between zero transport and several million cubic meters per year. Average values fall between 150,000 m³/yr and 1,500,000 m³/yr; 150,000 m³ is more than 30,000 dump truck loads. Once the enormous volume of the naturally moving sand is recognized, it becomes apparent why poorly designed harbors fill very quickly, beaches disappear, and spits migrate a considerable distance during a year. It also becomes obvious that the effort required to keep pace with the supply by dredging or pumping operations is enormously expensive and in many cases quite impossible.

The forces that supply the energy for the movement of beach materials come from the waves and the wave-produced currents. Surface winds supply about 10^{14} watts of power to the ocean surface to produce the waves and currents. There are about 440,000 km (264,000 mi) of coastal zone in the world, and approximately half of this coast is exposed directly to the ocean waves. Ocean waves average 1 m (3.3 ft) in height and produce power that is equivalent to 10^4 watts for each meter of exposed coastline. Under storm conditions, the wave height averages about

3 m (10 ft), and a meter of coastline receives 10^5 watts. It is this supply of energy that causes the beach erosion, the suspension of beach material, and the migration of beach sand along the narrow coastal zone.

Artificial Processes

People can change beaches by damming rivers to control floods and generate power. When a dam is built, a lake is formed behind the dam, and sand, gravel, and rock that had once moved down the river to join the sediments at the coast are now deposited in the lake behind the dam. An important source of material to the continued balance of a beach has been removed. Although the sediment supply is reduced, the longshore current continues to carry away suspended beach material. The net result is a loss of sediment and sand to the beaches.

Coastal zone engineering projects also result in changes to a beach. **Breakwaters** and **jetties** are built to protect harbors and coastal areas from the force of the waves. The areas behind jetties and breakwaters are quiet; materials suspended by the wave action and carried by the

Figure 11.25

Santa Barbara harbor as seen from the south. In the foreground a dredge removes the sediment that has built up in the protected, quiet water behind the jetty.

longshore current settle out here. The result is that less material is available for a beach farther down the coast, and as the longshore current continues to flow, this next beach begins to disappear. Small-scale examples of this process are seen when **groins,** rock or timber structures, are placed perpendicular to the beach to trap sand being carried in the longshore transport (see fig. 11.24).

Our beaches are disappearing into the sea. According to a U.S. Army Corps of Engineers study, more than 40% of the U.S. continental shoreline is losing more sediment than it receives. Of this 24,000 km (15,000 mi) of shoreline, 4300 km (2700 mi) are considered critical areas meriting public protection. Most of the areas designated critical are along the Atlantic and Gulf coasts; the majority are on barrier islands. Time and time again, coastal facilities to control erosion are constructed in response to the demands of one group, only to trigger a chain of events that results in newly created problems and more engineering projects. In many cases eroding beaches are maintained by supplying the beach with sand and gravel mined from upland areas or dredged from offshore sandbars. These "beach feed programs" are expensive but necessary to compensate for the losses of beach sediment.

11.6 Practical Considerations: Case Histories of Two Harbors

Designing coastal structures that correct or solve one problem without creating more problems is as much an art as it is a science. Measurements and calculations are made, but more is required, for there must be a thorough understanding of the natural processes in an area before the consequences of the new structure can be foreseen. A scale working model is a powerful aid in understanding the present state of a shore area and the results of future changes.

Natural processes such as tides, waves, and currents are reproduced on a scale model and then the proposed structure is introduced into the system. Its effect is observed, and design modifications can be made. Models of this type are expensive, but they are much less expensive than the costs of continually correcting the results of poorly designed structures that disturb the natural environment in an undesired manner.

The Santa Barbara Story

The Santa Barbara harbor project is a classic example of interference with coastal zone processes. At Santa Barbara, a jetty and breakwater were constructed to form a boat harbor. The jetty at the north side of the harbor juts out into the sea before it turns southward, forming a breakwater that runs parallel to the coast (see fig. 11.25). The longshore current and sediment transport move down the coast of southern California from north to south. This jetty-breakwater system creates a wave-sheltered area on the jetty's north side and also blocks the longshore current. Sand was deposited on this north side, where the longshore current was blocked, and the beach began to grow. The beaches to the south began to disappear, because they were starved of the sand that had been deposited to the north.

When the beach to the north had grown until it reached the seaward limit of the jetty, the longshore current could again move beach sediments southward along the ocean side of the breakwater. When the longshore current and the suspended sediment reached the end of the breakwater and the entrance to the harbor, the current formed an eddy that spiraled into the quiet water of the harbor. The sediment settled out and began to fill the harbor, forming a spit connected to the end of the breakwater. The sediment settling in the harbor deprived the beaches farther south of their supply but did not alter the forces acting to remove the sediment from these same beaches.

Figure 11.26
Ediz Hook is a long spit forming a natural breakwater at Port Angeles, Washington. The protected harbor behind the spit is deep enough for mooring the largest supertankers (infrared false color photo).

Today, a dredge pumps the sediments deposited in the harbor through a pipe and back into the longshore current on the south side of the harbor. In this way, the harbor remains open and the beaches to the south receive their needed supply of sand. Interference with a natural process requires the expenditure of much time, effort, and money to do the work nature did for nothing.

The History of Ediz Hook

Fifteen hundred miles north of Santa Barbara, another harbor has been altered through interference with the supply of beach material. The harbor of Port Angeles, located on the Strait of Juan de Fuca in Washington State, is protected from the storm waves in the strait by a naturally occurring 3½-mile-long curving spit known as Ediz Hook (fig. 11.26). The hook is composed mainly of sand and gravel, and protects an area large enough and deep enough to accommodate the largest and most modern supertankers. In recent years, however, the hook has undergone considerable erosion, and the danger of waves breaking through the hook near its base has become severe.

To understand how the hook came to exist at all and how its present condition has come about, it is necessary to go back about 14,000 years, to the time when the glaciers of the last ice age retreated from this area. At that time, sea level was lower than it is now, due to the amount of water on land in the form of ice. The Elwha River west of Port Angeles carried loose glacial deposits to the sea, forming a delta that curved to the east under the influence of local currents and waves in the Strait of Juan de Fuca. As sea level began to rise, the river, currents, and waves continued to carry sediment eastward, and the waves began to erode the cliffs in the area, adding still more sediment to the longshore current. As a result, a hook-type spit was built up at a point where the shoreline makes an abrupt angle with the strait. Over time, Ediz Hook was produced and was kept constantly supplied with sand and gravel from the cliffs and the river.

In 1911, a dam was constructed across the Elwha River to provide power and a freshwater reservoir. The sediments that had been flowing toward the hook (estimated at 30,000 m^3 per year) were now being deposited behind the dam. In 1930, a pipeline was constructed to deliver fresh water from the Elwha Reservoir to Port Angeles, and the route chosen was around the cliffs at beach level. To protect the pipeline, bulkheads were built along the face of the cliffs, and the bulkheads cut off the source of cliff-eroded sediments, estimated at 380,000 m^3 per year. At present, it is believed that only one-seventh of the total sediments once available to feed the spit remain in the longshore current. Sediment removal from the spit no longer equals sediment supply. Studies by the Army Corps of Engineers show a possible loss of 270,000 m^3 of sand each year from the outside of the hook; the hook is now sediment-starved.

Ediz Hook is not merely protection for the harbor; it is also the site of a large pulp mill, a U.S. Coast Guard Station, and a number of small harbor facilities. All are connected by a road running the length of the hook. In the 1950s, the occupants of the hook and the city acted to protect the seaward side of the hook by applying large boulder armoring called rip-rap as well as steel bulkheads. Cliff material was blasted into the water in hopes of supplying the required sediments. A constant battle between the people and the sea began. Nearly as fast as the armor was applied to the base of the spit, the waves tore away the protective barriers in places and crashed over the spit.

In 1971 the Corps of Engineers completed a planning study for further corrections. The 1973–1974 winter storms were severe, and the hook was again storm-damaged. In 1974 funds were appropriated, and the work of strengthening the seaward side of the hook began. Over a fifty-year period, annual maintenance costs are projected to be about $30 million, while revenue from the harbor is projected at $425 million. Since the benefit:cost ratio is about 14:1, it appears to justify the cost of the project.

Summary

The coast is the land area that is affected by the ocean. The shore extends from the low-tide level to the top of the wave zone. The beach is the accumulation of sediment along the shore.

Primary coasts are formed by nonocean processes (for example, land erosion; river, glacier, or wind deposition; volcanic activity; and faulting). Primary coasts include fjords, drowned river valleys, deltas, alluvial plains, dune coasts, lava and cratered coasts, and fault coasts. Secondary coasts are modified by ocean processes (for example, ocean erosion; deposition by waves, tides, or currents; and modification by marine plants and animals). Erosion produces regular and irregular coastlines. Deposition of eroded materials creates sandbars, sand spits, and barrier islands. Reef coasts are formed by marine organisms; mangrove swamps and salt marshes modify coastlines.

A typical beach has features that include an offshore trough and bar and a beach area comprising a low-tide terrace, beach face, beach scarp, and berms. Winter and summer berms are produced by seasonal changes in wave action.

Beaches are described by their shape; the size, color, and compostion of beach material; and their status as eroded or depositional beaches.

Beaches exist in a dynamic equilibrium, in which supply balances removal of beach material. The gentle summer waves move the sand toward the shore during onshore transport. In winter, high-energy storm waves scoop the sand off the beach and deposit it in a sandbar during offshore transport. In the surf zone, the breaking waves produce a longshore current. The longshore current moves the sediment down the shore in a process known as longshore transport. If the beach accumulates as much material as it loses, it is in a state of equilibrium. Drift sectors and coastal circulation cells are defined by the path followed by beach sediments from source to region of deposition.

Rip currents are narrow, fast, seaward movements of water and sediment through the surf zone. A series of small drift sectors forms major transport patterns along the coast, which are known as coastal circulation cells. The volume of naturally moving sand in coastal circulation patterns along a beach is enormous. The energy for the movement comes from the wind-driven waves and the wave-produced currents.

Artificial structures also change beaches (for example, the building of breakwaters, groins, and jetties in coastal waters). Santa Barbara harbor and Ediz Hook are examples of human interference with natural processes in the coastal zone.

Key Terms

coast	foreshore
coastal zone	offshore
shore	berm
beach	berm crest
primary coast	winter berm
secondary coast	summer berm
sill	scarp
moraine	low-tide terrace
drowned river valley	beach face
ria coast	trough
delta	tombolo
alluvial plain	lag deposit
dune coast	armored beach
lava coast	dynamic equilibrium
cratered coast	onshore current
fault bay	onshore transport
fault coast	longshore current
sea stack	longshore transport
bar	swash
barrier island	drift sector
sand spit	accretion
hook	coastal circulation cell
reef coast	breakwater
salt marsh	jetty
backshore	groin

Study Questions

1. The processes that form a stretch of flat, uniform coast depend on whether the land is rising or sinking. Discuss the processes that form a flat coastal area under these two conditions.

2. Why are multiple berms more likely to be seen on a beach between March and August than between September and February?

3. Fjord coasts and drowned river valleys, or ria coasts, are primary coasts. Explain why their appearances are distinctly different. Give examples of each.

4. Why is a resort hotel that is located on one of the barrier islands along the southeast coast of the United States a poor long-term investment?

5. Describe the processes required to create a tombolo. Consider and discuss (a) the distribution of wave energy, and (b) the longshore transport.

6. What conditions are required to maintain a beach with a constant profile and compostion? Consider both a static and a dynamic environment.

7. In order to create a small boat harbor along a wave-exposed sandy beach, a breakwater is built parallel to the shore. What effect will this structure have on the beach behind it?

8. Why does an eroding beach that is supplied with material from cliffs and the banks of an old glacier deposit become an armored beach, while an eroding beach that is supplied by river sediments does not?

9. How does the profile of a sand beach change during alternating seasonal periods of storm waves and more gentle small waves?

10. Sketch a beach section that is stable with respect to the supply and removal of beach sediments. Indicate the source of the sediments and the final deposition of sediments in the system. Add breakwaters and jetties to your beach and indicate what changes each of these structures will produce in your system.

11. What is causing the apparent rise of sea level around the world? Is the rise similar in all places?

12. What processes move beach sediments in a coastal circulation cell?

13. What are the lessons to be learned from the stories of Santa Barbara harbor and Ediz Hook?

14. How would the profile of a sand beach after a heavy winter storm differ from that of the same beach during the summer?

15. How would you expect the coastline of Africa adjacent to the western Sahara to change with time?

Suggested Readings

Bascom, W. 1960. Beaches. In *Ocean Science,* readings from *Scientific American* 203 (2): 171–81 (1977).

Bascom, W. 1980. *Waves and Beaches: The Dynamics of the Ocean Survey,* rev. ed. Doubleday, Garden City, N.J. 366 pp.

Bird, E. C. F. 1985. *Coastline Changes: A Global Review.* Wiley & Sons, New York. 219 pp.

Fox, W. T. 1983. *At the Sea's Edge: An Introduction to Coastal Oceanography for the Amateur Naturalist.* Prentice-Hall, Englewood Cliffs, N.J. 317 pp.

Gable, F., and D. Aubrey. 1989/90. Changing Climate and the Pacific. *Oceanus* 32 (4): 71–73.

Jacobson, J. 1989. A Really Worst Case Scenario. *Oceanus* 32 (2): 36–39.

Kaufman, W., and O. Pilkey. 1979. *The Beaches Are Moving.* Doubleday, Garden City, N.J. 326 pp.

Kemper, S. 1992. "If you can fish from your condo, you're too close." *Smithsonian* 23 (7): 72–86.

Lowenstein, F. 1985. Beaches or Bedrooms—The Choice as Sea Level Rises. *Oceanus* 28 (3): 20–29.

MacLeish, W. H., ed. Winter 1980–81. *Oceanus* 23 (4). (Articles on coastal zone management, barrier islands, and conservation.)

Milliman, J. 1989. Sea Levels: Past, Present, and Future. *Oceanus* 32 (2): 40–43.

Mitchell, J. 1986. Coastal Management since 1980: The U.S. Experience and Its Relevance for Other Countries. In *Ocean Yearbook 6,* Borgese, E. M., and N. Ginsburg, eds. Univ. of Chicago–Chicago. pp. 319–45.

Open University. 1989. *Waves, Tides and Shallow Water Processes.* Pergamon Press, Oxford, England; Open University, Milton Keynes, England. 186 pp.

Pilkey, O. H. 1990. Barrier Islands. *Sea Frontiers* 36 (6): 30–36.

Reid, W. V., and M. C. Trexler. 1991. *Drowning the National Heritage: Climate Change and U.S. Coastal Biodiversity.* World Resources Institute, Washington, D.C. 49 pp.

Ringold, P. L., and J. Clark. 1980. *The Coastal Almanac.* Freeman, San Francisco. 172 pp. (Data reference.)

Shepard, F. P. 1973. *Submarine Geology,* 3rd ed. Harper & Row, New York. 517 pp. (Classic text in geological oceanography.)

Wanless, H. R. 1989. The Inundation of Our Coastlines. *Sea Frontiers* 35 (5): 264–71.

12

Bays and Estuaries

12.1   **Estuaries**
        Types of Estuaries
        Circulation Patterns
        Temperate Zone Estuaries
12.2   **Embayments with High Evaporation Rates**
12.3   **Flushing Time**
12.4   **Degradation of Coastal Environments**
        Water and Sediment Quality
        The Plastic Trash Problem
        Oil Spills
12.5   **Marine Wetlands**

Box: Going, Going, Gone

12.6   **Practical Considerations: Case Histories**
        The Development of San Francisco Bay
        The Situation in Chesapeake Bay
Summary
Key Terms
Study Questions
Study Problems
Suggested Readings

There were many fish moving in through the deep water of the channel that night. They were full-bellied fish, soft-finned and covered with large silvery scales. It was a run of spawning shad fresh from the sea. For days the shad had lain outside the line of breakers beyond the inlet. Tonight with the rising tide they had moved in past the clanging buoy that guided fishermen returning from the outer grounds, had passed through the inlet and were crossing the sound by way of the channel.

As the night grew darker and the tides pressed farther into the marshes and moved higher into the estuary of the river, the silvery fish quickened their movements, feeling their way along the streams of less saline water that served them as paths to the river. The estuary was broad and sluggish, little more than an arm of the sound. Its shores were ragged with salt marsh, and far up along the winding course of the river the pulsating tides and the bitter tang of the water spoke of the sea.

Rachel Carson,
from *Under the Sea-Wind*

Galveston Bay estuary on the Texas coast.

A *long the coasts of the world, there are embayments or indentations where salt water from the oceans and fresh water from the land meet. Sometimes, heavy freshwater runoff floods and dominates these areas; at other times, the seawater is driven landward by severe storms and submerges the low-lying coasts. In some ways, embayments behave like small oceans, but the frequent addition of fresh water gives them characteristic circulation patterns of their own.*

We call embayments bays, harbors, gulfs, inlets, sounds, channels, and straits, names given historically but according to no defined rules. Usually bays and harbors are smaller than gulfs; inlets and sounds are elongate; straits connect two large bodies of water, while channels are narrow water connections between smaller areas. Each coastal embayment is unique, but together they share certain characteristics of circulation that are used to group them.

People are also a part of bay and estuary systems. They change these areas more than they realize, and their needs burden these complex and sensitive portions of the ocean. In this chapter we will investigate the relationships between these areas and the increasing populations that live near them and use them.

12.1   Estuaries

An **estuary** is a portion of the ocean that is semi-isolated by land and is diluted by freshwater drainage. All estuaries are coastal embayments, but not all embayments are estuaries unless they are isolated and diluted with fresh water. An estuary's circulation, water exchange with the ocean, and circulation patterns are determined by the tides, the river flow, and the geometry of the inlet. Circulation and distribution of salinity are used to distinguish between four basic types of estuaries.

Types of Estuaries

The simplest type of estuary is the **salt wedge estuary.** Salt wedge estuaries occur within the mouth of a river flowing directly into salt water. The fresh water flows rapidly out to sea at the surface, while the denser seawater attempts to flow upstream along the river bottom. The seawater is held back by the flow of the river, which causes a sharply tilted boundary to form between the intruding wedge of seawater and the fresh water moving downstream. This pattern is shown in figure 12.1. The net seaward surface flow is almost entirely fast-moving river water, because the large discharge of fresh water is forced into a thin layer above the salt wedge. The salt wedge moves upstream on the rising tide or when the river is at a low flow stage; the wedge moves downstream on the falling tide or when the river flow is high. The boundary between the salt water and the overriding river water is kept sharp by the rapidly moving

river water. The moving fresh water erodes seawater from the face of the wedge, mixing it upward into the turbulent river water, and raising the salinity of the seaward-moving fresh water. This one-way mixing process is called **entrainment;** very little river water is mixed downward into the salt wedge. Seawater from the ocean is continually added to the salt wedge to replace the salt water entrained into the seaward-flowing river water. In a salt wedge estuary, the circulation and mixing is controlled by the rate of river discharge; the influence of tidal currents is generally small in comparison to the river flow. Good examples of salt wedge estuaries are found in the mouths of the Columbia, Hudson, and Mississippi rivers. In the Columbia River, the salt wedge moves as far as fifteen miles upstream at times of high tide and low river flow, maintaining its identity with a sharp boundary between the two water types. Salt wedges also occur in the mouths of rivers where the rivers enter other types of estuaries, for example, the mouth of the Sacramento River in San Francisco Bay.

Estuaries that are not of the salt wedge type are divided into three additional categories on the basis of their net circulation and the vertical distribution of salinity. These categories are the well-mixed estuary, the partially-mixed estuary, and the fjord-type estuary.

Well-mixed estuaries (fig. 12.2) have strong tidal mixing and low river flow, creating a slow net seaward flow of water at all depths. Note that the mixing due to strong tidal turbulence is so complete that the salinity of the water is uniform over depth, and decreases from the ocean to the river. There is little or no transport of seawater inward at depth; instead, salt is transferred inward by turbulent diffusion. The nearly vertical lines of constant salinity move seaward on the falling tide or when river flow increases and landward on the rising tide or when river flow decreases. Shallow estuaries, including the Chesapeake and Delaware bays, are well-mixed estuaries.

Partially-mixed estuaries (fig. 12.3) have a strong net seaward surface flow of fresh water and a strong inflow of seawater at depth. The seawater is mixed upward and combined with the river water by tidal current turbulence and entrainment to produce a seaward surface flow that is larger than that of the river water alone. This two-layered circulation acts to rapidly exchange water between the estuary and the ocean. Salt moves into a partially-mixed estuary by turbulent diffusion, but more importantly, by **advection,** the inflow of seawater at depth. Examples of partially-mixed estuaries include deeper estuaries, for example, Puget Sound, San Francisco Bay, and British Columbia's Strait of Georgia.

Fjords or **fjord-type estuaries** (fig. 12.4) are those deep estuaries that have a moderately high river input but little tidal mixing. This pattern occurs in the deep and narrow fjords of British Columbia, Alaska, the Scandinavian countries, Greenland, New Zealand, Chile, and other glaciated coasts. In these estuaries, the river water tends to remain at the surface and move seaward with little mixing of the underlying salt water. Most of the net flow is in the surface layer, and there is little influx of seawater at depth. Salt is

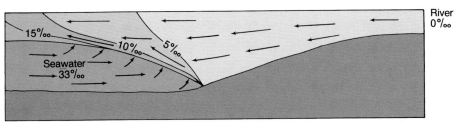

Figure 12.1

The salt wedge estuary. The high flow rate of the river holds back the salt water, which is drawn upward into the fast-moving river flow. Salinity is given in parts per thousand (‰).

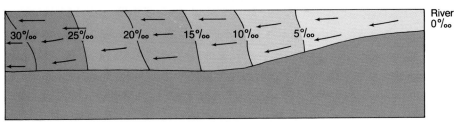

River
0‰

Figure 12.2

The well-mixed estuary. Strong tidal currents distribute and mix the seawater throughout the shallow estuary. The net flow is weak and seaward at all depths. Salinity is given in parts per thousand (‰).

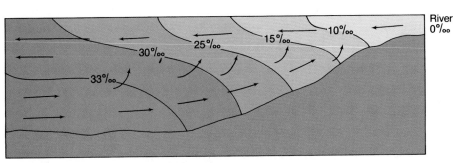

River
0‰

Figure 12.3

The partially-mixed estuary. Seawater enters below the mixed water that is flowing seaward at the surface. Seaward surface net flow is larger than river flow alone. Salinity is given in parts per thousand (‰).

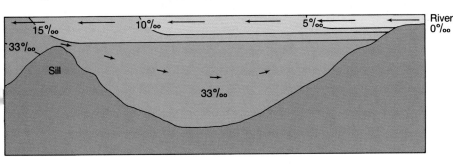

River
0‰

Figure 12.4

The fjord-type estuary. River water flows seaward over the surface of the deeper seawater and gains salt slowly. The deeper layers may become stagnant due to the slow rate of inflow. Salinity is given in parts per thousand (‰).

supplied at a slow rate by advective transport. Because of the low inflow at depth and the isolation of the deeper water by the entrance sill, the deeper water may stagnate.

The salt wedge estuary exists only as a strongly stratified estuary. These estuaries are never very well mixed vertically, except above the boundary between the layer of seaward-moving river water and the salt wedge. The other estuary types have various degrees of vertical stratification between the highly stratified, or poorly mixed, and the weakly stratified, or well-mixed. The processes that govern the degree of vertical mixing and stratification are the strength of the oscillatory tidal currents, the rate of freshwater addition, the roughness of the topography of the estuary over which these currents flow, and the average depth of the estuary. Tidal flow that oscillates with the rise and fall of the tide produces the energy for mixing; it does not usually result in a preferred directional flow and should not be confused with the net flow in each estuary.

Not all estuaries fit neatly into one of these categories. There are estuaries that fit between the types, and

some estuaries change from one type to another seasonally with changes in river flow or weekly as the tides change from springs to neaps. Studying how an individual estuary relates to examples of the different types helps the oceanographer understand the processes that govern an estuary and its water exchange with the ocean.

Circulation Patterns

This discussion of estuary circulation uses the partially-mixed estuary, with its inflow from the ocean at depth and its surface outflow of mixed river and seawater. As the water volume of the estuary increases on the rising tide and decreases on the falling tide, tidal currents are produced. Water in these currents moves back and forth in the estuary, but it does not necessarily leave the system and enter the open sea. However, the surface water moves farther seaward on the falling tide than it moves inward on the rising tide, and a parcel of surface water is moved progressively farther seaward on each tidal cycle. In the same way, the water from the sea moves into the estuary on the rising tide

Figure 12.5

Water and salt enter and exit a partially-mixed estuary.

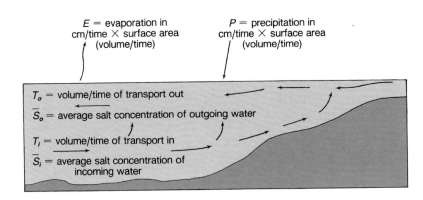

E = evaporation in cm/time × surface area (volume/time)

P = precipitation in cm/time × surface area (volume/time)

T_o = volume/time of transport out

$\overline{S}_o$ = average salt concentration of outgoing water

T_i = volume/time of transport in

$\overline{S}_i$ = average salt concentration of incoming water

farther than it drops back seaward on the falling tide. Averaged over many tidal cycles, this pattern produces a net movement seaward at the surface and a net movement landward at depth.

The **net circulation** of an estuary out (or seaward) at the top and in (or landward) at depth is of great importance, for it is this circulation that carries wastes and accumulated debris seaward and disperses them in the larger oceanic system. It is also through this circulation that organic materials and juvenile organisms produced in the estuaries and their marsh borderlands are moved seaward, while nutrient-rich water is brought inward at depth and replenishes the estuary's water and salt.

Understanding the net circulation of an estuary and evaluating the flows of surface water seaward and the underlying ocean water landward can be a long and expensive process. To make this determination directly requires the installation of recording current meters at various depths and at several cross-channel locations in an estuary. The currents are measured over many tidal cycles for both spring and neap tides, and at times of low, intermediate, and high freshwater discharge. The resulting current records at each depth and channel location are averaged to determine the net current distribution over depth for a given cross section in the estuary, yielding the net flow of the estuary. Data are averaged over one-week periods to determine the importance of spring and neap tides, over monthly cycles to find changes in patterns due to seasonal river and climate fluctuations, and over several years to produce an annual mean pattern. The cost of installing and maintaining equipment, as well as the costs of processing the collected data, make this direct approach a very expensive one.

A less expensive, indirect approach is to assume a **water budget** of inflow equal to outflow and a constant estuary volume averaged over time; all processes adding water to the estuary equal all processes removing water. A **salt budget** is also assumed; salt added is equal to salt removed. At the entrance to the estuary, measurements of salinity are made and averaged over space and time, establishing the average salinity of the outflowing surface water, $\overline{S}_o$, and the average salinity of the deeper inflowing seawater, $\overline{S}_i$. The formula

$$\frac{\overline{S}_i - \overline{S}_o}{\overline{S}_i}$$

is used to determine the fraction of river water in the seaward-moving surface layer at the estuary entrance.

If the rate of total river water inflow, R, is known for this same time period, T_o, the volume rate of seaward flow of the surface layer at the estuary entrance is found by using the following formulas without direct measurement:

$$\frac{\overline{S}_i - \overline{S}_o}{\overline{S}_i}(T_o) = R \text{ or } T_o = \frac{\overline{S}_i}{\overline{S}_i - \overline{S}_o}(R).$$

In a partially-mixed estuary, T_o is always larger than R, and both the river inflow, R, and the saltwater inflow, T_i, combine to produce the seaward flow, T_o, and maintain the water budget. Inflow equals outflow.

$$T_o = T_i + R.$$

Both evaporation, E, and precipitation, P, also remove and add water at the surface, as shown in figure 12.5. If both E and P were large, then R in the previous equations becomes $(R - E + P)$.

This method assumes that, over the time period for which the calculations are used, the average water volume and the total salt content of the estuary remain constant.

Temperate Zone Estuaries

In the middle latitudes of North America, most estuaries gain their fresh water from rivers, and the evaporation of water from the estuary surface is minor or is nearly balanced by direct precipitation. The salt concentration of the entering seawater, $\overline{S}_i$, is about 33‰, and $\overline{S}_o$ is about 30‰. Using the preceding equation and the example in figure 12.6, T_o is calculated to be about eleven times the rate of river inflow, or R. T_i in this case is ten times R. The influx of seawater at depth decreases as it moves into the estuary and the surface flow increases due to mixing and incorporating seawater from below as it moves seaward. While this mixing is occurring, the average salt concentration of the seaward-moving surface flow is also increasing, and T_o is formed when 10 unit volumes of seawater have combined with the 1 unit volume of river water. These relatively large values of T_o and T_i compared to R make them very important in the exchange of estuary water with ocean water.

In some fjordlike estuaries, the surface freshwater layer is as deep as the shallow sill at the entrance, and the average salt content of the surface layer at the entrance is very low. In this case, T_o is approximately equal to R; T_i is negligible, and the deep water of the fjord can stagnate.

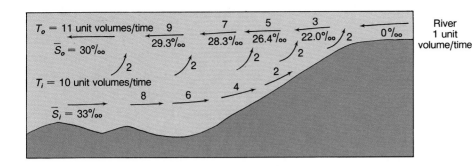

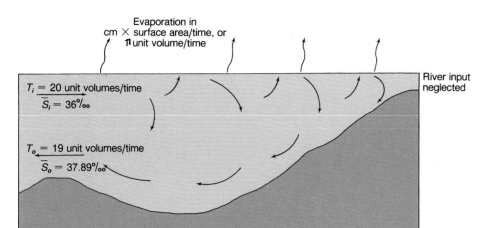

Figure 12.6

The seaward-moving surface water (odd-numbered arrows) increases in flow volume as the inflowing seawater (even-numbered arrows) moves upward. Upward mixing due to tidal turbulence is indicated by the vertical arrows. For every 2 units of seawater mixed upward, the inflow decreases and the outflow increases by the same 2 units. Salinity is given in parts per thousand (‰).

Figure 12.7

The evaporative sea or inverse estuary. Evaporation at the surface removes water (1 unit) and river inflow is negligible. Seawater flows inward at the surface (T_i = 20 units), and the seaward flow (T_o = 19 units) is at depth. Salinity is given in parts per thousand (‰).

12.2 Embayments with High Evaporation Rates

Embayments located near 30°N and 30°S latitudes have low precipitation and high evaporation rates. Although they may have rivers, these areas are not estuaries if they do not experience net dilution. The contribution of their rivers is usually minor when compared to the loss of water due to evaporation over the entire basin. The Red Sea and the Mediterranean Sea are examples of such areas. In these seas evaporation increases the surface-water salinity directly, causing the surface water to sink and accumulate at depth where it flows seaward to exit into the ocean. The ocean water is less dense than the outflowing high-salinity deep water, and ocean water flows into the sea at the surface. The T_o and T_i flows are present, but they have reversed their depth of occurrence, with T_o at depth and T_i at the surface. Because of this reversal, these embayments are sometimes called **inverse estuaries** (fig. 12.7).

If the excessive evaporation rate removing water from the surface of an evaporative embayment is equal to the rate of river inflow to an estuary, the T_o values for the evaporative estuary can be compared to those for the temperate estuary. Compare figures 12.6 and 12.7. In figure 12.7, a typical $\overline{S}_i$ value (36‰) is given for ocean-surface conditions existing at latitudes of 30°N and 30°S. The evaporative loss from the embayment is set at 1 unit volume per time, resulting in a T_i flow of 20 units and a T_o flow of 19 units. The resulting $\overline{S}_o$ value is 37.89‰. In figure 12.6, the river input of 1 unit volume per time results in

T_o and T_i flows of 11 units and 10 units, respectively. In general, if E in an inverse estuary is comparable to R in a temperate estuary, the inverse estuary has a better rate of exchange with the ocean than a true estuary. Not only is T_o likely to be larger, but because the inverse estuary has a continuous overturn as surface water sinks to depth and moves seaward, the exchange of water is more complete.

12.3 Flushing Time

If the mean volume of an estuary is divided by T_o, an estimate of the length of time required for the estuary to exchange its water is obtained; this is known as **flushing time.** If the net circulation is rapid and the total volume of an estuary is small, flushing is rapid. A rapidly flushing estuary has a high carrying capacity for wastes, because the dissolved or suspended wastes are moved rapidly out to sea and diluted. A slowly flushing estuary risks accumulating wastes and building up high concentrations of land-derived pollutants. Understanding the circulation of our estuary systems is essential to maintaining them as healthy, productive, and useful bodies of water. Population pressures on estuaries are heavy, for these areas are used as seaports, recreational areas, and industrial terminals as well as for their fishing resources.

If the waste products are associated only with the fresh water entering the estuary, it is possible to determine the ability of the fresh water to carry these products through the estuary without considering the action of the entire estuary. This requires determining how much fresh water is in the estuary at any one time, and the freshwater volume is estimated from the salinity of the estuary, which is the result of mixing fresh water and ocean water. For example, an estuary with an average salt concentration of 20% is one-third fresh water if the adjacent ocean salinity is 30%. Dividing the freshwater volume by the rate of river water addition yields the flushing time for fresh water and its associated pollutants.

Some bays and harbors do not have a freshwater source for dilution; these areas are flushed by tidal action. On each change of the tide, a volume of water equal to the area of the bay multiplied by the water level change between high and low water exits and enters the bay. When this volume, known as the **intertidal volume,** leaves the bay and enters the ocean, it may be displaced down the coast by a prevailing coastal current. On the next rising tide, an equivalent volume of different ocean water enters the bay. In this case, the flushing time is measured by the number of tidal cycles required to remove and replace the bay volume. This is determined by dividing the mean bay volume by the intertidal volume exchanged per tidal cycle. This flushing is often not complete, because currents along the coast do not always displace the intertidal volume a sufficient distance to prevent some portion of it from cycling back into the bay. Therefore, the number of tidal cycles required for flushing may be greater than the calculated estimate.

The same holds for those estuaries that have a two-layered flow. The seaward-moving surface layer, T_o, may experience some mixing with the T_i flow due to turbulence present at the entrance of the estuary. This mixing may incorporate some of the exiting estuary surface water into the deeper inflowing water, which results in a partial recycling in the estuary. Recycling of this type mixes the surface water to depth, aerating the deeper parts of the estuary.

Occasionally, the net circulation in an estuary is altered by ocean conditions. For example, if coastal upwelling occurs, the inflow of denser water to the estuary increases, and this increased inflow accelerates the estuary's circulation and causes the outflow to increase. When upwelling ceases and downwelling occurs, less-dense seawater is present at the estuary entrance, and the inflow is reduced. In these circumstances, the estuary's circulation may temporarily reverse as the denser, deeper water flows back to sea and the less-dense seawater moves inward at the surface.

Although each estuary is unique and must be judged accordingly, the knowledge gained from studying one estuary can be used to understand another system. Understanding the circulation patterns and the processes that form these patterns allows us to judge the degree to which an estuary can be modified and used while still preserving its environmental and economic value. Because estuaries lie at the contact zone between the land and the sea, and because humans are inhabitants of the land, alterations have been made in these areas and probably cannot be entirely avoided in the future. But we owe it to ourselves and to those who come after us to make the most intelligent and knowledgeable uses possible of these areas, for they represent a great natural resource that can renew itself if treated with care.

12.4 Degradation of Coastal Environments

Historically, people have depended on access to salt water for trade and transport and access to fresh water for their living requirements. This has resulted in major population centers becoming established along the shores of estuaries and coastal seas. In the United States today, over one-half of the population lives within 80 km (50 mi) of the coasts (including the Great Lakes). This population, with its necessary industries, energy-generating facilities, and waste-treatment plants, has created a tremendous burden on coastal zones. In the past, these natural waters were assumed to be infinite in their ability to absorb and remove the by-products of human populations. However, too much use and too many wastes discharged into too small an area at too rapid a rate have produced problems that can no longer be overlooked. These discharges can and do exceed the ability of the natural systems to flush themselves and disperse their wastes into the open ocean.

Recently, the realization has come that changes in the environment of these coastal regions were seriously degrading them, decreasing their aesthetic and economic value and, in some cases, endangering public health and safety as well as threatening their living resources. In the United States the 1960s and 1970s produced the laws and agencies that presently monitor and control the uses of these fragile areas. Today, there is a greater concern to act with care and judgment, but today we also have a legacy from past practices that is usually difficult and expensive to correct. The following discussions consider some of the major problems associated with our coastal environment.

Water and Sediment Quality

Dumping solid waste and pouring liquid pollutants into shore waters are inexpensive solutions to industrial and domestic waste problems. However, these practices have led to environmental degradation, with its hidden costs. The expense of cleaning up and of building plants to process these wastes is high and is increasing. The costs are paid by all of us as increased taxes and increased product prices. In times of high costs and high unemployment, it becomes more difficult to justify the short-term costs required to conserve natural marine resources, no matter how high the potential for eventual economic gain. When living is easy, we are more apt to consider this potential and to pay the costs of protecting the environment.

Off the mouth of the Hudson River lies the New York Bight, between New York and New Jersey. Street sweepings, garbage, dredge spoils, cellar dirt, and waste chemicals have been dumped into this area since the 1890s. Increasing quantities of floating debris found their way back to the beaches until 1934, when laws were passed to

Table 12.1

Contaminate Concentrations in Fine Sediments Sampled 1984–87

Location	Chlorinated pesticides* (non-DDT)	DDT* (total)	PCB* (total)
Boston Harbor	17.88–47.87	34.0–62.39	329.11–1128.76
Long Island Sound	7.27–28.94	7.41–76.83	122.86– 460.65
Chesapeake Bay	0.71– 9.63	1.43–14.11	6.29– 124.49
Tampa Bay	2.43–52.93	3.49–45.34	107.17– 747.5
Mobile Bay	0.18– 1.45	4.06–15.32	12.76
Galveston Bay	0.56– 4.18	0.42– 3.95	5.11– 53.32
San Diego Bay	10.14–13.01	4.6– 14.18	18.34– 788.55
San Francisco Bay	0.62– 7.12	5.28–32.94	65.59– 82.84
Puget Sound	0.23– 5.3	3.5– 12.97	25.82– 902.34
Port Valdez, AK	0.5	0.86	3.25
Honolulu Harbor	1.29	2.45	64.18

*parts per billion.

Source: National Oceanic and Atmospheric Administration Technical Memorandum *NOS OMA 44, 1988. Progress Report for Marine Environmental Quality.*

prohibit the dumping of "floatables." However, refuse from building and subway construction, toxic wastes from industry, acids, and sewage sludge continued to be dumped. It is estimated that between 1890 and 1971 the volume of solid waste dumped into the water of the New York Bight was 1.4 million m³. M. Grant Gross, in his NOAA monograph on waste disposal in the New York Bight, estimates this waste as the equivalent of a layer covering all of Manhattan Island to the height of a six-story building, or 20 m (66 ft). The dumping in the New York Bight area has degraded the quality of the water and the sediments. Many of the materials added to the water are both toxic and oxygen-demanding. Occasionally, this degraded water upwells along the coast and causes the death of marine organisms in shallower waters.

In 1987 the Environmental Protection Agency (EPA) closed this offshore dump, but agreed to let New York City and several New Jersey communities continue to dump sewage sludge, a product of secondary sewage-treatment plants, at a new site 171 km (106 miles) out to sea, at the edge of the continental shelf. Over 8 million tons of sludge have been dumped annually. In 1988, after a summer of severe pollution and waste-strewn beaches along the northeast coast, Congress voted to end the dumping of sewage sludge in the ocean by January 1, 1992. The bill also banned the dumping of industrial waste in the ocean. However, problems remain. Although the ban on sludge dumping benefits offshore organisms, it has little effect on the beach litter problems, and the sludge must still be dumped somewhere on land where it has the potential to leach its contaminants into freshwater drainage systems. Much of the debris washed onto the beaches was found to be the product of overflowing storm sewers that carried materials collected in streets and drains, and debris from marshes, which were flushed out to sea by storms and high tides, as well as refuse improperly dumped offshore. The increasing population of these coastal areas, accompanied by uncontrolled residential and commercial development, is considered a greater long-term threat to the area than these episodic events.

Refer back to chapter 3, section 3.8 for a review of a recent proposal to study the effects of deep-sea sludge dumping. Scientists making this proposal recognize that serious problems are developing as it becomes more difficult to find adequate land disposal sites, and they are suggesting a reassessment of deep-water ocean dumping.

Although the indiscriminate industrial dumping of materials in estuaries and coastal areas has been substantially reduced in response to recent legislation, there are other routes that pollutants or toxicants follow to reach the coastal zone. Pesticides such as DDT (dichloro-diphenyl-trichloro-ethane); long-lived toxic organic compounds such as polychlorinated biphenyls (PCBs); and heavy-metal ions such as lead, mercury, zinc, and chromium are still making their way through the environment, even though many of these materials are closely controlled or no longer used or manufactured. In the United States, PCB manufacture and sale stopped in 1978 and 1979, but electric transformers charged with PCBs are still in use. Although decontamination techniques have been developed to facilitate disposal, PCBs are a wide-spread pollutant in waters all around the world.

Concentrations of these toxicants always tend to be higher in the sediments than in the overlying water. A comparison of contamination of sediments in various U.S. bays and harbors is presented in table 12.1.

Despite international efforts to tighten controls on manufacturing processes and to reclaim toxic materials, accidents do happen. In November 1986 a fire in Basel, Switzerland, consumed the warehouse of a chemical firm. Water used to extinguish the blaze washed 30 tons of herbicides, pesticides, and mercury into the Rhine River. A contaminated slug of river water 25 miles (41.5 km) long moved downstream to the North Sea. Nearly all the plant and animal life over a 185-mile (307-km) stretch of this 820-mile (1361-km)-long river was destroyed. Over half a million fish and eels were estimated to have died. Towns dependent on the river for their water supply had to have water trucked in from other sources.

One week later, contaminated water from the same plant spilled into the Rhine again. While checking for the chemicals from these two spills, it was discovered that another, unreported spill of pesticides from a different source had entered the river two days prior to the fire and the initial spill. Ten years of effort to clean and restock the river had been destroyed. From the disaster have come new efforts in the nations that border the river to force the reconsideration of laws that regulate the chemical industries of the region.

Surface runoff from rural and agricultural lands finds its way through lakes, streams, and rivers to the estuaries. The runoff supplies pesticides and nutrients, which can poison or overfertilize the waters. In the latter case, the resulting overpopulation of plants eventually dies and decays, removing large quantities of oxygen. Lack of oxygen then kills other organisms, which in turn decay and continue to remove oxygen from the water. Animal wastes and failed septic systems also contribute contamination.

Metropolitan areas have surface runoff via storm sewers that add a wide mix of materials, including hydrocarbons from oil, lead from gasoline, residues from industry, pesticides and fertilizers from residential areas, and coliform bacteria from fecal material. Cities also add treated sewage effluent, which contributes its share of contaminants. Even the chlorine added first to municipal drinking water and then again to the treated sewage effluent as a bactericide may form a complex with organic compounds in the water to produce chlorinated hydrocarbons, which may be toxic in the marine environment. Considering all the pathways, sources, and types of materials that can be classed as pollutants, their management and eventual exclusion from the coastal zone is an extremely difficult and complex problem, since it cannot be solved by sewage treatment alone.

The annual average mass of treated sewage components delivered to the coastal region by seven southern California municipal discharge systems is shown in table 12.2. Because of the area's increasing population, the

	1980	1981	1982	1983	1984	1985	1986	1987	1988	1989
	1078	1080	1094	1122	1129	1143	1175	1179	1178	1200
	1,493	1,492	1,511	1,549	1,565	1,579	1,623	1,629	1,632	1,658
	232,100	224,900	224,200	244,700	197,700	204,500	184,900	148,500	97,000	83,400
	255,100	260,900	266,100	251,800	230,100	253,500	181,900	166,500	168,800	161,100
	38,400	36,700	37,300	35,700	30,000	34,300	29,000	25,700	25,300	22,600
	42,000	40,500	41,800	40,100	40,500	44,500	42,900	44,500	44,600	45,500
	30	28	25	26	24	26	22	15	11	11
	11	12	5.8	10	18	16	12	11	8.9	7.4
	39	32	21	23	16	16	14	9.0	3.4	1.9
	275	187	203	163	140	110	88	57	29	22
	335	337	284	272	251	239	202	125	76	68
	1.8	1.8	1.2	1.1	0.9	0.9	0.7	0.4	0.4	0.4
	224	167	168	163	133	118	127	76	63	54
	175	130	122	98	87	118	105	61	50	27
	11	15	6.4	6.5	6.5	5.6	8.2	7.2	6.7	7.6
	729	538	545	497	369	375	336	260	151	146
	671	480	290	223	310	48	51	53	26	20
	1,127	1,252	785	628	1,209	46	37	5	0	0

number of millions of gallons per day (MGD) of treated waste effluent has been steadily increasing. However, the mass of sewage components discharged each year has been decreasing, due to increased efforts that prevent harmful materials from entering waste streams at their sources and to increases in the level of wastewater treatment.

The Mediterranean has long been an example of a high-population area along a badly polluted inverse estuary. Of the Mediterranean pollutants, 85% entering the sea come from the land, and 85% of these come by way of the rivers. Sewage is a major component of river water, and 80% of the sewage flowing into the Mediterranean is untreated. Under the auspices of the United Nations, the Med Plan to improve the situation was adopted in 1976. By 1992 contaminated swimming beaches dropped by 30%, but few nation-participants as yet have established adequate guidelines to meet the Med Plan's requirements.

U.S. efforts to reduce the discharge of toxic materials into the marine environment are showing some signs of success. Improving the quality of treated sewage discharged into the nation's rivers is improving the levels of dissolved oxygen in the waters around discharge sites. While the increased use of agricultural fertilizers contributes nutrients to the freshwater runoff, improved sewage treatment decreases the nutrient level, and in general nutrient levels appear to be unchanged. The increase in the use of unleaded gasoline appears to be directly related to a reduction in the amount of lead entering the marine environment. Concentrations of lead in the surface waters of the Sargasso Sea have dropped about 30% between 1980 and 1984. Lead carried by the Mississippi River to the Gulf of Mexico has been reduced about 40% in the last decade.

Many toxicants reaching the estuaries do not remain in the water but become adsorbed onto the small particles of matter suspended in the water column. These particles clump together and settle out due to their increased particle size, concentrating the toxicants in the sediments. Analyses of estuary sediment core samples show changes in bottom sediment toxicant concentration over time. Figure 12.8a shows concentrations of heavy metal ions in sediments laid down between 1845 and 1990. Concentration levels for sediments deposited prior to 1845 are assumed to be at natural levels; those since 1845 are considered the products of human activities. In the last two decades the discharge to the sediments of these heavy metals has decreased with the exception of copper, which is still commonly used for water pipes. Figure 12.8b shows DDT and PCB concentrations between 1940 and 1980; widespread use of DDT didnot begin until the 1940s and was followed by PCBs in the 1950s. Sediment values indicate that their greatest use and discharge into the environment came in the early 1960s.

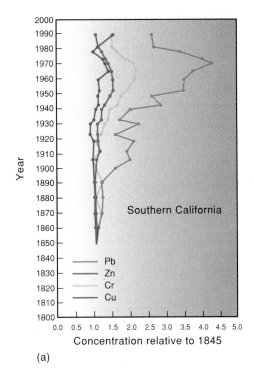

(a)

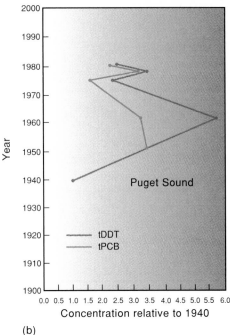

(b)

Figure 12.8

Toxicant concentrations in the sediments. (a) Heavy metal ion concentrations increased between 1845 and 1960; values have decreased during the environmentally conscious 1960s, 1970s, and 1980s. Pb = lead, Zn = zinc, Cr = chromium, Cu = copper. (b) DDT and PCB concentrations show increases until the control of DDT in the 1960s and the banning of PCBs in the 1970s.

Some of the particulate matter adsorbing toxicants has a high organic content and forms a food source for marine creatures. In this way, heavy metals and organic toxicants associated with the particles find their way into the body tissues of organisms, where they may accumulate and be passed on to predators. Throughout the United States, toxic residues are found in estuarine bottom fish. Shellfish concentrate heavy metals at levels that are many thousands of times over the levels found in the surrounding waters. The scallop can elevate levels of cadmium in its tissues about two million times over the water concentration, and oysters can concentrate DDT ninety thousand times over this concentration.

In 1967, a study was made of DDT sprayed on Long Island marshes to control mosquitoes. Although spray concentrations were not directly lethal to fish and birds, the effect of this long-lived toxic material as it was concentrated by the food chain is shown in table 12.3. DDT use in the United States has decreased, but there have been increases in the use of other long-lived toxicants such as dieldrin and aldrin, which are also concentrated by fish and shellfish.

An organism's ability to accumulate substances not only increases its concentrations of toxicants to levels that are sometimes injurious or fatal to the accumulating organism, but also produces organisms that are a hazard to humans, who may use them for food. A tragic example of the results of humans' ingesting organisms that had accumulated a toxin occurred between 1953 and 1960 in Minamata, Japan. Mercury from a local industrial source was released at high levels into the coastal embayment, from which much of the shellfish harvest for a local village was gathered. The mercury formed a complex that was readily taken up by the marine life, which led to severe

Table 12.3
Food Chain Concentration of DDT

	DDT residues[1]
water	0.00005
plankton	0.04
silverside minnow	0.23
sheepshead minnow	0.94
pickerel (predator)	1.33
needlefish (predator)	2.07
heron (small animal predator)	3.57
tern (small animal predator)	3.91
herring gull (scavenger)	6.00
osprey egg	13.8
merganser (fisheater)	22.8
cormorant (fisheater)	26.4

Source: From George M. Woodwell, et al. "DDT Residues in an East Coast Estuary." in *Science,* vol. 156, pp. 821–24, 12 May 1967. Copyright © 1967 by the American Association for the Advancement of Science. Reprinted by permission.
[1]Given in parts per million (PPM).

mercury poisoning and death among those eating the shellfish. The physical and mental degenerative effects of this mercury poisoning were especially severe on children whose mothers had eaten large amounts of shellfish during pregnancy. The condition produced has been named Minamata disease.

Figure 12.9

The impact of plastics on the marine environment: (a) Persistent litter includes plastic nets, floats, and containers. (b) A common mure entangled in a six-pack yoke. (c) A young grey seal caught in a trawl net. (d) A sea turtle with a partially ingested plastic bag. Photos (a), (b), and (c) are from Sable Island in the North Atlantic Ocean, approximately 240 km (150 miles) east of Nova Scotia, Canada.

The Plastic Trash Problem

Walk any beach in any estuary or along any open coast and see the tide of plastics being washed ashore. It is estimated that every year more than 135×10^6 kg (77 tons) of plastic trash have been routinely dumped by naval and merchant ships. The National Academy of Sciences estimates that the commercial fishing industry yearly loses or discards about 298 million pounds of fishing gear (nets, ropes, traps, and buoys) made mainly of plastic, and dumps another 52 million pounds of plastic packaging materials. Recreational and commercial vessels and oil and gas drilling platforms added their share, and a total of 14 billion pounds of cargo and crew wastes were added during each year of the 1980s. Plastics are a worldwide problem, with many sources and effective distribution to even the most remote areas by the currents. See figure 12.9a.

In 1960 U.S. production of plastics was 6×10^9 pounds; by 1988 the production was 60×10^9 pounds. Plastic is inexpensive, strong, and durable; these characteristics make it the most widely used manufacturing material in the world today and a major environmental problem. Nobody knows how long plastic stays in the marine envi-

ronment, but an ordinary plastic six-pack ring could last 450 years, and 16,000 six-pack rings were collected in three hours along 490 km (300 mi) of Texas beaches in 1989. Thousands of marine animals are crippled and killed each year by these materials. As many as thirty thousand fur seals a year are estimated to become entangled in lost or discarded plastic fishing nets and choked to death in plastic cargo straps. Lost lobster and crab traps made entirely or partially of plastic continue to trap animals; 25% of the 96,000 traps set off Florida's west coast were lost in 1984. Seabirds die entangled in six-pack rings and plastic fishing line. Seabirds and marine mammals swallow plastics. Porpoises and whales have been suffocated by plastic bags and sheeting, and the sheeting also clings to coral and to rocky beaches, smothering plant and animal life. Fish are trapped in discarded netting. Sea turtles eat plastic bags and die. Plastics are now considered as great a source of mortality to marine organisms as oil spills, toxic wastes, and heavy metals. See figure 12.9b, c, and d.

The Marine Plastic Pollution Research and Control Act of 1987 is a U.S. law that implements an international convention for the prevention of pollution from ships. Most of this law's provisions became effective at the end of 1988. Dumping plastic debris is prohibited everywhere in the oceans; other types of trash may be dumped at specific distances from shore. Ports and terminals are required to provide waste facilities for debris. The U.S. Coast Guard is the agency that must develop regulations for enforcement and compliance. Enforcement by this agency, plagued by ever-expanding duties and an ever-dwindling budget, is difficult. There is no international enforcement.

In the United States some manufacturers have begun to make six-pack rings of biodegradable plastic. Manufacturers add light-absorbing molecules that break down the plastic after a few months' exposure to sunlight; there is no evidence that this will solve the problem at sea. Even at the surface, the water keeps the plastic cool, and it becomes coated with a thin film of organisms that shade it from the light. As the problem worsens, it may be that only people educated to act in a responsible manner will reduce the tide of plastics rising around the world.

Oil Spills

The twentieth century runs on oil, petroleum, and petroleum products. This dependence requires the bulk transport of oil by sea to bring the crude oil to the land-based refineries and centers of use. This transport process exposes the world's coasts and estuaries to the hazard of oil spills associated with vessel casualties and transfer procedures. Because oil is found below the seafloor, the drilling of offshore wells exposes these areas to the risks of blowouts and accidental spills. Also, because industry, agriculture, and private and commercial transportation require petroleum and petroleum products, oil is constantly being released into the environment, to find its way directly or indirectly down to the sea. The average discharge of oil into the world's oceans and navigable waters during each of the last ten years has been 117×10^4 metric tons (340×10^6 gallons), not including the occasional megaspill of 3.8×10^4 metric tons (11×10^6 gallons) or more.

The transport of oil and oil products on the high seas creates the potential for accidents that release large volumes of oil. Spills far at sea are difficult to assess, because direct visual and economic impact on coastal areas does not occur. Damage to marine life cannot be accurately evaluated in such cases. Spills occurring due to the grounding of vessels or due to accidents in the storing and transferring processes are evaluated in coastal areas, where the environmental degradation and loss of marine life can be readily observed. Spills that occur in estuaries and along coasts affect regions that are oceanographically complex, biologically sensitive, and economically important. Three oil spills have occurred in the last 25 years that have become ecological landmarks; these are the sinking of the *Amoco Cadiz,* the grounding of the *Exxon Valdez,* and the wartime discharge of crude oil into the Persian Gulf.

In March 1978 a severe tanker spill occurred when the *Amoco Cadiz* lost her steering in the English Channel and broke up on the rocks of the Brittany coast of France (fig. 12.10a and b). Gale winds and high tides spread the oil

over more than 300 km (180 mi) of the French coast; more than three thousand birds died; oyster farms and fishing suffered severely. Of the approximately 2.1×10^5 metric tons of oil spilled, it is considered that 7×10^4 metric tons evaporated and were carried over the French countryside, $3–4 \times 10^4$ metric tons were cleaned up by the army and volunteers, another $3–4 \times 10^4$ metric tons penetrated down into the sand of the beaches to stay until winter storms washed it away, $2–3 \times 10^4$ metric tons (the lighter, more toxic fraction) dissolved in the seawater, and $4–5 \times 10^4$ metric tons sank to the seafloor in deeper water, where it continued to contaminate the area for an unknown period of time.

The United States experienced a large-volume oil spill in March 1989 when, 25 miles out of Valdez, Alaska, early on Good Friday morning, the *Exxon Valdez,* loaded with 170,000 metric tons (1.2 million barrels) of crude oil, ran onto a reef, tearing five huge gashes in her hull and spilling more than 240,000 barrels—ten million gallons—of oil into Alaska's Prince William Sound. Local contingency plans to clean up oil spills were not prepared to handle a situation of this magnitude; delays, caused in part by out-of-service equipment as well as a lack of equipment and personnel, the rugged coastline, and the tidal currents of the enclosed area in which the spill occurred combined to intensify the problem. The oil spread quickly, distributing itself unevenly over more than 900 square miles of water, taking a predictable and terrible toll of seabirds, marine mammals, fish, and other marine organisms. The oil moving out of Prince William Sound was not washed out to the open sea but was captured by the local currents and moved westward parallel with the coast, repeatedly oiling the rocky wilderness beaches in the weeks that followed (figure 12.10c and d). In the subarctic conditions of Prince William Sound, the photochemical and microbial degradation of the oil due to the cold temperatures proceeds more slowly than in temperate regions. The recovery of plant and animal populations also occurs more slowly in the cold water, where organisms tend to live longer and reproduce more slowly. Researchers inspecting the area during 1992, three years after the disaster, reported that the area was recovering and repopulating with organisms as the oil aged and degraded. They also noted that the areas most intensively cleaned to remove oil are recovering more slowly and with less-balanced populations than the areas in which the oil has been allowed to degrade naturally.

Each year a quarter of a million barrels of oil are routinely spilled into the shallow waters of the Persian Gulf. Much of the world's crude oil comes from the area's wells and is transported the length of the Gulf on its way to refineries around the world. It is one of the world's most polluted bodies of water; however, it is also an area of vigorous plant growth and supports fisheries of shrimp, mackerel, mullet, snapper, and grouper. During the eight-year Iran-Iraq war the bombing of oil facilities produced major spills, including one from an oil rig that poured out 172 metric tons (50×10^3 gallons) per day for nearly three months, but the greatest catastrophe came in the 1991 Gulf War when an estimated $0.86–1 \times 10^6$ metric tons ($250–300 \times 10^6$ gallons) of

Figure 12.10

(a) The tanker *Amoco Cadiz* aground and broken in two off the coast of France, March 1978. (b) A bay along the French coast was fouled with oil. (c) Hot water under high pressure is used to clean Prince William Sound beaches after the 1989 *Exxon Valdez* oil spill. (d) Booms collect the oil washed from the beach and hold it for pickup. (e) Booms are set in place to protect the saltwater intakes of a Saudi desalination plant on the Persian Gulf during the 1991 Gulf War. (f) Oil remains along the Persian Gulf coast after the Gulf War.

crude oil gushed into Gulf waters, the world's greatest oil spill to date. Some of this oil was deliberately released, some came from a refinery at a Gulf battle site, and bombing contributed additional quantities. This spill was almost twice as large as the estimated output of Ixtoc 1, an offshore Mexican well that blew out in 1980 and was the previous largest spill on record.

The average depth of the Persian Gulf is 40 m (131 ft) and the circulation of its high-salinity water is sluggish. Although changes in the wind in the weeks after the spill stopped the oil from drifting the entire length of the Gulf, some 570 km (350 mi) of Saudi Arabia's shoreline was oiled. Oil spill experts from the U.S., the U.K., the Netherlands, Germany, Australia, and Japan rushed to help.

Figure 12.11

A catamaran oil skimmer. The vessel cruises at about 3 knots, guiding surface oil between the twin hulls. The oil adheres to a moving belt and is lifted into on-board storage tanks.

They were able to protect the water-intake pipes of desalination plants and refineries, but other cleanup efforts were less successful (fig. 12.10e and f). Some areas have been bulldozed clean; others are being used as test plots to assess natural recovery. About half of the oil evaporated and about 3×10^5 metric tons have been recovered and stored in pits in the desert where some may be reclaimed. Much has sunk to the bottom of the Gulf, and still more oil may be seeping into the Gulf from sunken tankers. Scientists will continue to monitor the area and work to accelerate its restoration, but it is expected to take decades for the damaged areas to recover.

The tragedy of these spills demonstrated once again that there is no adequate technology available to cope with large oil spills, particularly under difficult weather and sea conditions, along irregular coastlines, or far from land-based supplies. Very little of the oil spilled under these conditions is recovered, on the average between 8% and 15%, which may be an inflated estimate because the recovered oil has a high water content. The technology for oil cleanup at sea includes oil booms and oil skimmers (see fig. 12.11. These devices are useful in confining and recovering small spills in protected waters.

The toxicity of weathered crude oil is low, and many oil spill experts believe that cleanup efforts should be concerned not with removing oil from the beaches but with moving oil seaward. They believe that many of the cleanup methods used along shorelines cause more immediate and long-term ecological damage than leaving the oil to degrade naturally. For example, habitats treated with hot water take longer to recover than those left untreated; oil loosened by hot water is washed down from the sparsely populated upper stretches of beaches into the crowded populations of more sensitive organisms along the lower shore. Backhoeing and high-pressure washing (100 lb/in²) destabilize gravel and sand beaches, killing any remaining live animals and working the oil into the sediments; refer back to figure 12.10c.

To prevent oil from reaching the beaches, some cleanup experts support newly developed, low-toxicity dispersants. These are most effective when applied in the

first 2–3 hours following a spill; the dispersed oil moves into deeper water where it is diluted and its toxicity is lessened.

Researchers are divided over the effectiveness of attempts to speed up natural degradation by adding nutrients to the beaches in order to increase populations of oil-degrading microorganisms. More data is needed before it can be determined how much if any difference in degradation occurs between treated and untreated beaches. Releasing other strains of microorganisms has been tested with limited success; the new organisms may not compete effectively with the natural populations.

The immediate damage from a large spill is obvious and dramatic; by contrast, the effect of the small but continuous additions of oil that occur in every port and harbor are much more difficult to assess, because they produce a chronic condition from which the environment has no chance to recover. Refined products such as gasoline and diesel fuel are more toxic to marine life than crude oil, but they evaporate rapidly and disperse quickly. They are less visible and so appear less offensive. Crude oil is slowly broken down by the action of water, sunlight, and bacteria, but the portion that settles on the seafloor moves down into the sediments, which it continues to contaminate for years.

The world's major ports and harbors are usually associated with the world's major bays and estuaries. Since most petroleum is transferred from vessel to shore at these locations, these sensitive, productive, and valuable areas will continue to be particularly vulnerable to oil damage in the foreseeable future.

12.5 Marine Wetlands

The value of estuarine areas as centers of productivity and nursery areas for the coastal marine environment is well known to oceanographers and fisheries biologists. Along the fringes of these estuaries the saltwater and brackish marshes, known as marine **wetlands,** provide nutrients, food, shelter, and spawning areas for marine species, including such commercially important organisms as crabs, shrimp, oysters, clams, and many species of fish. In the past these wetlands have been filled to produce usable ground for industry and port facilities; shorelines have been bulkheaded to protect them from erosion, and channels have been dredged to permit the entry of deep-draft vessels into the ports and harbors that support growing coastal centers of commerce and industry (fig. 12.12).

In 1960 the population density of U.S. coastal counties was 248 persons per square mile, four times the U.S. average. By 1988 population density in coastal counties reached 341 persons per square mile; by 2010 this coastal population is expected to be approaching 400 persons per square mile. To satisfy our desires to live near the shores in what were once marshes bordering estuaries, we destroy habitats and their populations. Modern Venices of recreation and retirement replace these areas with waterfront lots and homes, each with its own individual pleasure-craft moorage (fig. 12.13).

Figure 12.12
The industrialized estuary of the Duwamish River at Seattle, Washington.

Figure 12.13
The wetlands of Barnegat Bay, New Jersey, were replaced by a housing and recreational complex.

Going, Going, Gone

Louisiana is shrinking. Each year 40 to 50 square miles of Mississippi River wetland marsh disappear under the Gulf of Mexico. Twelve thousand years ago, Louisiana's shoreline extended 120 miles further into the Gulf of Mexico. In the mid-1800s, Bailize was a busy river town at the tip of the delta; today, it is under 15 feet of water. The wetlands of coastal Louisiana provide more than 4 million acres of marshes, ponds, lakes, bayous, and shallows, habitat for fish, shellfish, birds, and other animals. This great natural resource, important to the economy, biology, and history of the region, is dwindling.

Under natural conditions, the deposition of sediment by a river gradually raises the riverbed, decreasing its steepness and speed of flow. During a period of high flow, a river spills over its natural banks or levees and finds a new course where the steepness of the new channel is greater. In this way, a river wanders, leaving its old channel for a new one. This wandering changes the portion of the delta to which sediments and fresh water are being supplied. Older portions of a delta that have lost their sediment supply subside and erode, decreasing in area, while the new river channel creates delta wetlands in a new area.

The delta of the Mississippi River has been building over millions of years. Over this time, the sediments deposited by the river have added weight to the earth's crust, causing the crust to sink or subside; at the same time, the sediments have been eroded and dispersed by coastal waves and currents. If the supply of sediments exceeds the rate of subsidence and the rate of erosion, the delta grows and the wetlands increase in area. If the reverse occurs, the wetlands are lost to the Gulf of Mexico.

The Mississippi River has most recently been maintained in a fixed channel by raising the levee heights on each side of the river. This prevents loss of life and property due to flooding, and gives the port of New Orleans access to the river.

Box figure 12.1

The Mississippi River delta is disappearing into the Gulf of Mexico. Because the river is being kept in its channel by levees, riverborne sediments are continuing to the river's mouth and being deposited in deep water instead of spreading out and replenishing the delta. The loss of sediments also prevents compensation for the subsidence of the delta.

However, confining the river to a fixed channel cuts off much of the delta from sediments and fresh water. The sediments continue to the river's mouth, where they are discharged into deeper water. See Box figure 12.1. The lack of replenishing fresh water allows the water in the sediments to ooze out, and the slow process of soil compaction as well as subsidence results in the sinking of the land. The loss of sediment prevents the addition of land to subsiding areas, and the reduction of fresh water allows the intrusion of seawater over portions of the delta now protected from the river. Data from 1986 indicate that the Louisiana delta region accumulates sediments at a rate of 5.9–7.5 mm per year but the relative rise of sea level due to land subsidence and increasing water level is 11–13 mm per year. The result is a gradual loss of marshland and the conversion of vegetation from freshwater- to saltwater-tolerant species. Salt marshes eventually become bays as the sinking and erosion continues.

Other human activities and population growth contribute to the problem. Extraction of fresh water from the soil lowers the water table. Extraction of gas, oil, and sulfur leaves voids into which the land settles. Added weight from construction results in sediments consolidating and subsiding at faster rates than found in natural environments. All act to lower the land level, increasing the flooding and erosion. In the past, farming of the rich delta soils was accomplished by constructing extensive drainage systems to remove surplus fresh water. The sinking of these areas converts the drainage channels into access routes for seawater, which ultimately ruins the land for crop plants. Diking and draining or filling for agricultural, industrial, or residential use eliminate natural wetland areas. Marsh soils high in organic material exposed by these processes shrink, decay, and sink.

Hurricanes contribute to the situation. These storms damage barrier islands and drive salt water into the marshes, damaging the freshwater vegetation.

We may be able to save some marshland by rechanneling and directing sediments to sediment-starved areas, building small tidal dams to prevent saltwater flooding of freshwater marshes, and zoning coastal areas to prevent draining and filling. However, we must also recognize that upstream changes that are destructive to the marsh areas have often been designed to protect people and property in that upstream area. The result is a conflict with no simple solution. In addition, subsidence and erosion are natural processes often beyond our control, and change is a natural part of any environment. There will be changes in the Louisiana wetlands with or without the human factor; it is the rate of change that human activities accelerate.

Between the mid-1950s and the mid-1980s approximately 20,000 acres of coastal wetlands were lost per year in the contiguous United States. In the mid-1970s 4.4 million acres of coastal wetlands remained in the United States. Estuarine wetland losses have been greatest in six states: Louisiana, Florida, Texas, New Jersey, New York, and California. Much of the Louisiana loss is due to accelerated erosion and subsidence of Louisiana's coastal marshes; see Box: "Going, Going, Gone."

Public concern led to the passage of tidal wetland-protection laws in many coastal states and to stricter enforcement of federal laws in the seventies and eighties. In 1972 the federal government enacted the first nationwide wetlands regulatory program in Section 404 of the Clean Water Act and focused attention on coastal wetlands by enacting the Coastal Zone Management Act. These acts have reduced the rate of loss of coastal wetlands, but conflict occurs between the Environmental Protection Agency's monitoring policies and the U.S. Army Corps of Engineers' permit granting. In addition, various levels of state and county wetland management can and do cause confusion. In the northeast U.S., coastal wetlands are now better protected by state laws. Along the West and Gulf coasts, wetlands, although protected by state and federal laws, are still under heavy pressure for urban and industrial development. Where human use requires the permanent removal of estuarine land, there is now an attempt to mitigate the loss by restoring other degraded aquatic land or to assure that remaining natural areas be perpetually preserved. See the discussion of mitigation in chapter 13.

Each marine wetland is as unique as the estuary it borders, and the singular and distinctive nature of each is a part of the problem, for the difficulty of protecting a wetland is related to its legal definition. The law is subject to interpretation; under different conditions some wetlands escape protection while neighboring areas are included. The federal government's definition of wetlands has been under revision for more than five years, leading to uncertainty and confusion among monitoring agencies at all levels.

12.6 Practical Considerations: Case Histories

The Development of San Francisco Bay

The San Francisco Bay estuary (fig. 12.14) has a surface area of 1240 km^2 (480 mi^2). Its river systems drain 40% of the surface area of California. When visited by the early Spanish missionaries in 1769, the bay was surrounded by wetlands, where the 10,000–15,000 native peoples gathered much of their food. Development came slowly until 1848, when gold was discovered, and the population of San Francisco increased from approximately 400 to 25,000 in 1850. Within fifty years of the initial gold boom, the

Figure 12.14

The San Francisco Bay system. The city of San Francisco lies at the bottom center. The large, light-colored northern portion of the San Francisco Bay system is known as San Pablo Bay. The Sacramento River is joined by the San Joaquin River at the "Y" in the upper center of the photograph.

marshes were nearly gone, the bay was shallowed, fresh water had been siphoned off for irrigation, and nonnative species of animals had been introduced to the bay's waters, displacing native species.

Early in the bay's development the fisheries' resources (salmon, sturgeon, sardines, flatfish, crabs, and shrimp) were heavily exploited to feed the rapidly expanding population. By 1900, many of the fish, shellfish, and wildfowl stocks had been overharvested and depleted. Crabs were fished out in the 1880s, and the fishery was moved offshore, until its collapse in the 1960s. When the transcontinental railroad was completed in 1869, carload after carload of oysters were shipped from the east to mature on the bay's mud flats. The oysters did not become a self-reproducing stock, but many other small marine animals introduced with the oysters did naturalize, including the eastern soft-shell clam, the Japanese littleneck clam, and the marine pests: the oyster drill and the shipworm. Even the striped bass, the bay's best-known sport fish, was introduced from the East Coast in 1879. The only commercially harvested fish left in the bay are herring and anchovy.

In 1986 the clam *Poramocorbula amurensis* was discovered in the Carquinez Strait, an area north of San Francisco Bay. This clam was previously unknown in the area and appears to have arrived in the ballast water of cargo ships from Asian ports. During the next six years the clam spread southward into the bay and formed dense colonies, as many as 10,000 clams per square meter. The food requirements of these huge populations reduce amounts available for native species and further stress the system's species balance.

Until 1884, hydraulic mining for gold in the surrounding watershed washed huge amounts of silt and mud directly into the streams and rivers. The sands and muds in the rivers destroyed the salmon-spawning areas. Much of the sand and mud reached the bay, where it shallowed some areas, expanded the marshes, and altered the tidal flow. The rivers were flooded with the runoff from the barren land, and the resulting winter and spring floods produced changes in the estuaries, where the rivers enter the bay.

The marshlands of the bay and its river deltas, particularly the deltas of the Sacramento and the San Joaquin rivers, were diked, first to increase agricultural land and then for homes and industries. The conversion of wetland to dry land has reduced the tidal marshes from an area of about 2200 km^2 (860 mi^2) to 125 km^2 (49 mi^2). The increased cropland required increased irrigation, and the state built a series of dams, reservoirs, and canals with a water-storage capacity of 20 km^3 (.5 mi^3). Today, 40% of the Sacramento and San Joaquin rivers' flow is removed for irrigation, and another 24% is sent by aqueduct to central and southern California. By the year 2000, it is expected that San Francisco Bay will receive only 30% of its 1850 fresh-water input.

The diversion of water to the fields so reduces the amount of river water during periods of low flow in the summer that the irrigation pumps cause the water to flow upstream from the bay, carrying with it hundreds of juve-

nile salmon and striped bass, which are drawn into the pumps and die on the fields or in the ditches. It is thought that this has been a major cause of a drop in the population of striped bass since the 1960s. The loss of freshwater flow also changes the distribution and abundance of small, floating species on which the juvenile fish feed.

Intensive agricultural practices using fertilizers and pesticides, plus the runoff of water high in salts leached from the irrigated fields, has changed the quality of the fresh water entering the bay. In an effort to lessen this effect, runoff water from agriculture was directed into holding reservoirs, which in some cases were also wildlife refuges. In 1982, dead vegetation, deformed wildlife, and depressed reproduction in bird species were discovered at the Kesterson Reservoir. Levels of the element selenium, 130 times that of normal, were found in plant and animal tissues. The selenium was being leached from the irrigated soils and concentrated in the reservoir. Kesterson Reservoir was closed in 1986, but the water from the fields continues to flow into the San Joaquin and on into the bay. Recently, high concentrations of selenium have also been found in south bay ducks.

Domestic and industrial wastes from urban areas also enter the bay. Residence time of water in north San Francisco Bay fluctuates between one day during peak river flow in the winter to two months during low summer flow. Winter flows carry fresh water into the south bay and increase its exchange with the central bay, decreasing south bay residence time from months to weeks at this period of the year. In the south bay the city of San Jose has increased its freshwater sewage-treatment discharge as its population has grown. In addition to the pollution problems of the increased effluent, over the last ten years this freshwater increase has converted hundreds of acres of salt marsh to freshwater and brackish marsh, producing loss of habitat for native birds and small animals. San Jose has been ordered to create new habitat to offset this loss and to reclaim water from its sewage effluent in order to decrease the freshwater flow. If these improvements fail, a limit will be imposed on the volume of discharge, requiring the city to place limits on its population growth. Total wastewater discharge is expected to be 8% of the total freshwater inflow to the bay by the year 2000.

San Francisco Bay is considered the most modified major estuary in the United States. The city of San Francisco's population is now more than 700,000 in a metropolitan area of 5 million. Despite all the factors involved in this increasingly crowded area, the bay appears to have suffered less in total water-quality degradation than other major estuaries. In part, this is because the greatest urbanization has occurred near the mouth of the bay, allowing wastes to exit quickly through the Golden Gate. Improvements in sewage treatment since the 1960s have also improved conditions. There are patches of contamination from PCBs, oil, and chemical spills, and at this time the bay does not seem to be suffering from an overproduction of marine plants or a depletion of oxygen. Shellfish collection has recently been permitted for the first time in decades. The pressure for increased growth, the recent years of drought, and the loss of fresh water to irrigation continue to impact the bay's fish stocks. The population of delta smelt has declined by 90% from past levels, while adult striped bass have decreased from over 3 million in the past to 500,000 in 1991. In 1969, 118,000 winter chinook salmon made their way up the Sacramento River; in 1992, 191 were counted. As fresh water is extracted from the system, brackish water marshes have disappeared or have converted to saltwater systems, displacing the native organisms. As the area continues to develop and change, active research, effective monitoring, and intelligent management are needed to understand and protect the bay system.

The Situation in Chesapeake Bay

The Chesapeake Bay estuary (fig. 12.15) is a shallow, drowned river valley with a surface area of 11,500 km^2 (7000 mi^2) and an average depth of 6.5 m (21 ft). Chesapeake Bay flushes slowly; the residence time of the water is 1.16 years. The Chesapeake estuary has a long history as a major provider of oysters, blue crab, wildfowl, rockfish, and shad. It has also been under continuously increasing population pressure since colonial times, and has been absorbing more and more wastes from the expanding urban centers of Washington, D.C., Baltimore, Norfolk, and dozens of towns in Maryland, Virginia, and Delaware. After heavy harvesting in the 1950s, the oyster catch dropped by two-thirds between 1960 and 1983. In these same years the annual rockfish catch decreased from 6 million to 600,000 pounds, commercial fishing for shad has become nearly extinct in the upper Chesapeake, and the wildfowl population drop has been equally dramatic.

Although present harvests are low in comparison to those of the early years, the blue crab harvest for 1992 increased, and some areas report increased oyster catches. The bay's present total harvest of seafood is approximately 45 million kilograms per year (99 × 10^6 lb) with a dockside value of nearly $1 billion. The total value of the bay to Maryland and Virginia is estimated at $678 billion.

Partially treated and untreated sewage was recognized as the cause for outbreaks of typhoid fever among those drinking water and eating shellfish taken from the system in the early 1900s. Then, decaying untreated sewage robbed the water of oxygen; now plant growth, stimulated by nutrients from the effluent of sewage treatment plants and the runoff from fertilized fields, decays and consumes the dissolved oxygen. Increased plant growth also reduces the clarity of the water and shades the bottom. The low oxygen values and the reduction in bottom plant life due to decreased light together alter the bottom habitat. In addition, wastes from some five thousand factories, from military bases, and from sewage plants located between Virginia and New York find their way into the Chesapeake, adding conventional pollutants and toxic compounds that are trapped in the bottom sediments.

Efforts to improve water quality in Chesapeake Bay have produced nearly four thousand studies since the 1970s. In 1972, the Federal Clean Water Act provided en-

forcement standards and a system of permits governing the amount of pollutants that individual dischargers could dump into any body of water. Gradually, industrial and municipal discharges were regulated, but at the same time populations increased and industries expanded, increasing the volume and rate of discharge to the estuary.

At the present time, it is estimated that industries and sewage plants along the shores of Maryland and Virginia discharge about 4 trillion gallons of waste water annually, or about 20% of the water in the bay at any one time. An EPA study costing $27 million over seven years launched the "Save the Bay" campaign in 1983. Scientists working for the campaign reported that industries in Maryland dump more than 2700 tons of heavy metals into the bay each year, and that Virginia industries dump more than 400 tons during the same period. Much of the money has been spent to manage soil and fertilizer runoff from surrounding fields. Critics charge that the program ignores industrial and municipal discharges in excess of the levels allowed under the current permit system. The federal permit system is a self-policing effort, with dischargers setting their own limits depending on available technology and cost. The permit system has not responded to new technologies and enforcement and, until recently, has not been independently monitored. Many industries trying to avoid the discharge permit process direct their discharges to municipal sewage systems that are principally equipped to treat domestic and organic wastes. Chesapeake Bay is shallow with a large surface area; it has a very large ratio of watershed drainage area to volume of water. River systems that carry pollutants to the bay drain parts of New York, Pennsylvania, Delaware, Maryland, Washington, D.C., Virginia, and West Virginia. An unusual degree of cooperation must be achieved to solve the bay's problems.

While trying to understand the effect of human degradation of the Chesapeake, researchers have discovered that the natural changes in rainfall and temperature have large effects as well. At irregular intervals, tropical storms and hurricanes produce flood conditions that have catastrophic effects on the water quality and the organisms. Cyclic variation in survival of eastern oysters and striped bass have been related to such weather changes. In 1972, 1975, and 1979, the annual freshwater inflow was double the average. In 1972 in particular, tropical storm Agnes dropped 25.5 cm (10 in) of rainfall in a short period of time. This sudden surge of fresh water reduced the bay's salinity to about one-fourth its normal minimum level, and rinsed out the small floating organisms on which the larger animals feed. Severe winters, occurring at six- to eight-year intervals, promote heavy ice formation, which depresses the oxygen levels beneath the ice. These irregular but significant events superimposed on the increasingly degrading human influences make it difficult to determine exactly which factors are responsible for which changes in the Chesapeake estuary system.

Figure 12.15
Chesapeake Bay is a shallow estuary located on the heavily populated East Coast of the United States. The entrance to the estuary is at lower right.

Summary

The oceanographic classification of coastal embayments is based on the relationships between freshwater input, tidal flow, and net circulation. An estuary is a semi-isolated portion of the ocean that is diluted by fresh water. Estuary types include the salt wedge estuary, in which seawater becomes a sharply defined wedge that moves under the fresh water with the tide. Circulation and mixing in the salt wedge estuary are controlled by the rate of river discharge. In a shallow, well-mixed estuary, there is a net seaward flow at all depths due to strong tidal mixing and low river flow. Salinity is uniform over depth but varies along the estuary. A partially-mixed estuary has strong tidal turbulence, seaward surface flow of mixed fresh water and seawater, and an inflow of seawater at depth. Salt is transported in this system by advection and mixing. Fjord-type estuaries are deep estuaries in which the fresh water moves out at the surface and there is little tidal mixing or inflow at depth.

In a partially-mixed estuary, there is progressive movement of surface water seaward and deep water inward during each tidal cycle. The circulation of the estuary can be measured directly with current meters, which is a long and expensive process, or it can be measured indirectly by determining the water and salt budgets of the estuary.

In temperate latitudes, the volume transport of water between the estuary and the sea is much greater than the addition of fresh water. In fjords, the inward flow is small and the deep water tends to stagnate. In semienclosed seas with high evaporation rates, there is a net removal of fresh water; the seaward flow is at depth and the ocean water enters at the surface.

Estuaries that flush rapidly have a higher capacity for dissipating wastes than those that flush slowly. We can calculate the time period it takes for wastes associated with fresh water to pass through an estuary. In bays and harbors that are flushed only by tidal action, the recycling of resident water and wastes can occur. Partial recycling can also happen in estuaries with two-layered flow, as exiting surface water is mixed with incoming seawater.

Multiple uses of estuaries can overload them with wastes. Water quality is affected by dumping solid waste and liquid pollutants into coastal waters. Examples of practices leading to the degradation of the coastal marine environment include the dumping of solid waste in the New York Bight, the movement of pesticides and long-lived toxicants through the environment, chemical spills, the runoff of fertilizers and pesticides from agricultural lands, and storm sewer runoff from urban areas. Toxicants are adsorbed onto silt particles and become concentrated in the coastal sediments. Organisms concentrate toxicants and pass them on to other members of marine food chains. Plastic trash is an increasing problem to marine animal life. Oil spills are a special problem in inshore waters, for which there is no adequate cleanup technology.

Wetlands border estuaries; they are important as areas of nutrients, food, and shelter for many marine species. Many wetlands have been filled, dredged, developed, and lost. Case histories of San Francisco Bay and Chesapeake Bay are presented.

Key Terms

estuary	net circulation
salt wedge estuary	water budget
entrainment	salt budget
well-mixed estuary	inverse estuary
partially-mixed estuary	flushing time
advection	intertidal volume
fjord	wetland
fjord-type estuary	

Study Questions

1. If contaminated sediments are dredged from the floor of a harbor to use in a landfill, what hazards to the environment should be considered during both the dredging, disposal, and storage of the contaminated material?
2. Estuaries that receive silt-laden water from rivers and toxicants from urban sources tend to have lower toxicant levels in their waters than estuaries that receive clear river water and urban pollutants. Explain why.
3. Compare the circulation of a semienclosed basin at 30° N with that of an estuary located at 60°N. Which is less likely to accumulate waste products at depth? Why?
4. Estuaries are classified by their net circulation and salt distribution in this chapter. What other features could be used to describe and classify them?
5. Why are estuaries with short flushing times less apt to degrade when used as the receiving water for urban runoff than estuaries with long flushing times?
6. What happens to oil once it has been spilled into coastal waters?

7. Compare the oil spills from the Gulf War and the *Exxon Valdez*. Consider the geography and climate of each area, the dispersal of the oil, the effect of the oil on the beaches and the organisms, the effect on those who gain their living from the sea, and the effectiveness of the cleanup.

8. Study table 12.2. What general trends can be seen in wastewater contaminants between 1971 and 1989?

9. Identify several problems associated with the degradation of water quality in U.S. coastal waters. Are the solutions to these problems of equal difficulty? Why or why not?

10. Why is the concentration of toxicants in sediments higher than the concentration of toxicants in the overlying water?

11. Why have plastics become such a tremendous problem in marine waters?

12. Why are coastal wetlands important and why are we losing them?

13. Compare the circulations and histories of San Francisco Bay and Chesapeake Bay. How do these estuaries differ; how are they similar? What do you see in the future for each?

14. How has the damming of rivers for flood control altered sediment deposits in estuaries and coastal zones?

15. Why does a large but very shallow estuary support a different plant population than a very deep estuary of equal volume?

Study Problems

1. Determine the flushing time of an estuary in which $T_o = 9 \times 10^7$ m³/day and the volume of water is 30×10^8 m³.

2. An estuary has a volume of 50×10^9 m³. This water is 5% fresh water, and the fresh water is added at the rate of 6×10^7 m³/day. Why does the rate of addition of fresh water from the estuary to the ocean equal the input of fresh water from the land to the estuary? Consider the average salinity of the estuary as constant. What is the residence time of the fresh water?

3. Water entering an estuary at depth has a salinity of 34.5‰. The water leaving the estuary has a salinity of 29‰. The river inflow is 20×10^5 m³/day. Calculate T_o, the seaward transport.

4. A bay with no freshwater input can flush only by tidal exchange. If on each tidal cycle (ebb and flood) 10% of the bay's water volume is exchanged with ocean water, how much of the original water will still be in the bay after four tidal cycles?

5. If the flushing time of the bay in Problem 4 is calculated by dividing the volume of the bay by the intertidal volume (10% of the bay volume), the flushing time is ten tidal cycles. Compare this method to that used in Problem 4. In which case does water entering on the flood tide mix completely? In which case does the entering water move to the head of the bay without mixing? Which method best duplicates natural conditions?

Suggested Readings

Estuaries and Circulation

Lacombe, H. 1990. Water, Salt, Heat and Wind in the Med. *Oceanus* 33 (1): 26–36.

Nichols, F., J. Cloern, S. Luoma, and D. Peterson. 1986. Temporal Dynamics of an Estuary: San Francisco Bay. *Science* 231 (4738): 567–73.

Pritchard, D. W. 1967. What Is an Estuary: Physical Viewpoint. In *Estuaries,* Lauff, G. H., ed. Publication No. 83, American Association for the Advancement of Science, Washington, D.C. pp. 3–5.

Reid, W. V., and M. C. Trexler. 1991. *Drowning the National Heritage.* World Resources Institute, New York. 48 pp.

Water Quality and Pollution

Canby, T. V. 1991. After the Storm. *National Geographic* 180 (2): 2–32. (Oil spill cleanup in the Persian Gulf.)

Capuzzo, J. E. M. 1990. Effects of Wastes on the Ocean: The Coastal Example. *Oceanus* 33 (2): 39–44.

Hass, P. M., and J. Zuckman. 1990. The Med is Cleaner. *Oceanus* 33 (1): 38–42.

Hawley, T. M. 1990. Herculean Labors to Clean Wastewater. *Oceanus* 33 (2): 72–75.

Hodgson, B. 1990. Alaska's Big Spill. *National Geographic* 177 (1): 5–43.

Holloway, M. 1991. Soiled Shores. *Scientific American* 265 (4): 102–16. (Spill prevention and cleanup technology.)

Horgan, J. 1991. Muddled Cleanup in the Persian Gulf. *Scientific American* 265 (4): 99–110.

Horton, T. 1993. Chesapeake Bay. *National Geographic* 183 (6): 2–35.

Land, T. 1990. European Scientists Cooperate to Save the North Sea. *Sea Frontiers* 36 (1): 52–55.

National Oceanographic and Atmospheric Administration. 1990. *Coastal Environmental Quality in the United States.* National Oceanic and Atmospheric Administration, Rockville, Maryland. 34 pp.

Parker, P. A. 1990. Clearing the Oceans of Plastics. *Sea Frontiers* 36 (2): 18–27.

Wacker, R. 1991. The Bay Killers. *Sea Frontiers* 37 (6): 44–51.

Wilbur, R. J. 1987. Plastic in the North Atlantic. *Oceanus* 30 (3): 61–68.

13

Oceans: Environment for Life

13.1   **Buoyancy and Flotation**
13.2   **Osmotic Processes**
13.3   **Temperature**
13.4   **Pressure**
13.5   **Gases**
13.6   **Nutrients**
13.7   **Light and Color**
       Sunlight
       Bioluminescence
       Color
13.8   **Circulation**
13.9   **Barriers and Boundaries**
13.10 **Bottom Types**
13.11 **Environmental Zones**
13.12 **Classification of Organisms**

> *Box:* **Spartina:** *Valuable and Productive or Invasive and Destructive?*

13.13 **Practical Considerations: Modification and Mitigation**

Summary
Key Terms
Study Questions
Suggested Readings

The river is within us; the sea is all about us;
The sea is the land's edge also, the granite
Into which it reaches, the beaches where it tosses
Its hints of earlier and other creation.
The starfish, the hermit crab, the whale's backbone;
The pools where it offers to our curiosity
The more delicate algae and the sea anemone.
It tosses up our losses, the torn seine,
The shattered lobsterpot; the broken oar
And the gear of foreign dead men. The sea has many voices,
Many gods and many voices.

T. S. Eliot,
from "The Dry Salvages," *The Four Quartets*

A school of small fish (*Parapriacanthus ransonneti*) greet a diver in Indonesia.

A complex variety of environments for living organisms is provided by the oceans and coastal seas of the world. At the surface, conditions range from polar to tropical; over depth, from light to total and constant darkness; and around the world, from open ocean to sheltered bays. Currents and waves, rising and falling tides, sandy bottoms and rocky shores all contribute to the variation. Light–dark cycles as well as the tides modify an environment over the day; seasonal cycles modify it over the year, and other cycles take much longer. Some changes are gradual and extend over long distances; other changes occur quickly in a small area. This chapter introduces the ocean as a habitat by reviewing the properties of the world oceans that produce this special environment for life. Since the topics discussed here are considered in other chapters, their presentation is brief and keyed to their interaction with the marine organisms.

Organisms of the marine environment have much in common with the organisms of the land, but they also have very different problems and have developed unique solutions to cope with them. This chapter also explains some of the characteristics that enable organisms to live in the sea.

13.1   Buoyancy and Flotation

Organisms that make their home on land require structural strength to support their bodies in air and against gravity. Trees must be able to hold up their canopies of leaves, and animals need skeletons and muscles to give their bodies shape and the ability to move. The salt water surrounding marine organisms has a density similar to the tissues of many of the organisms. The **buoyancy** of objects in seawater is due to its density, and helps to keep the floating organisms at the surface. It supports the bodies of the bottom-living creatures and lessens the energy expended by swimmers.

Many organisms have ingenious adaptations to help them stay afloat. Some jellyfish-type animals, for example the Portuguese man-of-war and the by-the-wind-sailor (chapter 15, fig. 15.13), secrete gases into a float that enables them to stay at the sea surface. Some seaweeds secrete gas bubbles and form gas-filled floats, which help them keep their fronds in the sunlit surface waters while they are anchored to the seafloor. One floating snail produces and stores intestinal gases; another forms a bubble raft to which it clings. The chambered nautilus (fig. 13.1a), a relative of the squid, continually adds chambers to its shell and moves to the last chamber as it grows. A specialized tissue removes the ions from the empty chambers, causing water to diffuse out, then the chamber fills with gas, mainly nitrogen, from tissue fluids. The cuttlefish, another relative of the squid, has a soft, porous, internal

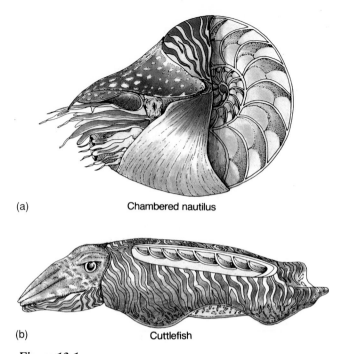

(a)   **Chambered nautilus**

(b)   **Cuttlefish**

Figure 13.1

(a) Chambered nautilus. (b) Cuttlefish. The chambered shell that provides buoyancy is shown in each organism.

shell or "bone"; it regulates its buoyancy by controlling the relative amounts of gas (also mainly nitrogen) and liquid within the shell (see fig. 13.1b).

Many fish have gas-filled swim bladders that keep them neutrally buoyant. Some fill their swim bladders by gulping air at the surface; others release gas from their blood through a gas gland to the swim bladder. When a fish changes depth, it adjusts the gas pressure in its swim bladder to compensate for the pressure change in the water, limiting its vertical swimming speed. If a fish with a swim bladder is forced suddenly to swim deeper, the increased external water pressure compresses the air, the bladder shrinks, and the fish sinks. If the fish is not able to readjust its system, it tires as it is forced to swim upward continually to compensate for the loss of buoyancy. A deep-swimming fish brought quickly to the surface with a fishing line will show bulging eyes and a distended body, as the gas in the swim bladder expands with decreasing pressure. Active, continuously swimming predatory species such as the mackerel, some tuna, and the sharks do not have swim bladders; bottom fish also lack swim bladders.

Small members of floating plant and animal populations store their food reserves as oil droplets that decrease their density and retard sinking. Large surface area-to-volume ratios that slow sinking are characteristic of small spheres, and single-celled organisms, particularly plants, take advantage of this by keeping their size small. Many have developed spines, ruffles, and feathery appendages that increase surface area and decrease their sinking rate, allowing them to more easily remain at or near the sea surface.

The tissues of most large marine animals are more dense than seawater. To reduce the density of its body tissues, the giant squid excludes the more-dense ions from its body fluids and replaces them with less-dense ions. Another squid species has one pair of its arms filled with low-density body fluids. Whales and seals decrease their density and increase their flotation by storing large quantities of blubber, which is mainly low-density fat. Sharks and some other varieties of fish store oil in their liver and muscle.

Seabirds float by using fat deposits in combination with light bones and air sacs developed for flight. Their feathers are waterproofed by an oily secretion called preen, which acts as a barrier to seal air between the feathers and the skin. This is important in keeping the birds warm, and it also helps to keep them afloat.

Mechanical strength is not a prerequisite for success in the marine environment. Marine organisms float and move with the water, and except where breaking waves are encountered at the surface or along the shore, the mechanical stresses of the water are slight. Many marine organisms are exceedingly delicate and fragile but survive and function well as long as they are surrounded by water.

13.2 Osmotic Processes

Special problems are posed for living creatures if the salt content of their body fluids differs from the salinity of the water that surrounds them. The body fluids of living plants and animals are separated from the seawater by membrane boundaries that are semipermeable, allowing some molecules to move across the membrane boundaries while other molecules cannot. Molecules that move across these membranes do so along a gradient from a region of high concentration of a substance to a region of low concentration of that substance, a process called diffusion. Water molecules pass across the membranes in a special type of diffusion known as **osmosis,** which was discussed in chapter 5.

Most fish have body fluids with a salt concentration that is about halfway between that of fresh water and seawater. In salt water, their tissues tend to lose water as it moves along an osmotic gradient from an internal high water concentration and low salt concentration to the lower water concentration and higher salt concentration of the ocean environment. Fish must constantly expend energy to prevent dehydration and an increase in the salt concentration of their tissues. Fish stay in fluid balance by drinking seawater nearly continually and excreting its salt across their gills. Sharks and rays do not have this problem, because their body fluids have the same approximate salt content as seawater, so that there is no osmotic gradient. These fish maintain a high concentration of urea in their tissues. The urea allows their tissue to retain water, prevents the movement of salt into their bodies, and keeps body fluid at approximately the same salt content as seawater.

The body fluids of many bottom-dwelling organisms, such as sea cucumbers and sponges, are also at the same salt concentration as the seawater. There is no concentration gradient; the water diffuses equally in both directions

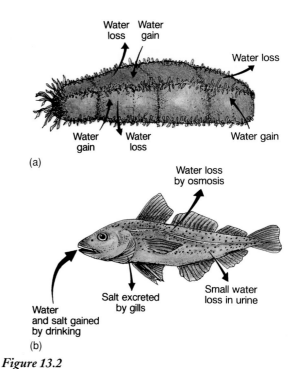

Figure 13.2

(a) The salt concentration of the seawater is the same as the salt concentration of the sea cucumber's body fluids (35‰). The water diffusing out of the sea cucumber is balanced by the water diffusing into it. (b) The salt concentration in the tissues of the fish is much lower (18‰) than that of the seawater (35‰). To balance the water lost by osmosis, the fish drinks salt water, from which the salt is removed and excreted.

across the membranes, and the salt content remains the same on both sides of the membranes. The fish, which does have to overcome an osmotic gradient, and the sea cucumber, which does not, are compared in figure 13.2.

Species may be limited in their geographic distribution by changes in salinity, for many organisms can maintain their salt-fluid balance over only limited salinity ranges. Since there is little change in salinity in deep water, species living below the surface layers are dispersed over large areas with respect to salinity. The surface-dwelling forms are more likely to find salinity barriers in coastal waters, since the salinity varies in bays and estuaries. Successful estuarine animals such as crabs are able to stabilize their osmotic processes by regulating the intake of salt ions. In some areas, a few species are able to survive in high-salinity lagoons and salt marshes, but they are unlikely to reproduce, and so the population must be replenished by new recruits from the sea. Some animals have an extraordinary ability to adapt to large changes in salinity over their life history. Salmon spawn in fresh water but move down the rivers as juveniles to live their adult lives in the sea. After several years (the time depends on the species), the salmon return to their home streams. The Atlantic common eel reverses this process by migrating downstream to spawn in the Sargasso Sea. The new generation of eels spends one to three years at sea, then returns to fresh water to live for up to ten years before migrating seaward. Other fish and crustaceans use the low-salinity coastal bays and estuaries as breeding grounds and nursery areas for their young, then as adults they migrate farther offshore into higher-salinity waters.

13.3 Temperature

The temperature of the deep oceans is low and nearly constant. At the surface and close to shore the water temperature varies with seasonal climate changes and geographic latitude zones. Temperature, like salinity, affects the density of the seawater and also affects its viscosity; density and viscosity are discussed in chapter 4. At polar latitudes, the surface water is cold, more dense, and more viscous, and organisms float more easily. In tropic latitudes, the warm, less-dense, less-viscous water is home for species with more appendages, larger surface areas, and greater gas-bubble production, for the water is less buoyant and offers less resistance to sinking.

When surface conditions produce a warm, low-density surface layer overlying denser water, the water column is stable, and floating organisms stay in the sunlit upper layers. Under these conditions, floating plant cells increase their rates of photosynthesis and reproduction. When surface waters cool and increase their density, they become unstable; the waters mix, overturn, and take the floating organisms with them into deeper water and away from the sunlight. The photosynthesis and reproduction of plants are decreased.

Plants and marine animals other than birds and mammals do not control their body temperatures. They are cold-blooded and their body temperatures vary with environmental conditions. The physiology of marine organisms is regulated by the temperature of the water, and within limits metabolic processes proceed more rapidly in warm water than in cold water. Cold-water forms frequently grow more slowly, live longer, and attain a larger size. To some species, changes in temperature act as signals to spawn or to become dormant. Although the heat capacity of the water restricts temperature fluctuations, the geographic distribution of seawater temperatures is sufficient to affect the distribution of marine organisms.

Seabirds and mammals are warm-blooded and maintain nearly constant body temperatures that are well above the temperature of the seawater. Because these animals are less restricted by the temperature of the water, they often have wider geographic ranges, for example, the annual migration of whales between polar and tropical waters. Some fish, although cold-blooded, are able to conserve heat in their swimming muscles, elevating their body temperatures. Because their muscles work more efficiently at higher temperatures, these fish are able to swim rapidly and cruise long distances in the coldest water, making them efficient predators. Fish of this type include some tuna, mackerel-sharks, the great white shark, and the dolphin fish.

At the greater depths, the uniformity of temperature with latitude creates an oceanwide environment that is unaffected by seasonal changes. At the sea surface, the temperature changes with latitude much the same as the climate changes on land. Annual changes in open sea-surface temperatures are small at the very high and very low latitudes; at the middle latitudes the annual changes in sea-surface temperature show a seasonal fluctuation. Ocean areas close to land undergo still greater changes in surface temperature, due to their shallowness and the influence of the greater annual temperature changes over the adjacent landmasses. Seasonal fluctuation in surface temperature at the middle latitudes is reflected in periods of spring and summer reproduction and growth and in winter dormancy.

On land, the general pattern of climate zones encountered by approaching the poles is also observed by increasing the altitude; the climate at sea level in polar zones is similar to the climate at the top of a high mountain peak at a lower latitude. In the ocean, conditions in surface water at polar latitudes are similar to those found at deeper depths at lower latitudes. Some shallow-water species of the polar seas are found at greater depths at the lower latitudes.

13.4 Pressure

Deep-living organisms, such as worms, crustaceans, and sea cucumbers, are unaffected by the pressure, because they do not have gas-filled cavities or lungs that must be maintained at high pressure or mechanically protected against the pressure of the overlying water. It is possible that pressure may alter metabolic rates and growth at greater depths, but little is known about such effects.

When humans descend into the sea, they need either protection from the pressure that will collapse their chest cavities and lungs or air supplied to the lungs at a pressure equal to the outside water pressure. Submarines and submersibles provide the first type of protection, while scuba (*S*elf-*C*ontained *U*nderwater *B*reathing *A*pparatus) and other commercial diving equipment supply the second type. After breathing gases under pressure, humans may experience the serious problems of decompression sickness and nitrogen narcosis. Divers breathing air under high pressure for extended periods absorb large quantities of gas, particularly nitrogen, into their blood and tissues. If they return to shallow depths of less pressure too quickly, these excess gases form bubbles in the body tissues and blood vessels, causing extreme pain, paralysis, and sometimes death. Excess nitrogen dissolved in the bloodstream also has a narcotic effect, which confuses the diver and restricts his or her ability to function normally. The 1990 depth record for an open dive in seawater using air is 137 m (452 ft).

Air-breathing marine mammals make dives of spectacular depth and duration without encountering such difficulties, because of their abilities to adjust their physiology. Record diving depths and duration for some of the whales and seals are given in table 13.1. These animals have a physiology that permits their blood to absorb more oxygen and to tolerate higher concentrations of carbon dioxide than land mammals. They also have a larger blood volume than nondiving mammals and a vascular shunt system that directs the blood flow through only the brain and heart

Table 13.1		
Diving Depths and Durations for Some Marine Mammals		
Mammal	**Diving record (m)**	**Duration record (minutes)**
sea lion	168	30
porpoise	300	6
bottle-nosed dolphin	450	120
fin whale	500	30
Weddell seal	600	75
elephant seal (male)	1530	77
sperm whale	2250	90

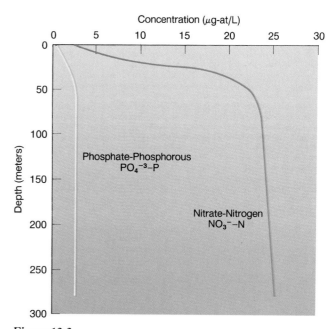

Figure 13.3
Nitrate and phosphate distribution in the main basin of Puget Sound in the late summer. The low surface values are the result of nutrient utilization by unicellular marine plants.

while underwater. Their muscles store additional oxygen and are able to tolerate the buildup of waste products from exertion to a greater degree than those of other animals. During their dives their lungs collapse completely, forcing the air out and preventing the blood from absorbing compressed gases at high pressure; therefore they do not suffer from diving illnesses and are able to rapidly change their depth.

13.5 Gases

Life in the water requires carbon dioxide and oxygen, as does life on land. Carbon dioxide is required by the plants for photosynthesis; it is contributed by the animals and by decay processes, and it is absorbed by the water from the atmosphere. Because seawater has the capacity to absorb large quantities of carbon dioxide, there is no shortage of carbon dioxide for the plant life. Also, carbon dioxide's role as a buffer limits the ocean's pH range, which keeps it a stable environment for living organisms (see chapter 5).

Oxygen is required by all organisms to liberate energy from organic compounds. Oxygen is available only at the ocean surface as a by-product of photosynthesis and from the atmosphere. Life below the surface depends on the vertical circulation processes (discussed in chapter 6) to replenish the oxygen at depth. The amount of oxygen in the water influences the distribution of organisms and is influenced by the temperature, salinity, and pressure of the water. Shallow tidal pools and bays on warm, quiet days increase in temperature and salinity, decreasing the ability of the water to hold oxygen, forcing motile animals out, and limiting these areas to the organisms that can successfully tolerate these changes. The bottoms of deep, isolated basins may be so low in oxygen that only non-oxygen-requiring, or **anaerobic,** bacteria can survive there (refer back to chapter 5).

13.6 Nutrients

Nitrate (NO_3^-) and phosphate (PO_4^{-3}) nutrients are required by the sea's plant life. They are the fertilizers of the sea and are stripped from the surface layers by the plants, which incorporate them into their tissues. These nutrients are liberated at depth by the decay of plant as well as animal tissues, or they are returned to the water in the form of waste products of herbivores and carnivores (see fig. 13.3). Vertical circulation and mixing transport the nutrients back to the surface in upwelling areas where life is abundant. Estuaries and coastal waters, where nutrients are supplied by land runoff and mixing from the continental shelf's shallow seafloor, are also rich with organisms. Plant populations are limited by the lack of any essential nutrient; if the concentration of such a nutrient falls below the minimum required, the population's growth ceases until the nutrient is replenished. Nutrients were introduced in chapter 5 and nutrient cycles will be discussed in chapter 14.

13.7 Light and Color

Sunlight

Without light there are no plants; therefore the distribution of plants in the oceans is light limited. Plant life is restricted to the **photic zone** where there is sufficient light energy for the process of photosynthesis. The depth of the photic zone, about 200 m (660 ft) in clear ocean water, is controlled by factors discussed previously in chapters 1, 4, and 6, including (1) the angle at which the sun's rays hit the

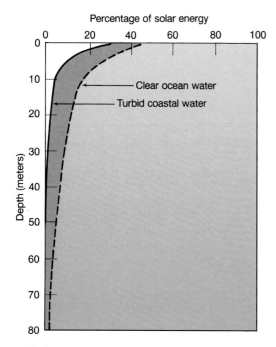

Figure 13.4

The percentage of solar energy available at depth in clear and turbid water.

earth's surface, which is related to latitude and change of season; (2) the different rates at which the wavelengths of light are absorbed, which is determined by the properties of water; and (3) the suspended particulate material present, which affects the rate of absorption. Below the photic zone is the **aphotic zone,** the zone in which there is no photosynthesis.

However, the presence of light does not guarantee plant life; nutrients must be available. It is because of a lack of nutrients that so much of the open ocean, exposed to high-intensity sunlight, is considered for all practical purposes a biological desert. Life is more abundant along the coasts and over the continental shelves because of the larger quantities of dissolved nutrients. Waves and currents of the coastal zone stir the bottom and mix up silt with the nutrients. The silt particles absorb and scatter light, reducing the depth to which it penetrates. As the single-celled plants reproduce, their increased numbers also act to limit light penetration, so that the photic depth may be reduced to less than 50 m (167 ft). The penetration of sunlight in clear and turbid seawater is compared in figure 13.4.

Bioluminescence

Another source of light is present in the oceans, the organisms themselves. On a dark night, when the wake of a boat is a glowing ribbon, and disturbed fish leave a trail of light, or the water flashes as oars dip and a person's hands glow briefly as a net is hauled in, living organisms are producing the light. The light is **bioluminescence** produced by the interaction of the compound **luciferin** and the enzyme **luciferase.** This phenomenon is often incorrectly referred to as phosphorescence; it has nothing to do with phosphorus or with the absorption of radiation, but is a chemical

reaction that produces light with a 99% efficiency. The same phenomenon is seen on land in the flashing of a firefly or the ghostly glowing of a fungus in the woods.

In the sea, the agitation of the water disturbs microscopic bioluminescent organisms, causing them to flash and produce glowing wakes and wave crests. Animals that feed on these organisms often concentrate the chemicals in their tissues and also glow. Jellyfish glow in this way and so do one's hands if they come in contact with crushed tissue. Other bioluminescent organisms in the sea include squid, shrimp, and some fish. Many middepth and deep-water fish carry light-producing organs, or **photophores;** some have patterns on their sides, possibly for identification. Others show photophores on their ventral surfaces, making them difficult to see from below against the light surface water, and still others have glowing bulbs dangling below their jaws or attached to flexible dorsal spines, acting as lures for their prey. The flashlight fish, found in the reefs of the Pacific and Indian oceans, has a specialized organ below each eye that is filled with light-emitting bacteria. These fish are known to use the light to see, communicate, lure prey, and confuse predators.

Color

Some sea animals are transparent, allowing them to blend with their water background, for example, jellyfish and most of the small floating animals in the surface layers, while other animals, particularly the fish, use color in many ways. In the clear waters of the tropics, where light penetrates to deeper depths, bright colors play their greatest role. Some brightly colored fish match the colors of the corals so well they become nearly invisible, and bold coloration increases during the breeding periods of some species. Other fish conceal themselves with bright color bands and blotches that disrupt the outline of the fish and may draw the predator's attention away from a vital area to a less important spot, for example, a black stripe over the eye and an eye spot on a tail or fin. Another use of bright color is to send a warning; organisms that sting, taste foul, have sharp spines, or poisonous flesh are often striped and splashed with color, for example, sea slugs and some poisonous shellfish. Among fish that swim near the surface in the well-lighted surface water, for example, herring, tuna, and mackerel, dark backs and light undersides are common. This color pattern allows the fish to blend with the bottom when seen from above and with the surface when seen from below. See figure 13.5.

In temperate regions, coastal waters are more productive and more turbid, and there is less light penetration. Drab browns and grays are concealing against the kelp beds of temperate waters, and cold-water bottom fish are usually uniform in color with the bottom, or speckled and mottled with neutral colors. The flatfish are well known for their ability to change their color, having skin cells that expand and contract to produce color changes (see fig. 13.6). Their extraordinary color-changing ability enables them to conceal themselves by matching the bottom type on which they live (see fig. 13.7). Squid may be the ocean's masters of

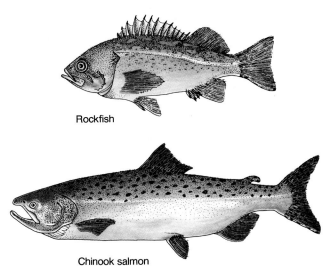

Figure 13.5
Viewed from above, the dark dorsal surface of the fish blends with the seafloor; viewed from below, the light ventral surface blends with the sea surface. This type of coloration is known as countershading.

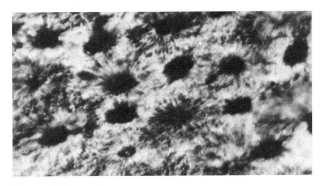

Figure 13.6
Pigment cells from a section of fish skin.

Figure 13.7
The winter flounder resting on a checkerboard pattern shows its use of camouflage.

color change; they are able to flash and change color patterns with great rapidity, expanding and contracting pigment cells. Many squid also have bioluminescent cells, and some have colonies of light-emitting bacteria covered by a flap of skin. When combined, these allow the squid to display hundreds of different and complex color patterns and sequences, allowing them to change color and disappear almost instantaneously.

Some species of deep-water shrimp are red when seen at the surface, but below the level of penetration of red wavelengths of light their red pigment absorbs blue and green light; little is reflected and the animals are dark and inconspicuous. In the deep ocean without light, color is of little importance except perhaps when combined with bioluminescence. Of course, we do not know how the animals see the colors, and there may be roles that color plays in the sea we do not know or understand. Color is thought to be important in species recognition, courtship, and possibly in keeping schools of fish together.

13.8 Circulation

The ocean's water is in constant motion. It is moved and mixed by currents (chapter 8), waves (chapter 9), and tides (chapter 10). Below the surface layers, the ocean environment is very uniform, providing marine organisms with similar conditions of temperature and salinity in any ocean at any time. Oceanic circulation brings food and oxygen, replenishes nutrients, and removes waste; it disperses floating organisms and scatters the reproductive stages of swimmers and attached forms.

Those plants and animals that drift rather than swim are carried along by the currents and run the risk of being carried out of a suitable habitat by either vertical or horizontal movement. However, analysis of their remains on the seafloor and observations of living populations show that this does not always happen. Populations of drifting animals appear to take advantage of their ability to move in the vertical direction either by swimming or by changing their buoyancy. They are able to maintain their place horizontally in ocean space by moving away from the surface during the day to depths where a current flows in a direction opposite to the surface current. They then move upward at night to be carried back to their starting position. Organisms also appear to maintain their position by adding to their population on the upstream side of a current to balance losses on the downstream side. This pattern is related to the new supply of food and nutrients brought into the population by the current upstream, in contrast to the food-depleted water downstream.

Vertical water motions in the sea are much slower than horizontal motions, but small displacements of organisms in the vertical direction can mean substantial changes in light, salinity, temperature, and nutrient supply. If the vertical flow is upward, it counteracts the tendency of organisms and other particulate matter to sink. In this way, light-dependent organisms are held in the photic zone. The upward motion of the water also supplies nutrients to the photic zone to promote plant growth. At the same time, these upward flows decrease the temperature of the surface waters and return water with a low oxygen content to the surface, where oxygen is replenished by photosynthesis and atmospheric exchange.

A downward vertical flow under an area of surface convergence accumulates a population of organisms as the surface flows move toward the area of downwelling. If the organisms cannot increase their buoyancy to compensate for the downward current, they are carried down to changes in light, temperature, salinity, nutrients, and gases. Areas of downwelling are usually regions of low plant growth, but at the surface convergence, the accumulation of organisms provides a rich feeding ground for carnivores.

13.9　Barriers and Boundaries

In the sea, species and populations are isolated from one another by barriers where the properties of the water change abruptly. Near-surface boundaries in the water column may be sharp; for example, rapid changes with depth in temperature (thermocline), density (pycnocline), and salinity (halocline) (refer to chapter 6). Light intensity also changes rapidly in the vertical direction (refer to chapter 4). These boundaries are barriers for marine organisms, because they do not survive if displaced through such a boundary. The effects of these barriers decreases at deeper depths, where water properties become more homogeneous.

Similar barriers exist in the horizontal direction, where surface water of one type is adjacent to surface water of another type (refer again to chapter 6). These boundaries are often associated with zones of surface convergence and divergence in the ocean and also occur in estuaries, between land-derived fresh water and seawater in the coastal zone. When they are associated with a rapidly flowing current of one type of water moving through or adjacent to another type, the boundaries are sharp; for example, populations that do well in the warm, saline waters of the Gulf Stream may die if they are displaced into the cold, less-saline Labrador Current water between the coast and the Gulf Stream.

Other boundaries are controlled by the topography of the seafloor (see chapter 2). Ridges that isolate one deep-ocean basin from another prevent the deeper water in the basins from freely exchanging, and water of dissimilar characteristics and populations may exist on either side of a submarine ridge. In other cases, the water and the populations in the two basins may be similar, but the populations are not able to move between the basins because the elevation of the ridge that separates them forces the animals to change their depth and pass upward into water with properties that they cannot tolerate. Isolated seamounts with their peaks in shallow water may support specific isolated communities of sea life in much the same way that mountain tops on land support widely separated arctic and alpine communities.

Lateral topographic barriers also isolate populations. Near-surface and surface species of the tropic regions are prevented from moving between the Atlantic and Pacific oceans by the land barrier of Central America. Tropical surface species such as sea snakes cannot migrate around the continental landmasses of North and South America, because to do so they must pass through regions of much colder water. Africa also acts as a barrier, keeping the tropical species of the Indian Ocean from communicating freely with the tropical species of the Atlantic, because the water south of Africa is too cold for the tropical species of either ocean.

13.10　Bottom Types

Although the properties of the seawater are critical for the survival of marine organisms, for many plants and animals the type of ocean bottom—rock, mud, sand, or gravel—is equally important. A seaweed that requires rock for attachment is unable to live in sand, and a burrowing worm from a mud flat or a shrimp from a sand beach cannot survive on a rocky reef. The material of the seafloor, or **substrate,** provides food, shelter, and attachment sites, each substrate type providing suitable living space for a different group of organisms. Substrates show greater variety along the shallow coastal areas; sandbars, mud flats, rocky points, and stretches of gravel and pebble are frequently found along the same strip of coastline. Further seaward, as the seafloor becomes more distant from the sea surface, the particle size of the sediments and the amount of organic matter associated with the substrates decrease; the substrate becomes more uniform. The decline in the variety of the substrate is matched by a decrease in the animal mass.

Organisms living attached to the seafloor often modify their habitats, providing food, shelter, and additional surfaces for the attachment of still other organisms. Forests of large seaweed attached to rocky bottoms in 20 m (66 ft) of water and eel grass beds in shallow, quiet, sandy, or muddy bays both provide such environments, but for quite different populations of organisms. Some crabs moving across the seafloor carry anemones attached to their backs, and some sea anemones harbor specialized fish within their tentacles. The most outstanding example of biological modification of a substrate is the tropical coral reef; here the organisms create a specialized environment over the stony skeletons of other organisms (see chapter 17).

13.11　Environmental Zones

Because the marine environment is so large and complex, biological oceanographers, marine biologists, and ecologists interested in the sea divide the marine environment into subunits called zones. The zone classifications used here are based on the system developed by Joel Hedgpeth in 1957; they are shown in figure 13.8. The water environment is the **pelagic zone,** and the seafloor environment is the **benthic zone.** The pelagic zone is divided into the coastal or **neritic zone** above the continental shelf, and the **oceanic zone,** or deep water away from the influence of land. The oceanic zone is then subdivided by depth as follows. The surface waters to 200 m (660 ft) are the **epipelagic zone; the mesopelagic zone** is the twilight zone between 200 m (660 ft) and 1000 m (3300 ft); the **bathypelagic zone** extends to 4000 m (13,200 ft), and the **abyssopelagic zone** extends to the deepest depths. All

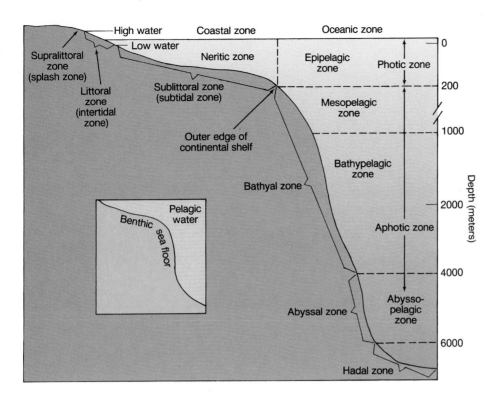

Figure 13.8
Zones of the marine environment.

oceanic subdivisions, except for the epipelagic zone, are aphotic, or without light. The photic zone coincides roughly with the epipelagic and neritic zones. Locate each zone in figure 13.8.

The seafloor, or benthic environment, is subdivided into comparable zones. Tidal fluctuations at the shoreline define the **supralittoral zone,** or **splash zone,** which lies just above the high-water mark and is covered by the sea only during the highest spring tides or by wave spray, and the **littoral zone,** or **intertidal zone,** which lies between high and low water and is covered and uncovered once or twice each day. The **sublittoral zone,** or **subtidal zone,** extends out along the continental shelf. The supralittoral, littoral, and inner portion of the sublittoral zones occupy the same area as the benthic photic zone because sufficient light is present to support single-celled plants and large benthic plants.

Within the aphotic zone are the deeper portions of the subtidal zone; the **bathyal zone,** extending from 200 m (660 ft) to 4000 m (13,200 ft) and coinciding with the continental slope, and the **abyssal zone,** between 4000 m and 6000 m, or roughly the area over the abyssal plain. The **hadal zone** lies below 6000 m (19,800 ft) and is associated with the trenches and deeps (see again fig. 13.8).

The properties of the littoral and epipelagic zones are keyed to latitude; keep in mind that these zones differ markedly, depending on whether they are at polar, temperate, or tropic latitudes. Substrate plays a basic role in benthic zones, and there is much less variety in substrates in the deeper zones. Life in all the zones is influenced by variations in temperature, light, dissolved gases, nutrients, and all the factors discussed in this chapter.

13.12 Classification of Organisms

All of these ocean zones together make up the varied marine environment, which is inhabited by a wide variety of organisms uniquely adapted to it. These organisms are divided into groups to promote ease of identification and to increase our understanding of the relationships that exist among them. Classic taxonomic categories are listed in table 13.2. A simple and practical method divides all marine organisms into three groups, based on where and how they live. Plants and animals that float or drift with the movements of the water are the **plankton.** Those that live attached to the bottom or on or in the bottom are the **benthos,** and the animals that swim freely and purposefully in the sea are the **nekton.** Chapters 15, 16, and 17 use this three-part system.

13.13 Practical Considerations: Modification and Mitigation

At the same time that we have been acquiring greater understanding of how the physical, chemical, and geological factors discussed in this chapter interact to produce the many and different environments of the ocean, we have been bringing greater and lasting changes to these ocean environments, especially in coastal bays and estuaries, where environments are typically small-scaled and varied. In some cases, using our knowledge of organisms and their preferred habitats, we have altered an area to enhance the

Table 13.2

Taxonomic Categories of Some Marine Organisms

	Killer whale	Northern fur seal	Pacific (Japanese) oyster	Giant octopus	Sea lettuce	Giant kelp
kingdom	Animalia	Animalia	Animalia	Animalia	Plantae	Plantae
phylum	Chordata	Chordata	Mollusca	Mollusca	Chlorophyta	Phaeophyta
class	Mammalia	Mammalia	Bivalvia (Pelecypoda)	Cephalopoda	Chlorophycae	Phaeophycae
order	Cetacea	Carnivora (Pinnipedia)	Anisomyaria	Octopoda	Ulvales	Laminariales
family	Delphinidae	Otariidae	Ostreidae	Octopodidae	Ulvaceae	Lessoniaceae
genus	*Orcinus*	*Callorhinus*	*Crassostrea*	*Octopus*	*Ulva*	*Macrocystis*
species	*Orcinus orca*	*Callorhinus ursinus*	*Crassostrea gigas*	*Octopus dofleini*	*Ulva lactuca*	*Macrocystis pyrifera*

populations of organisms we consider desirable over those we consider undesirable. We build artificial reefs using old car bodies, bags of old shells, or chunks of fractured concrete to encourage the growth of organisms that not only do well in a reef's protected nooks and crannies, but are well suited to recreational and sport fishing as well as commercial harvesting.

Development and modification of estuaries and bays is often done at the expense of wetlands and mud flats; refer back to chapter 12. Our engineering techniques allow us to move sediment, change the slope of the bottom, and alter substrates. We deepen channels, build marinas and protective breakwaters, and bulkhead the land to prevent its erosion. These activities do not remove marine habitats as much as they change them, substituting a new habitat and its biological communities for the previous habitat complex. In some cases our activities result in the creation of nonmarine uplands, and in this case there is a true loss of marine environment.

Our ability to modify or reengineer the oceans' shallow-water environments also allows us to create new marshes, tide flats, and special habitats from less-productive or less-desirable areas and to reestablish communities of marine plants and animals. We can and do choose which habitats and which organisms are to be conserved and which are to be sacrificed.

Present policies at national, state, and local levels, at a minimum seek to conserve our existing marine habitats, and to improve them when possible, while at the same time increasing populations promote development of coastal zones. Attempting to balance human and natural needs has led planners and developers to the coastal-management concept of **mitigation.**

When development projects alter or destroy an environment, management authorities at the local, state, or federal level may choose to enforce mitigation of these effects by requiring the developer to purchase an area of the same type as that to be developed and to arrange for it to be held in its natural state, or the developer may be required to reengineer an area to resemble what has been lost. Such required projects are examples of compensatory mitigation; compensation is paid for change and destruction. Projects that are voluntarily undertaken to improve coastal and shore areas are considered noncompensatory mitigation.

A successful mitigation project requires a thorough knowledge of the physical requirements needed to support the communities of plants and animals that the mitigation seeks to enhance. Any characteristics of the new environment that interfere with the mitigation process must be changed if the mitigated area is to sustain itself in the future. After the area has been restructured it must be monitored, and changes must be made as needed to maintain the new environment. If these preproject and postproject studies are not made, the mitigation effort is likely to fail.

While mitigation does preserve some habitats and species, it can only approximate the lost environment, not duplicate it. The result is still a shift or a change in habitat produced by human pressures.

Development pressures in deep-sea areas associated with mining and energy projects are still in the future. Tropical OTEC plants (chapter 6) will force cold, nutrient-rich water to the sea surface, changing the surface productivity; and manganese nodule mining (chapter 2) will increase deep-sea turbidity, changing the environment for bottom organisms in mined areas. However, because of the vast size of deep-sea areas with common properties, changes in deep water are likely to be less significant than changes in the coastal zone. How mitigation might be undertaken in the open ocean has not yet been considered.

Spartina: *Valuable and Productive or Invasive and Destructive?*

Spartina alterniflora, known as smooth cordgrass, many spiked cordgrass, and salt marsh cordgrass, is a deciduous, perennial flowering plant native to the Atlantic and Gulf coasts of the United States. It is the dominant native species of the lower salt marshes along the Atlantic seaboard from Newfoundland to Florida, and on the Gulf Coast from Florida to East Texas. It grows in the intertidal zone from mean higher high water to 1.8 m (6 ft) below mean higher high water.

These natural salt marshes are among the most productive habitats in the marine environment. Nutrient-rich water is brought to the wetlands during each high tide, making a high rate of food production possible. As the seaweed and marsh grass leaves die, bacteria break down the plant material, and insects, small shrimplike organisms, fiddler crabs, and marsh snails eat the decaying plant tissue, digest it, and excrete wastes high in nutrients. Numerous insects occupy the marsh, feeding on living or dead plant tissue, and red-winged black-birds, sparrows, rodents, rabbits, and deer feed directly on the cordgrass. Each tidal cycle carries plant material into the offshore water to be used by the subtidal organisms.

Spartina is an exceedingly competitive plant. It spreads primarily by underground stems; colonies form when pieces of the root system or whole plants float into an area and take root, or when seeds float into a suitable area and germinate. It establishes itself on substrates ranging from sand and silt to gravel and cobble and is tolerant of salinities ranging from near fresh to salt water (35‰). *Spartina* is able to tolerate high salinities because salt glands on the surface of the leaves remove the salt from the plant sap, leaving visible white salt crystals. Because of the lack of oxygen in marsh sediments, they are high in sulfides that are toxic to most plants. *S. alterniflora* has the ability to take up sulfides and convert them to sulfate, a form of sulfur that the plant can use; this ability makes it easier for the grass to colonize marsh environments. Another adaptive advantage is its bio-chemical photosynthetic pathway that uses carbon dioxide more efficiently than most other plants.

These characteristics make *Spartina alterniflora* a valuable component of the estuaries where it occurs naturally. The

Box figure 13.1
A naturally occurring *Spartina* marsh.

plant functions as a stabilizer and sediment trap and as a nursery area for estuarine fishes and shellfishes. Once established, a stand of *Spartina* begins to trap sediment, changing the substrate elevation, and eventually the stand evolves into a high marsh system where *Spartina* is gradually displaced by higher-elevation, brackish-water species. As elevation increases, narrow, deep channels of water form throughout the marsh (see Box fig. 13.1). Along the East Coast *Spartina* is considered valuable for its ability to prevent erosion and marshland deterioration; it is also used for coastal restoration projects and the creation of new wetland sites.

Spartina alterniflora has been introduced to and naturalized in Washington, Oregon, California, England, France, New Zealand, and China. *Spartina* was carried to Washington State in packing material for oysters transplanted from the East Coast in 1894. Leaving its insect

predators behind, the cordgrass has been spreading slowly and steadily along Washington's tidal estuaries, crowding out the native plants and drastically altering the landscape by trapping sediment. It turns tidal mud flats into high marshes inhospitable to many fish and waterfowl that depend on the mud flats. By 1988 it covered 1650 acres in Washington's Willapa Bay; it had spread to 2500 acres by 1991, and state officials predict that by 2010 it will cover most of the bay's 30,000 acres if left unchecked (see Box fig. 13.2). It is already hampering the oyster harvest, and it interferes with recreational use of beaches and waterfronts.

Spartina has been transplanted to England and New Zealand for land reclamation and shoreline stabilization. In New Zealand the plant has spread rapidly, changing mud flats with marshy fringes to extensive salt meadows and reducing the number and kinds of birds

continued

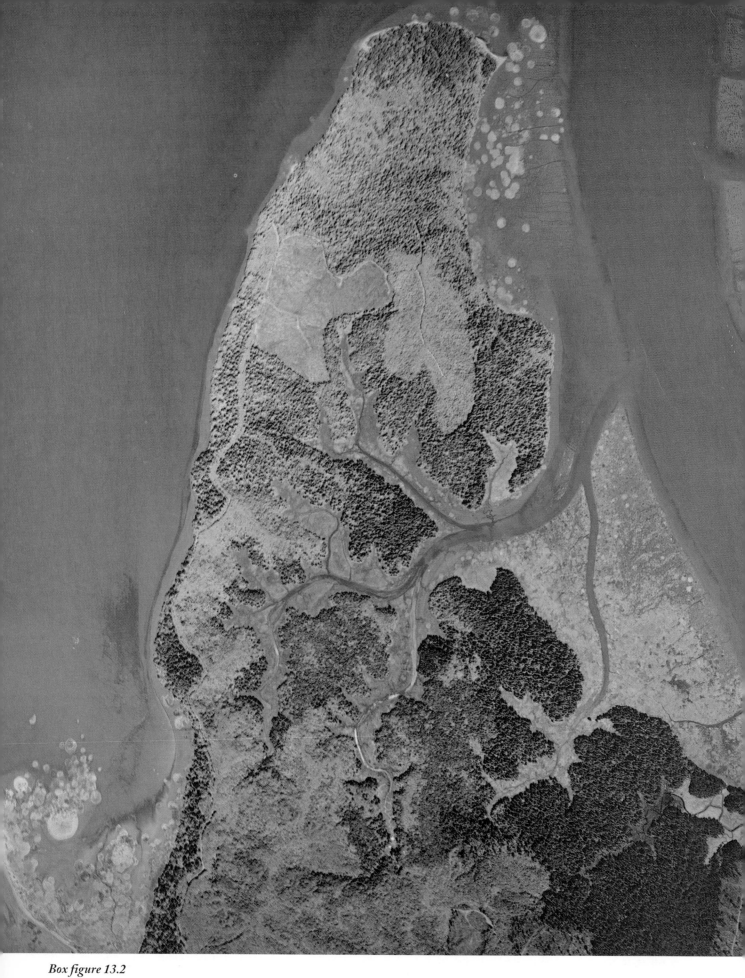

Box figure 13.2

Circular patches of *Spartina* spread along the mud flats of Willapa Bay, Washington. The large circle at lower left is thought to be the original colony. This photograph was taken with infrared film and is a false color image.

and animals that use the marsh. Another species of *Spartina* (*S. maritima*) occurs in marshes along the coasts of Europe and Africa. *S. alterniflora* was introduced into Great Britain from eastern North America in about 1800 and spread to form large colonies. The native species and the introduced species existed together throughout the nineteenth century, and by 1870 a sterile hybrid that reproduced by underground stems appeared. About 1890, a vigorous seed-producing form was derived naturally from this hybrid and spread rapidly along the coasts of Great Britain and northwestern France.

Efforts to control *Spartina* outside its natural environment have included burning, flooding, shading plants with black canvas or plastic, smothering the plants with dredged materials or clay, applying herbicide, and repeated mowings. Little success has been reported in New Zealand and England; Washington State has declared *Spartina* a "noxious weed" and set up tests using mowing and herbicide applications to control its spread.

Preliminary work has begun to determine the feasibility of using insects as biological controls, but effective biological controls are considered ten years away. Even with a massive effort it is doubtful that complete eradication of *Spartina* from nonnative habitats is possible, for it has become an integral part of these shorelines and estuaries during the last 100 to 200 years. A management plan that includes mowing for small, accessible patches and biological controls for long-term regulation may be the most realistic approach.

Summary

Organisms living in the sea are buoyed and supported by the seawater. Adaptations for staying afloat include low-density body fluids, gas bubbles, gas-filled floats, swim bladders, oil and fat storage, and extended surface areas and appendages.

Most marine fish lose water by osmosis. They drink continually and excrete salt to prevent dehydration. Sharks have the same concentration of salt in their tissues as there is in seawater. They therefore do not have a water-loss problem. Salinity is a barrier to some organisms; others can adapt to large salinity changes.

Temperature affects density, viscosity, and the water's buoyancy, as well as the stability of the water column. The body temperature and metabolism of all marine organisms, except for birds and mammals, are controlled by the sea temperature. Some fish conserve heat in their body muscles and elevate the temperature in these muscles. Temperatures at depth are uniform; sea-surface temperatures change with latitude and seasons.

Changes in pressure affect organisms with gas-filled cavities. Marine mammals have a unique ability to undergo large pressure changes due to their physiology and body chemistry. The swim bladders of fish are affected by pressure changes, and the fish must change depth slowly.

The carbon dioxide–oxygen balance in the oceans influences the distribution of all organisms. The availability of nutrients and light limits plant populations. The depth of light penetration in the oceans is controlled by the angle of the sun's rays, the properties of the water, and the material in the water. Light limits plant life, but nutrients are also required. Some organisms produce chemical light, known as bioluminescence. Animals use color for concealment and camouflage and also to warn predators of poisonous flesh and bitter taste.

Winds, tides, and currents mix the water. Moving water carries food and oxygen, removes waste, and disperses organisms. Floating populations are not necessarily scattered, but keep their place due to movements between surface currents and deeper currents. Upwellings supply nutrients and hold plants in the surface layers. Downwellings are regions of low plant growth.

Barriers for marine organisms include water properties, light intensity, zones of convergence and divergence, seafloor topography, and geography. Different substrates provide food, shelter, and attachment for different groups of organisms.

The marine environment is subdivided into zones. The major environments are the benthic and pelagic zones; there are numerous subdivisions of each of these zones.

The organisms of the sea are classified for identification and relationship. Organisms are also grouped as plankton, nekton, and benthos.

Development of marine areas, with the consequent loss of habitat and therefore populations, has led to the concept of mitigation, under which developers are required to preserve or replace habitats in an effort to maintain and preserve species.

Key Terms

buoyancy	epipelagic zone
osmosis	mesopelagic zone
anaerobic	bathypelagic zone
photic zone	abyssopelagic zone
aphotic zone	supralittoral zone/splash zone
bioluminescence	littoral zone/intertidal zone
luciferin	sublittoral zone/subtidal zone
luciferase	bathyal zone
photophore	abyssal zone
substrate	hadal zone
pelagic zone	plankton
benthic zone	benthos
neritic zone	nekton
oceanic zone	mitigation

Study Questions

1. How do so many delicate and fragile organisms exist in the oceans without damage?
2. What will happen to the body fluids of a frog placed in seawater? A sea cucumber placed in fresh water?
3. Discuss the effect of temperature on the distribution of organisms. Consider changes with latitude and with depth.
4. Although seals and whales are mammals, they do not suffer from either decompression sickness or nitrogen narcosis during deep dives of long duration. Explain why.
5. How does the role of bioluminescence differ from the role of sunlight in the sea?
6. Compare the flotation problems of a many-armed organism with those of an organism without arms but of the same density. Consider both organisms in 4°C water and in 20°C water.
7. In what ways does upwelling contribute to increasing the populations of surface organisms? What properties of seawater act as barriers for marine organisms?
8. Explain why the substrate of the seafloor becomes less diversified as one moves from the shore to the deep ocean.
9. What characteristics determine whether a plant or an animal belongs to the plankton, the nekton, or the benthos?
10. What properties of seawater act as barriers for marine organisms?
11. Why are *Spartina* marshes along the east coast considered productive while those along the west coast are considered destructive?
12. Why is the neritic zone of primary importance to the world's commercial fishing industries?
13. Find an example of mitigation being used in your community and discuss its effect.
14. How does countershading aid the survival of fish in nearshore areas?
15. How do rapidly increasing populations of single-celled plants limit their own growth?

Suggested Readings

Bertness, M. D. 1992. The Ecology of a New England Salt Marsh. *American Scientist* 80 (3): 260–68.

Fitzgerald, L. M. 1990. Seven Underwater Wonders of the World. *Sea Frontiers* 36 (6): 8–21.

Lerman, M. 1986. *Marine Biology, Environment, Diversity and Ecology.* Benjamin-Cummings, Menlo Park, Calif. 535 pp.

Life in the Sea, readings from *Scientific American,* 1982. Freeman, San Francisco. 248 pp. (Articles on marine organisms, the condition in which they live, and their food resources.)

MacLeish, W. H., ed. Fall 1980. *Oceanus* 23 (3). (Issue includes articles on the various senses of organisms of the sea.)

Nealson, K., and C. Arneson. 1985. Marine Bioluminescence: About to See the Light. *Oceanus* 28 (3): 13–18.

Nybakken, J. 1989. *Marine Biology, An Ecological Approach,* 2d ed. Harper & Row, New York. 514 pp.

Siezen, R. 1986. Cuttlebone: The Buoyant Skeleton. *Sea Frontiers* 32 (2): 115–22.

Sumich, J. L. 1992. *An Introduction to the Biology of Marine Life,* 5th ed. Wm. C. Brown, Dubuque, Ia. 449 pp.

Ward, P., L. Greenwald, and O. E. Greenwald. 1980. The Buoyancy of the Chambered Nautilus. *Scientific American* 243 (4): 190–203.

14

Production and Life

14.1 Primary Production
Gross and Net
Standing Crop
14.2 Controls on Primary Production
Light
Nutrients
Nutrient Cycles
14.3 Global Primary Production
14.4 Measuring Primary Production
14.5 Total Production
Food Chains and Food Webs

Box: Satellite Measurements

Trophic Pyramids
Other Systems
14.6 Practical Considerations: Human Concerns

Summary
Key Terms
Study Questions
Study Problems
Suggested Readings

One's first reaction to a close view of the life of the sea is confusion, unease. How can things be so beautiful, with such intricate balance and symmetry in their infinitely varied forms and, at the same time, seem so hostile, so threatening, ready to lunge, to snatch life from each other? So many of the moving parts of sea life seem to be designed, exquisitely tooled, for nothing but destruction and devouring. It is disconcerting, almost as though we have had the wrong idea about beauty and harmony: living things so evidently aimed at each others' throats should not have, as these things do, the aspect of pure, crystalline enchantment.

Perhaps something is wrong in the way we look at them. From our distance we see them as separate, independent creatures interminably wrangling, as a writhing arrangement of solitary adversaries bent on killing each other. Success in such a system would have to mean more than mere survival: to make sense, the fittest would surely have to end up standing triumphantly alone. This, in the conventional view, would be the way of the world, the ultimate observance of nature's law. It was to delineate such a state of affairs that the hideous nineteenth-century phrase "Nature red in tooth and claw," was hammered out.

What is wrong with this view is that it never seems to turn out that way. There is in the sea a symmetry, a balance, and something like the sense of permanence encountered in a well-tended garden.

Lewis Thomas,
from "Sensuous Symbionts of the Sea"

Surf grass produces oxygen, a by-product of photosynthesis.

*I*n all the waters of all the oceans, organisms prey on each other: big fish eat little fish, and little fish eat littler fish. These series of prey-predator relationships are called food chains, and each food chain has a beginning and an end. We are generally familiar with the predators at the upper ends of the chains (for example, salmon, tuna, swordfish, and seals), but without the beginnings of the chains, such large carnivores could not exist. The plankton are at the lower ends of the open-ocean food chains, and the microscopic plant plankton are the first link. The biological oceanographer devotes much time and study to this first link, for understanding the variation in abundance of the phytoplankton is the key to understanding the ocean's productivity, or the rate at which organic material is produced in the sea.

14.1   Primary Production

Gross and Net

Like plants on land, single-celled plant plankton, or **phytoplankton,** require sunlight, nutrients or fertilizers, carbon dioxide gas, and water. The nutrients and carbon dioxide are dissolved in the water surrounding and supporting the phytoplankton. These plant cells contain the pigment **chlorophyll,** which traps the sun's energy for use in **photosynthesis.** The photosynthetic process converts the carbon dioxide and water to high-energy organic compounds, which form new plant material. The production of new plant material by photosynthesis is termed **primary production.** The total amount, or mass, of organic material produced by photosynthesis is the **gross primary production** of the sea.

Photosynthesis is represented by the equation

$$6\,CO_2 \;+\; 6\,H_2O \xrightarrow[\text{chlorophyll}]{\text{solar energy}} C_6H_{12}O_6 + \; 6\,O_2$$

6 molecules carbon dioxide	+	6 molecules water	$\xrightarrow[\text{chlorophyll}]{\text{solar energy}}$	1 molecule sugar	+	6 molecules oxygen

The sugars produced by photosynthesis are broken down by the cells with the addition of oxygen to yield energy, carbon dioxide, and water in the process known as **respiration.** Respiration provides both plants and animals with energy for their life processes.

Respiration is represented by the equation

$$C_6H_{12}O_6 \;+\; 6\,O_2 \;\rightarrow\; 6\,CO_2 \;+\; 6\,H_2O + \text{life support energy}$$

1 molecule sugar	+	6 molecules oxygen	→	6 molecules carbon dioxide	+	6 molecules water	+	life support energy

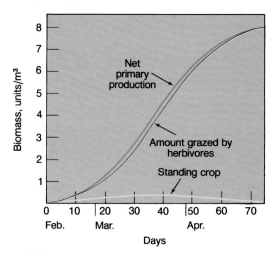

Figure 14.1

Net primary production is balanced by the grazing of herbivores. Both populations increase during the spring; the standing crop remains nearly constant throughout the year.

After the organic material required for respiration has been deducted, the gain in new organic material by a plant population is the population's **net primary production.** The net primary production is available for consumption to the ocean's plant eaters or herbivores and for decomposition to its bacteria.

Net and gross primary production are usually reported for a unit volume of water or the volume of water under a fixed area of the sea surface and over a given period of time. The organic matter produced can be expressed as the number of organisms or as their weight, known as **biomass.** Because organic substances are based on the element carbon, primary production is most often given as the mass of dry weight organic carbon in grams, produced under a square meter of sea surface over a time of years, months, weeks, or days, or $gC/m^2/time$, the rate of change of biomass.

Standing Crop

The total plant biomass under any area of sea surface at any instant in time is known as the **standing crop.** It is the result of growth, reproduction, death, and grazing. If the standing crop on two successive days shows an increase, then there is net production, but if the standing crop shows no change over the two days, the net production has not necessarily remained at zero but is the result of herbivores grazing the phytoplankton population at the same rate as the net primary production increases it. This relationship is shown in figure 14.1.

14.2   Controls on Primary Production

Light

At the polar latitudes, changes in the duration of daylight and darkness mark the seasonal cycle. During the summer days of nearly constant daylight, the light intensity and the

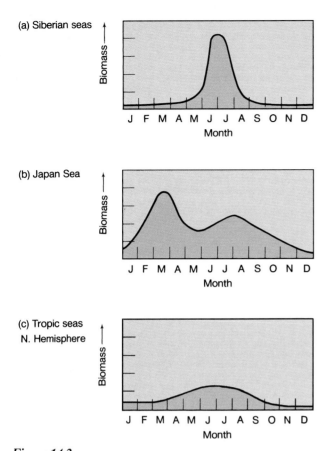

(a) Siberian seas

Biomass

J F M A M J J A S O N D
Month

(b) Japan Sea

Biomass

J F M A M J J A S O N D
Month

**(c) Tropic seas
N. Hemisphere**

Biomass

J F M A M J J A S O N D
Month

Figure 14.2

Phytoplankton growth cycles vary with the light available at different latitudes. At high latitudes (a), growth is limited to a brief period in midsummer. Seasonal light changes at the middle latitudes (b) increase growth in early spring and continue it through the summer. Sunlight levels vary little at tropic latitudes (c), where phytoplankton growth is nearly uniform throughout the year.

penetration of light below the surface are low, because of the angle at which the sun's rays strike the earth above 66 1/2°N and 66 1/2°S. For the remainder of the year, little sunlight is available at these latitudes. This results in a short period of rapid plant growth during the midsummer, shown in figure 14.2a, which presents the phytoplankton biomass with reference to the months of the year.

At the middle latitudes, the intensity of the sunlight and the duration of the daylight periods vary with the seasons. The extended growing period at these latitudes is demonstrated in figure 14.2b, the graph showing an initial increase in phytoplankton biomass associated with increasing light intensity and hours of daylight in the spring. A second increase in the population occurs during the midsummer.

Compare figure 14.2a and b with figure 14.2c, which shows the extended but lower-level growth in the tropics. There is little if any population increase associated with the slight seasonal change in sunlight; the year-round sunlight provides abundant high-intensity solar energy at all times.

In order to take advantage of the available sunlight, the phytoplankton must remain in the ocean's lighted surface layers. In the tropics, the oceans have a warm surface layer of low-density water. This layer exists all year long,

and the phytoplankton are not easily displaced into the denser underlying water. At the middle latitudes, a low-density, warm oceanic surface layer is found only during the summer. This layer appears gradually and is dependent on the water's absorption of the increasing solar radiation. If the warm weather is "late," or if it is interspersed with cool periods or strong storms, the formation of this layer is delayed and the active growth of the phytoplankton does not occur until the water is sufficiently stable and stratified to allow the plankton to remain in the upper layers. In the wintertime, the low-density surface layer of the middle latitudes is mixed with deeper water by surface cooling and winter storms. Vertical mixing, as well as the decrease in available sunlight, restricts phytoplankton growth. At polar latitudes, the low level of sunlight during the summer months forms a weakly stable density layering that is often associated with the fresh water that has returned from the melting sea ice. These factors combine with the short growing season to determine phytoplankton growth.

If the annual distribution of solar energy over the earth's surface were the sole factor influencing plant growth, we would expect the level of phytoplankton growth in the tropics (fig. 14.2c) to be maintained year-round, but at biomass levels approaching the summer values shown for the polar and middle latitudes (fig. 14.2a and b), and we would expect the peak values for the polar biomass to occur at much lower levels. However, the nutrient supply in these areas must also be taken into account. Remember that land drainage, mixing, overturn, and upwelling supply nutrients to the surface layers, and that these processes occur seasonally and in specific areas of the oceans. Both light and nutrients must be available if the phytoplankton population is to increase.

Nutrients

In the polar seas winter overturn resupplies the surface with nutrients; the growing season is short and the nutrients rarely fall below the level that is necessary for population increase. Here, the availability of light controls phytoplankton growth. In the tropics, the phytoplankton remain in the surface layers, because of the stability of the water column. The phytoplankton extract the nutrients from the low-density surface waters, but the upwelling and mixing processes for renewing the nutrient supply are weak. Phytoplankton production in the tropics and subtropics is limited by the nutrient supply to the photic layers.

In temperate latitudes, if winter storms are severe, or if the cooling of the surface water produces a shallow overturn, the surface nutrients are replenished. Nutrients are available to the plant cells when the spring increase in solar radiation warms the surface and increases the stability of the water column, and a spring bloom begins. Grazing organisms decrease the phytoplankton and release nutrients for a second bloom during midsummer. The rate at which nutrients are recycled or added to the surface layer controls the continued growth of the population. As autumn approaches, either the supply of nutrients or the level of sunlight limits phytoplankton growth at these latitudes.

Figure 14.3

Phytoplankton biomass, nutrient supply, and surface-water stability respond to solar energy changes at the middle latitudes in the Northern Hemisphere.

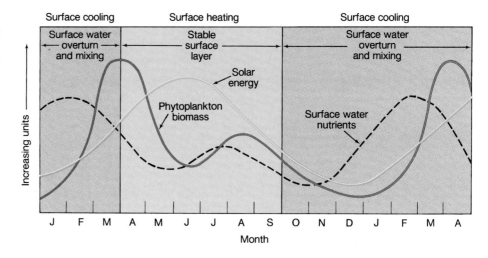

Figure 14.4

Lack of surface mixing and overturn at low latitudes results in a depressed phytoplankton biomass. This pattern is related to solar radiation levels that produce a year-round stable water column.

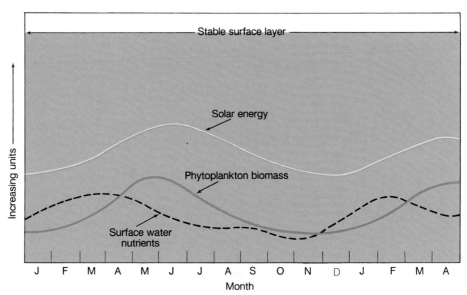

The interaction between the sunlight, the stability of the surface water, the nutrients in the surface water, and the phytoplankton biomass are presented in figures 14.3 and 14.4. Note the replenishment of nutrients to Northern Hemisphere middle-latitude surface waters during periods of overturn and mixing, and the decrease in nutrients during periods of increased sunlight and plant growth in figure 14.3. The lack of surface mixing and overturn at low latitudes is reflected in the low levels of nutrients and plant biomass despite the constant high level of solar energy; review figure 14.4.

Nutrient Cycles

If a plant or an animal dies naturally, or if portions of an organism remain uneaten, organisms of decay, known as **decomposers** (bacteria and fungi), release the energy within the organism's body to the environment as heat and break down the organism's organic molecules to basic molecules, such as carbon dioxide, nutrients, and water. Since no significant amount of new matter comes to the earth from space, living systems must recycle inorganic molecules to form the organic compounds of living organisms. Organisms contain many elements, but among the most important are nitrogen and phosphorus.

Nitrogen in the form of nitrate and phosphorus in the form of phosphate are two of the nutrients required for life (refer back to chapter 5). Nitrogen is essential in the formation of proteins, and phosphorus is required in energy reactions, cell membranes, and nucleic acids. Nitrates and phosphates are removed from the water by the primary producers as the plant populations grow and reproduce. Both are cycled into the animal populations as the animals feed on the plants and are returned to the water as the organisms die and decay. Excretory products from the animals are also added to the seawater, to be broken down and used again by a new generation of plants and animals. The cyclic nature of the nutrient pathway for nitrogen and phosphorus is illustrated in figures 14.5 and 14.6.

To cycle nitrogen, a number of different microorganisms must participate. Nitrogen gas is useless to plants; it must be converted to nitrate to be incorporated into plant tissue. A few species of microorganisms are able to convert nitrogen gas to ammonia, and then other species convert ammonia to nitrite. Still another species is required to convert nitrite to nitrate, the form most easily absorbed by the plants.

It is possible to follow the cyclic path of any molecule required for life. Water is a requirement for life; it is

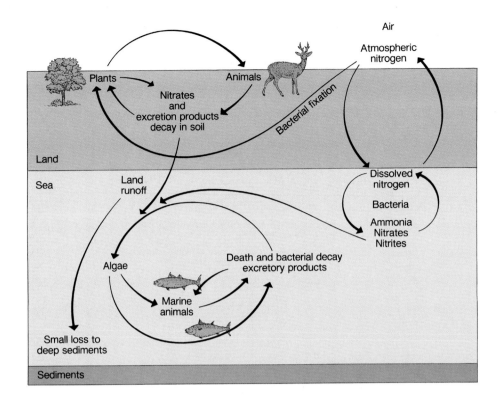

Figure 14.5

The nitrogen cycle. Although nitrogen makes up 78% of the earth's atmosphere, only a few microorganisms are able to change nitrogen gas to the nitrate that is used by land and sea plants, which are in turn eaten and recycled by animals.

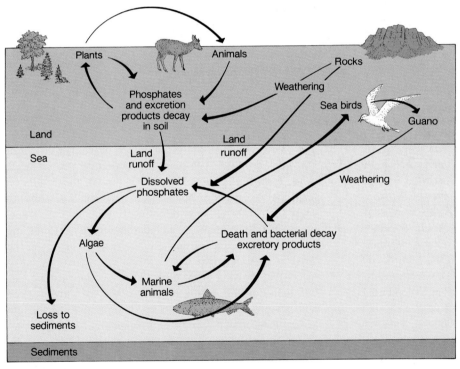

Figure 14.6

The phosphorous cycle. Dissolved phosphorus is carried to the sea by runoff from the land, where it is taken up by plants and recycled through animals until it is released from dead tissue by bacterial action.

part of all life processes and is cycled over and over through life-forms, time, and space (refer back to the hydrologic cycle described in chapter 1).

14.3 Global Primary Production

The geographic distribution of global primary production is shown in figure 14.7. There are a few narrow areas of very high productivity scattered around the world; they are

located against the northwest coasts of North and South America, the west coast of Africa, and along the west side of the Indian Ocean. These special areas are the product of upwellings that transport nutrients to the photic zone, and, as we shall see in chapter 16, these are the areas of the world's greatest fish catches.

Coastal areas are generally more productive than open ocean because along the coasts and in the estuaries, rivers and land runoff supply nutrients. Also, tidal currents cause mixing and turbulence that bring nutrients from shallow

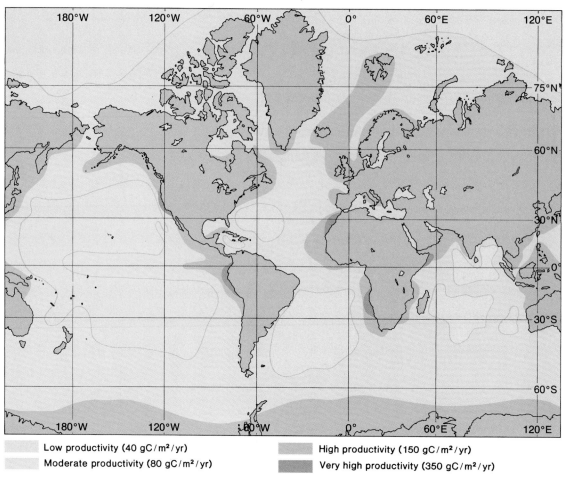

Low productivity (40 gC/m²/yr)	High productivity (150 gC/m²/yr)
Moderate productivity (80 gC/m²/yr)	Very high productivity (350 gC/m²/yr)

Figure 14.7

The distribution of primary production in the world's oceans.

depths to surface layers. Despite the turbulence, the fresh water creates a stable, low-density surface layer, which keeps the plant cells up where sunlight is plentiful.

If equal areas are considered, the quantities of organic material that are produced by upwelling, the coastal ocean, and the open sea can be compared. On the average, upwelling areas are about 2.5 times more productive than coastal areas and 6 times more productive than the open-ocean for the same units of area and time. This relationship is shown in the first column of table 14.1. The second column shows that the total area of each type of oceanic region is inversely related to its primary production. There is more than one hundred times more coastal area than upwelling area, and nearly six times more open-ocean area than coastal area. In the third column, the total primary production of each ocean area type per year is shown. These values demonstrate that most of the organic carbon produced by the oceans' plants is scattered in low concentrations over large areas of the open sea. Smaller total amounts are produced along the coasts and in upwelling regions, but these amounts are concentrated in smaller areas, making these areas very rich in primary production on a per unit area basis. A comparison of gross production in major upwelling areas is given in table 14.2.

To compare primary production on land with that at sea, study table 14.3. Primary production per square meter in the open sea is about the same as that of the deserts on land; the vast areas of open ocean are productive only because of their size, covering 90% of the total ocean's surface. Upwelling areas are comparable to pastureland and lush forestland, and certain estuary systems approach the most heavily cultivated land. Areas of intensively cultivated, fast-growing crops on land produce much more carbon than do most of the regions of the sea; however, on land people put large quantities of time and energy into raising their crops to produce the same high yields that natural processes provide in shallow estuaries. Keep in mind that from the human standpoint, the plants on land are often used directly for food, while those of the sea are not.

14.4 Measuring Primary Production

One method to determine the amount of plant material present in a water sample is to filter out the phytoplankton, count the cells, and multiply the number counted by the average mass per individual cell. Other less direct but less tedious methods are also available. Extracting the chlorophyll

Table 14.1
World Ocean Primary Production

Area	Primary production (gC/m²/yr)	World ocean area (km²)	(%)	Total primary production (metric tons of carbon/yr)
upwellings	350	0.36×10^6	0.1	0.13×10^9
coasts	150	54×10^6	15.0	8.1×10^9
open oceans	40–80	307×10^6	85.0	18.4×10^9

After: Berger, Fischer, Lai, and Wu, 1987; Sumich, 1992

Table 14.2
Gross Primary Production in Major Upwelling Areas

Amount (gC/m²/yr)	Upwelling areas	Amount (gC/m²/yr)	Upwelling areas
400–500	Peru, Canary Islands, Benguela, South Arabia (Yemen), Vietnam, Gulf of Thailand	100–300	California, Costa Rica, Chile, Benguela, New Guinea, Andaman Islands, Northwest Australia
200–400	Peru, Canary Islands, Benguela, Somalia, Malagasy (Madagascar), Orissa coast, Java, Sri Lanka (Ceylon), Flores, Banda, East Arafura	40–80	open ocean

After: Raymont, 1980

Table 14.3
Gross Primary Production: Land and Ocean

Ocean area	Amount (gC/m²/yr)	Land area	Amount (gC/m²/yr)
open ocean	40–80	deserts, grasslands	50
coastal ocean	100–200	forests, common crops, pastures	25–150
upwelling zones deep estuaries	200–500	rain forests, moist crops, intensive agriculture	150–500
shallow estuaries	500–1250	sugarcane and sorghum	500–1250

from a sample of phytoplankton and determining the concentration of pigment present is used to estimate the total quantity of plant material or biomass. Repeated sampling and chlorophyll determination in a column of water or under a fixed area of sea surface over time yield data on changes in biomass with time or net primary production. Another method exposes the chlorophyll in the phytoplankton cells to certain wavelengths of light, which cause the chlorophyll pigment to fluoresce. The strength or intensity of the fluorescence is read electronically to give a direct measure of the chlorophyll and phytoplankton biomass present in a given volume of water.

When new plant material is produced by the phytoplankton, dissolved carbon dioxide is converted to organic carbon compounds; nitrogen from nitrate and phosphorus from phosphate (the nutrients dissolved in seawater) are also required. A ratio exists between the oxygen gas and organic compound produced and the nitrogen and phosphorus removed from the seawater. These fixed ratios by weight are:

$$O_2 : C : N : P = 109 : 41 : 7.2 : 1.$$

These ratios are used to estimate primary carbon production by measuring either the rate of nutrient uptake by the plant population or the rate of production of dissolved oxygen. If 10 g of phosphorus is removed by the plants from a volume of water in a given time, then 410 g of carbon have been produced and incorporated into the phytoplankton biomass. If the rate at which the nitrogen or phosphorus is carried into a region by upwellings and currents and the rate at which it is removed by other currents is known, the rate at which it is being incorporated into the phytoplankton can be calculated, and primary production determined. This allows the oceanographer to estimate primary production over large areas of the ocean and to associate it with large-scale water movements and chemical cycles.

Notice that 7.2 times more nitrogen than phosphorus is needed. If the mechanisms supplying these nutrients to the photic zone are similar, then the available nitrogen will normally be used up before the phosphorus. Depletion of the seawater's nitrate level tends to be the limiting nutrient factor for the ocean's primary production, because nitrogen is critical to living organisms for the production of proteins and nucleic acids. Without nitrogen from nitrates, phytoplankton growth and reproduction will lag, and primary production will decrease.

A simple technique with light and dark bottles demonstrates the relationship between net and gross primary production over depth in an area. Seawater samples with their natural phytoplankton population are collected at selected depths, and the depths are keyed to the percentage of surface solar radiation available: for example, 100% (surface), 75%, 50%, 20%, and 10%. A water sample from each depth is divided into three subsamples: Subsample A is treated immediately to determine the amount of dissolved oxygen in the water at the beginning of the experiment; Subsample B is placed in a tightly sealed **dark bottle** covered with black tape or aluminum foil so that no light can penetrate the bottle; Subsample C is placed in a tightly sealed clear glass bottle, a **light bottle.** The light and dark bottles are then attached to a line and returned to the depth at which they were collected. They are left there for several hours and then brought back to the surface to have the amount of oxygen in each bottle determined as quickly as possible. The changes in the dissolved oxygen content with time can be used to describe the primary production at the sampled depths and light levels by subtracting oxygen values, as follows:

Bottle A – Bottle B$_{dark}$ = measure of oxygen used in respiration

Bottle C$_{light}$ – Bottle A = measure of net oxygen produced by photosynthesis, or net primary production

Bottle C$_{light}$ – Bottle B$_{dark}$ = measure of total oxygen produced, or gross primary production.

The carbon:oxygen ratio for photosynthesis, the volumes of water in the bottles, and the duration of the experiment are used to convert the calculated changes in dissolved oxygen to changes in gC/volume/time.

A light and dark bottle experiment is illustrated in figure 14.8. Note that the primary production varies with depth and that the values for respiration remain nearly constant. Where net primary production is zero, the oxygen produced by photosynthesis exactly meets the demand for oxygen by phytoplankton respiration. This is the **compensation depth** (refer back to section 5.2 in chapter 5).

If it is not convenient to return the light and dark bottles to the sea, the experiments can be carried out by wrapping or covering the clear light bottles with perforated screens to reduce the sunlight penetrating the bottle by the desired percentage, and then placing the bottles in a controlled temperature bath and exposing them to light equivalent to that available at the sea surface. In this way, light and dark bottle values can be obtained while the ship is on its way from station to station.

Similar experiments are made with light and dark bottles to which the radioactive isotope carbon-14 is added. After an incubation time the phytoplankton are filtered from the water, and the amount of carbon-14 incorporated into the plant cells in the light bottle is measured; the net primary production is determined by subtracting any radioactivity incorporated into cellular material in the dark bottle sample to account for nonphotosynthetic effects.

14.5 Total Production

Food Chains and Food Webs

Primary production forms the first link in the **food chains** that link plants, herbivorous animals, and carnivorous animals; together they make up the total production of the sea. Where there is high primary production by phytoplankton, seaweeds, or other marine plants, there are large populations of animals. Animals are consumers; they feed on the primary producers or on other consumers. The **herbivores** eat plant life directly; the **carnivores** feed on the herbivores or on other carnivores. Notice that in either case the carnivores are still linked to the primary producers. The most numerous and the greatest biomass of herbivores are the plant-eating **zooplankton** or animal plankton. These animals are the primary consumers that convert plant tissue to animal tissue; in turn they become the food for the other zooplankton, the flesh-eating carnivores or secondary consumers. See figure 14.9 for the relationship between herbivorous zooplankton and their phytoplankton food source. Again, the response varies with latitude; the zooplankton peak lags behind the phytoplankton production at the polar latitudes (fig. 14.9a) and middle latitudes (fig. 14.9b). The zooplankton reproduce when sufficient phytoplankton are present to support an increase in herbivores, and as grazing by herbivores increases, the phytoplankton biomass decreases. The herbivore population then decreases in its turn as the phytoplankton population declines. The zooplankton population in the tropics (fig. 14.9c) remains nearly constant, with a lower but stable phytoplankton biomass.

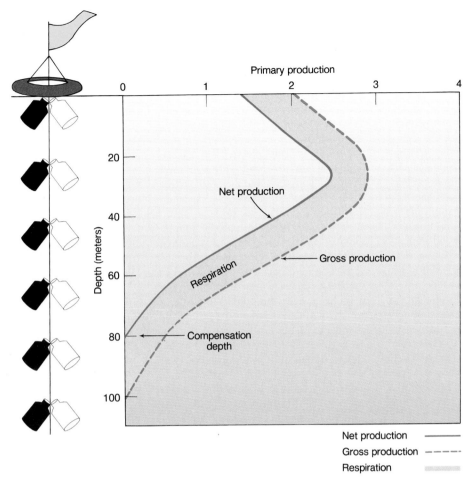

Figure 14.8

A light and dark bottle experiment provides values for respiration, net primary production, and gross primary production with depth. Respiration and production are measured in milligrams of O_2 per liter per six-hour interval. The compensation depth occurs where net production is zero. Light, nutrients, and stability of the water combine to provide conditions for the highest primary production at the depth of 30 meters.

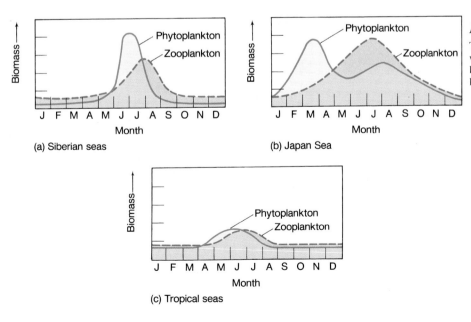

(a) Siberian seas

(b) Japan Sea

(c) Tropical seas

Figure 14.9

The zooplankton biomass varies directly with the phytoplankton biomass in (a) polar latitudes, (b) middle latitudes, and (c) tropical latitudes.

Satellite Measurements

If we are to make wise use of the ocean's living resources, we need to predict global primary production and understand the interactions between primary production, the standing stock abundance of producers, and the marine environment. Finding the relationships that will allow estimates of global primary productivity based on standing stock data is a priority of biological oceanographers seeking to understand the world oceans' total productivity, both plant and animal.

Our ability to make direct measurements of primary production using water samples collected by research vessels along all the world's coasts and in all the world's oceans is limited and extremely expensive. However, the continuous satellite measurements of sea-surface chlorophyll made in the years since satellites began to observe the world's oceans can be used to prepare maps of the global distribution of phytoplankton standing

stock. Box figure 14.1a, b, c, and d were prepared from satellite data collected between 1978 and 1986; each of these images shows chlorophyll concentration, representing biomass, averaged over three months. The color bar key is calibrated in units of milligrams of chlorophyll pigment per cubic meter of surface water. Changes in chlorophyll concentration from one image to the next are related to annual seasonal cycles of light intensity, photic zone depth, nutrients, water temperature, currents, grazing by herbivores, and population die-off. When studied in sequence these four images depict changes in chlorophyll or biomass, which relate to productivity. Plankton abundance is low in the central oceanic gyres (magenta to deep blue) and higher along coastal and upwelling areas (yellow, orange, and red); areas of moderate population are green-blue. Black ocean areas indicate insufficient data; the data

loss may be due to cloud cover, failure to extract the data from the satellite, or loss of the satellite signal. Compare the distribution of chlorophyll and its abundance in Box figure 14.1a, b, c, and d to the distribution of primary production from direct measurements in figure 14.7.

Images of the type seen in Box figure 14.1a, b, c, and d are the product of technology and cooperation between researchers and institutions. These images were prepared by David English, School of Oceanography, University of Washington, from data of the NASA Global Ocean Color Data Set, which was implemented by Gene Feldman of NASA's Goddard Space Flight Center. The Coastal Zone Color Scanner satellite data was processed using the University of Miami's DSP system developed by Otis Brown and Robert Evans with maintenance support from NASA.

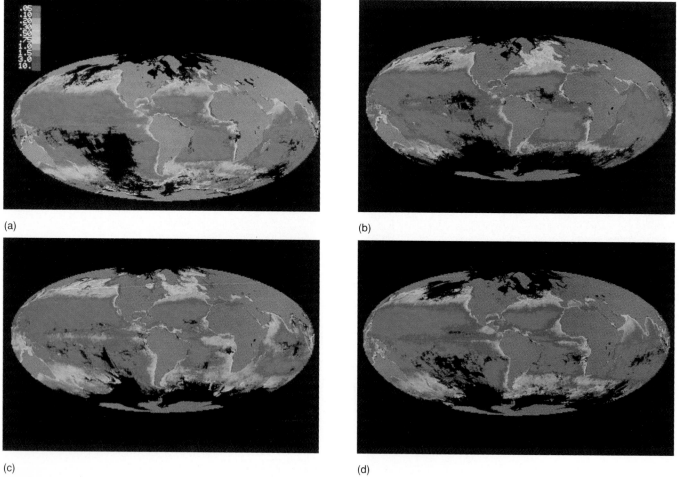

(a)

(b)

(c)

(d)

Box figure 14.1

Satellite observations of global chlorophyll concentration. (a) January, February, March. (b) April, May, June. (c) July, August, September. (d) October, November, December. Color is equivalent to milligrams of chlorophyll pigment per cubic meter of surface water; see color bar in (a).

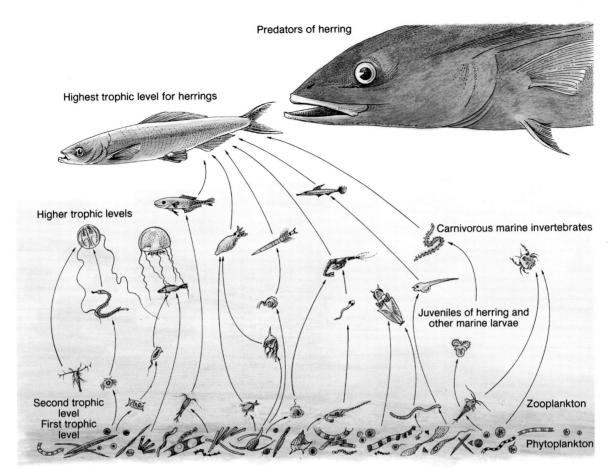

Predators of herring

Highest trophic level for herrings

Higher trophic levels

Carnivorous marine invertebrates

Juveniles of herring and
other marine larvae

Second trophic
level

First trophic
level

Zooplankton

Phytoplankton

Figure 14.10

The food web of the herring at various stages in the herring's life.

There may be many or few links in the food chain that lead to the top carnivore, a predator on which no other marine organism preys (for example, sharks and killer whales). Food chains are rarely simple and linear; they are more likely to show complex branching or interrelationships among organisms, in which case it is more appropriate to call the interconnecting pattern a **food web.** See the herring food web diagram (fig. 14.10) in which organisms change their prey-predation levels as they mature. See also the overall ocean food web in figure 14.11.

Trophic Pyramids

Food chains and food webs represent the pathways followed by nutrients and food energy as they move through the succession of plants, grazing herbivores, and carnivorous predators. These relationships are often demonstrated in the form of **trophic levels** representing links in the food chain; these levels form a **trophic pyramid** in which the trophic levels are numbered from the bottom to the top. See figure 14.12. The primary producers are always the first trophic level, the herbivorous zooplankton are the second trophic level, and the carnivores form the upper levels, up

Table 14.4		
Relative Abundance and Size of Marine Organisms		
Organic form	**Size range**	**Relative abundance**
fish	10 cm–100 cm	0.01
zooplankton	1 mm–10 cm	1.0
phytoplankton	0.001 mm–0.02 mm	10.0

to the top carnivore on which no other organisms prey. Figure 14.12 includes the sun, the energy source directly necessary to the primary producers and indirectly necessary to all other layers, and the decay and decomposition processes that recycle nutrients to the primary producers.

In general, moving upward from the first trophic level, the size of the organisms increases and the numbers and biomass of organisms decrease. The larger numbers of small organisms at the lower trophic levels collectively have a much larger biomass than the smaller numbers of large organisms at the upper levels. Table 14.4 relates the abundance of organisms to their size.

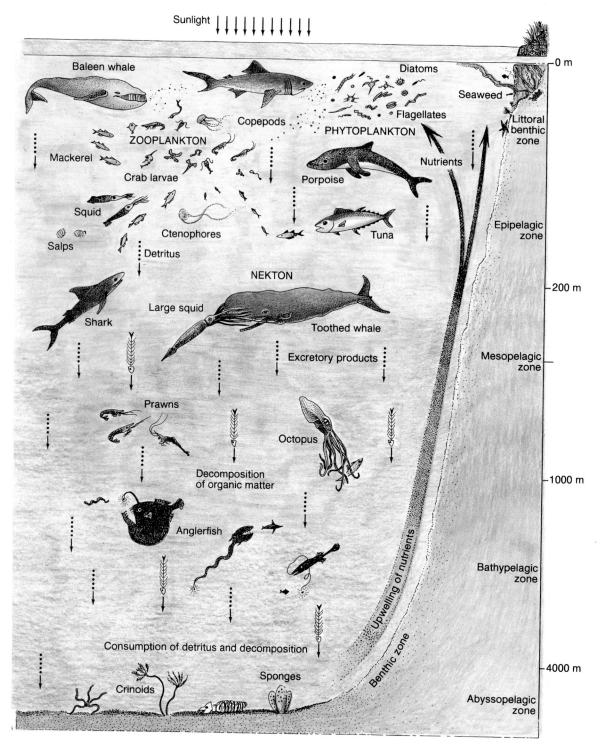

Sunlight

0 m

Baleen whale

Diatoms

Seaweed

Flagellates

Copepods

Littoral benthic zone

ZOOPLANKTON

PHYTOPLANKTON

Mackerel

Porpoise

Nutrients

Crab larvae

Epipelagic zone

Squid

Tuna

Ctenophores

Salps

Detritus

NEKTON

200 m

Large squid

Shark

Toothed whale

Mesopelagic zone

Excretory products

Prawns

Octopus

Decomposition of organic matter

1000 m

Anglerfish

Upwelling of nutrients

Bathypelagic zone

Benthic zone

Consumption of detritus and decomposition

4000 m

Crinoids

Sponges

Abyssopelagic zone

Figure 14.11

The ocean food web. Plant, animal, and bacterial populations are dependent on the flow of energy and the recycling of nutrients through the food web. The initial energy source is the sun, which fuels the primary production in the surface layers. Herbivores graze the phytoplankton and benthic algae and are in turn consumed by the carnivores. Animals at lower depths depend on organic matter from above. Upwelling recycles nutrients to the surface, where they are used in photosynthesis.

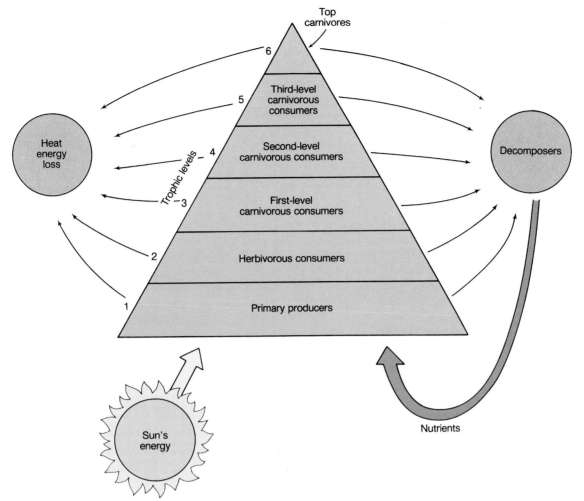

Figure 14.12

A trophic pyramid. Trophic levels are numbered from base to top. The first trophic level requires nutrients and energy. Nutrients are recycled at each level; energy is lost as heat at each level.

The overall efficiency of energy transfer up each layer of an open-ocean trophic pyramid is estimated at about 10%. If, in order to add 1 kg of weight, a person ate 10 kg of salmon, to attain that weight the salmon had to consume 100 kg of small fish, and the fish needed to consume 1000 kg of carnivorous zooplankton, which in turn required 10,000 kg of herbivorous zooplankton, needing 100,000 kg of phytoplankton to supply the eventual 1 kg gain at the top of the pyramid. The 90% energy loss at each trophic level goes to the metabolic needs of the organisms at that level. These needs include energy required for moving, breathing, feeding, and reproduction, as well as heat loss. In other words, an organism that consumes 100 units from the level directly below will use 90 units for its own metabolic needs and will convert only 10 units to body tissue available to predation from the level above. Therefore, feeding at high trophic levels is less energy efficient than feeding at low trophic levels.

Other Systems

In some areas of the ocean, total productivity is not tied to the primary productivity of the phytoplankton. The shallow waters above a tropic coral reef are clear, indicating a lack of phytoplankton. The reef, however, is a rich and varied community living in an independent, complex association that includes algae, herbivores, and carnivores. (See also the discussion of coral reefs in chapter 17.)

Shallow coastal areas in which the sea bottom is exposed to sunlight support masses of attached seaweed and bottom-living, single-celled plants. These function as primary producers in addition to the phytoplankton in the water column. The rapid growth and relatively large size of many seaweeds provide a large amount of organic material to the animal population. Animals graze the living plants or feed from the fragments left after the battering of winter storms and the populations' natural mortality.

The exceedingly high primary production of estuaries has various sources. In deep fjordlike estuary systems, the majority of the primary production comes from the phytoplankton in the water column. In these estuaries, although

Table 14.5
Oceanic Food Production

Area	Plant production (metric tons of carbon/yr)	Efficiency of mass and energy transfer per trophic level	Trophic level harvested	Estimated fish production (metric tons/yr)
open ocean	18.4×10^9	10%	5	1.84×10^6
coastal regions	8.1×10^9	15%	4	27.3×10^6
upwelling areas	0.13×10^9	20%	2	26.0×10^6

the concentration of organic carbon per cubic meter of water is highest near the surface, the water below the photic layer may contain a greater total amount of organic carbon because of the active vertical mixing, which displaces plant cells downward, into the water below the photic zone.

In broad, shallow estuaries, active production of organic matter by primary producers occurs at all depths, on the estuary floor as well as in the water column. Organic material is also produced in the bordering marshlands, some of which produce two crops of plant growth per year. All of these sources contribute organic matter as food to the animal population in a rather small volume of water. These shallow estuaries enjoy an extraordinary rate of primary productivity, supporting the food chains and webs within the estuary and also contributing organic matter to the coastal waters. Refer back to table 14.3 to compare the gross primary production of deep and shallow estuaries.

In recently discovered deep-water communities that surround hot-water vents on the ocean floor, the primary production is not of plant material based on solar energy; rather, the primary producers are bacteria that take their energy from the chemistry of the hot water flowing from cracks in the sea floor. These richly productive communities and their energy source are discussed with coral reefs and shallow-water communities in chapter 17. Other communities based on oil and gas seeping onto the surface of the seafloor are also discussed in chapter 17.

14.6 Practical Considerations: Human Concerns

When human activities supply additional nutrients, the primary production may exceed the ability of the local herbivores to consume it. A continuous and excessively high rate of primary production eventually results in an unnaturally high rate of decomposition, as unconsumed plant matter accumulates at depth, especially in an estuary. The decaying material consumes dissolved oxygen at a rapid rate, and a mat of organic debris may form over the sediments. The waters and the sediments become anoxic, which adversely affects the organisms at the higher levels in the food webs and chains. The section on Chesapeake Bay in chapter 12 illustrates this problem.

Custom, culture, economics, and availability influence the harvesting of the oceans by humans. Commercial harvests are made from both the higher trophic levels (for example, salmon, tuna, halibut, and swordfish), and the lower levels (for example, herring, shellfish, and anchovy). Harvesting high in the trophic pyramid is less energy efficient than harvesting at low trophic levels. Overharvesting any level endangers levels both above and below it by removing food resources from higher levels and preventing recycling of nutrients from higher to lower levels.

The highest yields of food resources for humans occur when the harvest of marine species is conducted at the lowest usable trophic level. The relationship between total plant production and theoretical production of fish is shown in table 14.5 for three basic areas of the ocean. The third column in this table gives the average efficiency of energy conversion between the trophic levels in each area, and the fourth column shows the trophic level at which humans usually harvest their food. The well-mixed, nutrient-rich waters of the coasts and estuaries support short, efficient food chains, shown in pyramid form in figure 14.13a and b. Fish production in these areas is greater because fish obtain food with less effort due to the high rates of plant production. Compare the efficiency of these food chains with the open-ocean food chain in figure 14.13c and the open-ocean fish production in table 14.5.

We need to learn how to harvest and use the lower trophic levels without depleting them. An understanding of the trophic levels and their relationship to each other is necessary, so that harvesting is controlled to ensure sufficient stocks for reproduction and the maintenance of the trophic pyramid.

Summary

Plants use photosynthesis to produce organic compounds from carbon dioxide and water in the presence of sunlight and chlorophyll. Oxygen is formed as a by-product. Respiration breaks down organic compounds with the addition of oxygen to yield energy, water, and carbon dioxide.

Gross primary production is the total amount of organic material produced by photosynthesis per volume per unit of time; net primary production is the gain in organic material per volume per unit of time after the

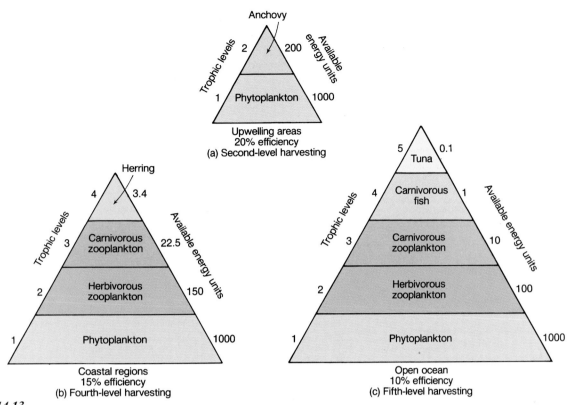

Figure 14.13

Trophic level efficiency varies among (a) upwelling areas, (b) coastal regions, and (c) the open ocean. The number of trophic levels and the level at which humans harvest differs with location.

organic material used in plant respiration is deducted. The amount of organic carbon produced by primary production in a volume over a time period is measured as the rate of change in biomass. The biomass available at a location at a specific time is the standing crop.

Primary production is controlled by the interaction of sunlight, nutrients, and the stability of the surface water in an area. At polar latitudes, the availability of light controls phytoplankton growth; nutrients are not limiting. In the tropics, the sunlight is available year-round, and the stability of the water column helps hold the phytoplankton at the surface. However, the nutrient supply is poor and it limits production. At temperate latitudes, light, nutrients, and water column stability vary with the seasons. Nutrients cycle through the land and sea, the plants and animals, and are returned to the water by death and bacterial decomposition.

Where sunlight is not limiting, coastal waters are always more productive than the open ocean. Good mixing, the nutrients in the land runoff, and a water column made stable by the addition of fresh water combine to make shallow coastal regions very productive.

Upwelling areas are 2.5 times more productive than coastal water, and 6 times more productive than the open ocean, but there is 6 times more open ocean than coastal water and more than 100 times more coastal water than upwelling area. Primary production is concentrated in coastal areas and at upwellings.

Primary production may be estimated by (1) counting the increase in plant cells; (2) determining the change in concentration of chlorophyll in a sample; (3) measuring the rates of nutrient supply and nutrient removal and using the known ratio of nutrients to calculate carbon production; (4) measuring the rate of incorporation of carbon-14 into the plant cells, or measuring the rate that oxygen is produced by phytoplankton growing in light and dark bottles.

The phytoplankton, or primary producers, are preyed upon by herbivorous zooplankton. These are the primary consumers that are preyed upon by the secondary consumers, or carnivorous zooplankton. The relationship between the phytoplankton and the herbivorous zooplankton varies with latitude. The term food web is more appropriate than food chain to describe the interconnecting patterns. Trophic pyramids present the relationship between producers and consumers in terms of the transfer of biomass and energy. Open-ocean transfer efficiency is approximately 10% between trophic levels. Efficiency is higher in coastal and upwelling areas. Nutrients are recycled; energy is not.

Tropic coral reefs, very shallow coastal waters and estuaries, and deep-ocean vents are productive areas in which productivity is not keyed to the phytoplankton.

If human activities supply additional nutrients to seawater, primary production increases, followed by decomposition and anoxia. Harvested food value is increased if the lower trophic levels are used, but care is required to conserve the stock for future harvests and for the requirements of other organisms.

phytoplankton
chlorophyll
photosynthesis
primary production
gross primary production
respiration
net primary production
biomass
standing crop
decomposer

dark bottle
light bottle
compensation depth
food chain
herbivore
carnivore
zooplankton
food web
trophic level
trophic pyramid

Study Questions

1. Why is the efficiency of energy transfer between trophic levels only about 10% in the open sea? Compare this efficiency of energy transfer with that found in upwelling regions.
2. Explain the general relationship between the abundance and size of organisms shown in table 14.4.
3. Distinguish between the terms in each pair:
   a. Standing crop, biomass.
   b. Photosynthesis, respiration.
   c. Producer, consumer.
   d. Food web, food chain.
   e. Net productivity, gross productivity.
4. Compare the productivity of polar, temperate, and tropic ocean regions. What factor or factors generally limit the productivity of each area?
5. The rate of primary production in the open ocean is less than the rate of primary production along the coasts, but the total primary production of the open ocean exceeds that of the coasts. Explain this apparent contradiction.
6. Which areas of the oceans are most productive? How does the productivity of these areas compare with the productivity of the land?
7. Why are estuaries less important than cultivated land as producers of human foods, although the primary production rate of estuaries approximately equals the most intensively cultivated land?
8. How do the surrounding land areas contribute to an estuary's productivity?
9. Draw a general diagram to explain the movement of (a) a gas and (b) a nutrient through the ocean environment and its plant and animal populations.
10. Explain why the compensation depth is deeper in tropical areas with low productivity than it is in high temperate or subarctic areas with primary production.
11. Consider figure 14.10 and explain how the herring food chain changes with time.
12. Why is a shallow-water estuary more productive than either a deep-water estuary or an upwelling zone?
13. Explain the changes in phytoplankton pigment concentrations in Box figure 14.1a, b, c, and d.
14. Why does the respiration of phytoplankton appear nearly constant with depth in figure 14.8?
15. If the ratios of O_2:C:N:P are 109:41:7.2:1 in phytoplankton organic matter, why is nitrate-nitrogen limiting to production when the concentrations of nitrates and phosphates are equal in seawater?

Study Problems

1. A sample of water from a depth of 5 m showed 8 mg of dissolved oxygen per liter. Part of this water was placed in a light bottle and part of it was placed in a dark bottle, and the bottles were returned to the 5-m depth. After six hours the oxygen content of the light bottle was found to be 8.9 mg of oxygen per liter; the dark bottle showed 7.4 mg of oxygen per liter. Calculate (a) the respiration rate, (b) net primary production, and (c) gross primary production. How many grams of new carbon were produced per liter in six hours?
2. If the nitrogen available as nitrate is removed from the water of an inlet at the rate of 3.6 mg of nitrogen per liter of water every eight hours, what is the rate of new carbon production by the phytoplankton?

Suggested Readings

Berger, W. H., K. Fischer, C. Lai, and G. Wu. 1987. *Ocean Productivity and Organic Carbon Flux, Part I.* SIO Reference Series 87-30. Scripps Institution of Oceanography. University of California, San Diego. 67 pp.

Isaacs, J. D. 1969. The Nature of Ocean Life. In *Life in the Sea,* readings from *Scientific American,* 1982. Freeman, San Francisco. pp. 4–17.

LaBrecque, M. 1988. A Global Chemical Flux. *Mosaic* 19 (3/4): 90–101.

Lerman, M. 1986. *Marine Biology, Environment, Diversity and Ecology.* Benjamin-Cummings, Menlo Park, Calif. 535 pp.

Parsons, T. R., M. Takahashi, and B. Hargrave. 1984. *Biological Oceanographic Processes,* 3d ed. Pergamon, Elmsford, N.Y. 330 pp.

Pomeroy, L. R. 1974. The Ocean's Food Web, A Changing Paradigm. In *Oceanography, Contemporary Readings in Ocean Sciences,* 2d ed. Pirie, R. G., ed. (1977). Oxford Univ., New York. pp. 105–15.

Raymont, J. E. G. 1980. *Plankton and Productivity in the Oceans,* 2d ed., Vols. 1 and 2: Phytoplankton. Pergamon, Elmsford, N.Y. 489 pp.

Rhyther, J. H. 1969. Photosynthesis and Fish Production in the Sea. *Science* 166 (3901): 72–76.

Sherman, K. 1991. *Food Chains, Yields, Models and Management of Large Marine Ecosystems.* Westview Press, Boulder, Co. 300 pp.

Sumich, J. L. 1992. *An Introduction to the Biology of Marine Life,* 5th ed. Wm. C. Brown, Dubuque, Ia. 450 pp.

15

The Plankton: Drifters of the Open Ocean

15.1   The Kinds of Plankton
Phytoplankton
Zooplankton

Box: A Krill-Based Ecosystem

15.2   Bacteria
15.3   Classification Summary of the Plankton
15.4   Sampling the Plankton

Box: Viruses in the Oceans

15.5   Practical Considerations: Marine Toxins
Red Tides
Other Toxic Blooms
Ciguatera Poisoning
Summary
Key Terms
Study Questions
Suggested Readings

In a sudden awakening, incredible in its swiftness, the simplest plants of the sea begin to multiply. Their increase is of astronomical proportions. The spring sea belongs at first to the diatoms and to all the other microscopic plant life of the plankton. In the fierce intensity of their growth they cover vast areas of ocean with a living blanket of their cells. Mile after mile of water may appear red or brown or green, the whole surface taking on the color of the infinitesimal grains of pigment contained in each of the plant cells.

The plants have undisputed sway in the sea for only a short time. Almost at once their own burst of multiplication is matched by a similar increase in the small animals of the plankton. It is the spawning time of the copepod and the glassworm, the pelagic shrimp and the winged snail. Hungry swarms of these little beasts of the plankton roam through the waters, feeding on the abundant plants and themselves falling prey to larger creatures.

Rachel Carson,
from *The Sea Around Us*

A drifting Scyphomedusa jellyfish.

*T*he word plankton comes from the Greek term **planktos,** *meaning to wander, and the plant and animal plankton are the wanderers and drifters of the sea. They exist in vast swarms, limited in their mobility, moving with the currents. The diversity of planktonic organisms is so great that it is not possible to discuss all of their life-forms here. Instead, representative animals, plants, and groups of animals and plants have been chosen for discussion. Because it is the plant life that is able to use the sun's energy to form the basis of life for all the animals, we begin by considering the ocean's floating plant life, the primary producers of chapter 14, and then go on to describe the animals that drift with the plant plankton, grazing upon it and upon each other.*

15.1 The Kinds of Plankton

Although many plankton have a limited ability to move toward and away from the sea surface, they make no purposeful motion against the ocean's currents and are carried from place to place suspended in the seawater. Some plankton are quite large; jellyfish may be the size of a large washtub, trailing 15 m (50 ft) of tentacles. But the phytoplankton and many zooplankton are generally too small for our unassisted vision and must be observed under a microscope.

Bacteria and very small phytoplankton cells are called **ultraplankton;** they are less than 0.005 mm in diameter and can only be collected using special filtering techniques. Slightly larger phytoplankton, called **nannoplankton,** have a size range between 0.005 and 0.07 mm. Zooplankton and phytoplankton between 0.07 and 1 mm are called **microplankton** or **net plankton,** because they are usually captured in tow nets made of very fine mesh nylon.

The microscopic phytoplankton are the "grasses of the sea." Just as a land without grass and herbs could not support the insects, small rodents, and birds that serve as food for the larger, meat-eating carnivores, a sea without phytoplankton could not support the zooplankton and the other larger animals. As the British biological oceanographer Sir Alister Hardy has said, "All flesh is grass."

Phytoplankton

The phytoplankton are mainly unicellular (or single-celled) plants known as algae. Each phytoplankton cell is **autotrophic** (or self-feeding) by the process of photosynthesis (refer back to chapter 14). Each cell is an independent individual, and even in the species in which the cells attach together in **filaments** (long chains) or other aggregations, there is no division of labor between the cells. There is only one large planktonic alga, the seaweed *Sargassum* that is found floating in the area of the North Atlantic known as the Sargasso Sea. *Sargassum* reproduces vegetatively by

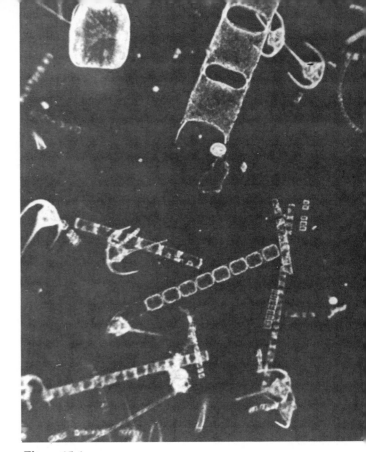

Figure 15.1
Phytoplankton. Diatoms are in chains; dinoflagellates are irregular single cells (enlarged).

fragmentation to form large mats, which provide shelter and food for a wide variety of organisms, including fish and crabs. The specialized organisms found living in the *Sargassum* mats occur nowhere else.

Groups of organisms belonging to the phytoplankton include the **diatoms, dinoflagellates,** and **coccolithophores;** silicoflagellates, cryptomonads, chrysomonads, green algae, and cyanobacteria or blue-green algae are present but less numerous. A generalized phytoplankton sample is shown in figure 15.1. In the following discussion the emphasis is placed on the diatoms and the dinoflagellates, as they are the most abundant and most important members of the marine phytoplankton.

Diatoms are single-celled plants found in areas of cold, nutrient-rich water. They are sometimes called golden algae, because their characteristic yellow-brown pigment, **fucoxanthin,** masks their chlorophyll. Some diatoms have **radial symmetry;** they are round, shaped like pillboxes, and are called **centric diatoms.** Others have **bilateral symmetry,** are elongate, and are called **pennate diatoms.** Centric diatoms float better than pennate diatoms; therefore pennate diatoms are often found on the shallow seafloor or attached to floating objects, while centric diatoms are more truly planktonic. Some common diatoms from temperate waters are shown in figures 15.2 and 15.3.

Around the outside of each diatom is a **frustule** (or cell wall) of pectin, a jellylike carbohydrate, impregnated with silica. The frustule is hard, rigid, transparent, and

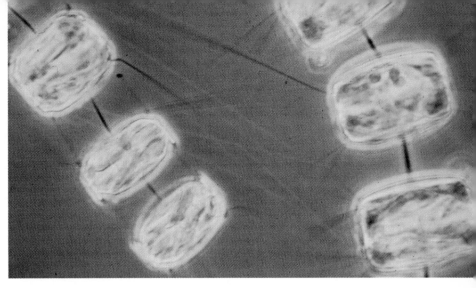

(a)

Figure 15.2

The diatoms are the most important members of the phytoplankton; they are found in a wide variety of forms. (a) Two species of *Thalassiosira*. (b) The large centric diatom *Arachnoidiscus*. (c) An enlarged detail of (b) showing the intricate details of the diatom's frustule. (d) The diatoms arranged in star-shaped groups are members of the genus, *Asterionella*.

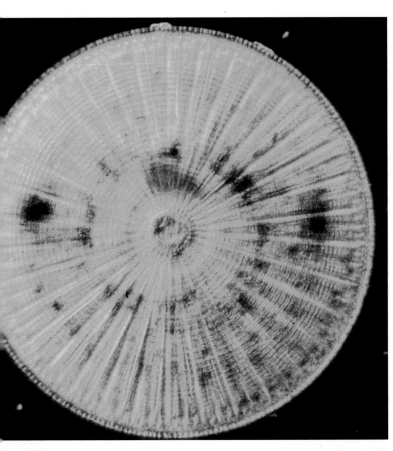

(b)

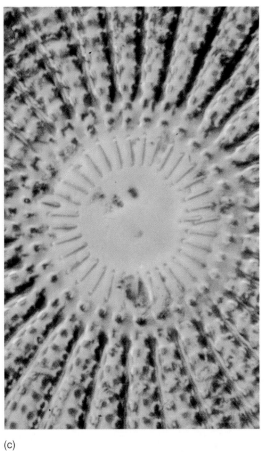

(c)

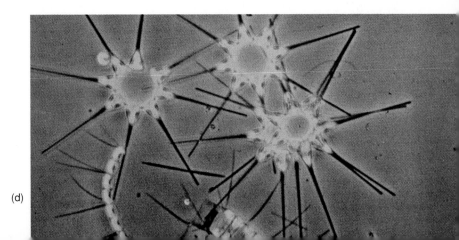

(d)

Figure 15.2 *continued*
(e) These *Nitzschia* cells show the
characteristic yellow-brown color of diatoms;
(f) *Chaetoceros* forms chains of cells with long
spines.

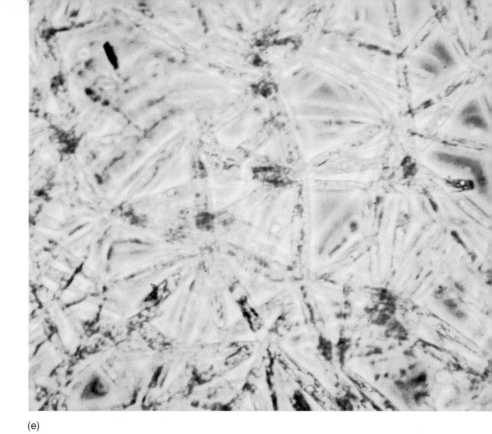

(e)

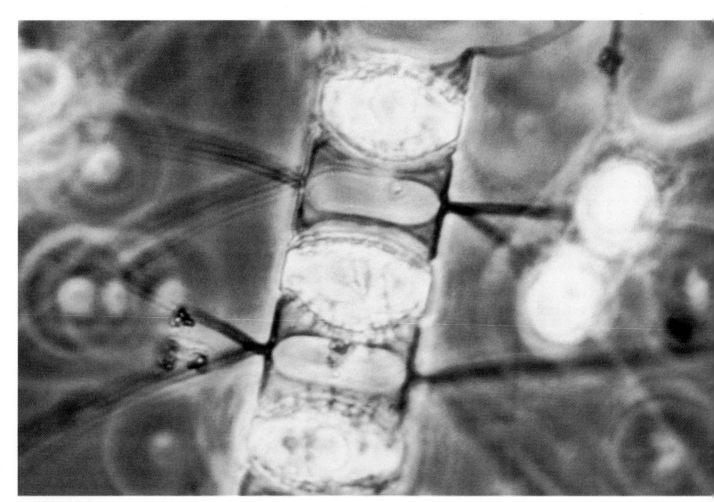

(f)

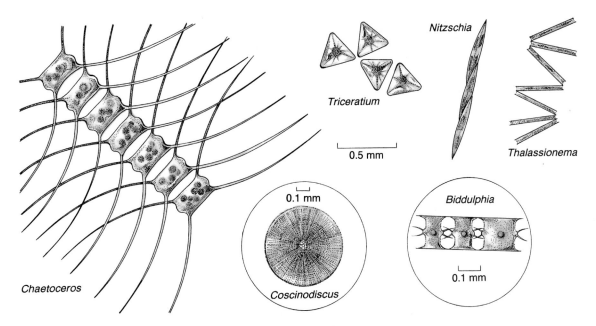

Triceratium

Nitzschia

0.5 mm

Thalassionema

0.1 mm

Coscinodiscus

Biddulphia

0.1 mm

Chaetoceros

Figure 15.3

Centric and pennate diatoms. Diatoms exist as single cells or in chains.

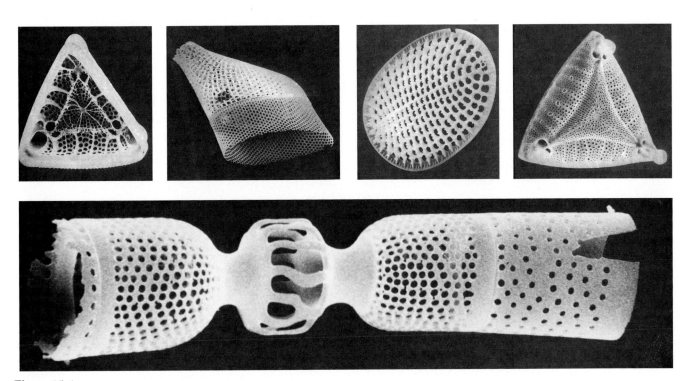

Figure 15.4

Stereoscan micrographs of diatom frustules.

delicately marked with pores that connect the living portion of the cell inside to its outside environment (see fig. 15.4). The two halves of a centric diatom's frustule fit together like a pillbox, and when the cell has grown sufficiently large, the cell inside divides, the two halves of the pillbox separate, a new inner half to each pillbox is formed, and two new daughter diatoms are produced. By this process, one of the new cells will remain the same size as the parent, while the other is always smaller, as its larger pillbox half is formed by the smaller bottom half of the parent pillbox.

This process is shown in figure 15.5. When a cell reaches a size level of about 25% of the original parent's size, it stops dividing and begins a sexual cycle. In this cycle it produces a naked **auxospore,** which increases in size, forms a new frustule, and begins to divide again. Diatoms divide very rapidly, every twelve to twenty-four hours under conditions of plentiful sunlight and nutrients. When this rapid division increases the population so that the water becomes discolored by the presence of millions and millions of cells, it is called a **bloom.**

Figure 15.5

The division of a parent centric diatom into two daughter diatoms. The two halves of the pillboxlike cell separate, the cell contents divide, and a new inner half is formed for each pillbox. One daughter cell remains the same size as the parent; the other is smaller.

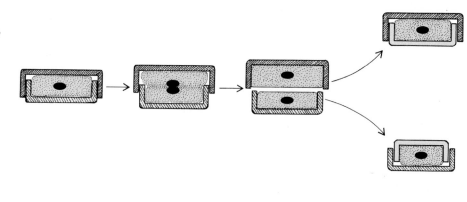

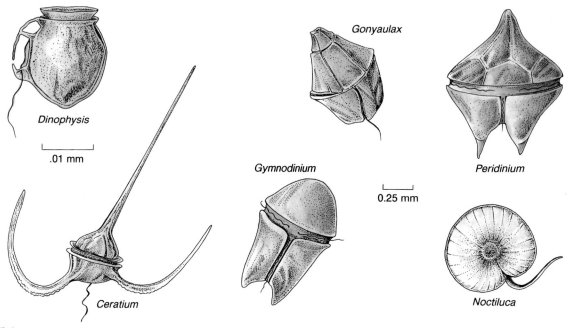

Figure 15.6

Dinoflagellates. *Noctiluca, Gymnodinium,* and *Gonyaulax* produce red tides. *Noctiluca* is a bioluminescent, nontoxic dinoflagellate. *Gymnodinium* and *Gonyaulax* produce toxic red tides and paralytic shellfish poisoning.

Frustules made with silica are more dense than seawater, but the diatom must stay afloat in the sunlit surface waters to survive. The internal cell material has a low density and increases its buoyancy by the production of oil as a storage product; fish that feed on large quantities of diatoms may have a distinctly oily taste. The diatom's small size also helps it to stay afloat, because small particles that are only slightly more dense than the liquid in which they are suspended sink very slowly. A small, as opposed to a large, spherical particle has a large surface area in comparison to its volume or mass, and this large surface area to volume ratio helps keep the cell afloat. Some diatoms possess spines, wings, or other projections that increase their surface area still more. A large surface area also provides the diatoms with increased exposure to sunlight and to water containing the necessary gases and nutrients for photosynthesis and growth.

Diatoms are most important as the first level of food production. Those that are not consumed by herbivores eventually die and sink to the ocean floor. In shallow areas of the oceans, the cells reach the seafloor with some

organic matter still locked inside. Deposits made long ago in this manner have formed petroleum, or oil. Diatom frustules sinking to the greater depths of the oceans build up siliceous sediments under areas of abundant diatom populations. (Refer back to the section on biogenous sediments in chapter 2.) Sometimes geologic processes lift these silica-rich sediments above the sea, where they are mined as **diatomaceous earth,** which is used in industrial filtration systems, in the filtering of wine as well as swimming pools, and as an abrasive in toothpaste and silver polish.

The dinoflagellates differ from the diatoms in several respects. Dinoflagellates usually have two **flagella,** or whiplike appendages, that beat within grooves in the cell wall. One groove encircles the cell like a belt, and the other lies at right angles to it. The beating of these flagella makes the cells motile and causes them to spin like tops as they move through the water. They also tend to migrate vertically in response to sunlight, but this ability to move is limited, and they are still at the mercy of the waves and currents. Representative dinoflagellates are shown in figure 15.6.

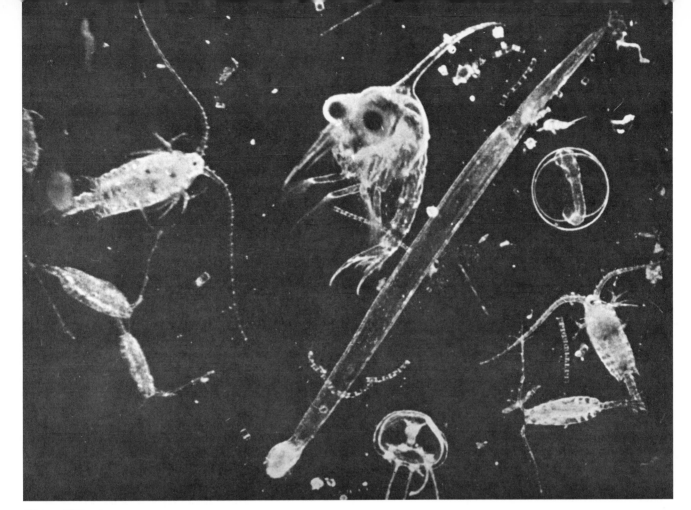

Figure 15.7
Zooplankton sample (enlarged).

Dinoflagellates are red to green in color and can exist at lower light levels than diatoms, because they can both photosynthesize like a plant and ingest organic material like an animal. They have both autotrophic and **heterotrophic** abilities. Heterotrophic organisms feed on other organisms or on organic substances. Their external walls do not contain silica, but may be armored with plates of cellulose, giving them the appearance of spinning armored helmets. Other dinoflagellates have a smooth, flexible outer surface showing no such plate structures. Some dinoflagellates are called fire algae, because they glow with bioluminescence at night. (Refer back to chapter 13 for a discussion of this phenomenon.)

Dinoflagellates are found over most of the oceans but do not contribute to the bottom sediments, because both the cell walls and the soft parts decay completely. Although they do make up a substantial portion of the phytoplankton, dinoflagellates are not as important as the diatoms as a primary ocean food source. The cells reproduce by a division process similar to that found in diatoms, but without the reduction in size. They can, under favorable conditions, multiply even more rapidly than diatoms to form blooms.

Coccolithophores and silicoflagellates are relatives of the diatoms. The coccolithophores are single-celled plants with **coccoliths,** or outer calcareous plates, that are deposited

as sediment when the cells die (see fig. 2.19b). These coccoliths are soluble under the conditions of low temperature and high pressure found in the deep oceans; therefore, calcareous coccolith deposits are limited to shallow regions of less than 4000 m (13,200 ft). Coccolithophores are also limited by the surface-water temperatures to the warm tropic regions, and so coccolith deposits are to be found in shallow areas at the lower latitudes. Like the dinoflagellates, they possess two flagella, and like both diatoms and dinoflagellates, they may reproduce by simple division, although some species do have a form of sexual reproduction. Silicoflagellates are small autotrophic cells with flagella and an internal hollow skeleton of silica. They are not believed to occur in large numbers.

Zooplankton

The animal members of the plankton, or the zooplankton, are either grazers on phytoplankton (herbivores), feeders on other members of the zooplankton (carnivores), or feeders on both plants and animals (omnivores). A general zooplankton sample is shown in figure 15.7. Many of the zooplankton have some ability to swim and can even dart rapidly over short distances in pursuit of prey or to escape from predators. They may move vertically in the water column, but they are still at the mercy of the waves and currents and so are considered planktonic (or drifting) organisms.

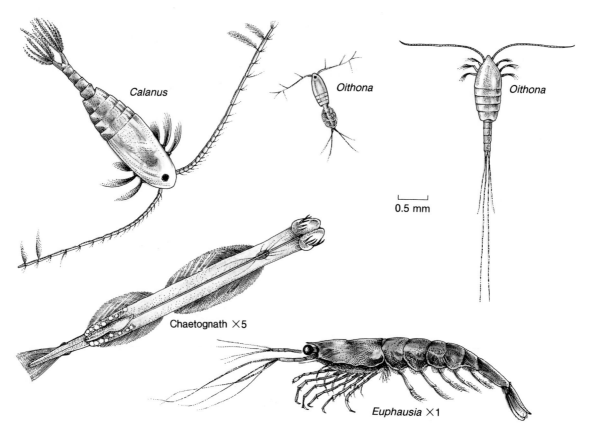

Figure 15.8
Crustacean members of the zooplankton and an arrowworm or *Chaetognath. Calanus* and *Oithona* are copepods. *Euphausia superba* is known as krill.

Representatives of nearly every animal phylum are found in the zooplankton. The life histories of organisms that make up the zooplankton are varied and show a variety of strategies for survival in a world where reproduction rates are high and life spans are short. These animals may produce three to five generations a year in warm waters, where food supplies are abundant and temperatures accelerate life processes. At high latitudes, where the season for phytoplankton growth is brief, the zooplankton may produce only a single generation in a year. The voracious appetites and rapid growth rates of the carnivorous zooplankton are responsible for liberating nutrients to be recycled by the phytoplankton.

Zooplankton exist in patches of high population density between areas that are much less heavily populated. The high population patches attract predators, and the more sparse populations between the denser patches preserve the stock, as fewer predators feed there. Turbulence and eddies disperse individuals from the densely populated patches to the intervening sparser areas. Convergence zones and boundaries between water types concentrate zooplankton populations, which attract predators.

Plankton accumulate at the density boundaries caused by the layering of the surface waters, and the variation of light with depth and the day-night cycle play additional roles. Some zooplankton migrate toward the sea surface each night and return to depth each day, either in an attempt to maintain their light level or in response to the movement of their food resource. This daily migration may be as much as 500 m (1650 ft) or less than 10 m (33 ft).

Accumulations of organisms in a thin band extending horizontally along a pycnocline or at a preferred light intensity or food resource level are capable of partially reflecting sound waves from depth sounders. The zooplankton layer is seen on a bathymetric recording as a false bottom or a deep scattering layer, the DSL. (Refer to chapter 4.) Echo-sounding studies are used to record the vertical migration of this layer of plankton and predators and to measure the vertical and horizontal extent of the layer.

Among the most common and widespread zooplankton types worldwide are the small **crustaceans** (shrimplike animals): **copepods** and **euphausiids** (fig. 15.8). These animals are basically herbivorous and consume more than half their body weight daily. Copepods are smaller than euphausiids; euphausiids move more slowly and live longer than the copepods. The euphausiids, because of their size, also eat some of the smaller zooplankton along with the phytoplankton that make up the bulk of their food. Both reproduce much more slowly than the diatoms, doubling their populations only three to four times a year. They may make up more than 60% of the zooplankton in any of the world's oceans and serve as a food source for small fish. In the Arctic and Antarctic, the euphausiids are the **krill,** occurring in such quantities that they provide the main food for the

Figure 15.9
The Antarctic krill, *Euphausia superba.*

baleen whales. Baleen (or whalebone) whales have no teeth; instead they have a netlike strainer of baleen suspended from the roofs of their mouths. After the whales gulp the water and plankton, they expel the water through the baleen, leaving the tiny krill behind. Whales of this type include the blue, right, humpback, sei, and finback whales.

The Antarctic krill, *Euphausia superba* (fig. 15.9), are present in enormous quantities. Estimates of total biomass have varied from 5 million to 6 billion metric tons, and are probably in excess of 900 million metric tons. Because of these large biomass estimates, the krill in the Southern Ocean are considered a potentially valuable international fishery. They are harvested commercially by fishing fleets from the former USSR, Japan, Korea, Poland, and Chile. In 1981 and 1982 Japan and the former Soviet Union, using large stern trawlers with nets 80 m (264 ft) wide and catching 8–12 tons of krill in a single haul, harvested 448,000 metric tons in 1981 and 529,000 metric tons in 1982.

Marketing the krill for human consumption has not been very successful. Fresh krill are almost flavorless, and when dried the flavor becomes strong and somewhat unpleasant. The shell has been found to contain extremely high amounts of fluoride and must be removed for human consumption. In the former Soviet Union, the catch was fed mainly to livestock and poultry; the Japanese use their krill as feed on their fish farms. In addition there is the expense of the long distance to the fishing grounds, and because the krill deteriorate rapidly, they must be processed between hauls, limiting the daily harvest. Because of these difficulties the krill catch dropped between 1983 and 1985 to less than 200,000 metric tons and since then has averaged less than 400,000 metric tons. The 1990 catch was 381,000 metric tons; after the breakup of the Soviet Union the catch dropped to 289,000 metric tons during the Antarctic 1991–92 season.

All the nations presently harvesting krill in the Southern Ocean are members of the Convention for the Conservation of Antarctic Marine Living Resources, and all have agreed not to overharvest this resource. However, any evaluation of the krill fishery requires us to remember that while the krill are available in huge numbers, they are the food of many whales, and they form a basic link in the food web for the seals, penguins and other birds, squid, and fish of the Antarctic region; see the box in this chapter.

Arrowworms, or **chaetognaths** (see fig. 15.8), are abundant in ocean waters from the surface to the great depths. These macroscopic (2–3 cm) (1 in), nearly transparent, voracious carnivores feed on other members of the zooplankton. Several species of arrowworms are found in the sea, and in some cases a particular species is found only in a certain water mass. The association between organism and water mass is so complete that the species can be used to identify the origin of the water sample in which it is found. In the North Atlantic, the arrowworm, *Sagitta setosa,* inhabits only the North Sea water mass and *S. elegans* is found only in oceanic waters.

Foraminiferans and **radiolarians** are microscopic, single-celled, amoebalike protozoans; they are shown in figure 15.10. Foraminiferans, such as the common *Globigerina,* are encased in a compartmented calcareous covering, or shell, while the radiolarians are surrounded by a silica **test,** or shell. The radiolarian tests are ornately sculptured and covered with delicate spines. Openings in the test allow a continuity between internal protoplasm and an external layer of protoplasm. Pseudopodia (false feet), many with skeletal elements, radiate out from the cell. Radiolarians feed on diatoms and small protozoa caught in these pseudopodia. Both foraminiferans and radiolarians are found in the warmer regions of the oceans. After death, their shells and tests accumulate on the ocean floor, contributing to the sediments. Calcareous foraminiferan tests are found in shallow-water sediments, while the siliceous radiolarian tests, which are resistant to the dissolving action of the seawater, predominate at greater depths, commonly below 4000 m (13,200 ft) (see chapter 2, figs. 2.19 and 2.20). **Tintinnids** (see fig. 15.10) are tiny protozoans with moving hairlike structures, or **cilia.** These organisms are often called bell animals and are found in coastal waters and in the open ocean.

Pteropods (fig. 15.11) are mollusks; they are related to the snails and slugs. They may or may not have a small calcareous shell, depending on the species, but all have a foot that is modified into a transparent and gracefully undulating wing. Their hard, calcareous remains contribute to the bottom sediments in shallow tropic regions. Some pteropods are herbivores and some are carnivores.

Transparent, delicate, and luminescent, the **ctenophores,** or comb jellies (fig. 15.12), float in the surface waters. Some have trailing tentacles; all are propelled slowly by eight rows of beating cilia. The small, round forms are familiarly called sea gooseberries or sea walnuts; by contrast, the beautiful, tropical, narrow, flattened Venus' girdle may grow to 30 cm (12 in) or more in length. A group of Venus' girdles drifting at the surface and catching the sunlight with their beating cilia is a spectacular sight from the deck of a ship. All ctenophores are carnivores, feeding on other zooplankton.

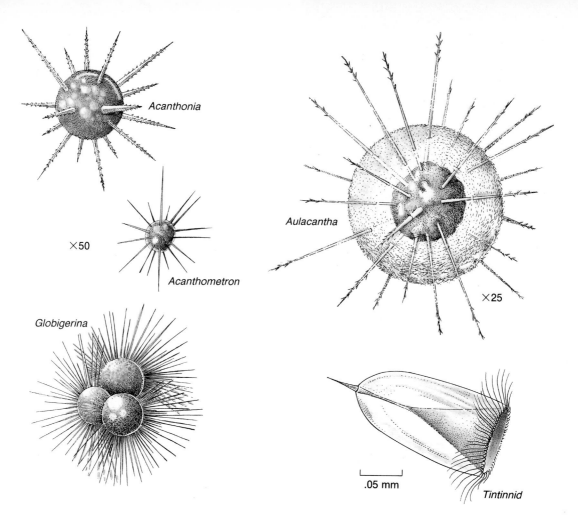

Figure 15.10

Selected members of the radiolaria (*Acanthonia, Acanthometron,* and *Aulacantha*), a foraminifera, (*Globigerina*), and a tintinnid.

The tunicate, another transparent member of the zooplankton, is related to the more advanced vertebrate animals (animals with backbones) through its tadpolelike larval form. **Salps** (see fig. 15.12) are pelagic tunicates that are cylindric and transparent; they are commonly found in dense patches scattered over many square kilometers of sea surface.

Both ctenophores and pelagic tunicates, although jellylike and transparent, are not to be confused with jellyfish (fig. 15.13). True jellyfish come from another and unrelated group of animals, the **Coelenterata** or **Cnidaria.** Some jellyfish, such as the common *Aurelia* and the colorful *Cyanea* with its trailing stinging tentacles, spend their entire lives as drifters. Others, such as *Gonionemus,* a small jellyfish of the Atlantic and Pacific oceans, and *Aequorea,* found in many temperate waters, are members of the plankton for only a portion of their lives, as they eventually settle and change to a bottom-dwelling, attached form similar to a sea anemone. Another group of unusual jellyfish are the

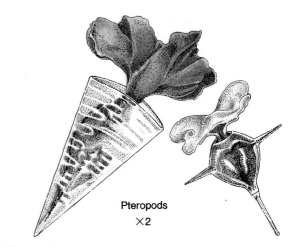

Pteropods
×2

Figure 15.11

The pteropods are planktonic mollusks.

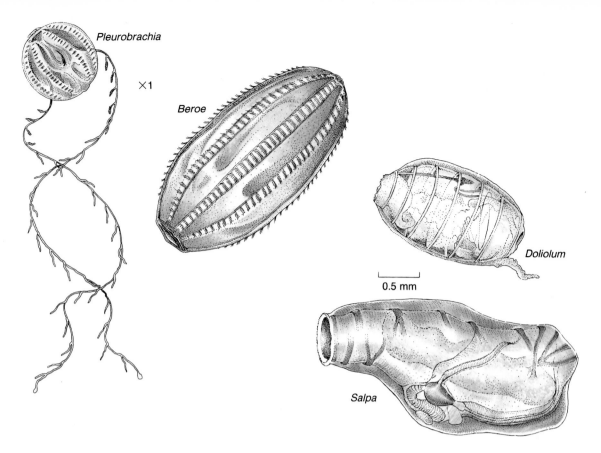

Figure 15.12
The comb jellies (ctenophores) *Pleurobrachia* and *Beroe*. *Pleurobrachia* is often called a sea gooseberry. *Salpa* and *Doliolum* are tunicates.

colonial forms, including the Portuguese man-of-war, *Physalia,* and the small by-the-wind-sailor, *Velella.* Both are collections of individual but specialized animals, some of which have the task of gathering food, reproducing, or protecting the colony with stinging cells, while some form a float.

All of the zooplankton discussed to this point, with the exception of certain jellyfish, spend their entire lives as plankton and are called **holoplankton** (see fig. 15.14). However, an important portion of the zooplankton spends only part of its life as plankton; these are members of the **meroplankton.** The meroplankton include the eggs, larval and juvenile stages of many organisms that spend most of their lives as either free swimmers (such as fish) or bottom dwellers (such as crabs and starfish). For a few weeks, the **larvae** (or young forms) of oysters, clams, barnacles, crabs, worms, snails, starfish, and many other organisms are a part of the zooplankton. The currents carry these larvae to new locations, where they find areas to settle and food sources.

jIn this way, repopulation of areas in which a species may have died out occurs, and overcrowding in the home area is reduced. Sea animals produce larvae in enormous numbers, and so these meroplankton are an important food source for other members of the zooplankton and other animals. The parent animals may produce millions of spawn, but only small numbers of males and females need survive to adulthood to guarantee survival of the stock.

Larvae often look very unlike the adult forms into which they develop (see figs. 15.15 and 15.16). Early scientists who found and described these larvae gave each a name, thinking they had discovered a new type of animal. We keep some of these names today, referring, for example, to the trochophore larvae of worms, the veliger larvae of sea snails, the zoea larvae of crabs, and the nauplius larvae of barnacles.

Other members of the meroplankton include fish eggs, larvae, and juvenile fish. The young fish feed on other larvae, until they grow large enough to hunt for other

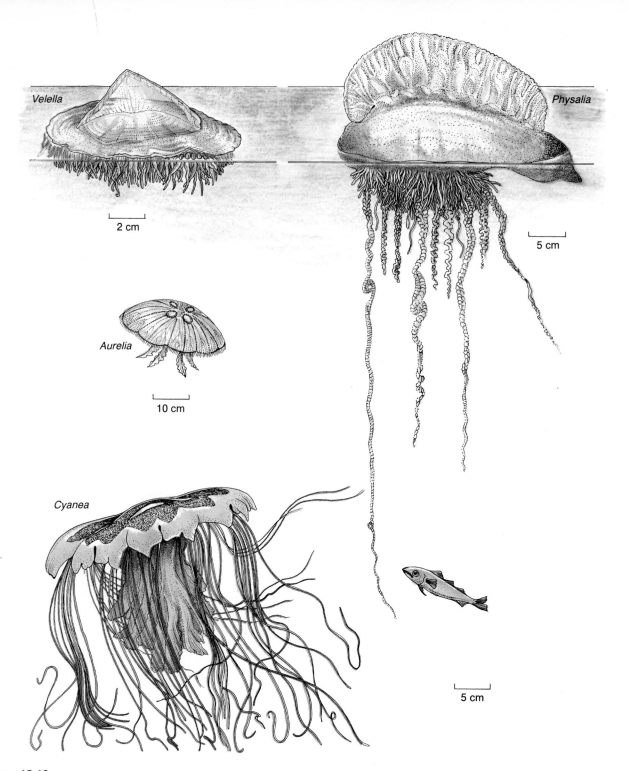

Figure 15.13

Jellyfish belong to the Coelenterata or Cnidaria. *Velella,* the by-the-wind-sailor, and *Physalia,* the Portuguese man-of-war, are colonial forms.

(a)

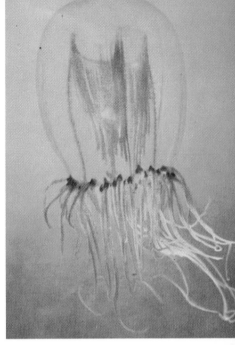

(c)

(b)

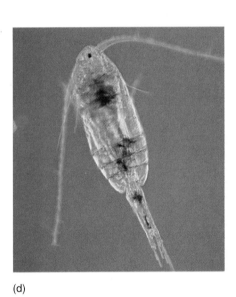

(d)

Figure 15.14

Animals that spend their entire lives in the zooplankton are known as holoplankton and include the single-celled (a) foraminiferans and (b) radiolarians, and the more complex organisms such as (c) the jellyfish (*Polyorchis penicellatus*) and (d) a copepod.

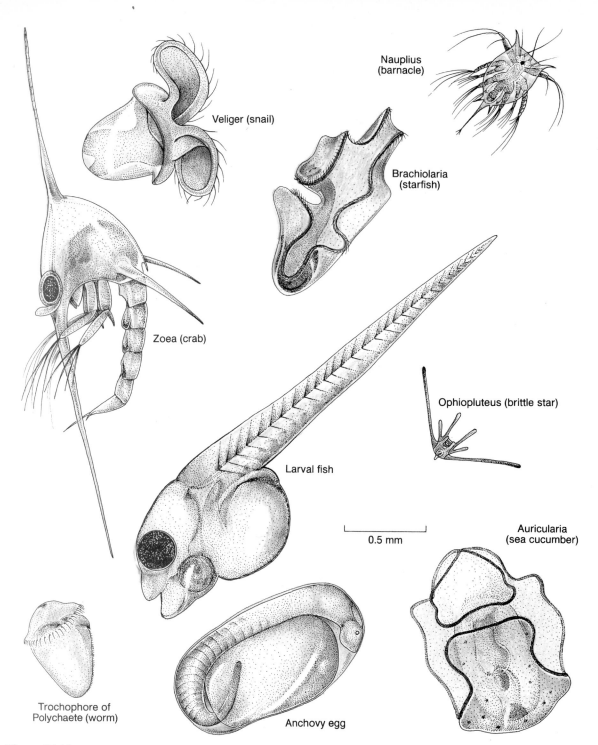

Veliger (snail)

Nauplius
(barnacle)

Brachiolaria
(starfish)

Zoea (crab)

Larval fish

Ophiopluteus (brittle star)

0.5 mm

Auricularia
(sea cucumber)

Trochophore of
Polychaete (worm)

Anchovy egg

Figure 15.15

Members of the meroplankton. All are larval forms of nonplanktonic adults.

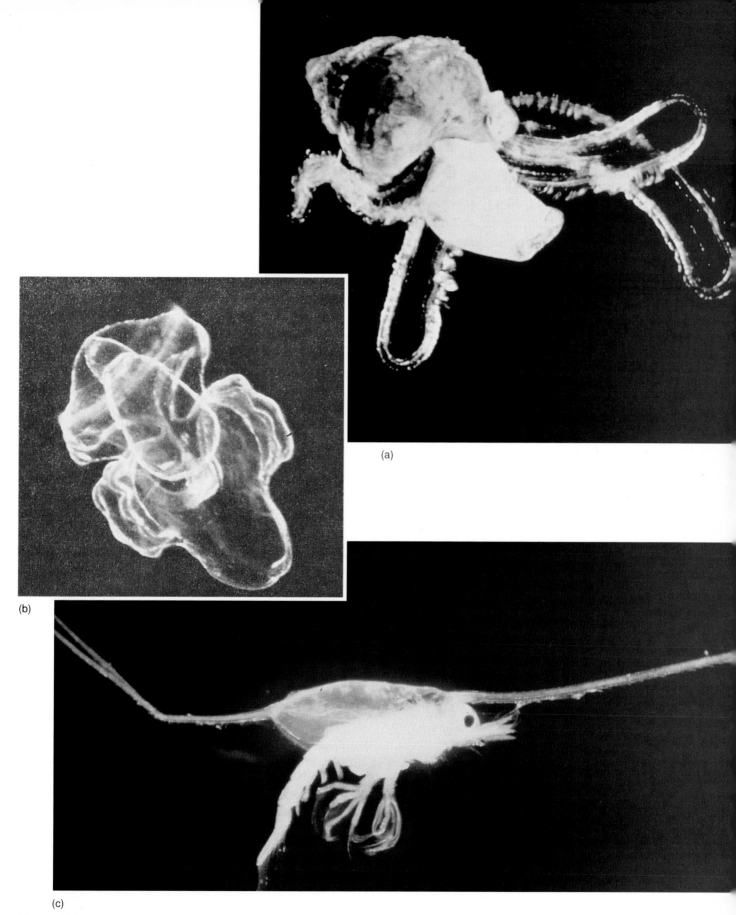

Figure 15.16

Larval forms of benthic marine organisms or meroplankton are an important part of the zooplankton: (a) The larva of a marine snail; (b) the larva of a starfish; (c) the zoea larva, a stage of crab development.

Almost all the life of Antarctica and the surrounding Southern Ocean depends on the sea. The food web of the Southern Ocean is based on the phytoplankton, mainly diatoms, that harness the energy of the sun. Estimates of primary production vary; in summer, in ice-free areas, the average lies between 20 and 100 g of carbon per square meter per year (gC/m2/year). Among the herbivorous zooplankton one species, Euphausia superba, or krill, probably amounts to half the total zooplankton biomass. Krill are the key organisms of the Southern Ocean ecosystem, a unit that includes the area's community of organisms and the environment with which it interacts. Although krill are circumpolar in distribution, their concentration is not uniform, and the greatest concentration occurs in the Weddell Sea, between 0° and 60°W. Some swarms have been estimated at more than 2 million tons.

In summer, the ice pack melts back, the primary production increases, and the krill move toward the surface to graze the phytoplankton over large areas. In winter, primary production is practically gone from the surface waters due to reduced light, ice-pack cover, and increased turbulence; at this time, the krill appear to descend into deeper water and feed on the phytoplankton detritus. The annual production of krill biomass in the Southern Ocean has been estimated at 750 million–1300 million tons.

Squid are an important part of the Antarctic food web. There may be more than twenty species, some depending on krill as their most important food. The total annual consumption of squid by whales, birds, and seals is calculated at about 35 million tons. Some species of Antarctic fish stay in the Southern Ocean year-round, feeding on krill; other species migrate into Antarctic waters each summer to feed on krill. It has been estimated that all Antarctic fish combined may consume as much as 100 million tons of krill each year.

There are few species of birds in Antarctica, but populations are usually very large. Birds of Antarctica feed principally on crustacea (mainly krill and copepods), squid, fish, and carrion. Krill amounts to 78% of all the food they eat. Recent estimates indicate about 115 million tons of krill are eaten annually by birds, either directly or indirectly. In winter, the birds either switch to a diet of squid or fish or they migrate northward. Most penguin species feed on krill supplemented, in some cases, by fish. The two largest penguins, the emperors and the kings, take fish and squid only.

Seven species of seals are found in the Southern Ocean; four species, the crabeater, leopard, Ross, and Weddell, are almost totally confined to the ice zones. Weddell seals eat mainly fish and squid; the crabeater's diet is about 94% krill; krill makes up 37% of the leopard seal's diet. The crabeater seal, now numbering about thirty million, is the most abundant seal in the world. The fur seals at South Georgia feed almost exclusively on krill. Stocks of Antarctic seals at present total about 33 million with a biomass of about 7 million tons. These seal populations annually consume over 130 million tons of krill (two or three times the current consumption by whales) and at least 10 million tons of squid.

Baleen, or plankton-feeding, whales of the Southern Ocean include the blue, fin, sei, minke, humpback, and southern right whales. Before whaling exploitation, they were probably about four times more abundant and had a biomass five times greater than at present. The major food of the baleen whales is krill. In 1904, the whale biomass was about 45 million tons, and whales consumed an estimated 190 million tons of krill.

Competition for food probably limited their size and their numbers. By 1973, the whales had declined to a biomass of about 9 million tons, and they ate about 43 million tons of krill. The reduction in the whale population means that some 150 million tons of krill formerly eaten by whales have become available to the remaining whales, other predators, and human harvesters.

In the past thirty years, there has been an increase in the pregnancy rate of fin and sei whales and an apparent, but controversial, decrease in their age at maturity due, presumably, to the greater availability of food. The minke population may be double what it was before whaling. The crabeater seal populations also have experienced a decrease in the age of sexual maturity (4 years in the 1950s to 2.5 years in the early 1960s), also thought to be the result of increased growth rates based on food abundance. The 14%–17% increase in the population of the South Georgia fur seals is unusually high, probably related to the abundance of krill.

The three most abundant penguins, the chinstrap, Adelie, and macaroni, have shown increases in population. Large increases in king penguins are thought to be due to their feeding on krill-eating squid. We do not know the response of the fish and squid stocks to the increase in available krill, but it is likely to be similar. The total quantity of krill taken by all predators in the Southern Ocean may be about 500 million tons per year. Populations and numbers may change, but directly or indirectly the Antarctic ecosystem depends on krill, which in turn depends on the primary production of the Southern Ocean waters.

Population figures in this section are from Laws, Richard M., *The Ecology of the Southern Ocean,* American Scientist, 1985.

---

foods. Some large seaweeds release **spores,** or reproductive cells, that drift in the plankton until they are consumed or settle out to grow attached to the sea bottom.

15.2 Bacteria

The bacteria are another important group of organisms found in the plankton. Bacteria are the smallest living organisms; they are microscopic, single cells and many have the ability to reproduce by cell division every few minutes when they inhabit a favorable environment. Autotrophic marine bacteria include the photosynthesizing **cyanobacteria,** most abundant in intertidal and estuarine areas and producing dense blooms in warm-water regions, as well as the bacteria associated with the hydrothermal vents of the deep seafloor where they are the primary producers for the animal communities surrounding the vents (refer to chapter 17). Heterotrophic marine bacteria are free living in seawater and exist on every available surface including the seafloor, decaying material, the surface of organisms, and pieces of floating wood and other matter.

Bacteria play an important role in the decay and breakdown of organic matter, returning it to the sea as basic chemicals and compounds to be used again by new generations of plants and animals; refer back to chapter 14. Some are also able to degrade pollutants such as oil; refer back to chapter 12.

They are a significant additional food source for planktonic larvae and a variety of single-celled protozoans. A film of bacteria is found on minute particles of floating organic material; the small size of these particles, with their attached bacterial population, makes them an ideal food for many small zooplankton. They are important in the food webs of the deep sea, where they are consumed by a large number of small animals.

Plant remains sinking into the deep sea have been a recent focus for sampling and research. Bacterial activity on this material decreases with depth and is minimal below 2000 m (6600 ft), but once it has reached the seabed it is subject to vigorous bacterial activity. Samples collected from the sediment surface 4500 m (14,800 ft) deep in mid-oceanic areas of the northeastern Atlantic show a rich community of active bacteria that are able to degrade and transform the plant material faster than had previously been assumed. Experiments showed that the surface water contained 5×10^5 bacteria per milliliter; seafloor samples contained $8-20 \times 10^6$ bacteria per milliliter. The deposit of this plant material shortly after the spring phytoplankton bloom represents an important food resource for the deep-sea animal community.

The extreme environment of the deep seafloor may contain bacteria with abilities that could be very useful to humans, and Japan is preparing to search for these organisms. Japan's *Shinkai 6500*, the world's deepest-diving manned submersible, will be used to bring back bacteria, under pressure, from the bottom of the deep oceans. Organisms will be transferred to an automated laboratory where they can be cultured and isolated in water under pressure and with the correct temperature and pH. Japanese scientists intend to search for bacteria with genes that can be useful in Japan's biotechnology industries.

15.3 Classification Summary of the Plankton

The types of plankton can be categorized in the kingdoms, phyla, and classes, as in the following lists.

I. *Kingdom Monera:* cells, simple and unspecialized; single cells, some in groups or chains.
   A. *Bacteria:* single cells, in chains or groups; autotrophic and heterotrophic, aerobic and anaerobic; important as food source, in decomposition.
   B. *Cyanobacteria:* blue-green algae; autotrophic single cells, in chains or groups, produce some red blooms in sea; phytoplankton.

II. *Kingdom Protista:* grouping of microscopic and mostly single-celled organisms; autotrophs (algae) and heterotrophs (protozoa).
   A. *Phylum Chrysophyta:* golden-brown algae; yellow to golden autotrophic single cells, in groups or chains; contribute to deep-sea sediments; phytoplankton.
      1. *Class Bacillariophyceae:* diatoms.
      2. *Class Chrysophyceae:* coccolithophores, silicoflagellates, and other flagellates.
   B. *Phylum Pyrrophyta:* fire algae; single cells with flagella; produce most red tides; bioluminescence common; usually considered phytoplankton.
      1. *Class Dinophyceae:* dinoflagellates.
   C. *Phylum Sarcodina:* radiolarians, foraminiferans, zooplankton.
   D. *Phylum Ciliophora:* ciliates, zooplankton.

III. *Kingdom Plantae:* plants; primarily nonmotile, multicellular, photosynthetic autotrophs.
   A. *Division Phaeophyta:* brown algae; *Sargassum* maintains a planktonic habit in the Sargasso Sea.

IV. *Kingdom Animalia:* animals; multicellular heterotrophs with specialized cells, tissues, and organ systems; zooplankton. For temporary members of the zooplankton (or meroplankton), see the Meroplankton listed in V.
   A. *Phylum Coelenterata* or *Cnidaria:* radially symmetrical with tentacles and stinging cells.
      1. *Class Hydrozoa:* jellyfish as one stage in the life cycle, including such colonial forms as Portuguese man-of-war.
      2. *Class Scyphozoa:* jellyfish.
   B. *Phylum Ctenophore:* comb jellies; translucent; move with cilia; often bioluminescent.
   C. *Phylum Chaetognatha:* arrowworms; free-swimming, carnivorous worms.
   D. *Phylum Mollusca:* mollusks; the snail-like pteropod is planktonic.
   E. *Phylum Arthropoda:* animals with paired, jointed appendages and hard outer skeletons.
      1. *Class Crustacea:* copepods and euphausiids.
   F. *Phylum Chordata:* animals, including vertebrates, with dorsal nerve cord and gill slits at some stage in development.
      1. *Subphylum Urochordata:* saclike adults with "tadpole" larvae; salps.

V. *Meroplankton:* larval forms from the phyla Annelida (segmented worms), Mollusca (shellfish and snails), Arthropoda (crabs and barnacles), Echinodermata (starfish and sea urchins), and Chordata (fish). See also the classification summaries of the nekton (chapter 16) and the benthos (in chapter 17).

15.4 Sampling the Plankton

The biological oceanographer needs to know what species of plants and animals make up the plankton in a given geographic area, the abundance of these organisms, and where

A **virus** is a noncellular particle made up of genetic material surrounded by a protein coat. Viruses are highly successful parasites; they infect plant, animal, and bacterial cells. Viruses are unable to carry on metabolic activities by themselves and can replicate themselves only inside a host cell. During replication the virus may acquire some of the host's genes; it may then transport these genes to a new cell, leading to exchange of genetic material between organisms.

Recently, microbiologists have discovered an unexpected abundance of viruses in marine waters (10^6–10^9 viruses/mL). One series of samples showed concentrations of viruses between 5×10^6 and 15×10^6 per milliliter during the spring and summer but dropping to less than 10^4 viruses per milliliter in the winter. Most of the viruses appeared to be free in the water, but some were associated with bacteria. It has been assumed that marine bacteria production is maintained in balance by the grazing of single-celled heterotrophs, but these high concentrations of viruses indicate that viral infection may be an important factor in the ecological control of planktonic bacteria and that viruses might actively exchange genes between host populations infected by the same viral strain.

Other researchers report high viral abundance in ocean water and also large numbers of bacteria infected with viruses. Up to 7% of the heterotrophic bacteria and 5% of the cyanobacteria from diverse marine locations contained mature viruses. Estimates of total viral infection from this data conclude that about 32% of the heterotrophic bacteria and about 15% of the cyanobacteria contain mature virus particles at any given time, indicating again that viral infection may be a significant mechanism of mortality in marine bacteria.

In laboratory experiments the addition of viral particles reduced primary productivity by as much as 78%, indicating that infection by viruses could be a factor regulating phytoplankton community structure and primary productivity in the oceans. Because viruses can infect a variety of marine phytoplankton, including diatoms, they may affect ocean food webs by restricting or terminating phytoplankton blooms.

The abundance of viruses in ocean waters indicates routine viral infection of aquatic bacteria, and this is likely to mean that natural genetic engineering experiments in bacterial populations have been occurring for a very long time. New research raises the possibility that marine bacteria, via viral exchange of genetic material, might be able to develop resistance to antibiotics used in aquaculture, or even acquire traits from artificially engineered bacteria released into coastal waters.

---

in the water column they are located. Traditionally, plankton are sampled by towing fine-mesh, cone-shaped nets (fig. 15.17) through the sea behind a vessel or by dropping a net straight down over the side of a nonmoving vessel and pulling it back up like a bucket. After the net is returned to the deck it is rinsed carefully, and the "catch" is collected in a labeled jar. If the net has been hauled vertically through the water in one place, the volume of water that has passed through the net can be calculated from the area of the net's mouth and the distance through which the net was pulled. If the net was towed behind the vessel, the volume can be calculated from the time of the tow and the speed of the tow, which will tell the distance towed. This information can then be used with the area of the net opening to find the volume of water that passed through the net. However, to measure the water volume directly and accurately, a flow meter is placed in the mouth of the net. The catch is then known for the total volume of water sample, but there is no indication of how the species are distributed within that volume.

It is also possible to raise and lower the net as it is being towed horizontally. This movement allows sampling to be averaged over both distance and depth. If the sample is to be taken at a specific depth, the net may be lowered closed and only opened when the desired depth is reached. After the towing operation, the net is again closed before it is brought up through the shallower water.

Today, oceanographers sampling zooplankton use multiple-net systems mounted on a single frame; the nets are opened and closed on command from the ship. The

Figure 15.17
A plankton net is retrieved after a near-surface tow.

frame also carries electronic sensors that relay data on salinity, temperature, water flow, light level, net depth, and cable angle to the ship's computer.

Plankton tows must be rapid enough to catch the organisms but slow enough to let the water pass through the net. If the tow is too fast, the water will be pushed away from the mouth of the net, and less water than expected will be filtered. Size of the mesh from which the net is made also influences the catch; a fine-mesh net clogs rapidly and a large-mesh net loses organisms.

Plankton may also be filtered from samples taken by water bottle or a submersible pump. Whether the sample is taken by net, bottle, or pump, the next step is to determine the number and kinds of plankton in the sample. Because the numbers of organisms are so large, in most cases it is not possible to directly inspect or count the total catch. Instead the catch may be precisely subdivided, and a subsample may be checked under the microscope. If the amount of plankton is of greater interest than the kinds of organisms, an electronic particle counter may be used to find the total count (population density), or a subsample may be dried and weighed to determine its mass.

Sonar may be used to determine the quantity of zooplankton directly; the echo returned to the ship is related to the density of the zooplankton but does not determine the species present. The abundance of phytoplankton can be determined by dissolving the chlorophyll pigment out of the sample and measuring the pigment concentration. New optical instrumentation has been designed to measure the natural fluorescent signal coming from chlorophyll pigment in phytoplankton cells. These measurements are made at sea, directly and instantaneously at depth, and then related to phytoplankton abundance.

All these methods allow scientists to record population densities or to inspect individual organisms, but they do not tell us how these organisms behave in their environment. A new video plankton recorder designed at the Woods Hole Oceanographic Institution uses a strobe light with four video cameras at four different magnifications to photograph the plankton. The strobe flashes sixty times a second, capturing images of the organisms that researchers hope will allow them to learn about the swimming, feeding, and reproduction of the plankton. Eventually they hope to program the system to recognize different kinds of plankton, enabling them to acquire species and population information while the system is running at sea.

15.5 Practical Considerations: Marine Toxins

Red Tides

Toxic blooms of single-celled organisms that discolor seawater are often known as **red tides;** most are produced by certain species of dinoflagellates. These blooms may or may not be poisonous to fish and other organisms and may or may not produce symptoms of paralytic shellfish poisoning (PSP), neurotoxic shellfish poisoning (NSP), or diarrhetic shellfish poisoning (DSP) in humans eating clams, mussels, and oysters that have ingested the dinoflagellates. In North American waters several different dinoflagellates produce red tides; *Gonyaulax, Alexandrium,* and *Gymnodinium* (previously *Ptychodiscus*) are toxic, but *Noctiluca* is not. See figure 15.6. Species of *Gonyaulax* and *Alexandrium* found in temperate latitudes are generally nontoxic to the shellfish themselves, but the toxins produced by these dinoflagellates are concentrated in the tissues of the shellfish as they feed on the bloom. It is the toxins that produce PSP, NSP, or DSP, in the humans eating the affected shellfish. *Gymnodinium* and a species of *Gonyaulax* found in the warmer waters of the Gulf of Mexico kill fish. *Gonyaulax* also kills shrimp and crab; *Gymnodinium* does not.

It appears to some scientists that the red tide problem is growing around the world. In 1986 a major outbreak of red tide caused by the dinoflagellate *Gymnodinium brevi* struck the Texas Gulf Coast beaches. It spread 500 km (300 miles) along the coast and killed more than 22 million fish over more than two months. Harvesting of shellfish was banned along three-quarters of the Texas coast south of Galveston. The state of Texas lost at least $1.4 million in oyster production, with a total economic loss of $3.7 million; businesses based on tourism and recreation suffered severely. That same year hundreds of bottle-nosed dolphins died along the coasts of New Jersey and Maryland. A bloom of *G. brevi* occurred in the warm waters off Florida's west coast. The winter was mild; the red tide survived the winter, and the current carried it around Florida and up the East Coast of the United States. The red tide arrived off the Carolinas at about the same time the menhaden fish began their annual summer migration up the East Coast. The menhaden fed on the *G. brevi* and accumulated the toxin in their livers. After the dolphins fed on the menhaden, the toxin attacked the heart muscles of the dolphins or caused the animals to lose heat uncontrollably, killing them. In 1987 and 1988 a red tide along the Gulf Coast of Florida again spread northward, this time to North Carolina, and again dolphins died. The shellfisheries were closed and economic losses reached $25 million. In 1990 the first confirmed outbreak of DSP appeared in North America and was traced to a dinoflagellate bloom in Canadian waters. In 1992 PSP toxin was detected for the first time in the guts of Dungeness crabs from Alaskan waters.

No one knows precisely what triggers the sudden bloom of these organisms, but red tides often happen in spring and summer after heavy rains have produced a land runoff of nutrient-rich water. Dinoflagellates produce a cyst or resting cell that settles to the bottom and mixes with the sediments. In the shallow waters of bays and harbors along the New England coast, temperature and light appear to play major roles in activating the cysts. In deeper coastal waters the cysts seem to have a natural biological clock or annual cycle of reactivation. It is thought that sudden disturbances of the bottom by either natural or artificial means (such as slumping of sediments or dredging) stimulate the cysts into activity and rapid reproduction.

Some scientists think that the continuous addition of nitrates and phosphates to coastal waters from both sewage and agricultural runoff may be partly to blame for the apparent increase in the number and severity of red tide episodes. Extra nutrients mean extra growth and may explain why instead of a series of small blooms, the world's coastal areas are experiencing such massive outbreaks. An additional factor may be the increased commercial ship traffic that brings hitchhiking cells to new environments. In a survey of cargo vessels entering Australian ports, 40% were found to have viable dinoflagellate cysts in their ballast tanks. Other scientists do not believe that there is a rise in toxic blooms; they point to better statistics, communication, and reporting as well as increased eating of shellfish and fish. However, the experience of the Japanese in the inland sea of Japan suggests otherwise: Red tides increased from forty a year in 1965 to more than three hundred a year in 1973. In 1972 authorities introduced controls to cut the nutrients entering the sea by half; the frequency of red tides peaked in 1975 and has been declining since.

The toxins that produce PSP are powerful nerve poisons that can cause paralysis and death if the breathing centers are affected. Some species of red tide dinoflagellates produce several toxins; one species of *Gymnodinium* can produce at least five different toxins. The saxitoxins that are found in butter clams but are produced by *Alexandrium* and *Gonyaulax* are fifty times more lethal than strychnine. The toxins are not affected by heat, so cooking the shellfish does not neutralize the poison. Even after the visible signs of red water due to a dinoflagellate bloom have disappeared, the shellfish can retain the toxin in their tissues for long periods, and so the beaches are kept closed to shellfish harvesting. NSP symptoms are similar to those of basic food poisoning; airborne, wind-spread cells may cause irritated eyes, runny noses, and persistent cough. DSP outbreaks may have been occurring throughout history and been attributed to bacterial contamination; only recently have researchers linked DSP outbreaks to the presence of a dinoflagellate. Between 1976 and 1982 more than thirteen hundred documented cases of DSP occurred in Japan; in 1981 over five thousand cases were reported from Spain. Other outbreaks have occurred in Norway and Sweden and along the French coast. Scientists speculating why the toxin is produced believe it is a defense against predators.

Just as not all red tides are toxic, not all red water is caused by dinoflagellates. The Red Sea received its name because of dense blooms of nontoxic cyanobacteria with large amounts of red pigment. The Gulf of California has been called the Vermilion Sea for the same reason. The Indian Ocean has red tides due to the presence of a toxic cyanobacterium, not a dinoflagellate.

Other Toxic Blooms

In 1987, at Prince Edward Island, Canada, three people died and more than one hundred others became sick from eating shellfish contaminated with domoic acid. The domoic acid

(a)

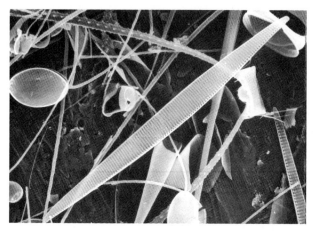

(b)

Figure 15.18

Electron photomicrographs of two species of the domoic acid producing diatom *Pseudonitzschia*. (a) *P. pungens* (1000×) responsible for amnesic shellfish poisoning in Prince Edward Island, Canada. (b) *P. australis* (1400×) closed U.S. West Coast shellfish and crab fisheries.

was traced to a bloom of *Pseudonitzschia pungens*, (fig. 15.18a)—a diatom, not a dinoflagellate; diatoms had not been known to produce toxins until this outbreak occurred. In 1991 pelicans eating anchovies off the California coast were dying from domoic acid produced by another species, *P. australis* (fig. 15.18b). Shellfish and crab fisheries along the California, Oregon, and Washington coasts were closed.

Domoic acid poisoning in humans may cause short-term memory loss and is called amnesic shellfish poisoning (ASP). Other symptoms include nausea, muscle weakness, disorientation, and organ failure.

The reasons for the sudden appearance of toxic levels of domoic acid are unknown, although speculations are similar to those proposed for the dinoflagellate outbreaks. One researcher has suggested that when excess phosphorus and nitrogen are present the diatom populations grow explosively, and when the phosphorus is depleted some

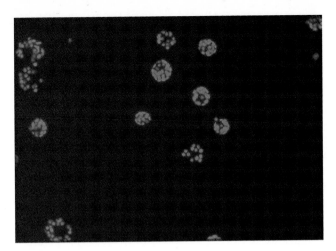

Figure 15.19

Heterosigma is associated with fish mortality in pen-reared fish stocks. Photo taken with fluorescent techniques.

species secrete toxins that kill animals, releasing the phosphorus in the animals' tissues and allowing the diatoms to continue their bloom.

Another diatom, *Chaetoceros,* has been involved in the mortality of pen-reared salmon. It is thought that the spiny cells of these organisms cause mechanical injury or clogging of gill tissue, leading to bacterial invasion and suffocation, but the exact mechanism is unknown. *Heterosigma,* neither a diatom nor a dinoflagellate but a member of a small group of green and golden-brown flagellates, blooms orange-brown in estuaries during periods of sunlight, warm water, and high nutrient concentrations; see figure 15.19. It is associated with fish mortality in Japan, Scotland, New Zealand, and the U.S. Pacific Northwest.

Ciguatera Poisoning

It is estimated that each year between ten thousand and fifty thousand people eating fish in the tropical regions of the world are affected by **ciguatera** poisoning. More than four hundred species of fish have been found to be affected, and the presence of ciguatoxin in the fish dramatically affects the development and growth of inshore fisheries in these areas. The situation is complex, because not all of the fish of the same species caught at the same time and in the same place are toxic.

There is no way to prepare an affected fish to make it safe to eat. Symptoms of ciguatera poisoning are extremely variable and may include headache, nausea, vomiting, abdominal cramps, possible irregular pulse beat, reduced blood pressure, and, in severe cases, convulsions, muscular paralysis, hallucinations, and death. Symptoms can occur in various combinations, and no proven antitoxin is known.

In the United States, the number of cases may be over two thousand per year. These are clustered in Florida, Hawaii, the Virgin Islands, and Puerto Rico. Over 80% of the resident adults in the United States and British Virgin Islands report having been poisoned at least once. Ciguatera poisoning is thought to be underreported in Hawaii and more serious than generally acknowledged. An outbreak involving twelve persons in Maryland and a single case in Boston involved consumption of grouper shipped from Florida to local restaurants.

Several dinoflagellates are associated with ciguatera poisoning, but the dinoflagellate *Gambierdiscus toxicus* is most often associated with the problem. This dinoflagellate was discovered only ten years ago by scientists at the University of Tokyo, but attempts to grow *G. toxicus* in the laboratory have been disappointing because lab-cultured organisms produce less toxin than those living in their natural surroundings. Ciguatoxic dinoflagellates live in close association with many types of seaweeds and appear to need nutrients exuded by the seaweeds. They flourish in areas of human or natural disturbance such as dredging, hurricanes, and destruction of coral reefs. Perhaps the seaweeds and the dinoflagellates are mixed into the water column during such disturbances and dispersed. In some areas it is considered likely that more than one organism produces a combination of toxins contributing to ciguatera poisoning.

The dinoflagellates are eaten by herbivores and the ciguatoxins move through the food web. Japanese researchers in the Gilbert Islands found an evolution of toxicity over the years. Initially only a few species are toxic. At the peak of the outbreak almost all reef fish become toxic, and in the final stages only large eels and certain snappers and groupers remain toxic. This cycle appears to take at least eight years and, in this case, points to a food-chain cycle in which the herbivorous fish become toxic first, followed by the carnivores. Exactly how this occurs is not known with certainty because there is no technology to analyze the toxins, and how ciguatera moves up a food chain without killing the fish that consume it is also unknown. Present testing can identify toxic fish only by bioassay that requires feeding suspected fish to test animals, but simple color tests are expected to be available in the near future.

Ciguatera is an international problem. It has delayed development of Egyptian Red Sea fisheries. Sri Lanka reports hundreds of cases each year. In New Guinea, it is believed that thousands are poisoned each year, but that most cases go unreported as they are attributed to magic. Some islands in the western Pacific have been abandoned because of local ciguatera problems. A bottom fishery in Samoa is required to discard all red snappers (as much as 50% of the catch). It hampers the fledgling Puerto Rico fishing industry, and the loss to the Floridean/Caribbean/Hawaiian seafood industry is estimated at $10 million annually. Other losses include export markets, the loss of use of the banned fish, the cost of treatment and the time lost from employment by victims, and the costs associated with monitoring and implementing fishing and marketing regulations.

Summary

The plankton are the drifting organisms. The microscopic plankton are divided into groups by size. Phytoplankton are autotrophic single cells or filaments. *Sargassum* is the only large planktonic seaweed. The diatoms are found in cold, upwelled water; they are yellow-brown, with a hard, transparent frustule, and they store oil, which increases their buoyancy. Centric forms are round; pennate forms are elongate. Diatoms reproduce rapidly by cell division and make up the first trophic level of the open sea.

Dinoflagellates are single cells with both autotrophic and heterotrophic capabilities. Their cell walls are smooth or are heavily armored with cellulose plates. They, too, reproduce by cell division. These organisms are responsible for much of the bioluminescence in the oceans. Coccolithophores and silicoflagellates are very small autotrophic members of the phytoplankton.

Some herbivorous zooplankton reproduce several times a year, while others reproduce only once, depending on the water temperature and phytoplankton food supply. Carnivorous zooplankton are important in the recycling of nutrients to the phytoplankton. Heavy concentrations of zooplankton are found at convergence zones and along density boundaries. Zooplankton migrate toward the sea surface at night and away from it during the day, forming the deep scattering layer.

Zooplankton members that spend their entire lives in the plankton are called holoplankton. The copepods and euphausiids are the most abundant members of the holoplankton. Euphausiids are also known as krill; they form a basic food of the baleen whales. Krill is the zooplankton base for all the Antarctic ecosystems; recently it has been harvested for human consumption with mixed success.

Other small members of the holoplankton are the carnivorous arrowworms, the calcareous-shelled foraminiferans, the delicate, silica-shelled radiolarians, the ciliated tintinnids, and the swimming snails, or pteropods. Large zooplankton include the comb jellies, salps, and jellyfish; all are nearly transparent, but each belongs to a different zoologic group.

The meroplankton are the juvenile (or larval) stages of nonplanktonic adults. This group comprises fish eggs, very young fish, and the larvae of barnacles, snails, crabs, starfish, and many other nonplanktonic animals. The spores of seaweeds and the marine bacteria are also planktonic.

Plankton sampling is done with a plankton net or with a water bottle. The kinds of organisms in a sample are determined microscopically; the numbers of organisms are counted, or samples are dried and weighed.

Heavy blooms of dinoflagellates and some other kinds of phytoplankton produce red tides. Some red tides are toxic; others are not. The toxin is concentrated in shellfish and produces paralytic and other types of shellfish poisoning in humans and sometimes in other animals. Red tides appear to be triggered by the particular combination of environmental factors, the disturbance of dormant dinoflagellates, and the addition of increasing amounts of nutrients to coastal waters. Domoic acid produced by diatoms is a newly discovered toxin. Ciguatoxin is another dinoflagellate product that affects humans and hampers fishery development around the world.

Key Terms

ultraplankton	euphausiid
nannoplankton	krill
microplankton/net plankton	baleen whale
autotrophic	chaetognath
filament	foraminiferan
diatom	radiolarian
dinoflagellate	test
coccolithophore	tintinnid
fucoxanthin	cilia
radial symmetry	pteropod
centric diatoms	ctenophore
bilateral symmetry	salp
pennate diatoms	Coelenterata/Cnidaria
frustule	colonial forms
auxospore	holoplankton
bloom	meroplankton
diatomaceous earth	larva
flagella	spore
heterotrophic	ecosystem
coccolith	cyanobacteria
crustacean	virus
copepod	red tide
	ciguatera

Study Questions

1. Why does a pycnocline located above the compensation depth promote a phytoplankton bloom?
2. Why are meroplankton produced in such large numbers?
3. When the discoloration has left the water after a PSP red tide, the shellfish may not be safe to eat. Explain why.
4. Patches with abundant populations of zooplankton are frequently found separated by patches with sparse populations. How does this help to assure survival from predators?
5. If the krill of the Southern Ocean were heavily harvested for human consumption, explain the possible effects on the rest of the organisms in that area.
6. Describe four ways to subdivide the plankton.
7. Why are there more planktonic centric diatoms and more benthic pennate diatoms?
8. Discuss what happens to a diatom population if no auxospores form.
9. Why is a planktonic stage important to a nonplanktonic adult?
10. Discuss how plankton may maintain themselves in a given region of the ocean even though there are currents flowing through that region.
11. When you are sampling plankton with a plankton net, how can you determine the quantity of the plankton in a volume of water? Assume that you know (1) the cross-sectional area of the net and (2) the length of time you towed the net and the speed at which you towed the net, or you know (1) the cross-sectional area of the net and (2) the distance you towed the net.

12. Distinguish between diatoms and dinoflagellates, between euphausiids and copepods. Which are heterotrophs and which are autotrophs?

13. Make a simple food web for the Southern Ocean around Antarctica. Include fish, whales, seals, squid, penguins, and sea birds.

14. Episodes of toxic phytoplankton blooms appear to be increasing along the world's coasts. (1) List possible reasons for this increase. (2) Distinguish among PSP, NSP, ASP, and ciguatera.

15. Toxic materials, such as oil, may form a thin layer at the sea surface where many plankton also accumulate. How might this affect populations of the nekton and the benthos?

Suggested Readings

General

Lerman, M. 1986. *Marine Biology, Environment, Diversity and Ecology.* Benjamin-Cummings, Menlo Park, Calif. 535 pp.

Life in the Sea, readings from *Scientific American,* 1982. Freeman, San Francisco, 248 pp.

Niesen, T. M. 1982. *The Marine Biology Coloring Book.* Harper & Row, New York. 218 pp.

Sumich, J. L. 1992. *An Introduction to the Biology of Marine Life,* 5th ed. Wm. C. Brown, Dubuque, Ia. 449 pp.

Plankton

Baden, D. G. 1990. Toxic Fish: Why They Make Us Sick. *Sea Frontiers* 36 (3): 8–14. (Ciguatera poisoning.)

Beddington, J. R., and R. M. Beddington. May, 1982. The Harvesting of Interacting Species. *Scientific American* 247 (5): 62–69. (Krill: harvesting and its effect on the ecosystem.)

Cherfas, J. 1990. The Fringe of the Ocean—Under Siege from Land. *Science* 248 (4952): 163–65. (Red tides and toxic blooms.)

Coleman, B. C., R. N. Doetsch, and R. D. Sjoblad. 1986. Red Tide: A Recurrent Marine Phenomenon. *Sea Frontiers* 32 (3): 184–92.

Culotta, E. 1992. Red Menace in the World's Oceans. *Science* 257 (5076): 1476–77.

Hammer, W. 1984. Krill, Untapped Bounty from Sea. *National Geographic* 165 (5): 627–43.

Hardy, A. 1958. *The Open Sea: Its Natural History,* part 1; The World of Plankton. Houghton Mifflin, Boston. 335 pp.

Laws, R. 1985. The Ecology of the Southern Ocean. *American Scientist* 73: 26–40.

Mistry, R. 1992. Lilliputian World of Plankton. *Sea Frontiers* 38 (1): 42–47.

Morse, A. N. C. 1991. How Do Planktonic Larvae Know Where to Settle? *American Scientist* 79: 154–68.

Nicol, S. 1987. Krill, Food of the Future? *Sea Frontiers* 33 (1): 12–17.

Nicol, S., and W. de la Mare. 1993. Ecosystem Management and the Antarctic Krill. *American Scientist* 81 (1): 36–47.

Rodgers, D. L., and C. Muench. 1986. Ciguatera: Scourge of Seafood Lovers. *Sea Frontiers* 32 (5): 338–46.

Smith, D. L. 1977. *A Guide to Marine Coastal Plankton and Marine Invertebrate Larvae.* Kendall/Hunt, Dubuque, Ia. 161 pp.

Wrobel, D. J. 1990. Transient Jewels. *Sea Frontiers* 36 (2): 8–17. (Jellyfish and combjellies.)

16

The Nekton: Free Swimmers of the Sea

16.1 The Mammals
Whales and Whaling
Dolphins and Porpoises

Box: Echolocation and Communication

Seals, Sea Lions, and Walruses
Sea Otters
Sea Cows
Marine Mammal Protection Act
16.2 The Reptiles
Sea Snakes
Sea Turtles
16.3 The Squid
16.4 The Fish
Sharks and Rays
Commercial Species of Bony Fish
Deep-Sea Species of Bony Fish
16.5 Classification Summary of the Nekton
16.6 Practical Considerations: Commercial Fisheries
Anchovies
Tuna
Salmon
Redfish
Sharks
Problems and Policies
Fish Farming
Summary
Key Terms
Study Questions
Suggested Readings

We need another and a wiser and perhaps a more mystical concept of animals. Remote from universal nature, and living by complicated artifice, man in civilization surveys the creature through the glass of his knowledge and sees thereby a feather magnified and the whole image in distortion. We patronize them for their incompleteness, for their tragic fate of having taken form so far below ourselves. And therein we err, and greatly err. For the animal shall not be measured by man. In a world older and more complete than ours they move finished and complete, gifted with extensions of the senses we have lost or never attained, living by voices we shall never hear. They are not brethren, they are not underlings; they are other nations, caught with ourselves in the net of life and time, fellow prisoners of the splendour and travail of the earth.

Henry Beston,
from *The Outermost House*

Butterfly fish, Molokini Island, Hawaii.

*T*he nekton are the free swimmers of the oceans, moving through the water independent of the motion of currents and waves. Approximately five thousand species of nekton swim freely through the pelagic and neritic regions of the oceans. Most of the nekton are fishes; others are marine mammals, ocean-living reptiles, and squid. Members of the nekton are able to move toward their food and away from their predators; many occupy the top trophic levels of the marine food webs, as either herbivores or carnivores. Sizes range from the smallest fish of the tropic reefs to the largest animal ever to have existed on this earth, the blue whale. This chapter examines representative swimmers in coastal waters and in the open sea; it contains additional information on their harvests, including quantities taken, methods used, and current population status.

16.1   The Mammals

Marine **mammals** are warm-blooded air breathers. They may spend all of their lives at sea, or they may return to land to mate and give birth. In either case, the young are born live and are nursed by their mothers. Included in this group are large and small whales (including porpoises and dolphins), seals, sea lions, walruses, sea otters, and sea cows.

Whales and Whaling

Whales belong to the mammal group called **cetaceans.** Some cetaceans are toothed, pursuing and catching their prey with their teeth and jaws (for example, the killer whale, the sperm whale, and the porpoises); others have mouths fitted with strainers of **baleen** or whalebone through which they filter the seawater and remove the krill. The blue, finback, right, sei, gray, and humpback whales are baleen whales. The mouths of toothed and baleen whales are compared in figure 16.1. The blue, finback, and right whales swim open-mouthed and engulf water and plankton. The tongue acts to push the water through the baleen, and the krill are trapped. The sei whale swims with its mouth partly open and uses its tongue to remove the organisms trapped in the baleen. The humpbacks circle an area rich in krill and expel air to form a circular screen, or a net of bubbles. The krill bunch together toward the center of this net, and the whales pass through the dense cloud of krill and scoop them up. The gray whale feeds mainly on small bottom crustaceans and worms.

Some whales migrate seasonally over thousands of miles; other whales stay in cold water and migrate over relatively short distances. The California gray whale and the humpback whale make long migratory journeys. In the summer, the California gray whale is found in the shallow waters of the Bering Sea and the adjacent Arctic Ocean,

(a)

(b)

Figure 16.1

(a) The killer whale (*Orcinus orca*) is a toothed whale. (b) Bowhead whale (*Balaena mysticetus*) baleen. This dead whale has been hauled onto the ice and is shown lying on its back.

where they feed all summer, building up layers of fat and blubber. In October, when the northern seas begin to freeze, the whales begin to move south, and in December, the first gray whales arrive off the west coast of Baja California. Here they spend the winter in the warm, calm waters of sheltered lagoons, where the gray whales calve and mate but find little food; by the time they leave for their northward migration in February and March, they have lost 20%–30% of their body weight. The animals move north

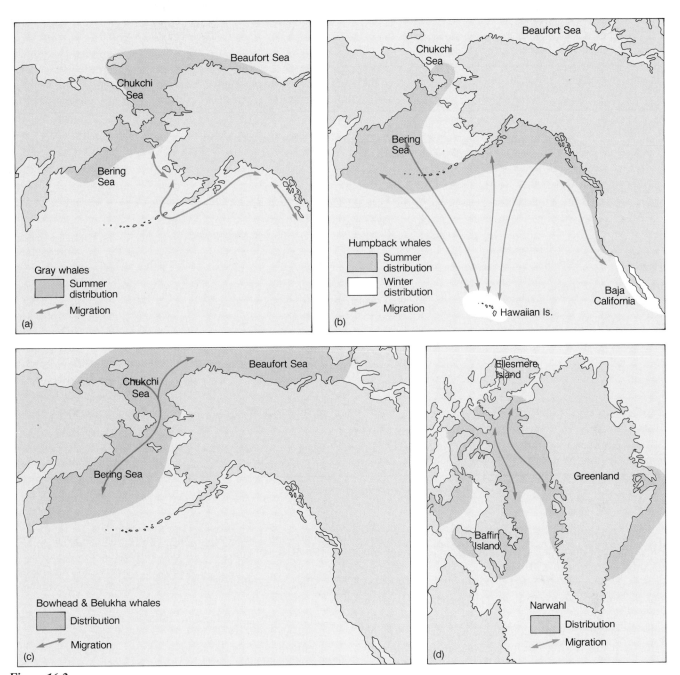

Figure 16.2

Migration paths and seasonal distribution of whales. (a) The California gray whale, (b) the humpback whale, (c) the bowhead and belukha or beluga whales, and (d) the narwhal.

singly or in twos or threes, sometimes in groups of ten to twelve up the west coast of the United States, Canada, and Alaska. Moving at about 5 knots day and night, the whales make their annual 18,000-km (11,000-mi) migratory journey to link areas that provide abundant food with areas that insure reproductive success. (See fig. 16.2a.) Protection of the gray whales' calving and winter grounds in Baja California has allowed the population to build, and in December, 1992, the gray whale became the first marine mammal to be recommended for removal from the U.S. endangered species list.

The humpback whale also has well-defined migration patterns. Humpbacks are found in three geographically and reproductively isolated populations: in the North Pacific, North Atlantic, and Southern Ocean. The North Pacific humpback spends the summer feeding in the Gulf of Alaska, along the northern islands of Japan, and in the Bering Sea; in winter, these North Pacific humpbacks migrate to the Mariana Islands in the west Pacific, the Hawaiian Islands in the central Pacific, and along the west coast of Baja California in the eastern Pacific. At these warmer latitudes, calves are born and mating takes place. (See fig. 16.2b.)

Table 16.1

Principal Characteristics of the Great Whales

	Distribution	Breeding grounds	Average weight (tons)	Greatest length (m)	Food
Toothed whales					
sperm	worldwide; breeding herds in tropic and temperate regions	oceanic	35	18	squid, fish
Baleen whales					
blue	worldwide; large north-south migrations	oceanic	84	30	krill
finback	worldwide; large north-south migrations	oceanic	50	25	krill and other plankton, fish
humpback	worldwide; large north-south migrations along coasts	coastal	33	15	krill, fish
right	worldwide; cool temperate	coastal	(50)	17	copepods and other plankton
sei	worldwide; large north-south migrations	oceanic	17	15	copepods and other plankton, fish
gray	North Pacific; large north-south migrations along coasts	coastal	20	12	benthic invertebrates
bowhead	Arctic; close to edge of ice	unknown	(50)	18	krill
Bryde's	worldwide; tropic and warm temperate regions	oceanic	17	15	krill
minke	worldwide; north-south migrations	oceanic	10	9	krill

Source: After K. R. Allen, *Conservation and Management of Whales,* Washington Sea Grant Program, 1980.
Key: ( ) = estimate.

The bowhead, the beluga or belukha, and the single-tusked narwahl are whales that remain in cold water but still migrate over short distances each year. The largest population of bowhead whales is found in the Bering, Chukchi, and Beaufort seas; this whale spends nearly all of its life near the edge of the Arctic ice pack. Bowheads, singly or in pairs, often accompanied by belugas, migrate north from the Bering Sea to feed in the Beaufort and Chukchi seas as the ice recedes in the spring, returning south to the Bering Sea in groups of up to fifty as the ice begins to extend in the winter. Their mating and reproductive cycles are not well known, but they probably mate during the spring migration and calve sometime during April and May. (See fig. 16.2c.)

The narwhal is the most northerly whale and is found only in Arctic waters, most commonly on both sides of Greenland. In summer, they move north along the coasts of Ellesmere and Baffin islands, and in the autumn, they return south to the waters along the Greenland coast. (See fig. 16.2d.)

The whales of stories and songs are the "great whales": the blue, sperm, humpback, finback, sei, and right whales (see fig. 16.3 and table 16.1); they are also the whales of the whaling industry. The earliest-known European whaling was done by the Norse between A.D. 800 and 1000. The Basque

people of France and Spain hunted whales first in the Bay of Biscay, and then in the 1500s the Basque whalers crossed the Atlantic to Labrador. They set up whaling stations along the Labrador coast to process the blubber of bowhead and right whales into oil for transport back across the Atlantic. In Red Bay, Labrador, the operation reached its peak in the 1560s and 1570s, when one thousand people gathered seasonally to hunt whales and produce 500,000 gallons of whale oil each year. By 1600, whaling had become a major commercial activity among the Dutch and the British, and at about the same time the Japanese independently began harvesting whales. In the 1700s and 1800s the whalers from the United States, Great Britain, the Scandinavian countries, and other northern European countries pursued the whales far from shore, hunting them for their oil and baleen. Kills were made using hand-held harpoons, and the whales were cut up and processed on land or on board the ships at sea. Long voyages, intense effort, and dangerous combat between whalers and whales characterized these whale hunts.

In 1868, Svend Føyn, a Norwegian, invented the harpoon gun with its explosive harpoon and changed the character of whaling. Ships were motorized, and in 1925, harvesting was increased further by the addition of great factory ships, to which the small, high-speed whale-hunting

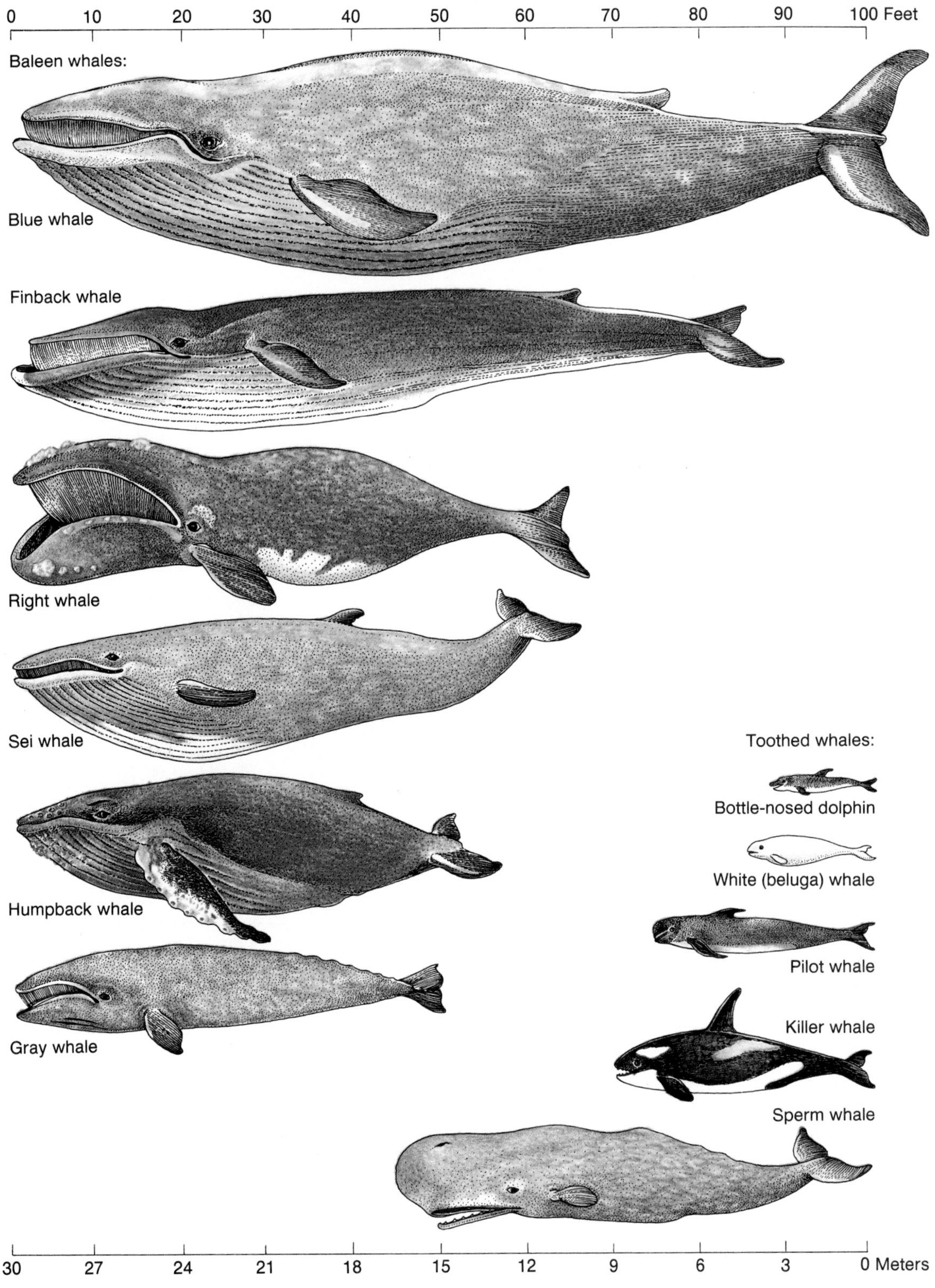
0
10
20
30
40
50
60
70
80
90
100 Feet
Baleen whales:
Blue whale
Finback whale
Right whale
Sei whale
Humpback whale
Gray whale
Toothed whales:
Bottle-nosed dolphin
White (beluga) whale
Pilot whale
Killer whale
Sperm whale
30
27
24
21
18
15
12
9
6
3
0 Meters

Figure 16.3
Relative sizes of baleen and toothed whales.

vessels brought the dead whales for processing. This system freed the fleets, now centered in the Antarctic, from dependence on shore stations. These methods continued into the twentieth century, greatly increasing the efficiency of the hunt and rapidly depleting the whale stocks of the world.

In the 1930s, the annual blue whale harvest was between 4% and 6% of the estimated total original population. Many of the captured females carried young, and harvesting at this rate reduced the population of blue whales to less than 4% of its original numbers, threatening the species with extinction. In 1946, representatives from Australia, Argentina, Britain, Canada, Denmark, France, Iceland, Japan, Mexico, New Zealand, Norway, Panama, South Africa, the Soviet Union, and the United States met in

Table 16.2

Estimated Population Status of Great Whales

Species	Original population (estimated)	Current population (estimated)
sperm	1,377,000	1,900,000
blue	166,000–226,400	8,555
finback	449,700–452,700	105,200–121,900
humpback	119,400	10,500
right	120,000	3,050–3,250
sei	108,100–109,400	25,110
gray	15,000–20,000	21,113
bowhead	54,680	>7,800
Bryde's	92,000	30,200–55,500
minke	320,000	331,800

Source: National Marine Fisheries Service, National Oceanographic and Atmospheric Administration, 1989.

Washington, D.C., to establish the International Whaling Commission (IWC). The regulations drawn up prohibited the killing of the remaining gray, bowhead, and right whales and of cows with calves. Opening and closing dates for whaling and minimum size data were set for each species harvested. Although each factory ship carried an observer, they could only report offenses and recommend disciplinary action; the government of the country registering the ship involved was responsible for any action. With IWC and its regulations in effect, 31,072 whales were killed in 1951, more than 50,000 in 1960, and in 1962 over 66,000 animals were killed. The annual harvest of whales between 1910 and 1977 is given in figure 16.4. To understand what these numbers mean to the whale populations, it should be remembered that it took New Bedford, Massachusetts, whalers over one hundred fifty years to kill 30,000 whales. The current population status of the great whales is given in table 16.2.

As the 1970s drew to a close, the era of commercial whaling appeared to be closing as well. In 1979, the IWC placed a moratorium on all whaling in the Indian Ocean and outlawed the use of factory ships as floating bases from which to send out hunter vessels, but whaling continued from land bases in Antarctica. In the spring of 1982, the IWC voted a ten-year moratorium on commercial harvesting of whales, except dolphin and porpoise, in order to study the whale populations and assess their ability to recover. The moratorium began in 1985–86 and continues in effect. The moratorium called for assessment of whale stocks to be completed by 1990, but no large-scale IWC sampling program has been set up.

Some countries (Japan, Norway, Iceland, Korea, and the former Soviet Union) filed objections to the moratorium in 1982 and have turned to "scientific" or "research" whaling to assess stocks. Korea abandoned its research whaling programs in 1987 after a series of negative reviews by the IWC, and Iceland announced the end of its research

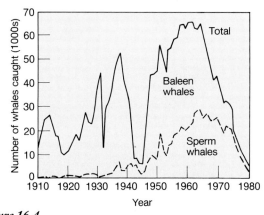

Figure 16.4

Total annual catches of baleen and sperm whales in all oceans from 1910 to 1980.

whaling in 1989. Japan took its last commercial catch of sperm whales in the spring of 1988 but has continued to harvest minke whales, as many as three hundred per year, under research proposals made to the IWC, although the IWC adopted resolutions against the harvests.

Under the IWC regulations, aboriginal/subsistence whaling is permitted. IWC regulations attempt to balance the needs of aboriginal peoples in Alaska, Denmark (Greenland), and the Soviet Union who depend on limited whaling for subsistence, cultural, and nutritional needs with the conservation needs of the whales.

The 1991 IWC meeting raised fears that commercial whaling will soon resume, because whale stocks are coming back as a result of the moratorium. For example, the bowhead whale population, which had been hunted from 20,000 to 1,000 whales during the last century, has rebuilt to 7,500, an average increase of 3% a year. At the meeting it was agreed that 41 bowheads may be taken each year by Alaskan Eskimos and 169 minke whales a year by Soviet Eskimos, and that Greenlanders may take 21 fin whales a

Echolocation and Communication

Target

When submerged marine mammals descend to a world in which little light is available, many of these animals use sound instead of sight to picture their environment as well as communicate with each other. The best-known communication between marine mammals is the "song" of the humpback whales, lasting up to thirty minutes. These songs are thought to be announcements of presence and territory, although some scientists believe the singing is a secondary sexual characteristic of males in the breeding season. Female gray whales stay in contact with their calves by a series of grunts, and Weddell seals are known to communicate by audible squeaks.

Toothed whales, particularly porpoises and the sperm whale, a few baleen whales, including the gray, blue, and minke, and the Weddell seal, California sea lion, and possibly the walrus are known to make the sharp sound required to produce the reflected echoes that allow certain marine mammals to orient themselves and locate objects. This method of using sound to picture the environment is known as **echolocation.** Although these animals can produce a range of sounds,

the most useful sound for echolocation appears to be clicks of short duration released in single pulses or trains of pulses. The bottle-nosed dolphin produces clicks in frequencies audible to the human ear and higher, each lasting less than a millisecond and repeated up to eight hundred times per second. When each click hits its target, part of the sound is reflected back; the animal continually evaluates the time and direction of return to learn the speed, distance, and direction of the reflecting target. Low-frequency clicks are used to scan the general surroundings, and higher frequencies are used for distinguishing specific objects.

Since many of these marine mammals have no vocal cords, how do they produce these sounds? Porpoises and dolphins appear to move air in their nasal passages, which vibrate certain structures to produce the clicks; the whistles and squeaks are made by forcing air out of nasal sacs. The bulbous, fatty, rounded forehead of the porpoise acts as a lens to concentrate the clicks into a beam and direct them forward. Sperm whales produce shorter, more powerful, long-range pulses at lower frequencies; these sounds travel

more slowly but carry several kilometers. Each pulse from the sperm whale is compound, lasting about twenty-four milliseconds and made up of up to nine separate clicks. The sperm whale's massive forehead is filled with oil and may be used to focus the sound pulses.

All marine mammals have good hearing. Humans hear in the range of 16–20,000 vibrations or cycles per second; the bottle-nosed porpoise responds above 150,000 vibrations per second. How do they pick up the faint incoming echoes of their own clicks and screen out the louder outgoing clicks and other sea noises? Sounds enter through the lower jaw and travel through the skull by bone conduction. Within the lower jaw, fat and oil bodies vibrate, and the sound is channeled through the oil directly to the middle ear. Areas on each side of the forehead are also very sensitive to incoming sound. The hearing centers in the brains of marine mammals are extremely well developed, presumably to analyze and interpret returning sound messages. Their vision centers are less developed, and they are believed to have no sense of smell.

---

year and 315 minke whales over three years. Whaling nations reminded the IWC that the original intent of the organization was to "conserve" whales in order to protect the stocks as a harvestable natural resource.

The 1992 IWC meeting adopted a Revised Management System to calculate catch quotas for individual species of whales. Sixteen nations voted to approve the RMS; Norway voted against it, and Japan and Britain with nine other countries abstained. The Japanese were allowed to continue limited research whaling, up to 330 minke

whales a year. The 1993 meeting continued the moratorium, and shortly after the meeting Norway announced its resumption of whaling.

While populations of California gray whales and the bowhead whales have increased, other populations, the blue whales, humpbacks, right whales and sperm whales, are still present only in very low numbers. Some researchers believe this is due to the difficulty of finding mates in such small populations; there is also the possibility that the noise produced by increasing ship traffic interferes with whale calls. Other scientists are concerned that krill harvesting and the global depletion of fish species is affecting the whale populations; pollution may also play a role.

The intertwined history of whales and people is not yet ended. In the words of Herman Melville from the pages of *Moby Dick*,

> *The moot point is whether the Leviathan can long endure so wide a chase and so remorseless a havoc; whether he must not at last be exterminated from the waters, and the last whale, like the last man, smoke his last pipe, and then himself evaporate in the final puff.*

Dolphins and Porpoises

Dolphins and porpoises are small, toothed whales (see fig. 16.5e). They are the clowns of the sea, gentle, friendly, and easily trained, and they figure in many folktales, stories, movies, and TV programs. In the open ocean, they are observed traveling at high speeds and in large schools; they occasionally leap clear of the water, in apparent fun and high spirits, and may even alter their course to keep a vessel company for hours, swimming easily just in front of the bow. Porpoises have been observed swimming at speeds in excess of 30 knots, a feat that interests scientists and researchers. Their abilities to communicate and their intelligence, as demonstrated by their learning and recall abilities, are under study. They have been trained to help divers and to act as messengers between those working at depth and their surface vessels; the U.S. Navy continues to investigate their potential as underwater assistants.

These smaller cetaceans are found in both tropic and temperate waters. Although marine, they will go up rivers and channels into shallow brackish waters. They have been seen moving across the very shallow lakes and canals of the Mississippi delta region in water barely deep enough to support their high-speed swimming.

The last two decades have placed great pressure on the world's dolphins and porpoises. A 1990 United Nations–sponsored symposium reported that more than a million of these mammals die each year in nets, usually the unwanted "bycatch" of those fishing for other species. Dolphin and porpoise populations are affected by certain fishing techniques used by the tuna industry; this situation is investigated further in the discussion of the tuna fishery at the end of this chapter. Two small species, the Mexican porpoise and the black dolphin of the Chilean coast, may be endangered. There are only a few hundred Mexican dolphins left in the Gulf of California, and the Chilean dolphins are reportedly being killed in large numbers for fish and crab bait.

Seals, Sea Lions, and Walruses

Seals and sea lions belong to the **pinnipeds,** or "feather-footed" animals, so named for their four characteristic swimming flippers. Representative pinnipeds are shown in figures 16.5 and 16.6. These animals are marine mammals that still retain their ties to land, spending considerable time ashore on rocky beaches, on ice floes, or in caves. They are found from the tropics to the polar seas, ranging from the

nearly extinct monk seal of the western Hawaiian Islands and Mediterranean area to the fur seals of the Arctic. The common harbor seal, the harp seal of the northwest Atlantic, the leopard seal of the Antarctic, and the 2-ton male elephant seal, with its great pendulous snout, are all true seals, or seals without external ears and with torpedo-shaped bodies that require them to use a wriggling, wormlike motion to move on land. The northern fur seal and the sea lion (the seal of the circus and amusement park) are eared seals with longer necks and supple forelimbs tipped with broad flippers, which are used for walking and hold the animal's body in a partially erect position on land. Seals may undertake long sea migrations, congregating in spring and summer at specific locations for breeding. For example, the northern fur seal ranges the North Pacific, Bering, and Okhotsk seas, coming ashore to breed in the Pribilof Islands. The habits and natural histories of each of the pinnipeds differ from those of the others, and much is still left to learn.

The walrus (fig. 16.5d) is placed in a separate subgroup. It has no external ears and is able to rotate its hind flippers so that it may walk on a hard surface. Its heavy canine teeth (or tusks) are unique and are found in both males and females. These tusks help the walrus to haul itself out of the water onto the ice, and are probably used to glide over the bottom, like sled runners, while it forages for clams with its heavy muscular whisker pads. It has been said that the tusks are used for digging clams, but the tusks show wear on the front surfaces, not the ends and back surfaces as they would if used for the clam-digging function.

Some seals and sea lions are currently enjoying a period of relative peace, compared to the sealing days of the nineteenth and early twentieth centuries. The Guadalupe fur seal of southern California was hunted to the brink of extinction, until, in 1892, only seven animals were thought to survive; its remarkable recovery to a current population of 1500 is due to protection by both the United States and Mexican governments and to luck. Between 1870 and 1880, hunters for furs and oil reduced the northern elephant seal population to one hundred and cut the Pacific walrus population in half. Recently, an increasing number of Pacific walrus have been killed for their ivory. Numbers are unreliable, and the thirty-year catch-monitoring program was canceled in 1990 due to lack of funds, but there are reports of yearly kills between eight thousand and twelve thousand animals since 1980. There is a growing concern that this kill rate, approaching yearly harvests of the late nineteenth and early twentieth centuries, may be more than the slowly reproducing walrus populations can sustain.

The northern fur seal's Pribilof population was 2.5 to 3 million in 1867; heavy hunting reduced it to only 200,000 to 300,000 by 1910. Although the population recovered from these excessive harvests and the present population is approximately 1.1 million, there has been a long-term downward trend since the mid 1950s. This has been blamed on entanglement with nets, lines, plastic strapping rings, and other debris, but recent studies suggest that the population's decline is primarily related to a reduced prey base

(a)

(b)

(c)

(d)

Figure 16.5

Marine mammals. (a) The harbor seal (*Phoca vitulina*) is a friendly, curious animal that coexists well with humans. (b) This large male northern fur seal (*Callorhinus ursinus*) stands vigilant watch over his harem of females in the Pribilof Islands. (c) Sea otter (*Enhydra lutris*) and her pup dine on crab. (d) The walrus (*Odobenus rosmarus*) feeds from the bottom and relaxes on the Arctic ice floes. (e) Pacific white-sided dolphins (*Lagenorhynchus obliguidens*) play in the bow wave of a research vessel.

(e)

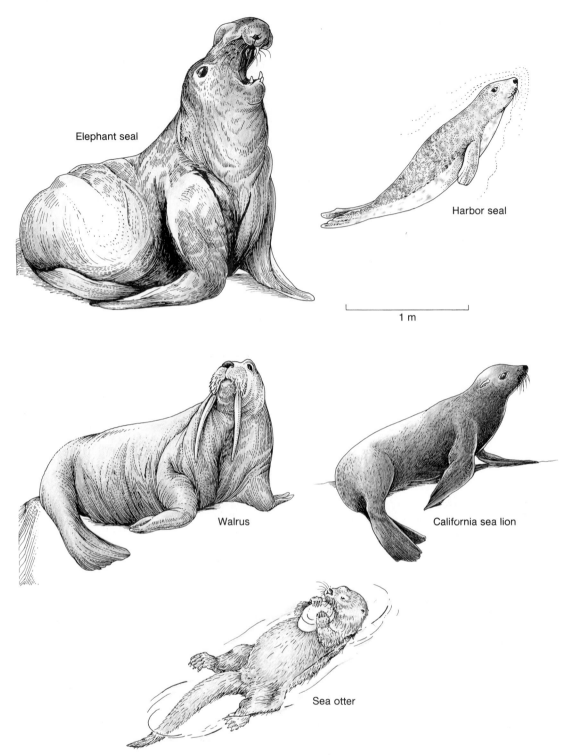

Figure 16.6
Examples of marine mammals. The elephant seal, harbor seal, and California sea lion are pinnipeds. The sea otter, related to river otters, feeds on clams and sea urchins while floating on its back.

related to increased large commercial fish catches. (See also the decline of Steller sea lions discussed in section 16.6, Problems and Policies.)

In the past, northern fur seal populations have been managed and harvested under international agreement between the United States, Canada, Japan, and the Soviet Union. At present, fur seals are covered by the Marine Mammal Protection Act, which is discussed at the end of this section. There has been no commercial harvest since

1984. A yearly subsistence harvest of less than two thousand seals is allowed for the Pribilof Islanders; the 1990 harvest was 1,241 seals.

The annual commercial harvest of white-fur harp seal pups in eastern Canada has dropped the Canadian harp seal population from an estimated initial stock of between 3 million and 4 million to its present level of 1 million to 1.5 million over the last 150 years. This once lucrative harvest has been banned by the Canadian government, after the

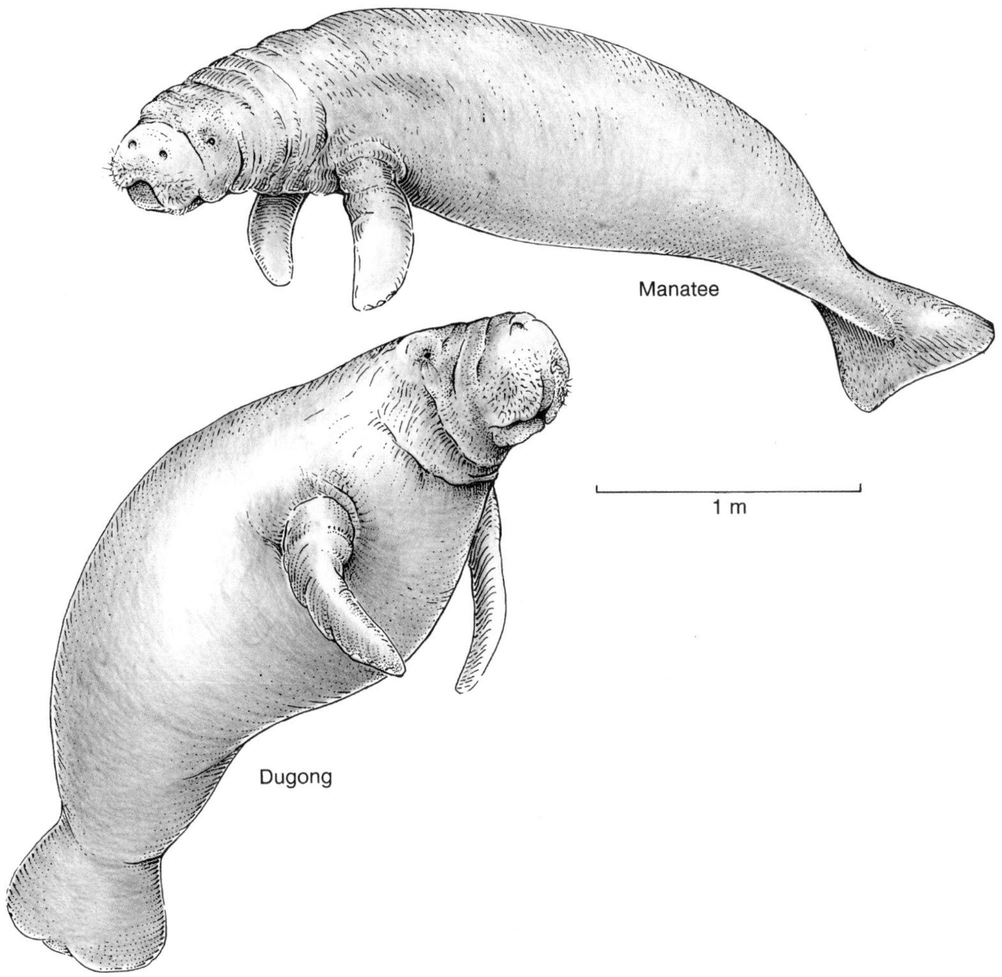

Figure 16.7
The tropic manatee of the
Caribbean and the dugong of
Southeast Asia are herbivorous
marine mammals.

Manatee

1 m

Dugong

harvest had become the focus of a public outcry against
sealing. The market for the once prized furs collapsed in
1983 when the European Economic Community, bowing to
international pressure, banned imports of the pelts.

The seals are not, however, completely at peace. In
the summer of 1988 European harbor seals began to die in
large numbers along the North Sea coasts of the
Netherlands, West Germany, Denmark, Sweden, and
Norway. More than nine thousand out of the fifteen
thousand to sixteen thousand harbor seals previously found
in this area died, as well as hundreds of seals along the
coast of Britain. The primary cause of the disease was
found to be phocine distemper virus; serological evidence
in Antarctic seal populations points to previous outbreaks
of this virus in Southern Ocean populations.

Sea Otters

Sea otters (figs. 16.5c and 16.6) are related to river otters but
are larger and live in salt water. This animal lives in coastal
areas, taking shellfish and other food from the bottom in
relatively shallow water. It differs from both seals and
whales in having no insulating layer of blubber beneath the
skin and so must depend entirely on its dense fur for
warmth. This soft, thick fur became the sea otters' death
warrant in the eighteenth and nineteenth centuries, when
they were hunted nearly to extinction; prime pelts sold for
more than $1000 each. The sea otter was brought under

protection in 1911, and many scientists doubted that the
species could survive. The Alaskan population in the mid-
1970s had increased to between 100,000 and 120,000; in
recent years overcrowded populations have been thinned.
In California in 1977, 1800 to 2000 animals were counted,
but the population there is still officially designated as
"threatened" due to increasing human activity, such as offshore oil development, and to the sea otter's appetite for the
commercially valuable abalone and clams. The species, although now reestablished, will need continued watching
and management in this area.

Sea Cows

Manatees and **dugongs** are also known as **sea cows** (see
fig. 16.7); they are members of the **Sirenia** and are thought
to be the possible source of the mythic mermaid. Manatees
are found in the brackish coastal bays and waterways of the
warm southern Atlantic coasts and in the Caribbean, and
dugongs are found in the seas of Southeast Asia, Africa,
and Australia. At present, the growth of human populations
and their need for protein is putting increasing hunting
pressure on the dugongs in the southern Pacific. Manatees
in the coastal waters of the Caribbean and South Atlantic
are frequently injured and killed by collisions with the propellers of large and small vessels. Although they are protected along the Florida coast, the known manatee mortality
between 1979 and June, 1992 was more than seventeen

hundred animals, 26% from boat collisions. Manatees that died at or around the time of birth set a record high in 1991 at 53, continuing a steep upward mortality trend.

In former times the Steller sea cow existed in the shallow waters off the Commander Islands in the Bering Sea. These sea cows were slow-moving, docile, totally unafraid, and present in only limited numbers. These characteristics, coupled with a low reproduction rate, made them unable to withstand the human hunting pressure. The last Steller sea cow was killed for its meat in about 1768.

Marine Mammal Protection Act

In 1972, the Congress of the United States established the Marine Mammal Protection Act, Public Law 92–522. This act includes a ban on the taking or importing of any marine mammals or marine mammal product. "Taking" is defined by the act as harvesting, hunting, capturing, or killing any marine mammal or attempting to do so. The act covers all United States territorial waters and fishery zones. It is also unlawful "for any person subject to the jurisdiction of the United States or any vessel or any convoy once subject to the jurisdiction of the United States to take any marine mammals on the high seas" except as provided under pre-existing international treaty.

The act effectively removed the animals and their products from commercial trade in the United States. Only under strict permit procedures and with the approval of the Marine Mammal Commission can a few individual marine mammals be caught for scientific research and public display.

Alaskan natives are exempted from the act for purposes of subsistence hunting and of creating and selling authentic native articles of handicraft and clothing. If it is determined that a species or stock being hunted under this exemption is being depleted, further regulations may be established to conserve the animals.

In some cases, the Marine Mammal Protection Act has been very successful in its mission of protecting marine mammal populations; some would argue that it has been too successful. Interactions between people and marine mammals competing for the same resource and/or habitat are creating problems that are not easily solved. Since passage of the act, harbor seal populations have increased 7%–10% per year along the U.S. West Coast. The seals feed on fish that are harvested commercially and increasingly rob fishing nets directly. Under the provisions of the act, those who fish can only attempt to discourage the animals, and this is not always successful. In Oregon, an acoustic device called a "sealchaser" was initially successful in keeping the seals away from the nets, but in a few weeks the seals became accustomed to the sound and used it to find the nets. Along the central California coast, the California sea lion population increases each year, and between 80,000 and 125,000 sea lions consume 100,000–250,000 tons of Pacific hake (whiting) annually along the West Coast of the United States.

16.2   The Reptiles

Although many land reptiles visit the shore to feed, mainly on crabs and shellfish, few reptiles are found in today's seas. Examples of marine reptiles are shown in figure 16.8. The only modern marine lizard is the big, gregarious marine iguana of the Galápagos Islands. It lives along the shore and dives into the water at low tide to feed on the algae. This iguana has evolved a flattened tail for swimming and has strong legs with large claws for climbing back up on the cliffs. It regulates its buoyancy by expelling air, allowing it to remain underwater. The large monitor lizards of the Indian Ocean islands, known as the Komodo dragons, are capable of swimming but do so only under duress.

Alligators may enter shallow shore water, and crocodiles are known to go to sea. The estuarine crocodile of Asia is found in the coastal waters of India, Sri Lanka (Ceylon), Malaya, and Australia. The Indian gavial and false gavial are slender-nosed, fish-eating crocodilians found in nearshore waters.

Sea Snakes

There are about fifty different kinds of sea snakes found in the warm waters of the Pacific and Indian oceans; there are no sea snakes in the Atlantic Ocean. Sea snakes are extremely poisonous, with small mouths, flattened tails for swimming, and nostrils on the upper surface of the snout that can be closed when the snakes are submerged. The sea snakes' skin is nearly impervious to salt but it is permeable to gases, passing nitrogen gas as well as carbon dioxide and oxygen. The snake is able to lose nitrogen gas to the water, allowing it to dive as deep as 100 m (300 ft), stay submerged as long as two hours, and surface rapidly with no decompression problems. (See chapter 13 for a discussion of the way in which marine mammals handle diving and decompression.) Sea snakes eat fish, and most reproduce at sea by giving birth to live young.

Sea Turtles

Sea turtles live in the ocean but nest on land. The four large sea turtles are the green, hawksbill, leatherback, and loggerhead turtles. Green sea turtles are herbivores; they weigh 140 kg (300 lb) or more. The hawksbill is a tropic species found on reefs and feeding primarily on sponges. The loggerhead, weighing between 70 and 180 kg (150 and 400 lb), is usually found around wrecks and reefs, feeding on crabs, mollusks, and sponges. The leatherback is the largest sea turtle, weighing up to 630 kg (1400 lb). This giant feeds only on jellyfish, following them over large areas of the ocean.

These animals are all migratory and make long sea journeys of many hundreds of miles between their nesting sites and their foraging areas. The green turtles of the Brazil coast migrate 2250 km (1400 mi) to Ascension Island in the South Atlantic to lay their eggs. The female turtle lays a hundred or more eggs in a scooped-out depression in the warm sand, covers them, and returns to the sea. While the eggs incubate, they are an easy prey for humans, dogs, rats, and other carnivores. In more and more cases no

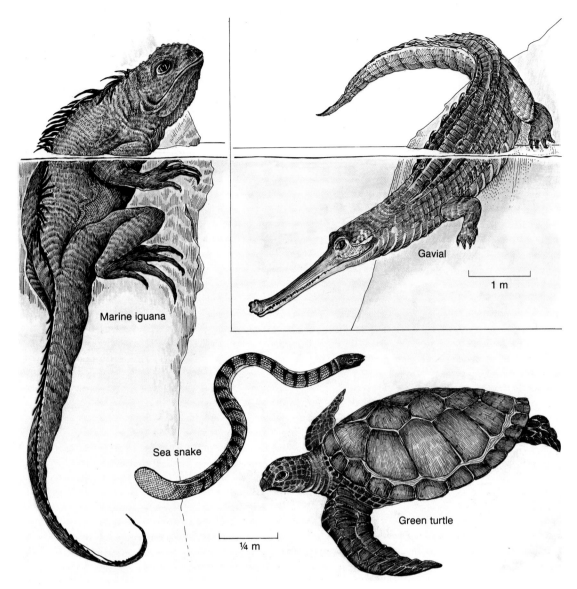

Figure 16.8
Marine reptiles.

Marine iguana

Gavial

1 m

Sea snake

Green turtle

¼ m

eggs survive to replenish the population. If the young turtles hatch, they must make their way across the sand while hungry birds attack and must then evade the waiting predatory fish when they do find the water.

The greatest threat to the survival of all sea turtles is the human. Turtle eggs and turtle meat are prized by humans throughout the Pacific, and turtle nests are regularly raided by poachers. Until recently, five thousand Caribbean sea turtles were used each year by a New York soup company to produce 600,000 liters of turtle soup. Hawksbill turtles are hunted for tortoiseshell, which is used to form combs, boxes, jewelry, and other ornaments. Japan imported 234,000 shells between 1981 and 1989. Turtles' skins are used for leather and their fat and oils for cosmetics.

The Kemp's ridley turtle is a small sea turtle found in the Gulf of Mexico and the South Atlantic. It is the rarest and most endangered sea turtle. No one knew where Kemp's ridley turtles nested until 1947, when their only natural nesting site was found at Rancho Nuevo on Mexico's eastern coast. That year, forty thousand females came ashore to lay their eggs. In spite of protection by the Mexican government, in 1985 there were less than five

hundred egg-laying females. The Kemp's ridley is under seige; their eggs are stolen to make aphrodisiac cocktails; they are killed for food and for their hides; they are suffocated in shrimp trawls; and they are preyed on by raccoons, skunks, and seabirds as the young try to reach the sea. Scientists have tried to start a new colony on Padre Island, Texas, by collecting the eggs, hatching the baby turtles, and then imprinting them with a chemical signature to assure their return to Padre Island. Eighteen thousand "head-started" turtles were released since 1978, but by 1992 not one of these turtles had been observed returning to any beach. The project has become controversial among scientists and conservationists; supporters include shrimpers who have indicated they prefer to pay for a turtle-restocking program (proven or unproven) rather than use the turtle excluding devices described in the next paragraph.

Worldwide, the slaughter of these sea creatures is enormous. Although placing species on the United States endangered species list prevents the legal importing of turtle products into this country, according to the U.S. Fish and Wildlife Service there is a vast international illegal trade in turtle products that rivals the illegal ivory trade in

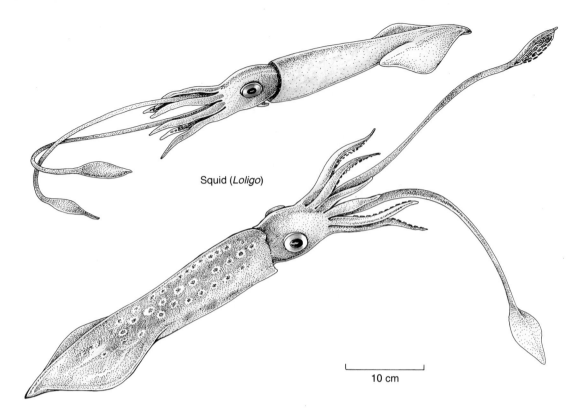

Squid (*Loligo*)

10 cm

Figure 16.9
The squid is a swimming mollusk and a member of the nekton.

dollar value. The problem of turtles drowning in shrimp nets is believed to play a significant role in the depletion of sea turtle populations. U.S. shrimpers accidentally catch an estimated 48,000 sea turtles each year in the Gulf of Mexico and South Atlantic. Regulations were passed in 1987 to require the shrimp fishermen to use turtle excluding devices (TEDs) that allow the turtles but not the shrimp to escape the nets. Shrimpers initially resisted the TED requirement, claiming it reduced shrimp yields, and thus appealed enforcement of the regulation. The regulation was placed under the Endangered Species Act in order to gain compliance and went into effect for offshore waters in 1989. Vacillation by federal officials, lawsuits by environmental groups, and blockades by shrimpers resulted in on-again, off-again use of TEDs and restricted net tow times. In December, 1992 new regulations were issued by the National Marine Fisheries Service (NMFS) requiring all trawls greater than 11 m (35 ft) wide to have TEDs in place whether fishing onshore or offshore. Trawls less than 11 m (35 ft) wide and used in inshore waters are exempt from TEDs until 1994 so that NMFS may conduct further research to prove that TEDs do exclude turtles and do not interfere with catch efficiency.

16.3   The Squid

Squid (fig. 16.9) are abundant in deeper water, travel in large schools, and, like many of the mesopelagic fish, migrate toward the surface at night. It is not unusual in the open ocean to see schools of squid rapidly swimming near the surface at night and often shooting out of the water at the wave crests. Giant squid (*Architeuthus*), more than 15 m (49 ft) long, are the food of the sperm whale. Long before a single giant squid was ever seen, bits and pieces were found in the stomachs of sperm whales, so that their existence was anticipated. Scientists know very little about these huge creatures but suspect that there may be large numbers of them in the mesopelagic zone.

16.4   The Fish

The fish dominate the nekton. Fish are found at all depths and in all the oceans, but their distribution patterns are determined directly or indirectly by their dependency on the ocean's primary producers. Fish are concentrated in upwelling areas, shallow coastal areas, and estuaries. The surface waters support much greater populations per unit of water volume than the deeper zones, where food resources are sparser.

Fish come in a wide variety of shapes related to their environment and behavior. Some are streamlined, designed to move rapidly through the water (tuna and mackerel); others are flattened for life on the seafloor (sole and halibut), while still others are elongate for living in soft sediments and under rocks (some eels). Fins provide the push or thrust for locomotion and occur in a variety of shapes and sizes. Fins are used to change direction, turn, balance, and brake. Flying fish use their fins to glide above the sea surface; mudskippers and sculpins walk on their fins. A sample of the variety found among marine fishes is shown in figure 16.10.

(a)

(b)

(c)

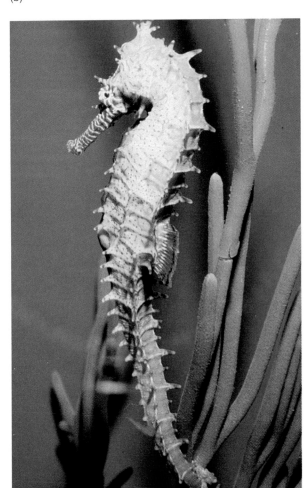

(d)

(e)

(f)

Figure 16.10

The bony fishes dominate the world's aquatic environments. Their diversity is enormous and they are found in almost every conceivable aquatic environment. Two of the thousands of species of reef fish: (a) an emperor angel and (b) a butterfly fish. (c) The wolf eel is a carnivore with strong jaws and long teeth. (d) The sea horse usually swims vertically but slowly using the fanlike motion of its fins. Its scales are modified to form protective armor. (e) The clown fish enjoys a symbiotic relationship with a sea anemone. (f) The red snapper inhabits rocky coastal areas; it is used as a food fish.

Schooling is common among certain types of fish (herring, mackerel, menhaden). Schools may consist of a few fish in a small area, or they may cover several square kilometers; for example, herring in the North Sea have been seen in schools 15 km long and 5 km wide (9 × 3 mi). Usually the fish are all of the same species and similar in size. Fish schools have no definite leaders, and the fish change position continually. Schooling fish keep their relationship to one another constant as the school moves or changes direction. Most schooling fish have wide angle eyes and the ability to sense changes in water displacement that allow them to keep their place with respect to their neighbors. Schooling probably developed as a means of protection; each fish has less chance of being eaten in the school than alone. The school may also keep reproductive members of the population together.

Ocean fish are divided into two groups: (1) fish with skeletons of cartilage and (2) fish with skeletons of bone. Cartilaginous fish—sharks and their relatives, the skates, rays, and ratfish—are considered more primitive than bony fish. The fish that are used for food are mainly those with skeletons of bone.

Sharks and Rays

The shark is an ancient fish; it predates the mammals, first appearing in the earth's oceans 450 million years ago. Sharks differ from other fish by their skeletons of cartilage rather than bone and by their toothlike scales. Shark scales have a covering of dentine similar to vertebrate teeth and are extremely abrasive; sharkskin has been used as a sandpaper and polishing material. The shark's teeth are modified scales; they are replaced rapidly if they are lost, and they occur in as many as seven overlapping rows in the jaw. Sharks are actively aware of their environment through good eyesight; excellent senses of smell, hearing, and mechanical reception; and electrical sense. A shark's vision is more important than previously thought; sharks see well under dim-light conditions. They have the ability to sense chemicals in their environment through smell, taste, a general chemical sense, and unique pit-organs distributed over their bodies. These pit-organs contain clusters of sensory cells resembling taste buds. The shark's sense of smell is acute; it has a pair of nasal sacs located in front of its mouth, and when water flows into the sacs as the shark swims, the water passes over a series of thin folds with many receptor cells. Sharks are most sensitive to chemicals associated with their feeding, and all are able to detect such chemicals in amounts as dilute as one part per billion. Receptors along the shark's sides are sensitive to touch, vibration, currents, sound, and pressure. The movement of water from currents or from an injured or distressed fish are sensed by the shark's lines of receptors that communicate with the watery environment by a series of tubes; water displacement stimulates nerve impulses along these systems. The shark is able to hear and uses its hearing in the location of its prey. Pores in the shark's skin, especially around the head and mouth, are sensitive to small electric fields. Fish and other small marine organisms produce electric fields

around themselves, and the shark uses its electroreception sense to locate prey and recognize food. As a shark swims through the earth's magnetic field, an electric field is produced that varies with direction, giving the shark its own compass. We still do not understand completely how all these senses function or how they affect the behavior of sharks; we do know that the shark is extremely well tuned to its environment.

There are some three hundred species of sharks widely spread through the oceans and found in rivers more than a hundred miles from the sea, as well as one species landlocked in fresh water, the Lake Nicaragua shark. Some of these sharks are shown in figure 16.11. The whale shark is the world's largest fish, reaching lengths of more than 15 m (50 ft). This graceful and passive animal feeds on plankton and is harmless to other fish and mammals. The docile basking shark, 5 to 12 m (15–40 ft) long, is another plankton feeder. It is found commonly off the California coast and in the North Atlantic, where it has been harvested for its oil-rich liver. A third species of plankton-feeding shark, 4 m (14 ft) long and weighing approximately 680 kg (1500 lb), nicknamed megamouth, is known from two specimens, one caught in 1976 and one in 1984.

Many sharks are swift and active predators, attacking quickly and efficiently, using their rows of serrated teeth to remove massive amounts of tissue or whole limbs and body portions. They also play an important role as scavengers, and, like wolves and the large cats on land, eliminate the diseased and aged animals. Sharks can and do attack humans, although the reasons for these attacks and the periodic frenzied feeding observed in groups of sharks is not understood. A human swimming inefficiently at the surface may look like a struggling, ailing animal and be attacked; a diver swimming completely submerged may appear as a more natural part of the environment and be ignored.

Studies of shark attacks on humans in South African waters indicate that most shark attacks occur in shallow waters and in rivers (where there are more swimmers), in murky water, and during the darker parts of the day, such as at dawn and at dusk. Shark attacks have increased along the rocky northern California coast during the last fifteen years. Fourteen attacks upon humans on surfboards have occurred since 1972 along this coast. In the summer of 1982, beaches were closed north of San Francisco. This increase in shark attacks is considered by many to be related to the increase in the sharks' prey, the seal and sea lion populations protected by the Marine Mammal Protection Act. Although surfing is more popular along the beaches of the southern California coast, the shark attacks occur along the rocky coasts where seals and sea lions congregate. It has also been observed that modern surfboard designs appear very similar in shape to a seal when viewed from below.

Researchers have spent considerable time trying to perfect a shark repellent. During World War II, aircraft and ship crew members were provided with a so-called shark repellent made from copper acetate and a soluble ammonium compound, which had a positive psychological effect on the men but had little effect on the sharks. Current

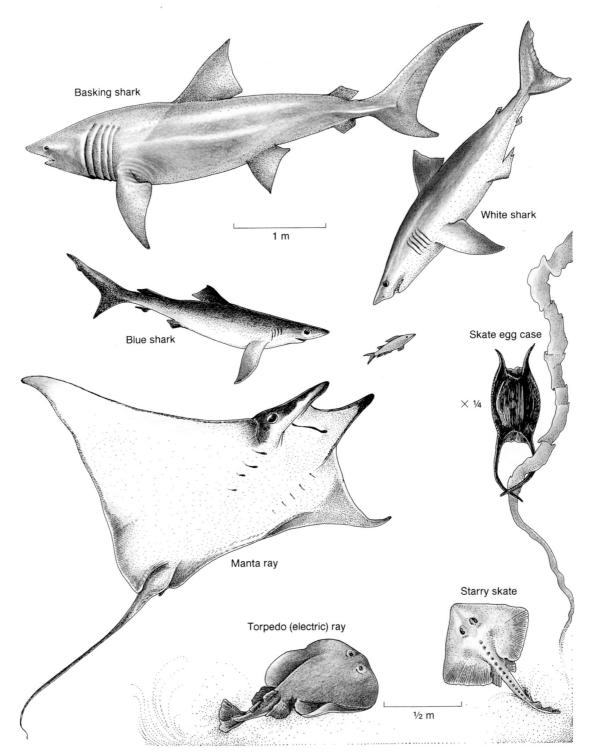

Figure 16.11

Representative members of the cartilaginous fish. The leathery egg case of a skate is known as a mermaid's purse.

research in shark repellents includes experiments with underwater sound as well as with chemicals. Recent studies indicate that an excretion found on the surface of a certain flat fish may act as a shark deterrent.

Skates and rays, also shown in figure 16.11, are flattened, sharklike fish that live near the seafloor. They move by undulating their large side fins, which gives them the appearance of flying through the water. The large manta rays are plankton feeders, but most rays and skates are carnivorous, eating fish but preferring crustaceans, mollusks,

and other benthic organisms. Their tails are usually thin and whiplike and, in the case of stingrays, carry a poisonous barb at the base. Some of the skates and a few of the rays have shock-producing electric organs that are located along the sides of the tail in skates and on the wings of the rays; their purpose appears to be mainly defensive. Like the sharks, most rays bear their young live. Skates enclose the fertilized eggs in a leathery capsule called a sea purse or mermaid's purse that is ejected into the sea and from which the young emerge in a few months; see figure 16.11.

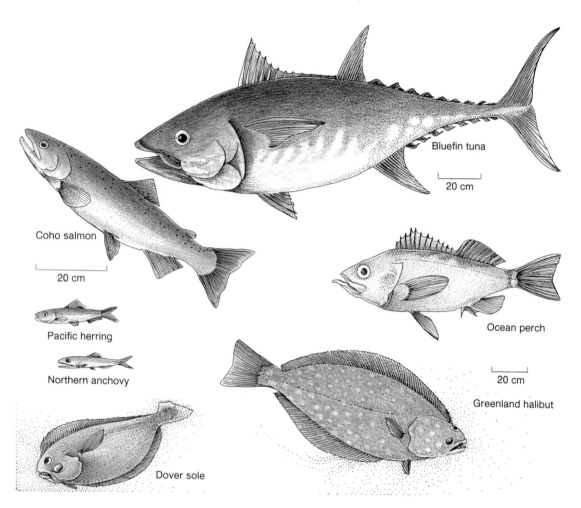

Figure 16.12
Commercially harvested members of the bony fish.

Commercial Species of Bony Fish

Most commercially valuable food fish are found between the ocean's surface layers and a depth of approximately 200 m (660 ft), or in the epipelagic zone. Most of these fish are streamlined, active, predatory, and capable of high-speed, long-distance travel. Among the most important species are the enormously abundant small, herring-type fish, such as sardine, anchovy, menhaden, and herring. These fish feed directly on plankton and are found in large schools in areas of high primary productivity. Other valuable fish that are commercially harvested include mackerel, pompano, swordfish, and tuna. These fish are caught at sea, out of sight of land.

Fish that live on or near the bottom, known as **demersal fish** or bottom fish, do not swim as rapidly as those that live in the water column above. The flounder, halibut, turbot, and sole are commercially valuable bottom fish. Perch and snapper tend to congregate along the seafloor in the shallower, nearshore areas. They are called rockfish because they hide among the rocks and live in the cracks of underwater reefs. Representative fish are shown

in figure 16.12. In a later section of this chapter, we will discuss the commercial exploitation and economic significance of some of these food fish.

Deep-Sea Species of Bony Fish

The fish of the deep sea are not well known, because the depths below the epipelagic zone are difficult and expensive to sample. No fish from these depths have been exploited commercially. A variety of deep-sea types is shown in figure 16.13.

In the dim, transitional mesopelagic layer, between 200 and 1000 m (660–3300 ft), the waters support vast schools of small luminous fish. *Cyclothone* is believed to be the most common fish in the sea. Each of its species lives at a relatively fixed depth; the deeper-living species are black while the shallower-living species are silvery to blend with the dim light. At this depth the lantern fish has a worldwide distribution; some two hundred species are distinguished by the pattern of light organs along their sides. They are a major item in the diet of tuna, squid, and porpoises. *Stomias,* a fish with a huge mouth, long, pointed teeth, and light organs along its sides, and the large-eyed hatchet fish prey on the great clouds of euphausiids and copepods found at this depth.

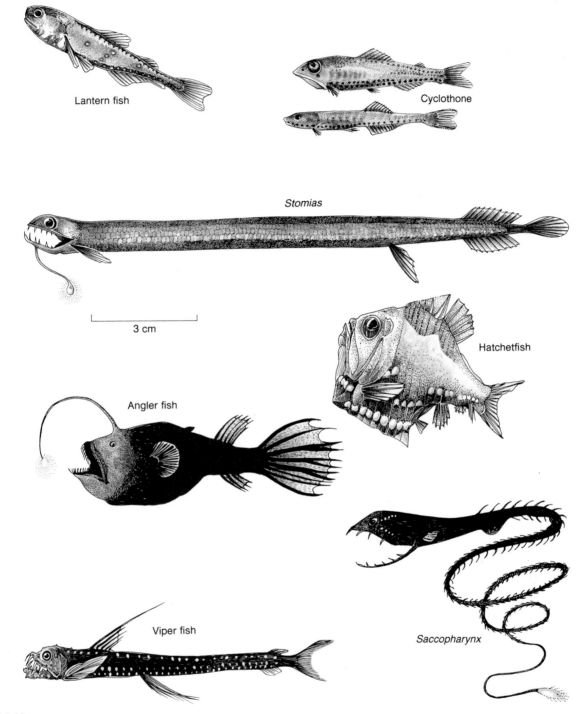

Figure 16.13

Fish from the deep sea.

Sharks, sablefish, and grenadiers move freely through these deeper waters. Often, species found at great depths at low latitudes are the same as the cold-water species found at shallow depths at high latitudes. The sablefish is caught commercially in the waters off Alaska, British Columbia, and Washington State; it is also found in large numbers off southern California, between 800 and 1500 m (2640–4950 ft).

At one time, it was assumed that although fish did exist at depths below 200 m (660 ft), there were very few of them. Researchers are now finding that life is more abundant at these depths than was previously believed. Submersibles, which allow direct observation and better underwater photography techniques, have helped to change our ideas. Underwater cameras, with attached bait, show populations of scavenging fish, including sharks, congregating at a baited target over a period of a few hours. Probably these species follow scentlike clues carried by the deep-water currents. Particularly abundant deep-water populations have been found in areas of the Pacific where the surface populations are sparse; more than forty large fish were attracted to the bait at 6000 m (19,800 ft) under the least productive waters of the Pacific gyre, while at only 4000 m under the most productive waters of the Antarctic the bait was visited by only a few small fish. A partial explanation may be that in the least populated, shallower areas food descends to the bottom, while in productive, highly populated regions it is consumed on the way down, leaving little to feed the deeper dwellers.

In the perpetually dark bathypelagic zone there are predators with highly specialized equipment for catching their prey. These fish are equipped with light-producing organs used as lures, large teeth that in some species fold backward toward the gullet so the prey cannot escape, and gaping mouths with jaws that unhinge to allow the catching and eating of fish larger than themselves. Among the most famous of these predators are *Macropharynx longicaudatus* and the related *Saccopharynx,* which have funnellike throats and tapering bodies ending in whiplike tails. When the stomach is empty the fish appears slender, but it expands to accept anything the great mouth can swallow. The female angler fish, *Ceratias halboelli,* has a dorsal fin modified into a fishing rod, which slides back and forth along a canal in the back of the fish; an illuminated lure dangles from the tip of the rod. Other fish are attracted to the lighted bait, which the angler fish moves forward along her back until the bait is just above her jaws.

Although fierce and monstrous in appearance, most of these fish are small, between 2 and 30 cm (1–12 in) in length. They breathe slowly, and the tissues of their small bodies have a high water and low protein content. The fish go for long periods between feedings and use their food for energy production rather than for increased tissue production.

16.5 Classification Summary of the Nekton

The nekton, all members of the kingdom Animalia, can be classified in the following phyla and classes list.

I. *Kingdom Animalia:* animals; multicellular heterotrophs with specialized cells, tissues, and organ systems.
   A. *Phylum Mollusca:* mollusks.
      1. *Class Cephalopoda:* squid; the octopus is a member of the benthos.
   B. *Phylum Chordata:* animals with a dorsal nerve cord and gill slits at some stage in development.
      1. *Subphylum Vertebrata:* animals with a backbone of bone or cartilage.
         a. *Class Agnatha:* jawless fish; lampreys and hagfish.

   b. *Class Chondrichthyes:* jawed fish with cartilaginous skeletons; sharks and rays.
   c. *Class Osteichthyes:* bony fish; all other fish, including commercial species such as salmon, tuna, herring, and anchovy.
   d. *Class Reptilia:* air breathers with dry skin and scales; young develop in self-contained eggs; turtles, sea snakes, iguanas, and crocodiles.
   e. *Class Mammalia:* warm-blooded; body covering of hair; produce milk; young born live.
      (1) *Order Cetacea:* whales.
         (a) *Suborder Mysticeti:* baleen whales, including blue, gray, right, sei, finback, and humpback whales.
         (b) *Suborder Odonticeti:* toothed whales, including the killer and sperm whale, the porpoise, and the dolphin.
      (2) *Order Carnivora:* sea otters, seals, sea lions, and walruses.
      (3) *Order Sirenia:* dugongs and manatees.

16.6 Practical Considerations: Commercial Fisheries

In 1950, the total world marine fish catch was approximately 21 million metric tons. During the next forty years, as human populations exploded, the fishing effort by all nations intensified, and the technology and gear used to hunt and catch the fish improved dramatically. In 1955, the catch was 28 million metric tons, and by 1960 at 40 million metric tons it was nearly twice what it had been ten years previously. By 1970 the catch was 70 million metric tons; it rose to 76 million metric tons in 1985 and rose again to 81 million metric tons in 1986. By 1989 the total world fish catch was 86 million metric tons, but in 1990, for the first time in forty years, the world's marine harvest dropped, to 83 million metric tons.

At the same time, there has been a shift in the use made of the catch. In 1950, 90% of the world fish catch was consumed as food by humans, and 10% was used to make fish-meal products to feed the poultry and livestock of the more-developed countries. By 1980, approximately 60% of the world catch was going to feed people, and 40% was being used as feed for domestic animals. The use of fish meal in this way is only about 20% efficient; that is, for every ton of fish meal fed, only 0.2 ton of additional animal protein is produced.

At the present time, total world fish catches continue to increase due to fishing effort, the opening of new fisheries, such as the North Pacific bottom fishery and the South African pilchard fishery, and an increased consumer

(a)

(b)

(c)

Figure 16.14

(a) Freshly caught pollock is poured from a trawl net onto the deck of a seafood-processing vessel. (b) Fish are cleaned and boned below decks in the processing plant. After washing to remove blood and oils and then drying, the fish meat becomes a tasteless, odorless fish product, surimi. (c) The surimi base is made into analogs such as crab shapes.

demand for fish and fish products. For example, Alaskan pollock, a bottom fish, is processed to remove the fats and oils that give the fish its flavor, and a highly refined fish protein called **surimi** is produced. Surimi is processed again and flavored to form artificial crab, shrimp, and scallops. The processing of pollock to form surimi is shown in figure 16.14. Surimi, a major fish product in Japan, is the fastest-growing product in the U.S. seafood market. Meanwhile, many traditional and valuable fisheries have declined, often as a result of overfishing. Five fisheries are discussed in the following sections; each illustrates the difficulties faced by those fishing, the fishery managers, the fishing nations, and the fish consumers as the catches decline and the costs rise.

Anchovies

The anchovy fishery, concentrated in the upwelling zone off the coast of Peru, has produced the world's greatest fish catches for any single species. Anchovies are small, fast-growing fish that feed directly on the phytoplankton in the

upwelling zone and travel in dense schools that are easy to net in large quantities. These fish are not caught for human consumption but are processed into fish meal that is exported as feed for domestic animals.

The fishery began in 1950 with a harvest of 7000 metric tons. By 1962, the harvest had risen to 6.5 million metric tons, and as the world demand for fish meal rose, the anchovy fishing intensified. In 1970, the fishery's peak year, about 12.3 million metric tons were taken. As the fishing increased, the average size of the fish taken decreased, and it took more and more fish (smaller and younger) to make up the same catch. During the period between 1950 and 1972, the Peruvian coast was visited three times by El Niño. (Refer back to chapter 7.) When the general wind pattern reverses, the winds become westerly, diminishing the upwelling and moving warmer, nutrient-poor surface water eastward into the areas of normally cold and productive upwelled water. The decrease in nutrients and the increase in temperature result in the destruction of the plant and animal populations; the decomposing organic

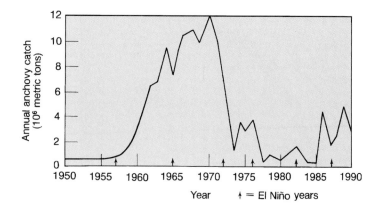

Figure 16.15

The anchovy catch of Peru and Ecuador by calendar year given in millions of metric tons. Years of El Niño are 1957, 1965, 1972, 1976, 1982, and 1986.

material strips oxygen from the water and further reduces the populations. El Niño devastates the fishing, and large numbers of seabirds that feed on the schools of anchovies starve.

The heavy fishing of 1970–71 and the severe effect of El Niño in 1972 resulted in only 2 million metric tons harvested in 1973. After 1973, government quotas were placed on the anchovy catch and it increased slowly in 1974 and 1975, but El Niño struck again in 1976, and 1982–83. The 1983 through 1985 catches were kept below 150,000 metric tons. Since then the catch has stabilized, with harvests between 1988 and 1990 being 3.6, 5.4, and 3.8 million metric tons, respectively; see figure 16.15. In this fishery, overfishing combined with environmental change to intensify the effect of overharvesting.

Tuna

The tuna fishery is an expensive, sophisticated business that involves many nations, including the United States. The United States tuna industry accounts for about 50% of the world tuna catch and operates out of southern California. A modern tuna boat costs $5 million or more and takes advantage of the tuna's schooling habit to catch the fish in huge seine nets 1100 m (3600 ft) long and 180 m (595 ft) deep; a single set of a net brings in over 150 tons of tuna. Although the modern open-ocean netting of tuna is an efficient method of harvesting, it has a significantly undesirable side effect. Tuna are accompanied in their movement across the oceans by porpoises that swim just above the schools of tuna and are used by the tuna boats to locate the schools. The net is set in a circle around the school of tuna and pulled up into a bag. If porpoises are caught in the net they can drown. The total cost of the tuna harvest in porpoises is unknown, but since the passage of the 1972 Marine Mammal Protection Act, conscious and legally enforced efforts have been made to reduce the porpoise kill. It is estimated that the largest kill occurred in 1970, with a loss of more than 500,000 porpoises.

Today many nations ban the import of tuna caught by encircling dolphins with nets, and pressure is mounting on other nations to abandon this technique; a voluntary program sponsored by the Inter-American Tropical Tuna Commission trains skippers to use dolphin-saving maneuvers. United States conservation groups monitor tuna seine fishing and canneries to insure that tuna eaten in the U.S. is "dolphin-safe" from seiners that set their nets around floating logs or other objects. These efforts have decreased the annual dolphin kill to 20,000 from more than 150,000.

Salmon

Another fishery in which the United States has an important role and investment is the salmon industry of the Pacific Northwest and Alaska (see fig. 16.16). These fish spawn in fresh water and remain there as juveniles for a year or less, depending on the species. Then they begin their long journey to the sea, where they will live and grow for one to four years. At the end of this time, the fully mature adults return to the same rivers, lakes, and streams in which they were born to spawn in their turn. As they return, the salmon are fished both commercially and for sport. The demand for salmon is high, and the fish bring a high price per pound, but like many other heavily exploited fish stocks, the salmon are becoming scarcer, and the fishing is now tightly controlled. The fishing season is shortened, and the catches decrease, causing profits for those who fish to fall; and management requires more and more regulations in attempts to ensure a population for the next year. The numbers of coho and chinook salmon caught off the Oregon coast in 1992 were 62% below the 1991 catch; a fishing season that once extended from May to October is now open for only two weeks.

An additional difficulty in maintaining wild salmon runs is the degradation of their freshwater environment. The salmon depend on high-quality, pollution-free fresh water and clean, gravel-bottomed, shady, cool streams for successful natural spawning. People have routinely dammed rivers for power usage and for flood control, cutting off the salmon from their home streams; people have harvested trees up to the stream edge, removing shade and allowing mud and silt eroded from the exposed land to cover the streambed. People have used streams to carry away their wastes, degrading the water quality and damaging both the juvenile fish and the returning runs. In many areas, these activities are being corrected, but the naturally spawning fish are now few in number, and people must pay both the cleanup costs and the costs of hatchery programs to supplement the dwindling natural stocks.

Figure 16.16
A salmon purse seiner hauling in its net.

Redfish

The red drum, or redfish, fishery of the Gulf of Mexico is an example of a local, small-scale fishery that suddenly expanded due to national demand and without adequate safeguards for the fish stock. Stocks of redfish occur in both the inshore and offshore waters of the Gulf of Mexico, with the larger fish, believed to be the spawning stock, gathering in large schools offshore. In 1983, 7.6 million pounds were caught in the recreational fishery and 3 million in the commercial fishery. Only 7% of the commercially caught fish came from offshore waters. In 1984, recreational fishing took 6.5 million pounds; commercial fishing took 4.3 million pounds, 25% from offshore waters. By 1985, the commercial catch had grown to 6.5 million pounds, 55% from offshore waters, and in the first half of 1986, 6 million pounds were fished commercially, nearly all from offshore waters.

The rapid growth of this commercial fishery is attributed to the sudden popularity of the Cajun dish blackened redfish. Commercial fishing responded to the increased demand for redfish by developing a purse-seiner fleet assisted by spotter planes and able to fish the offshore schools. Each set of the purse seine nets takes 50,000–150,000 pounds of redfish. In 1986, to forestall collapse of the fishery by overfishing, federal agencies placed a 1-million-pound emergency quota on redfish caught in federally regulated waters. The states of Texas, Louisiana, Alabama, Mississippi, and Florida passed laws to prevent the landing of fish caught by purse seiners. The commercial catch dropped to 4.8 million pounds in 1987. Further analysis of the stock led to a prohibition of direct targeted commercial fishing of redfish in some areas while allowing recreational fishing. In other areas all fishing was banned. In 1988 there was another federal emergency closure when the 1.7-million-pound commercial and recreational quota was exceeded. Only 287,000 pounds of redfish were caught by the commercial fishery in 1988. Recreational fishing was also closed by the states in 1988 to protect the existing redfish stocks and to allow the states time to develop regulations enhancing juvenile redfish survival and protecting offshore stocks. Whether the rage for Cajun cooking will dissipate and the demand for redfish will drop is unknown. However, while the demand is high, careful management of the redfish stocks is required to assure sustained yield of this resource.

Sharks

As overfishing takes its toll on familiar commercial species, the fishing pressure is transferred to other species. Worldwide, 370 species of shark have been identified in a population estimated at more than 4 billion. Until recently sharks were fished for sport and for local markets, but recent estimates are that more than 200 million sharks are being taken annually, twice the catch of five years ago. Annual U.S. sales of shark steaks and fillets have increased from 3000 metric tons in 1985 to 5000 metric tons in 1990. The Asian demand for shark fin soup has supported an expanding shark fishery that amputates the shark fins and throws the animals back into the ocean to die. When U.S. commercial shark fishing began in the late 1970s, 21 metric tons of thresher sharks were brought to California docks; by 1982 the thresher catch was 1100 metric tons. In 1989 the catch fell to 300 metric tons, and the thresher fishery collapsed. Overfishing appears to be decimating shark populations in all the world's oceans; the world catch was 710,000 metric tons in 1990. The problem is intensified by sharks' low reproductive rate, their slow growth, and the long period required for their maturation. Lack of shark-fishery management, lack of information concerning shark biology, and the possibility of continued worldwide expansion of shark fishing could be devastating to this link in the ocean food web.

Problems and Policies

Around the world too many fishing boats are taking too many fish, too fast. Diminishing fish populations are the victims of relentless overfishing, and harvest regulations fail to keep pace with declining stocks. Fifteen percent of the species currently fished in U.S. waters are near their maximum population potential, and, according to the U.S. Office of Fisheries Conservation and Management, 41% of the species are overfished. The Newfoundland cod fishery is in severe decline, and Iceland has warned its fishing industry that the cod stock could collapse unless the annual catch is cut back by 40% in 1993. North Atlantic swordfish landings in the U.S. have declined 70% from 1980 to 1990, and the average weight of the swordfish has fallen from 115 to 60 pounds. The western Atlantic adult bluefin tuna population dropped from 300,000 to 30,000 in twenty years, and many believe this fish is doomed, as a single fish sells for as much as $30,000 in Japan. These data mean trouble for people who depend on fish as their source of protein as well as for the hundred million people that make their living by catching, processing, or selling fish. Trouble may be coming to the other marine animals that compete for their share of the catch. Steller sea lion populations in Alaska have dropped 70% since the early 1970s with a 1% decline from 1990 to 1991 and a 5% decline from 1991 to 1992; in the mid-1970s and mid-1980s studies in the Gulf of Alaska and the Bering Sea showed that more than 50% of the Steller sea lion's diet was pollock, but the pollock they ate in the 1980s were about 37% smaller by weight, meaning the sea lions had to spend more time and energy obtaining the same amount of fish.

Problems for the United States fishing industry accelerated in 1977 when U.S. control over its fisheries was expanded to a zone that extends 200 miles out to sea. Within this zone, only U.S. boats are permitted to fish. Other countries may fish only under the terms of agreements negotiated with the United States. Since the establishment of the 200-mile economic zone, the U.S. fishing fleet has grown dramatically through a federal loan program encouraging the building of new boats; for example, the trawler fleet in New England expanded from 590 vessels in 1976 to more than 1000 in 1985. Boats built in the 1980s are extremely efficient, equipped with sonar and depth recorders and computers that remember the sites of previous catches and home in on these sites at a later date. Planes, helicopters, and even satellite data are used to find the fish. In addition, the fishing boat and gear, the crew's wages, and the fuel needed are all increasing the cost of commercial fishing. Each year the shrinking stocks mean greater regulation and ever smaller quotas; for example, a year's quota of Pacific halibut is now taken in two frantic twenty-four-hour periods.

Since 1987 the longline fishery fleet in Hawaii has grown from 37 vessels to more than one hundred in 1990. In 1985 this fleet landed less than 13.6 metric tons (30,000 lb) of Pacific swordfish; by 1990 the harvest had grown to more than 1136 metric tons (2.5 million lb). The success of the fishery is expected to spread to other Pacific Island areas such as Guam, the Mariana Islands, and American Samoa. Will this intense fishing by longliners impact other fisheries, seabirds, and marine mammals?

The oceanic drift-net fishery became a serious problem in the 1980s, laying out nets up to 40 miles long each night. These nets of almost invisible nylon hang like walls in the water, trapping and killing nearly everything that swims into them. In 1990 the U.S. Marine Mammal Commission estimated the aggregate length of these nets at 25,000 miles, enough to ring the earth. Three Asian nations, Japan, Taiwan, and South Korea, have used the nets to catch squid and fish in the Pacific and Indian oceans; in the Atlantic several European nations have drift-netted, mainly for albacore tuna. These nets do not rot; sections torn free in storms float in the ocean for months, even years, as ghost nets catching everything they encounter; there is no estimate on the animal life they destroy. Japan, Taiwan, and South Korea agreed to abide by a United Nations resolution to halt drift-net fisheries by the end of 1992, but compliance will be difficult, requiring alternative employment for many thousands of workers.

Large numbers of marine animals die each year only because they are caught incidentally while fishing for other species. **Incidental catch,** or bycatch, or what are often called "trash fish," represent a tremendous waste of marine resources. During a single year, more than 25,000 porpoise have been killed as incidental catch in the tuna fishery. The Asian high-seas drift-net fisheries in the North Pacific yearly is reported to claim 10,000–15,000 Dall porpoise and 500,000–700,000 seabirds trying to feed on the netted fish. It is estimated that shrimp trawlers catch more than 45,000 sea turtles and that more than 12,000 of them do not survive. The world's shrimp fishery is estimated to have an annual catch of 1.1 million–1.5 million tons; the associated bycatch of finfish is 5 million–21 million tons, of which 3–5 million tons are discarded. Alaskan trawlers for pollock and cod throw back to the sea some 25 million pounds of halibut, worth about $30 million, as well as salmon and king crab because they are prohibited from keeping or selling this bycatch. Another 550 million pounds of groundfish are discarded in Alaskan waters to save space for larger or more valuable fish.

The discard rate on bycatch varies from place to place; if incidental catch does not bring a high enough price, and if processors are not available, these "trash fish" will be returned to the sea, usually dead. In the Gulf of Mexico, 1 pound of shrimp results in 15 pounds of bycatch, nearly all discarded. In Southeast Asia, the need for protein is so great that most of the bycatch is marketed locally.

Fish Farming

A possible alternative way to increase the fish harvest from the sea is by farming it. An old Chinese proverb states, "If you give a person a fish, he will have food for one day, but if you teach him to raise fish, he will have food for a lifetime." Farming the water, known as **mariculture** or **aquaculture,** began in China some four thousand years ago. The Chinese were culturing common carp in 1000 B.C., and a book was written in 500 B.C. giving directions on fish farming, including methods for building the pond, selecting the stock, and harvesting it. In China, Southeast Asia, and Japan, fish farming in fresh and salt water has continued to the present as a practical and productive method of raising large quantities of fish. Carp, milkfish, *Tilapia,* and catfish are raised. The methods used are labor intensive, and most fish farms are small, family-run operations. In **monoculture,** fish are raised as a single species. In **polyculture,** they are raised in ponds they share with other species. Surface feeders and bottom feeders are raised together, making use of the total volume of the pond; this type of system may even include ducks.

Israel's requirements for protein to feed a large population concentrated in a small area have resulted in the development of acres of fish ponds, which produce annually 130 million pounds of carp, *Tilapia,* and mullet in polyculture, with plans for further development. Fish farming in the United States, however, produces only 2% of our fishery products, but that amount includes 50% of the catfish and nearly all the trout. In order to be successful in the United States, fish farming must be profitable, which requires a proven market and a large-scale operation, along with the development of a technology to keep costs down. Another factor to be considered is that a profitable operation requires careful choice of the species. To be economically productive, a species must reproduce in captivity, have juvenile forms that survive well under controlled conditions, gain weight rapidly, eat a cheap and available food, and fetch a high market price.

Salmon farming is on the rise worldwide. In 1973 the world's salmon pens produced 136 metric tons of salmon; by 1980, 910 metric tons were being raised. Norway and Scotland have been particularly successful; in 1988 Scotland produced 20,000 metric tons and Norway harvested 80,000 metric tons. The 1990 harvest of 282,000 metric tons accounted for 30% of the world's marketed salmon. The Canadian government has adopted a national policy of encouraging salmon farms, and the farms are proliferating rapidly along the British Columbia coast, north of Vancouver. Maine, Washington, California, and Oregon allow salmon farming, but the permit procedure is long and complex. Although pen-rearing of salmon has been tried and is successful in the protected waters of Puget Sound in Washington State (fig. 16.17), it is an expensive operation that raises problems in the Sound's multiple-use waterways. Salmon pens are not particularly attractive to those with homes along the shore; the pens get in the way of recreational boating and water sports, and the fish produce waste products into the water.

In Texas, the craze for redfish discussed earlier has opened the way to redfish mariculture. At present, there is one redfish hatchery in the state and several farms that are raising redfish in indoor and outdoor pools. Hatcheries are beginning to produce, but initial startup costs are estimated to be between $300,000 and $400,000, which may limit individual ventures and allow only corporate investment.

Hatchery rearing and release of seagoing fish (salmon and steelhead trout) is called **ocean ranching** or **sea ranching.** Federal and state hatcheries provide millions of fish, which are caught in both the commercial and recreational fisheries along the coasts of Oregon, Washington, and Alaska. Because pen culture of salmon is expensive, the release from private hatcheries of salmon that will return to the hatcheries to be netted has been proposed. This private sea ranching is being tried in Oregon, where a private sea-ranching facility is associated with the Weyerhauser Company's paper plant. The warm water from the plant is mixed with river water to accelerate the salmon's early growth. These juveniles are then transported to holding areas adjacent to the sea and imprinted with a chemical label in the water (fig. 16.18). The fish are released to the ocean. Upon maturing, the salmon return to the holding areas, seeking them by the chemical label. Here, they are harvested for the market. Natural survival is compared with hatchery survival and returns for spawning in table 16.3.

A population of natural spawners must produce at least two thousand eggs per pair to assure the survival of a spawning pair to sustain the natural run. Hatcheries increase the survival at the egg to fry stage eightfold. Thus they have the ability to assure continuance and even build a run of fish by spawning more than one pair of returnees. The greater return of mature fish to the hatchery because of the increased juvenile survival is what makes ocean ranching appear attractive as an enterprise.

The idea of private ocean ranching has not proved popular with those who fish commercially, because they fear the intrusion and control of the large companies and a possible drop in salmon prices. Most economists and fishery

Figure 16.17

An aerial view of salmon pens in which fish are raised to market size by Sea Farm Washington, Inc.

Figure 16.18

The ocean-ranching release and recapture facility at Coos Bay, Oregon, is operated by Anadromous, Inc., of Corvallis, Oregon.

biologists believe that an ocean-ranching program in the private sector would add fish to the harvest without any great effect on market prices.

Summary

The nekton swim freely and independently. The marine mammals include whales, seals, sea lions, walruses, sea otters, and sea cows. The whalers hunted baleen whales and the sperm whales, which are toothed whales. As hunting techniques changed, harvesting became more intensive, and some species of whales have been threatened with extinction. The International Whaling Commission regulates whaling on a voluntary basis; a ten-year moratorium on commercial whaling is in progress, but scientific whaling continues. The IWC is at present considering new catch quotas. Dolphins and porpoises are small toothed whales.

The fur seals and sea otters were hunted heavily during the nineteenth and early twentieth centuries. Walrus harvests have recently increased alarmingly, and the northern fur seal has been declining since the 1950s. The Steller sea cow is extinct; manatees and dugongs are found in the warm waters of the Indian and Atlantic oceans. The Marine Mammal Protection Act was passed in 1972 to protect all marine mammals and to prohibit commercial trade in marine mammal products.

Sea snakes, the marine iguana, seagoing crocodiles, and sea turtles are the reptile members of the nekton. Sea turtle populations are under great pressure by hunters and poachers. The squid also belong to the nekton.

Fish are found at all depths in all oceans. Sharks, rays, and skates are primitive fish with cartilaginous skeletons. All other fish have bony skeletons, including the commercially fished species and the highly specialized types of the deeper ocean.

Table 16.3
Survival Rates of Salmonid Spawners

	Eggs (produced)	Eggs to fry stage (survivors)	Fry to adult (survivors)	Caught in fishing	Left to spawn (survivors)
natural spawners	2000 100%	200 10%	6 0.3%	4 0.2%	2 0.1%
hatchery spawners	2000 100%	1600 80%	48 2.4%	32 1.6%	16 0.8%

Source: After K.R. Allen, *Conservation and Management of Whales,* Washington Sea Grant Program, 1980.

World fish catches increased dramatically between 1950 and the 1980s, due to the increased fishing effort and improved technology. Although stock abundance has declined, new fisheries have developed, and the fishing effort has continued high to keep the catch stable. At the same time, more of the fish is used as domestic animal feed and less is used for human consumption.

The problems of the traditional commercial fisheries include overfishing, increased regulation, increased costs, and large quantities of incidental catch. The Peruvian anchovy fishery had produced the world's greatest fish catches until it was hit by the effects of overfishing and El Niño. The United States tuna fishery has had to cope with the killing of porpoises as a side effect of fishing technology, and the Pacific Northwest and Alaskan salmon fisheries are experiencing difficulty due to the degradation of the salmon's freshwater environment and the loss of natural salmon runs. The Gulf Coast redfish fishery has been threatened with collapse due to its recent explosive growth, and unregulated shark fisheries are depleting shark populations worldwide.

Aquaculture has proved to be practical and productive in Asia and Israel; European and Canadian salmon farms are increasingly successful. In the United States, marine aquaculture is a small industry that requires large-scale operations, proven markets, and cost-cutting technology. Ocean ranching is the release of hatchery-raised seagoing fish to increase fishing stocks.

Key Terms

mammal	demersal fish
cetaceans	surimi
baleen	incidental catch
echolocation	mariculture
pinniped	aquaculture
manatee	monoculture
dugong	polyculture
sea cow	ocean ranching/sea ranching
Sirenia	

Study Questions

1. The Marine Mammal Protection Act of 1972 has allowed the recovery of certain populations to such an extent that the marine mammals are now competing for food resources of economic value to humans. Under these circumstances should any adjustments be made to this act? Explain why.
2. What sensory abilities help a shark to find its prey?
3. In the last twenty-five years the world's marine fish catch has grown from 40 million metric tons to more than 80 million metric tons. During this period the catch from many large fisheries has declined. How do you explain this contradictory situation?
4. Fish farming is growing slowly in the United States in comparison to other countries. What reasons can you give for this situation?
5. How has the United States' 200-mile economic zone affected U.S. fisheries?

6. Why are bottom fish more likely to show the first effects of a degraded nearshore environment, rather than fish that inhabit the water column?
7. How will the ability to predict El Niño help manage the Peruvian anchovy fishery?
8. Why are more toothed whales found at low latitudes and why do more baleen whales reside in temperate and polar latitudes? Consider food requirements only.
9. Why is it difficult to increase a pinniped population in the northern temperate latitudes after hunting has ceased? Do you think the situation would be the same in the southern temperate latitudes?
10. Sea snakes are thought to have their origin in the land snakes of India and Southeast Asia. Why are there no sea snakes in the Atlantic Ocean? What might be the effect of a sea level canal through the isthmus of Panama?
11. If equal numbers of natural-run fish and hatchery fish are spawned, the returning fish will be in the ratio of 6 natural run: 48 hatchery fish (see table 16.3). If a quota of 36 of the returning fish is assigned to a commercial fishery, why is there a risk of completely destroying the natural run? Discuss the impact of hatchery-enhanced fish runs on natural fish runs as hatchery production is increased and fish catch quotas are based on the total number of mixed stock returns.
12. Discuss the impact of humans on populations of walrus, fur seals, and sea otters.
13. Compare fishing as practiced in the United States and in less-developed countries.
14. Compare fish farming in the United States with fish farming in other countries.
15. What characteristics do deep-sea fish have in common? How do these characteristics enhance fish survival in this environment?

Suggested Readings

General

Lerman, M. 1986. *Marine Biology, Environment, Diversity and Ecology.* Benjamin-Cummings, Menlo Park, Calif. 535 pp.
Life in the Sea, readings from *Scientific American,* 1982. Freeman, San Francisco. 248 pp.
Niesen, T. M. 1982. *The Marine Biology Coloring Book.* Harper & Row, New York. 218 pp.
Sumich, J. L. 1992. *An Introduction to the Biology of Marine Life,* 5th ed. Wm. C. Brown, Dubuque, Ia. 449 pp.

Mammals

Ackerman, D. 1992. Last Refuge of the Monk Seal. *National Geographic* 181 (3): 128–44.
Allen, K. R. 1980. *Conservation and Management of Whales.* Washington Sea Grant, Univ. of Washington, Seattle. 107 pp.
Bonner, W. N. 1982. *Seals and Man, A Study of Interactions.* Washington Sea Grant, Univ. of Washington, Seattle. 170 pp.
Calkins, D. 1992. Steller Sea Lions: Still Threatened in Alaska. *Alaska's Wildlife* 24 (6): 21–27.
Chapan, D. G. 1988. Whales. *Oceanus* 31 (2): 64–70.
Darling, J. D. 1988. Whales, An Era of Discovery. *National Geographic* 174 (6): 872–909.
Gentry, R. 1987. Seals and Their Kin. *National Geographic* 171 (4): 475–501.
Golden, F., ed. Spring 1989. *Oceanus* 32 (1). (Issue devoted to whales.)
Haley, D., ed. 1978. *Marine Mammals.* Pacific Search, Seattle. 256 pp.
Hall, A. 1984. Man and Manatee. *National Geographic* 166 (3): 400–18.

Nelson, C., and K. Johnson. 1987. Whales and Walruses as Tillers of the Sea Floor. *Scientific American* 256 (2): 112–17.

Norris, K. S. 1992. Dolphins in Crisis. *National Geographic* 182 (3): 2–35.

Sanderson, S., and R. Wasseisug. 1990. Suspension Feeding Vertebrates. *Scientific American* 262 (3): 96–101. (Whale sharks and baleen whales.)

Scammon, C. M. 1874. *The Marine Mammals of the Northwestern Coast of North America.* Dover, N.Y. (1968). 324 pp.

Siniff, D. B. 1988. Seals. *Oceanus* 31 (2): 71–74.

Swartz, S. L., and M. L. Jones. 1987. Gray Whales Make a Comeback. *National Geographic* 171 (6): 754–71.

Trites, A. 1992. Northern Fur Seals: Why Have They Declined. *Aquatic Mammals* 18 (1): 3–18.

Tuck, J. A., R. Grenier, and R. Laxalt. 1985. Sixteenth Century Basque Whaling in America. *National Geographic* 168 (1): 40–57.

Whitehead, H. 1984. The Unknown Giants: Sperm and Blue Whales. *National Geographic* 166 (6): 774–89.

Würsig, B. 1988. The Behavior of Baleen Whales. *Scientific American* 258 (4): 102–7.

Zimmer, C. 1992. Portrait in Blubber. *Discover* 13 (3): 86–89. (Elephant seals.)

Zopal, W. 1987. Diving Adaptations of the Weddell Seal. *Scientific American* 256 (6): 100–105.

Squid

Roper, C. F. E., and K. J. Boss. 1982. The Giant Squid. *Scientific American* 246 (4): 96–105.

Reptiles

Carr, A. 1984. *So Excellent A Fishe: A Natural History of Sea Turtles.* Scribner, New York. 248 pp.

Lohmann, K. J. 1992. How Sea Turtles Navigate. *Scientific American* 266 (1): 100–106.

McClintock, J. 1991. Deep-Diving, Warm-Blooded Turtles. *Sea Frontiers* 37 (1): 8–13. (Leatherback turtles.)

Minton, S. A., and H. Heatwole. 1978. Snakes and the Sea. *Oceanus* 11 (2): 53–56.

Ritchie, T. 1989. Marine Crocodiles. *Sea Frontiers* 35 (4): 212–19.

Taubes, G. 1992. A Dubious Battle to Save the Kemp's Ridley Sea Turtle. *Science* 256 (5057): 614–16.

Fish

Eastman, J. T., and A. L. DeVries. 1986. Antarctic Fishes. *Scientific American* 255 (5): 106–14.

Hardy, A. 1959. *The Open Sea: Its Natural History,* Part 2: Fish and Fisheries. Houghton Mifflin, Boston. 322 pp.

Horn, M. H., and R. N. Gibson. 1988. Intertidal Fisheries. *Scientific American* 258 (1): 64–70.

Idyll, C. P. 1976. *Abyss—The Deep Sea and the Creatures that Live In It,* rev. ed. Crowell, New York. 428 pp.

Isaacs, J. D., and R. A. Schwartzlose. 1975. Active Animals of the Deep-Sea Floor. In *Ocean Science,* readings from *Scientific American* 233 (4): 202–9 (1977).

MacLeish, W. H., ed. Winter 1981–1982. *Oceanus* 24 (4). (Issue devoted to sharks: vision, feeding, reproduction, and behavior.)

Partridge, B. L. 1982. The Structure and Function of Fish Schools. *Scientific American* 246 (6): 114–23.

Perrine, D. 1989. Shark Attack! *Sea Frontiers* 35 (1): 31–41. (Why sharks attack people.)

VanDyk, J. 1990. The Long Journey of the Pacific Salmon. *National Geographic* 178 (1): 3–37.

Wheeler, A. 1975. *Fishes of the World, An Illustrated Dictionary.* Macmillan, New York. 366 pp.

Wood, L. 1986. Megamouth, New Species of Shark. *Sea Frontiers* 32 (3): 192–98.

Wu, N. 1990. Fangtooth, Viperfish and Black Swallower. *Sea Frontiers* 36 (5): 32–39. (Deep-sea fishes.)

Fisheries

Baldwin, R. F. 1990. Fundy Farming. *Sea Frontiers* 36 (5): 40–45. (Salmon farming.)

Bauereis, E. I., and J. N. Kraeuter. 1984. Power Plants and Striped Bass: A Partnership. *Oceanus* 27 (1): 40–45.

Borgese, E. M. 1980. *The Story of Aquaculture.* Abrams, New York. 236 pp.

Brown, L. R. 1985. Maintaining World Fisheries. In *State of the World 1985,* Brown, L. R., ed. Norton, New York. pp. 73–96.

Food and Agricultural Organization of the United Nations. 1990. *Yearbook of Fisheries Statistics, Catches, and Landings.* Vol. 70. Rome, Italy.

Iverson, E. S., and J. Z. Iverson. 1987. Salmon Farming Success in Norway. *Sea Frontiers* 33 (5): 354–61.

Lee, C. M. 1984. Surimi Gel and the U.S. Seafood Industry. *Oceanus* 27 (1): 35–39.

Ling, S. 1977. *Aquaculture in Southeast Asia, A Historical Overview.* Washington Sea Grant, Univ. of Washington, Seattle. 108 pp.

Michael, R. G., ed. 1987. *Managed Aquatic Ecosystems.* Elsevier, New York. 166 pp.

17

The Benthos: Dwellers of the Seafloor

17.1 **The Plants**
General Characteristics of Benthic Algae
Kinds of Seaweeds
Other Plants
17.2 **The Animals**
Animals of the Rocky Shore
Symbiotic Relationships
Tide Pools
Animals of the Soft Substrates
The Deep Seafloor
Fouling and Boring Organisms
17.3 **Classification Summary of the Benthos**
17.4 **The Tropical Coral Reefs**
17.5 **High-Energy Environments**
17.6 **Deep-Ocean Chemosynthetic Communities**
17.7 **Sampling the Benthos**

Box: *Mussel Watch*

17.8 Practical Considerations: Harvesting the Benthos
The Animals
The Algae
Kelp Bioconversion
Biomedical Products

Box: *Genetic Manipulation of Fish and Shellfish*

Summary
Key Terms
Study Questions
Suggested Readings

The exposed rocks had looked rich with life under the lowering tide, but they were more than that: they were ferocious with life. There was an exuberant fierceness in the littoral here, a vital competition for existence. Everything seemed speeded-up; starfish and urchins were more strongly attached than in other places, and many of the univalves were so tightly fixed that the shells broke before the animals would let go their hold. Perhaps the force of the great surf which beats on this shore has much to do with the tenacity of the animals here. It is noteworthy that the animals, rather than deserting such beaten shores for the safe cove and protected pools, simply increase their toughness and fight back at the sea with a kind of joyful survival. This ferocious survival quotient excites us and makes us feel good, and from the crawling, fighting, resisting qualities of the animals, it almost seems that they are excited too.

John Steinbeck,
from *The Log from the Sea of Cortez*

Coral reef, Caribbean Sea.

The animals and plants that live on the seafloor or in the sediments are members of the benthos. The plants are found in the sunlit shallow coastal areas and in the intertidal zones, while the benthic animals are found at all ocean depths. The benthos include remarkably rich and diverse groups of organisms, among them the luxuriant, colorful tropical coral reefs, the newly discovered organisms surrounding deep-sea hydrothermal vents, the great cold-water kelp forests, and the hidden life below the surface of the mud flat and the sandy beach. Benthic organisms are important food resources and provide valuable commercial harvests, for example, oysters, clams, crabs, and lobsters. In this chapter we first present an overview of the benthos by group and habitat, with special focus on some of the world's more intriguing benthic communities, and then we consider the harvesting of the benthos, its problems, and its potential. This chapter is intended to serve only as an introduction to this topic, for no single chapter can do justice to the profusion of organisms present in this group. If you wish to establish a greater appreciation and understanding of the benthos, we suggest additional readings in books and magazines devoted to marine biology and biological oceanography (see the selected references at the end of this chapter).

17.1 The Plants

Seaweeds are members of a large group called **algae.** The unicellular planktonic algae were considered in chapter 15, and in this chapter we consider the large benthic algae or seaweeds, conspicuous members of coastal intertidal and subtidal communities. Because the algae photosynthesize and contain chlorophyll pigments, some biologists consider them plants; others consider them plantlike but prefer to place them in other categories, because their body form, reproduction, accessory pigments, and storage products generally differ from those of land plants. In this discussion the large benthic algae (or seaweeds) will be considered primitive members of the plant kingdom. These algae have simple tissues; they do not produce flowers or seeds, and their pigments and storage compounds vary from group to group.

General Characteristics of Benthic Algae

Although seaweeds are often found floating at the water's edge or thrown up on shore after a storm, they are not floating or planktonic algae. These large algae are benthic organisms that have been dislodged from the bottom. Seaweeds have a basal organ called a **holdfast** that anchors the plant firmly to a solid bottom or substrate. The holdfast is not a root; it does not absorb water and nutrients. Because the seaweeds grow attached, they are found in

areas of rocky substrate and not in areas of mud or sand, where the holdfast has nothing to which it can attach. Above the holdfast is a stemlike portion known as the **stipe;** it may be so short that it is barely identifiable, or it may be up to 35 m (115 ft) in length. These long stipes belong to certain brown seaweeds commonly known as **kelps,** which are found in the temperate and colder waters of the world. The stipe in most algae does not have tissues that transport water and nutrients or organic compounds, in contrast with the stem of a land plant, but the stipes of some larger algae known as kelps form tissue that is similar to conductive tissue that transports photosynthetic products in land plants. The stipe acts as a flexible connection between the holdfast and the **blades,** the alga's photosynthetic organ. Blades serve the same purpose as leaves, but they do not possess veins, the specialized conducting tissue seen in leaves. These tissues are not needed by the algae, because their blades are thin and bathed on all sides by water, and the algal cells take up water directly and do not require a supply from the ground conducted up a stem and through veins to leaves. The blades may be flat, ruffled, feathery, or even encrusted with calcium carbonate. Because seaweed is surrounded by water, with its dissolved carbon dioxide and nutrients, and because it exposes a large blade area to both the water and the sun, it is an efficient producer. The general characteristics of a benthic alga are shown in figure 17.1.

Since the benthic algae are primary producers, dependent on sunlight, they are confined to the shallow, sunlit areas of the oceans. Sometimes during a storm, the plants are dislodged, taking with them the rocks to which the holdfasts are attached. The plants, carrying their rock anchors, often drift for some time before they sink back to the bottom. If they sink too deep for sufficient light, they die, but if their blades remain in the photic zone, they will continue to grow. This process is also responsible for spreading seaweeds into new areas.

In the sea, the quality of light as well as the quantity of light changes with depth (see chapter 4). The red end of the visible light spectrum is quickly absorbed at the surface, and it is the blue-green light that penetrates to the greatest depths. On land and at the sea surface, plants receive the full spectrum of visible light, and it is the green algae with the same chlorophyll pigments as land plants, able to absorb both long and short wavelengths of light, that are found in shallow water. Algae found at moderate depths have a brown pigment that is more efficient at trapping the shorter wavelengths of solar light. At maximum growing depths, the algae are red, for the red pigment can best absorb the remaining blue-green light. Both the brown algae and the red algae possess the green chlorophyll, but the red and brown pigments mask its color, so that the seaweeds appear red or brown. Therefore, the characteristic pattern for seaweed growing on a rocky shore is the green algae in shallow water, then the brown algae, and last the red algae at greater depths. Few large red algae are seen at low tide, for they are primarily sublittoral species.

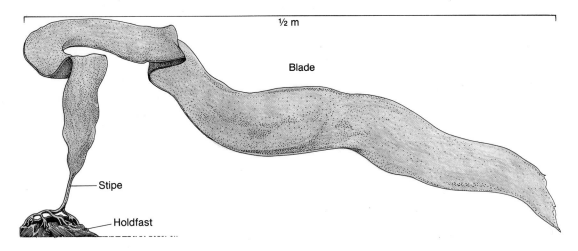

Figure 17.1
Benthic algae are attached to the seafloor by a holdfast. A stipe connects the holdfast to the blade. *Laminaria* is a kelp and a member of the brown algae.

Seaweeds provide food and shelter for many animals. They act in the sea much as the forest and shrubs do on land. Fish and other animals, such as sea urchins, limpets, and some snails, feed directly on the algae; other animals feed on shreds and pieces as they settle to the bottom. Some organisms use large seaweeds as a place of attachment; some of the smaller algae grow on the large kelps. Algae that produce calcareous outer coverings are important in the building of tropical coral reefs, which are discussed later in this chapter.

Kinds of Seaweeds

Algae can be divided into convenient groups based on their pigments. Green algae are moderate in size and may form fine branches or thin, flat sheets. They are mostly freshwater organisms, but a small number do occur in the sea, including *Ulva,* the sea lettuce, and *Codium,* known as dead man's fingers. Green algae are most similar to the land plants; they have the same green chlorophyll pigments as land plants and also store starch as a food reserve.

All brown algae are marine and range from simple microscopic chains of cells to the kelps, which are the largest of all the algae. Kelps have more structure than most algae but are much simpler than flowering land plants. The kelps have strong stipes and effective holdfasts that allow them to colonize rocky points in fast currents or heavy surf, a habitat favored by the sea palm, *Postelsia.* Other kelps grow with holdfasts well below the depth of wave action and float their blades at the surface supported by gas-filled floats; for example, the bull kelp, *Nereocystis,* is especially abundant along the Alaskan, British Columbian, and Washington coasts, and the great kelp, *Macrocystis,* is found along the California coast. Other species are found off Chile, New Zealand, northern Europe, and Japan. Their brown color comes from fucoxanthin pigment, which masks the green chlorophyll. Storage products include the carbohydrates laminarin and mannitol, but not starch.

The red algae are almost exclusively marine; they are the most abundant and widespread of the large algae. Their body forms are varied and often beautiful, flat, ruffled, lacy, or intricately branched. Their life histories are specialized and complex, and they are considered the most advanced of the algae. All contain the pigments phycoerythrin and phycocyanin that mask their chlorophyll. Their storage product is floridean starch.

Although the algae are commonly classified by color, the visible color can be misleading. Some red algae appear brown, green, or violet, and some brown algae appear black or greenish. Some representative algae are shown in figure 17.2.

We considered diatoms as part of the plankton (see chapter 15), yet there are also benthic diatoms. These diatoms are usually of the pennate type and grow on rocks, muds, and docks, where they produce a slippery brown coating that is treacherous for the walker.

Other Plants

Algae do not produce seeds or flowers, but there are a few flowering plants with true roots, stems, and leaves that have made a home in the sea. Eelgrass, with its strap-shaped leaves, is found growing on mud and sand in the quiet waters of bays and estuaries along the Pacific and Atlantic coasts, and turtle grass is common along the Gulf Coast. Surf grass flourishes in more turbulent areas exposed to waves and tidal action. Temperate-area salt marshes are dominated by marsh grasses able to tolerate the brackish water. Vegetation from the marshes is washed into the estuaries by tidal creeks, and its nutrients are released to be recycled by the algae and the sea grasses. These sea grasses are rich sources of food and shelter for animals and act as attachment surfaces for some algae. In the tropics mangrove trees grow in swampy intertidal areas. Their intertwining roots provide shelter and also trap sediments and organic material, helping eventually to fill in the swamps and move the shore seaward.

Figure 17.2

Representative benthic algae. *Ulva* and *Codium* are green algae. *Postelsia, Nereocystis,* and *Macrocystis* are kelps. The kelps and *Fucus* are brown algae. The red algae are *Corallina, Porphyra,* and *Polyneura. Corallina* has a hard, calcareous covering.

17.2 The Animals

Benthic marine animals, unlike benthic plants, are found at all depths and are associated with all substrates. There are more than 150,000 benthic species, as opposed to about 3000 pelagic species in the sea. About 80% of the benthic animals belong to the **epifauna;** these are the animals that live on or attached to rocky areas or firm sediments. Animals that live buried in the substrate belong to the **infauna** and are associated with soft sediments such as mud and sand. There are only about 30,000 species of infauna, as compared to 120,000 different species of epifauna. Some animals of the seafloor are **sessile,** or attached to the seafloor, as adults (for example, barnacles, sea anemones, and oysters), while others are motile all their lives (for example, crabs, starfish, and snails). Although most benthic forms produce motile larval stages that spend a few weeks of their lives as members of the plankton (refer to the discussion of the meroplankton in chapter 15), the planktonic larval stage is of particular importance to the sessile species. The mobility of the juvenile stages allows the species to relieve overcrowding and to colonize new areas. Sessile adults must wait for their food to come to them, either under its own power or carried in by waves, tides, and currents; motile organisms are able to pursue their prey, scavenge over the bottom, or graze quietly on the seaweed-covered rocks.

The distribution of benthic animals is not governed by any single controlling factor such as light or pressure. In fact, the increasing pressure that occurs with depth does not have much effect on animals that are more than 90% water and have no internal gas storage chambers or lungs. Animal distribution is controlled by a complex interaction of factors, creating living conditions that are extremely variable. The substrate may be solid rock, movable cobbles, shifting sand, or soft mud. Temperature is nearly constant in deep water but changes abruptly in shallow areas covered and uncovered by the daily tidal cycle; salinity, pH, exposure to air, oxygen content of the water, and water turbulence all change abruptly as well in the intertidal zone. Benthic animals exist at all depths and are as diverse as the conditions under which they live. Their life-styles are related to their varied habitats, and the following sections discuss both habitat and life-style.

Animals of the Rocky Shore

The rocky coast is a region of rich and complex plant and animal communities living in an area of environmental extremes. It is the meeting place between the more variable land conditions and the more stable sea conditions. As the water moves in and out on its daily tidal cycle, rapidly changing combinations of temperature, salinity, moisture, pH, dissolved oxygen, and food supply must be endured. At the top of the littoral zone, organisms must cope with long periods of exposure, heat, cold, rain, snow, and predation by land animals and seabirds as well as with the waves and turbulence of the returning water. At the littoral zone's lowest reaches, the plants and animals are rarely exposed but

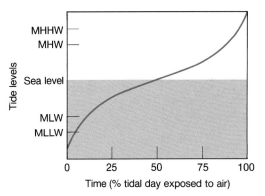

Figure 17.3
The time of exposure to air for intertidal benthic organisms is determined by their location above and below mean sea level and by the tidal range. (MLLW = mean lower low water range; MLW = mean low water; MHW = mean high water; MHHW = mean higher high water.)

have their own problems of competition for space and predation by other organisms. The exposure endured by marine life at different levels in the littoral zone is shown in figure 17.3.

The distribution of the plants and animals is governed by their ability to cope with the stresses that accompany exposure, turbulence, and loss of water. Along the rocky coasts of such areas as North America, Australia, and South Africa, biologists have noted the patterns formed as the plants and animals sort themselves out over the intertidal zone. This grouping is termed zonation; **vertical zonation,** or **intertidal zonation,** is shown in figure 17.4. The distribution of the seaweeds, with the green algae in shallow water, the brown algae in the intertidal zone, and the red algae in the subtidal area, is an example of such zonation.

The following is a general discussion of some of the more conspicuous life-forms found in these rocky intertidal zones. It is not a guide to the marine life of rocky shores, as substantial differences occur from region to region. If you wish to explore the rocky shores in your area, obtain a manual written specifically for your location.

In the supralittoral (or splash) zone, which is above the high-water level and is covered with water only during storms and highest tides, the animals and plants occupy an area that is as nearly land as it is ocean bed. At the top of this area, patches of dark lichens and algae appear as crusts on rocks and stones and are often nearly indistinguishable from the rock itself. Scattered vegetation, including tufts of green algae, provides grazing for the small herbivorous snails and limpets of the supralittoral zone. The snail *Littorina,* the periwinkle, is so well adapted to an environment that is more dry than wet that, unlike its marine relatives, it is an air breather, and some species will drown if caught underwater. Snails withdraw into their shells to prevent moisture loss and seal themselves off from the air with a horny disk called an operculum. Limpets are able to use their single muscular foot to press themselves tightly to the rocks to prevent drying out.

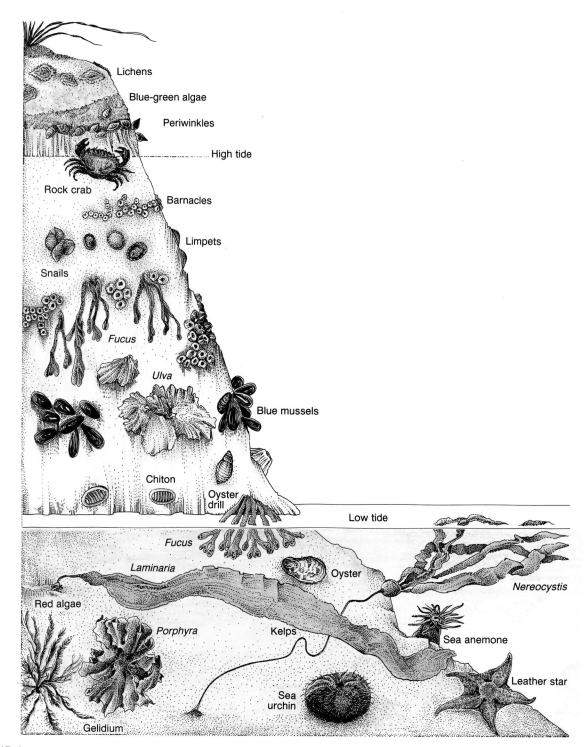

Figure 17.4

A typical distribution of benthic plants and animals on a rocky shore at temperate latitudes. Vertical zonation is the result of the relationships of the organisms to their intertidal environment.

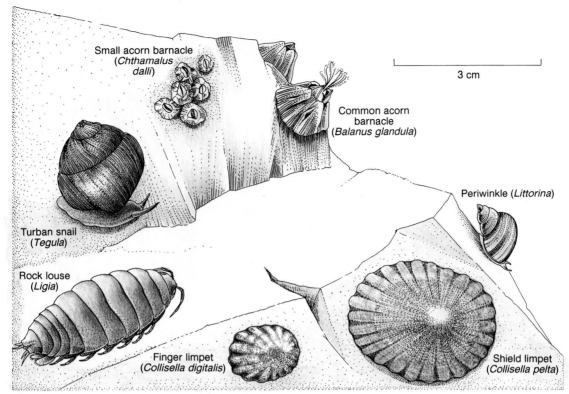

Figure 17.5

Organisms of the supralittoral zone. The limpets and the snail, *Littorina,* are herbivores. The barnacles feed on particulate matter in the water. *Ligia* is a scavenger.

Just below the zone of snails and limpets are found small acorn barnacles, which must filter food from the seawater and are able to survive even though they are covered with water only during the few days of high water or spring tides each month. Barnacles are crustaceans, related to crabs and lobsters but cemented firmly in place. They have been described as animals that lie on their backs and spend their lives kicking food into their mouths with their feet. In some areas, the rocky splash zone is the home of another crustacean, the large (3–4 cm [1–2 in]) isopod *Ligia.* Organisms of the supralittoral zone are shown in figure 17.5.

The width of the supralittoral zone varies. The slope of the rocks, variations in light and shade, exposure to waves and spray, tidal range, and the frequency of cool days and damp fogs all contribute to establishing its limits.

Conspicuous members of the upper midlittoral zone are illustrated in figure 17.6. These organisms include several other species of barnacles, limpets, snails, and two other **mollusks:** the bivalved (or two-shelled) mussels and the chitons, which may appear to resemble limpets but on closer inspection will be seen to have shells of eight separate plates. Chitons, like limpets, are grazers that scrape algae from the hard surfaces. Mussels are filter-feeders; food removed from the water is trapped in a heavy mucus and moved to the mouth by liplike palps. The animals are well anchored to the rocks by a variety of mechanisms: a muscular foot in chitons and limpets, strong cement in barnacles, and special threads in mussels. Their profiles tend

to be rounded, so as to present little resistance to the breaking waves. Tightly closed shells protect many of the organisms from drying out during periods of low tide. Species of brown algae, typically rockweed (*Fucus*), are in this zone. The algae of the area have strong holdfasts and flexible stipes.

Gooseneck barnacles are found attached to rocks where wave action is strong. They have evolved an interesting feeding style, facing shoreward and feeding by taking particulate matter from the runback of the surf rather than facing the sea, as might be expected.

Mussel beds promote shelter for less conspicuous animals, such as the segmented sea worm, *Nereis,* and small crustaceans. Shore crabs of varied colors and patterns are found in the moist shelter of the rocks. The crabs are active predators as well as important scavengers. Small sea anemones, huddled together in large groups to conserve moisture, are also found in the higher regions of the midlittoral zone.

The area is crowded; the competition for space appears extreme. The free-swimming juveniles (or larval forms) of these animals settle and compete for space, each with its own special set of requirements. New space in an inhabited area becomes available as the whelks, which are carnivorous snails, prey on some species of thinner-shelled barnacles, and the sea stars move up with the tide to feed on the mussels. This predation restricts the thinner-shelled barnacles to the upper levels of the littoral zone, which are

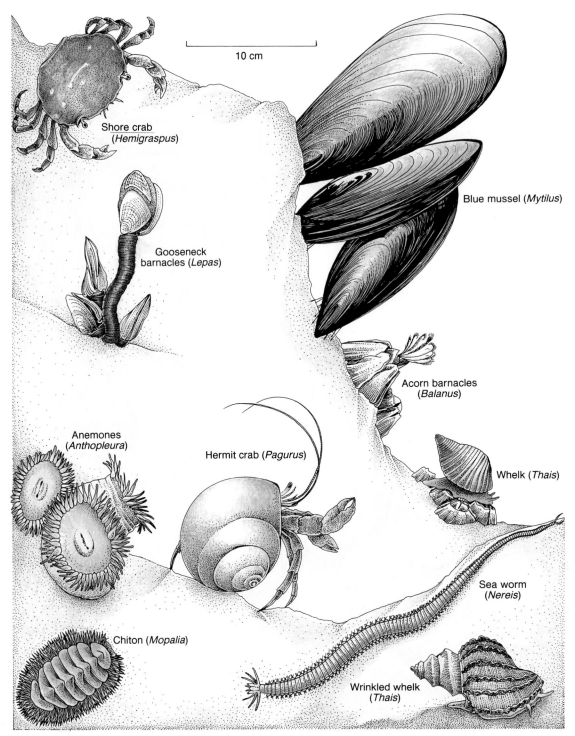

10 cm

Shore crab
(*Hemigraspus*)

Blue mussel (*Mytilus*)

Gooseneck
barnacles (*Lepas*)

Acorn barnacles
(*Balanus*)

Anemones
(*Anthopleura*)

Hermit crab (*Pagurus*)

Whelk (*Thais*)

Sea worm
(*Nereis*)

Chiton (*Mopalia*)

Wrinkled whelk
(*Thais*)

Figure 17.6

Representative organisms from the midlittoral zone. The mussels and barnacles filter their food from the water. The chitons graze on the algae covering the rocks. *Thais*, a snail, is a carnivore. Small shore crabs and hermit crabs are scavengers. *Balanus cariosus* is a larger and heavier barnacle than the barnacles of the supralittoral zone. *Nereis* is often found in the mussel beds. The anemones huddle together to keep moist when exposed.

too dry for the predator snails. Other species of barnacles inhabit the lower midlittoral zone in association with the predator snails, which cannot pierce the heavier plates of the mature individuals. Seasonal die-offs of algae and the battering action of strong seas and floating logs also act to clear space for newcomers.

A selection of organisms from the lower littoral zone is found in figure 17.7. The larger anemones are common inhabitants of the midlittoral and lower littoral zones. These delicate-looking, flowerlike animals attach firmly to the rocks and spread their tentacles, loaded with poisonous darts called nematocysts. The darts are fired when small fish, shrimp, or worms brush the tentacles. The prey is

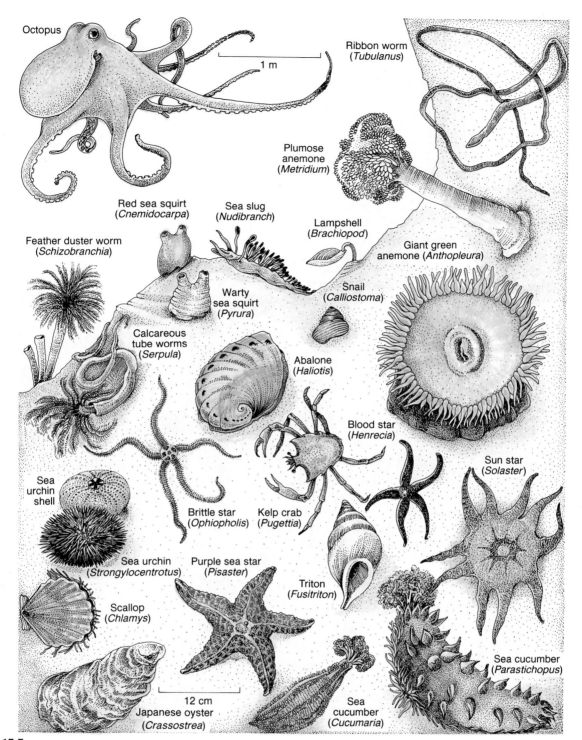

Figure 17.7

Lower littoral zone organisms. A variety of related organisms inhabit the area. The starfish feed on the oysters; the related sea urchins are herbivores; and the sea cucumbers feed on detritus suspended in the water. Among the mollusks are the oysters, scallops, snails, abalone, nudibranchs, and octopuses. The oysters and scallops are filter-feeders. *Calliostoma* and the abalone are grazers; the nudibranchs, the octopus, and the triton snail are predators. Note the size of the anemones in this zone.

paralyzed, the tentacles grasp, and the prey is pushed down into the anemone's central mouth. A finger inserted into the tentacles of most anemones will feel only a stickiness and a tingling sensation, but some large tropic species can produce a painful injury. Some snails and sea slugs prey on the anemones. They are unaffected by the nematocysts, and certain sea slugs store the anemone's nematocysts in their tissues and use them for their own defense.

Starfish of many colors and sizes make their home in the lower littoral zone; these slow-moving but voracious carnivores prey on shellfish, sea urchins, and limpets. Their mouths are on their undersides, at the center of their central disks, surrounded by strong arms that are equipped with hundreds of tiny suction cups, or tube feet. The tube feet are operated by a water-vascular system, a kind of hydraulic system that attaches the animal very firmly to a hard surface. In feeding, the tube feet attach to the shell of the intended prey and, by a combination of holding and pulling, open the shell sufficiently to insert the starfish's stomach, which can be extruded through its mouth; enzymes are released, and digestion begins. Shellfish sense the approach of a starfish by substances it liberates into the water, and some execute violent escape maneuvers. Scallops swim jerkily; clams and cockles jump away; even the slow-moving sea urchins and limpets move as rapidly as possible. Starfish are so effective as predators that oyster farmers must take care to exclude them from their oyster beds.

The filter-feeding sponges, some flat and some vase-shaped, encrust the rocks. On a minus tide, delicate, free-living flatworms and long **nemerteans,** or ribbon worms, armed with poison-injecting mouth parts, are found keeping moist under the mats of algae. Snails and crabs inhabit this zone; the scallop, another filter-feeding bivalve, and red algae are found as well, along with beds of kelp, eelgrass, and surf grass. Calcareous red algae encrust some rocks and are seen as tufts on others. Occasionally **brachiopods,** or lampshells, are found in the lower littoral zone. They resemble clams but are completely unrelated to them. The shell encloses a coiled ridge of tentacles used in feeding.

Beautiful, graceful, and colorful sea slugs, or **nudibranchs,** are active predators, feeding on sponges, anemones, and the spawn of other organisms. Although soft-bodied, they have few if any enemies, because they produce poisonous acid secretions. Herbivores are also present in the lower intertidal region; species of chitons and limpets as well as the sea urchin graze on the algae covering the rocks. Sea cucumbers are found wedged in cracks and crevices; some types are identifiable by their brightly colored tentacles, which act as mops to remove food particles from the water and thrust them into the animal's mouth. Tube worms secrete the leathery or calcareous tubes in which they live and extend only their graceful, feathered tentacles to strain their food from the water.

The octopus is an interesting but shy member of the benthos. Octopuses are seen occasionally from shore on very low tides. These eight-armed carnivorous animals are soft-bodied mollusks. They feed on crabs and shellfish and live in caves or dens identifiable by the piles of waste shells outside. They are known for their ability to flash color changes and move gracefully and swiftly over the bottom and through the water. The world's largest octopus is found in the coastal waters of the eastern North Pacific. It commonly measures 2 to 3 m (10–16.5 ft) in diameter and weighs 20 kg (45 lb), but specimens in excess of 7 m (23 ft) and 45 kg (100 lb) have been observed. They are not aggressive, although they are curious and have been shown to have learning ability and memory.

Symbiotic Relationships

Competition for food and space and predator-prey relationships are common in the intertidal region, but other, cooperative relationships are also present. The beautiful green anemone of the Pacific coast, *Anthopleura xanthogrammica,* is green because a small, single-celled alga grows in the animal's cells. This mutually beneficial relationship or **mutualism,** in which the alga receives protection as well as carbon dioxide and nitrogenous compounds from the anemone, and the anemone acquires organic compounds and some oxygen from the alga, is a type of **symbiosis.** A symbiotic relationship is one in which the two dissimilar organisms live in a close, intimate relationship.

Another interesting symbiotic relationship is found in tropical waters between the sea anemone and the clown fish; refer back to figure 16.10e. The clown fish acquires protection by nestling among the anemone's stinging tentacles while acting to lure other fish within the anemone's grasp. Shellfish of all types play host to various worms and small crustaceans. Rather than being mutually beneficial, these relationships are often beneficial to one partner and of no harm to the other, a relationship known as **commensalism.** For example, shrimp fish acquire protection by hovering, head down, among the spines of the sea urchin. The small fish *Nomeus* takes shelter within the tentacles of the Portuguese man-of-war.

A relationship in which one partner is harmed by the other is **parasitism,** and parasitic flatworms, roundworms, and bacteria similar to those that infect land organisms are found in the sea. These relationships are not confined to benthic organisms. Cleaner fish feed by removing the parasites and damaged tissues from other fish. Sea mammals are also often heavily parasitized.

Tide Pools

The zonation of benthic forms varies with local conditions; zones are generally narrow where the beach is steep and the tidal range is small, while in areas where the beach is flat and the range of tides is large the zones are wide. The orientation of the shore to sunlight and shade, wave action, and beach topography often results in the apparent displacement of benthic zones. A tide pool formed from water left by the receding tide in a rock depression or basin provides a habitat for animals common to the lower littoral zone. Some small, isolated tide pools provide a very specialized habitat due to evaporation and the resulting increase in salinity of the remaining water, as well as to the accompanying rise in temperature. On a summer day, the water in a tide pool may feel quite warm to the touch. Other tide pools act as catch basins for rainwater, lowering the salinity of the water and its temperature in the fall and winter. Isolated tide pools often support blooms of microscopic algae that give the water the appearance of pea soup, and in various parts of the world, a tiny, bright-red copepod, *Tigripus,* is also found.

The size and depth of a tide pool are also important. The deeper the tide pool and the greater the volume of water, the more stable its environment during its isolation by the receding tide. The larger the tide pool, the more slowly it will change temperature, salinity, pH, and carbon dioxide–oxygen balance. Subtidal animals such as starfish, sea urchins, and sea cucumbers cannot survive in pools such as that described in the preceding paragraph. These animals do not tolerate significant changes in their chemical and physical environment; they require large, deep pools. A few fish species, such as the small sculpins, can be found in tide pools. These fish are patterned and colored to match the rocks and the algae within the pool. They spend much of their time resting on the bottom, swimming in short spurts from one resting place to another. Each tide pool is a specialized environment populated with organisms that are able to survive under the conditions established in that particular pool.

The bottom of the littoral (or intertidal) zone merges into the beginning of the sublittoral (or subtidal) zone extending across the continental shelf. If the shallow areas of the subtidal zone are rocky, many of the same lower littoral zone organisms will be found. When soft sediments begin to collect in protected areas or deeper water, the population types change, and animals of the rocky bottom are replaced by those commonly found on mud and sand substrates.

Animals of the Soft Substrates

Mud, sand, and gravel shores are usually less-stable habitats than the rocky areas discussed in the previous section. A pebble or gravel beach does not support much plant or animal life, for the pebbles roll and shift under the waves, scrubbing off any attached growth. Open-coast beaches frequently lose much of their sand under winter storm conditions and regain it with the gentler waves of summer, making these beaches difficult areas for benthic organisms to colonize.

When currents deposit the sands and muds in quiet coves and bays, the energy for movement of the substrate is greatly reduced, and the habitat is mechanically more stable. In these areas, it is the size and shape of the sediment particles and the organic content of the sediment that determine the quality of the environment. The size and shape of the particles, and therefore the size of the spaces between the particles, regulate the flow of water between particles and also the availability of oxygen dissolved in the water. Beach sand is fairly coarse and thus porous to water, both gaining and losing it quickly, while fine particles of mud hold more water and replace the water more slowly.

Therefore, sand beaches exchange water, dissolved wastes, and organic particles more quickly than mud flats, which act as traps for organic material. The finer the mud particles, the tighter they pack together and the slower the exchange of water; oxygen is not resupplied quickly nor are wastes removed rapidly. The fragrance of mud flats is due to the decomposition of the organic material, and the odor of hydrogen sulfide may be quite strong. Digging into the mud will generally show a black layer 1 or 2 cm (.5–1 in) below the surface. Above this layer the water between the sediment particles contains dissolved oxygen; below the black layer, organisms (mainly bacteria) function without oxygen, producing hydrogen sulfide, the rotten egg smell. Lack of oxygen may restrict the depth to which infauna species can be found, but some animals, such as clams, are commonly found below the black layer. They use their siphons to obtain food and oxygen from the water above the sediments.

Waves along gravel and sandy shores produce an unstable environment. Few plants can attach, and therefore few grazing animals are found. In locations protected from waves and currents, eelgrass and surf grass help stabilize the small-particle sediments and provide shelter, substrate, and food, creating a special community of plants and animals. Most sand and mud animals are **detritus** feeders, living on bits and pieces of organic material washed in from the sea or off the land. Most detritus is formed from plant material that is degraded by bacteria and fungi. A familiar sand creature, the sand dollar, feeds on detritus particles found between the sand grains. Some animals, such as clams, cockles, and certain polychaete worms, are filter-feeders, feeding on the detritus and microscopic organisms suspended in the water. Other animals are deposit feeders that engulf the sediment and process it in their gut to extract organic matter in a manner similar to that of earthworms. Deposit feeders are usually found in muds or muddy sands with a high organic content; for example, burrowing sea cucumbers and the lugworm, *Arenicola,* which produces the coiled castings seen outside its burrow. Small crustaceans, crabs, and some worm species are scavengers, preying on any available plant or animal material, while still other worms and snails are carnivores. The moon snail is a clam eater that drills a hole in the shell of its prey and then sucks out the flesh.

Bacteria not only play the major role in the decomposition of plant and animal material; they also serve as a major protein source. It is estimated that 25% to 50% of the organic material the bacteria decompose is converted into bacterial cell material, which is consumed by microscopic protozoans, which in turn serve as food for tiny worms, clams, and crustaceans. Areas of mud that are high in organic detritus produce large quantities of bacteria.

The intertidal area of a soft-sediment beach shows some zonation of benthic organisms, but it is not nearly as clear-cut as the zonation along a rocky cliff. In temperate latitudes, small crustaceans called sand hoppers are found at the high intertidal region; they are replaced by ghost crabs in the tropics. Lugworms, mole crabs, and ghost

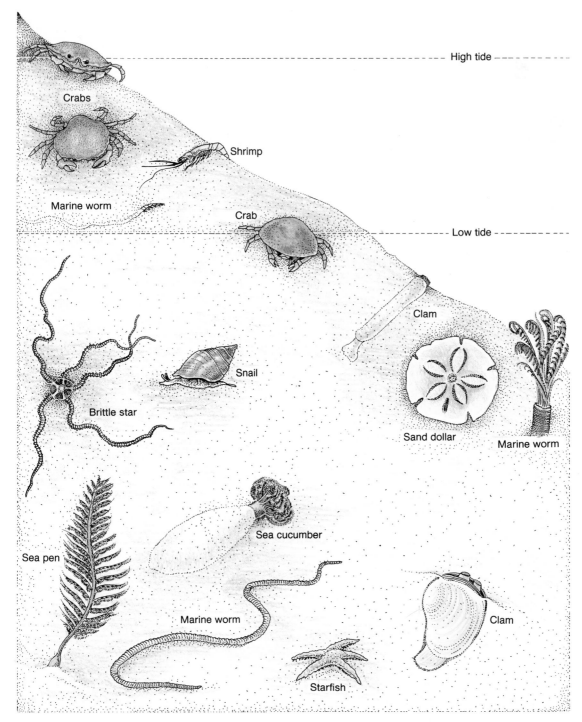

Figure 17.8

Zonation on a soft-sediment beach is less conspicuous than that found on a rocky beach. Animals living at the higher tide levels burrow to stay moist.

shrimp occupy the midbeach area, while clams, cockles, polychaete worms, and sand dollars are found in the lower intertidal region. The subtidal zone is home to sea cucumbers, sea pens, more crabs and clams, and some species of worms, snails, and sea slugs. The distribution of life in soft sediments is shown in figure 17.8. A selection of animals from this region is found in figure 17.9.

The Deep Seafloor

The deep seafloor includes the flat abyssal plains, the trenches, and the rocky bases of seamounts and midocean ridges. The seafloor sediments are more uniform and their particle size is smaller than those of the shallower regions close to land sources. The environment of the bathyal, abyssal, and hadal zones is uniformly cold and dark.

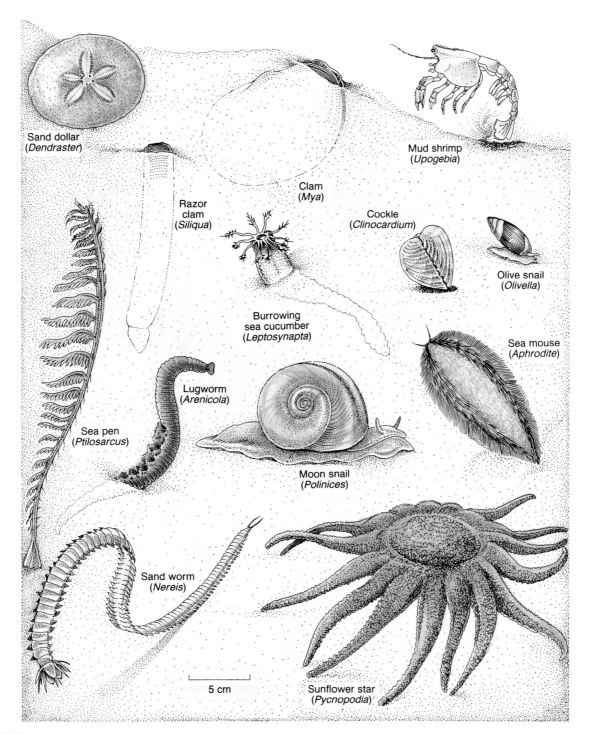

Figure 17.9

Organisms of the soft sediments. Infauna types include the shrimp, *Upogebia,* the lugworm, *Arenicola,* the clam, the cockle, and the burrowing sea cucumber. The sand dollar feeds on detritus; the moon snail drills its way into shellfish; the sea pens feed from the water above the soft bottom. The sea mouse, like *Nereis,* is a polychaete worm.

Many of the animals of the deep seafloor are very like their benthic relatives in the shallow waters of the continental shelf. But in general, it appears that while the population of an organism decreases with depth, the number of species within that population increases. That is, diversity of species, which is determined from the ratio of number of species to number of organisms in a sample, increases with depth. Samples show that few species of burrowing copepods are found to be duplicated between samples. This diversity of species between samples is also evident in the protozoan foraminiferans. Thirty species of planktonic foraminiferans are known, whereas one thousand species of benthic foraminiferans have been described.

The stable conditions of the seafloor appear to have favored deposit-feeding infaunal animals of many species. Many members of the deep-sea infauna are very small; they are known as the **meiobenthos** and measure 2 mm or less. The meiobenthos include nematode worms, burrowing crustaceans, and segmented worms. At 7000 m (23,000 ft), tusk shells lie buried in the ooze with tentacles at the sediment surface to feed on the foraminiferans. Acorn worms are found frequently in samples taken at 4000 m (13,000 ft). Hagfish burrow into the sediment at the 2000-m (6600-ft) depth. Detritus-eating worms and bivalve mollusks have been found on the seafloor in all the oceans.

Among the epifauna, protozoans are abundant and are widely distributed. Glass sponges attach to the scattered rocks on oceanic ridges and seamounts (see fig. 17.10); so do sea squirts and sea anemones. Their stalks lift them above the soft sediments into the water, where they feed by straining out organic matter. Stalked barnacles attach to the stalks of glass sponges and sea squirts as well as to shells and boulders. Tube worms are common, ranging in size from a few millimeters to 20 cm (8 in), and sea spiders with four pairs of very long legs that span up to 60 cm (27 in) are found at depths to 7000 m (23,000 ft). Snails are found to the greatest of depths; those in the deepest trenches frequently have no eyes or eye-stalks.

The beard worms, or **poganophora,** are found in more-productive areas at depths to 10,000 m (33,000 ft). They secrete a close-fitting tube and stand erect, with only their lower portion buried in the sediment. They have no mouth, no gut, and no anus and absorb their food through their skin. The poganophora associated with the vent communities are discussed in a later section of this chapter.

Horny corals, or sea fans, which resemble plants more than animals, grow at depths of 5000 to 6000 m (16,000–20,000 ft); so do solitary stone corals, which grow larger at these depths than do the coral organisms in the surface waters. Sea lilies, or crinoids, which are related to starfish, are also found at this depth, as are brittle stars and sea cucumbers. Sea cucumbers are found in areas of sediments that are rich in organic substances; they are a dominant and widely spread organism of the deep seafloor.

Bottom photographs and cores show that the deep sediments are continually disturbed and reworked by sea cucumbers and worms in the process of extracting organic matter. This process, called **bioturbation,** reworks the sediment and destroys layers, resulting in the uniform sediments that cover vast areas of the ocean floor.

Fouling and Boring Organisms

Organisms that settle and grow on pilings, docks, and boat hulls are said to foul these surfaces. Fouling organisms include barnacles, anemones, tube worms, sea squirts, and algae. When these organisms grow on a vessel's hull, they slow its movement through the water and add to the costs in shipping time and haul-out for cleaning. Fouling organisms also make it difficult to operate equipment in the

(a)

(b)

(c)

Figure 17.10

(a) A vase-shaped glass sponge 640 m (2100 ft) under the surface on the Brown Bear Sea Mount in the northeast Pacific Ocean. (b) A deep-sea crab photographed at a depth of 2000 m (6550 ft) on the Juan de Fuca Ridge, in an area of little life. (c) A group of deep-sea sponges and an anemone are seen at 684 m (2244 ft) on the Brown Bear Sea Mount.

(a)   (b)

Figure 17.11

Unprotected wood, such as driftwood, is subject to destruction by marine borers. (a) This beach log has been riddled by shipworm. (b) The small holes surrounding a shipworm hole are bored by a crustacean known as gribble.

ocean environment, and much research goes into experiments with paints and metal alloys to discover methods of discouraging and controlling these fouling organisms.

Other organisms naturally drill or bore their way into the substrate. Sponges bore into scallop and clam shells; snails bore into oysters; some clams bore into rock. Organisms that bore into wood are costly, for they destroy harbor and port structures as well as wooden boat hulls. Two organisms are responsible for most of the wood boring: the shipworm (*Teredo*) is a mollusk, and the gribble is a small crustacean. *Teredo* is a wormlike mollusk with one end covered by a bivalved shell. These mollusks secrete enzymes that break down and at least partially digest the wood fibers, which are scraped away by the shell's rocking and turning movement that bores deep and extremely destructive holes all through a piece of wood. The gribble gnaws more superficial and smaller burrows, but it, too, is very destructive. The destructive effects of these organisms are shown in figure 17.11.

17.3 Classification Summary of the Benthos

Benthic organisms include members of the following kingdoms and phyla.

I. *Kingdom Monera:* includes the bacteria, an important food source in mud and sand environments; also cyanobacteria or blue-green algae, especially abundant on coral reefs.

II. *Kingdom Protista:* convenience grouping of microscopic, usually unicellular plants and animals.
   A. *Phylum Chrysophyta:* golden-brown algae; mainly planktonic, but benthic diatoms are abundant.
   B. *Phylum Protozoa:* infauna include many members among sand and mud particles, including amoeboid and foraminiferan species.

III. *Kingdom Plantae:* plants; primarily nonmotile, multicellular, photosynthetic autotrophs.
   A. *Division Chlorophyta:* green algae; seaweeds such as *Ulva* and *Codium.*

B. *Division Phaeophyta:* brown algae; coastal seaweeds, including the large kelps *Nereocystis, Macrocystis,* and *Postelsia. Sargassum* maintains a planktonic habit in the Sargasso Sea.

C. *Division Rhodophyta:* red algae; littoral and sublittoral seaweeds of varied form.

D. *Division Tracheophyta:* plants with vascular tissue; true roots, stems, and leaves present.
   1. *Class Angiospermae:* flowering plants; seeds enclosed in fruit; eelgrass, surf grass, and mangrove trees.

IV. *Kingdom Animalia:* animals; multicellular heterotrophs with specialized cells, tissues, and organ systems.

A. *Phylum Porifera:* sponges; simple, nonmotile filter-feeders.

B. *Phylum Cnidaria (Coelenterata):* radially symmetric, with tentacles and stinging cells.
   1. *Class Anthozoa:* sea anemones, corals, and sea fans.

C. *Phylum Platyhelminthes:* flatworms; free-living and parasitic forms.

D. *Phylum Nemertina* or *Nemertea:* ribbon worms.

E. *Phylum Aschelminthes:* round worms, or nematodes.

F. *Phylum Brachiopoda:* lampshells.

G. *Phylum Annelida:* segmented worms.
   1. *Class Polychaete:* free-living marine carnivores, including tube worms, *Nereis,* and clam worms.

H. *Phylum Pogonophora:* deep-sea tube-dwelling worms; giant members found around hot vents on the seafloor.

I. *Phylum Mollusca:* mollusks.
   1. *Class Polyplacophora:* chitons.
   2. *Class Scaphopoda:* tusk shells.
   3. *Class Gastropoda:* single-shelled mollusks; snails, limpets, and sea slugs.
   4. *Class Pelecypoda* or *Bivalvia:* bivalved mollusks; clams, mussels, scallops, and oysters.
   5. *Class Cephalopoda:* octopus. The squid is a member of the nekton.

J. *Phylum Arthropoda:* animals with paired, jointed appendages and outer skeletons.
   1. *Class Crustacea:* crabs, shrimp, and barnacles.

K. *Phylum Echinodermata:* all marine, radially symmetric, spiny-skinned animals with water-vascular systems, including starfish, sea urchins, sand dollars, and sea cucumbers.

L. *Phylum Hemichordata:* acorn worms.

M. *Phylum Chordata:* animals, including vertebrates, with dorsal nerve cord and gill slits at some stage in development.
   1. *Subphylum Urochordata:* filter-feeding, saclike adults with "tadpole" larvae; sea squirt.

17.4 The Tropical Coral Reefs

Coral reefs are the most luxuriant and complex of all benthic communities (see fig. 17.12). Corals are colonial animals; the corals of the curio shop and jeweler are the calcareous skeletons of these animals. Individual coral animals are called **polyps.** A coral polyp is very similar to a tiny sea anemone with its tentacles and stinging cells, but unlike the anemone, a coral polyp extracts calcium carbonate from the water and forms a calcareous skeletal cup. Large numbers of these polyps grow together in colonies of delicately branched forms or rounded masses.

Although various corals are found in shallow and deep water and in temperate and tropic climates, reef-building corals have specialized requirements restricting their range. These reef-building corals require warm, clear, shallow, clean water and a firm substrate to which they can attach. Because the water temperature must not go below 18°C and the optimum temperature is 23° to 25°C, their growth is restricted to tropical waters between 30°N and 30°S and away from cold-water currents. Waters at depths greater than 50–100 m (150–300 ft) are also too cold for significant secretion of calcium carbonate. Most Caribbean corals are found in the upper 50 m (150 ft) of lighted water while Indian and Pacific corals are found to depths of 150 m (500 ft) in the more transparent water of these oceans. Reefs usually are not found where sediments limit water transparency. The largest coral reef in the world, the Great Barrier Reef, stretches more than 2000 km (1200 mi), from New Guinea southward along the east coast of Australia.

Clear, shallow water is required by the reef-building coral, because within the tissues of the polyps are masses of single-celled dinoflagellate algae called **zooxanthellae** that require light for photosynthesis and therefore must stay in the photic zone. Polyps and zooxanthellae have a symbiotic relationship in which the coral provides the algal cells with a protected environment, carbon dioxide, and nitrate and phosphate nutrients, and the algal cells photosynthesize, returning oxygen and removing waste. Experiments have shown that the zooxanthellae supply the corals with substantial amounts of the organic products of photosynthesis and also enhance the ability of the coral to extract the calcium carbonate from the seawater. The degree of interdependence is thought to vary from species to species. The polyps feed actively at night, extending their tentacles to feed on zooplankton, but during the day, their tentacles are contracted, exposing the outer layer of cells containing zooxanthellae to the sunlight.

The corals require a firm base to which they can cement their skeletons. The classic reef types of the tropic sea—fringing reefs and barrier reefs—are attached to existing islands or landmasses. Atolls are attached to submerged seamounts. (For a review of reef formation around a seamount, see chapter 2.) Delicate, finely branched corals inhabit the more protected areas of the reef, while the more massive rounded corals are found in regions of heavy surf. Corals are slow-growing organisms; some species grow less than 1 cm (0.4 in) in a year, and others add up to 5 cm

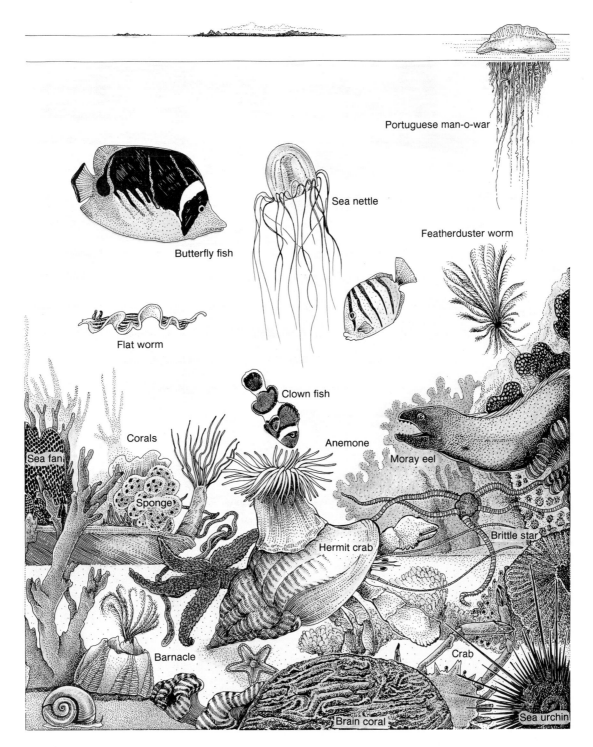

Figure 17.12

A coral reef is a complex, interdependent but self-contained community. Members include corals, clams, sponges, sea urchins, anemones, tube worms, algae, and fish.

(2 in) a year. The same coral may be found in different shapes and sizes, depending on the depth and the wave action of an area. Environmental conditions vary over a reef, forming both horizontal and vertical zonation patterns that are largely the product of wave action and water depth as shown in figure 17.13. On the sheltered (or lagoon) side of the reef, the shallow **reef flat** is covered with a large variety of corals and other organisms. Fine coral particles broken off from the reef top produce sand, which fills the sheltered lagoon floor. On the reef's windward side, the reef's highest point, or **reef crest,** may be exposed at low tide and is pounded by the breaking waves of the surf zone. Here the more massive rounded corals grow. Below the low-tide line to a depth of 10 to 20 m (35 to 65 ft) on the seaward side is a zone of steep, rugged buttresses, which alternate with grooves in the reef face. Masses of large

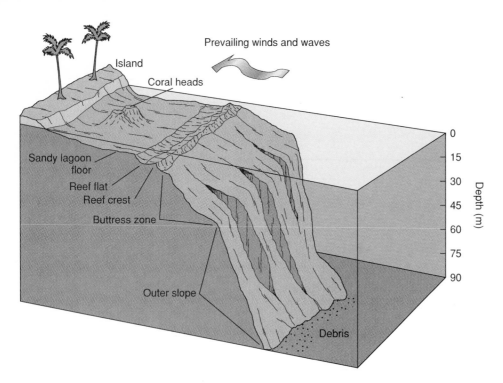

Figure 17.13
Coral reef zonation.

Prevailing winds and waves

Island

Coral heads

Sandy lagoon
floor

Reef flat

Reef crest

Buttress zone

Outer slope

Debris

Depth (m)

0

15

30

45

60

75

90

corals grow here, and many large fish frequent the area. The buttresses dissipate the wave energy, and the grooves drain off fine sands and debris, which would smother the coral colonies. At depths of 20 to 30 m (65 to 100 ft), there is little wave energy, and the light intensity is only about 25% of its surface value, although it is adequate to support reef algae and corals. The corals are less massive at this depth, and more delicately branched forms are found here. Between 30 and 40 m (100–130 ft), the slope is gentle and the level of light is very reduced; sediments accumulate at this depth, and the coral growth becomes patchy. Below 50 m (165 ft) the slope drops off sharply into the deep water. The reef exists as a balance between the growth of the organisms on the reef surface, as they build on top of old, dead, calcareous remains, and the wearing away of the reef by mechanical and biological forces.

Coral reefs are complex assemblages of many different types of plants and animals, and competition for space and food is intense. Algae, sponges, and corals are constantly growing over and killing each other. Some species are active only at night; some fishes, snails, shrimp, the octopus, fireworm, and moray eel. During the day, other species depend on color and vision to make their way. It has been estimated that as many as three thousand animal species may live together on a single reef. The giant reef clam, *Tridacna,* measures up to a meter in length and weighs over 150 kg (330 lb). It also possesses zooxanthellae in large numbers in the colorful tissues that line the edges of the shell. Crabs, moray eels, colorful reef fish, poisonous stonefish, long-spined sea urchins, sea horses, shrimp, lobsters, sponges, and many more organisms are all found living here together. The total reef community can be considered to be an autotrophic association based on the sun's energy as transformed by its photosynthetic members. On some reefs, the zooxanthellae have been shown to produce

several times more organic material per unit of space than the phytoplankton, probably due to the rapid recycling of nutrients between the corals and the zooxanthellae.

The reefs are not formed exclusively from the calcium carbonate skeletons of the coral. Encrusting algae that produce an outer calcareous covering contribute; so do the minute shells of foraminifera, the shells of bivalves, the calcareous tubes of polychaete worms, and the spines and plates of sea urchins. All are compressed and cemented together to form new places for more organisms to live. At the same time, some sponges, worms, and clams bore into the reef; some fish graze on the coral and the algae, and the sea cucumbers feed on the broken fragments, reducing them to sandy sediments.

Coral reefs around the world are showing signs of stress; from Florida to Fiji reefs are infested with algae, predators, and recurrent episodes of bleaching. Seventy percent of the coral reefs in the Philippines were reported in a critical condition at the 1992 International Coral Reef Symposium in Guam. Between 1984 and 1986 several sections of the Florida Keys reef decreased in area by 8%. In 1983 half of the coral colonies off the west coast of Panama were severely damaged.

Coral bleaching, in which the corals expel their zooxanthellae, have occurred from time to time, but unless the episode has been especially severe the corals regained their algae and recovered. Isolated instances of this bleaching, apparently in response to local stress such as heavy rains, decreased salinity, and pollutants have not been unusual. However, in the tropical eastern Pacific in 1983 large amounts of bleaching, thought to have been associated with the 1982–83 El Niño event, occurred in reefs off Costa Rica, Panama, and the Galápagos Islands. Mass bleaching in the Atlantic began in 1987 in the Florida Keys and eventually stretched from Bermuda south to the Bahamas, affecting coral reefs throughout the Caribbean. Elevated water temperatures were observed in the Caribbean in 1987–88, perhaps also due to a large-scale weather phenomenon. In 1990 the water warmed again over an 18-day period, and a week later the corals bleached; the 1990–91 bleaching seems to be killing corals that managed to recover from the 1987 episode. French Polynesia also reported large-scale bleaching in 1991. No one is sure why massive bleaching episodes have occurred; one possibility is that when the water warms it causes the algae to produce more oxygen (a by-product of photosynthesis) than the corals can stand; another idea has attempted to link the problem to global warming, but there is no evidence so far of a persistent warming trend in the oceans. Whatever the cause, researchers believe the corals are weakened by temperature extremes, leaving them susceptible to disease and other stresses.

Periodically there occurs a dramatic population increase of the sea star *Acanthaster,* or crown of thorns, which feeds on the coral polyps. No one knows why the *Acanthaster* population increases so rapidly, but evidence points to a correlation between rainy weather (low salinity) and runoff (increased nutrients), allowing larger numbers of *Acanthaster* larvae to survive. The clearing of land for agriculture and the development of coastal areas are probably also related. There is also concern that the harvesting of the large conchs that prey on the starfish may upset the population balance. A 1979 outbreak in Guam has continued, and there appears to be a chronic outbreak in the Ryukyu Islands south of Japan. The Great Barrier Reef has endured two population explosions of *Acanthaster* since the 1960s, and here fossil evidence points to outbreaks having occurred for at least the last eight thousand years. Most reefs have been able to regenerate when the *Acanthaster* populations die back, and the opening up of areas on the reef by the starfish may even allow slower-growing coral species to expand. Some researchers feel that we have not been studying the problem long enough to know whether these outbreaks are the result of human activities or whether they are a part of some unknown natural cycle.

Humans and their activities are among the greatest threats to the reefs. Reefs are damaged by careless sport divers trampling delicate corals, by boats grounding or dragging their anchors, and by pesticide runoff from fields and resort areas, but excess nutrients and sediments are two of the most important sources of decline. As the tropical rain forests are cleared, the trees are replaced by farms, mines, resorts, and other human enterprises. The result is increased runoff, carrying agricultural fertilizers and human sewage into coastal waters. Additional nutrients favor the growth of algae that grow faster than the corals, outcompete the corals for space, smother the existing corals, and prevent new coral colonies from starting. Along the coast of the small Central American nation of Belize lies the largest barrier reef in the Western Hemisphere. Clearing of mangroves and expanding citrus and banana plantations have brought silt, pesticides, and fertilizers to the Belize reefs, which now have serious algal overgrowth. Sewage from new resorts and hotels and septic tanks from an increasing population provide the nutrients for algal overgrowth in Belize and the Florida Keys.

In addition coral reefs are despoiled by shell collectors and mined for building materials. In French Polynesia, Thailand, and Sri Lanka, tons of coral are used as construction material each year. In the Philippines hammers and chisels are used to collect chunks of live coral for the U.S. aquarium market. Eighty percent of marine fishes sold for the worldwide aquarium trade come from the reefs of the Philippines. These fish are stunned with sodium cyanide or dynamite to make them easy to catch; the estimate is that for every fish that survives to reach an aquarium, twenty or more fish die. Explosives destroy large patches of coral, and "muroami," a technique in which fifty to three hundred swimmers pound weights on the reef to scare fish out into waiting nets, also has destroyed large areas of the Philippine reefs. The coral reefs of Indonesia and reefs off Tanzania in East Africa are also reported to have been extensively damaged by fishing with explosives.

Whatever the cause, if the reef's chemical balance is upset the results can be disastrous; too little oxygen and the animals die, too many nitrates and the algae overgrow the reef. If the grazers decline, the predators decline, upsetting the balance once again. Ensuring the continued existence of these beautiful and productive areas requires both an increased understanding of the complex nature of reef communities and the development of policies designed to protect them from human interference.

17.5 High-Energy Environments

Recent work has shown that intertidal communities constantly battered by the waves are more productive than the world's lush, green rain forests. On the average, waves deliver 0.335 watts of energy per square centimeter of coastline, about fifteen times more energy than comes from the sun. Even during calm periods, wave energy is 100% greater than solar energy. The algae have little woody tissues; instead, the kelps of the rocky intertidal have 2.5 times more photosynthetic area per square meter of growing surface and are from two to ten times more productive than rain forest vegetation. The mussels, which are consumers, have been found to match or exceed the rain forest productivity when growing in areas of high wave action.

The method of harnessing the energy is indirect. Wave action reduces predators, such as starfish and sea urchins, allowing more mussels and more kelp to live in a unit area. The moving water brings a constant supply of nutrients to the algae and keeps their blades in motion, so they are never in the shade for long. Lastly, the waves can dislodge mussels, allowing more kelp to move in. The waves allow the primary producers to become highly concentrated, and so the consumers can grow and expand their population as well.

Deep-Ocean Chemosynthetic Communities

Before 1977, deep-sea benthic communities were thought to be made up of small numbers of deposit-feeding animals living on and in the soft sediments. These animals grow slowly in the cold water and depend for food on the slow descent of decayed organic material from the surface layers. Since then, the communities of animals found along the Galápagos Rift and East Pacific Rise between 2500 and 2600 m (8000–8500 ft), on the Juan de Fuca Ridge at 1570 m (5000 ft), off the west coast of Florida at 3266 m (11,000 ft), in the central Gulf of Mexico at 700–800 m (2300–2600 ft), and along the Gorda Ridge west of Oregon at 3000 m (9840 ft) have been shown to be completely different. Animals in these areas include filter-feeding clams and mussels, in addition to anemones, worms, barnacles, limpets, crabs, and fish. The clams are very large and show the fastest growth rate of any known deep-sea animal, up to 4 cm (2 in) per year. The tube worms are startling in size, up to 3 m (10 ft) long. These worms belong to the phylum Pogonophora, which is a small and, until now, relatively obscure group related to segmented worms (see fig. 17.14). Many of these areas were revisited by several biological expeditions in 1990 and 1991. Vent populations along the East Pacific Rise were found to have remained stable. Although the abundance of tube worms decreased and the clams and mussels increased at the Galápagos Ridge sites between 1979 and 1985, little change had occurred between 1985 and 1990.

Dense clouds of bacteria are the base of the trophic system for these communities. These bacteria are able to utilize dissolved chemicals in the hot vent water to obtain the energy they require to live and grow in a process known as **chemosynthesis.** They have no link to the sun through photosynthesis, either directly or indirectly. The bacteria are able to do in the dark what plants do in the sunlight, that is, fix carbon from carbon dioxide into organic molecules such as sugars that can be used in the metabolism of other organisms. In these vent areas the bacteria are able to oxidize the sulfides, particularly hydrogen sulfide that issue from hydrothermal vents. Refer back to chapter 3 for a discussion of the hydrothermal vent environment.

The tube worms, which may reach up to 3 meters in length and several centimeters in diameter, have no mouth and no digestive system; the soft tissue mass of their internal body cavity is filled with bacteria. Tube worms have the ability to transport carbon dioxide, oxygen, and hydrogen sulfide in their blood in a way that prevents poisoning or interaction. The substances are delivered to bacteria held in the special body tissues, and the synthesis of organic molecules is done within the bacteria.

Both the clams and the tube worms have red flesh and red blood. The color is due to the oxygen-binding molecule, hemoglobin. The oxygen is needed for the oxidation of the hydrogen sulfide and to maintain body tissues and

Figure 17.14

Tube worms surround a vent area. This photograph was made from the submersible *Alvin.*

high growth rates. During studies of the large shrimp populations that surround some vent areas, a reflective spot located just behind the head of the shrimp was noticed. The reflective spot consists of two lobes, each connected to the brain by a large nerve cord; light-sensitive pigment related to visual pigment is also present. This system allows the shrimp to detect the radiation associated with the hot vents (refer back to figure 3.29) and may allow the shrimp to orient themselves with the vents and their food supply.

The vent plankton are believed to feed on the bacteria; the bacteria and the plankton provide nutrition for a variety of filter-feeders. Other bacteria form mats that surround the vents, and these bacterial mats are the food for snails and other grazers. The bacteria form the base of a broad-based, self-contained trophic system in which nutrition passes from animal to animal by symbiosis, grazing, filtering, and predation. These crowded communities have a biomass five hundred to one thousand times denser than the biomass of the normal deep sea.

Along the continental slope of Louisiana and Texas, faulting has fractured the sea bottom and produced environments where oil and gas seep up and onto the surface of the seafloor; these are the so-called cold seeps. In 1985 clams, mussels, and large tube worms were collected from these sites at depths between 500 and 900 m (1600–3000 ft); see figure 17.15. Bacteria living at these oil and gas seeps use

(a)

(b)

(c)

Figure 17.15

Photographs from Gulf of Mexico oil and gas seeps: (a) Mussels and tube worms, (b) tube worms, (c) seeping gas and oil, bacterial mats, and chemosynthetic communities in the Green Canyon oil lease area. Note sampling devices and gas bubbles.

methane (the principal component of natural gas) as a carbon source. The symbiotic relationship between these bacteria and the seep animals is unique in the animal world because the bacteria live not on but within the cells of their hosts. The bacteria live within the gill cells of the mussels and use methane as their sole energy source; laboratory experiments show that the mussels can extract some energy by filter-feeding from the water, but they are not able to grow without getting energy from the bacteria. Clams predominate in areas of oil seepage, while mussels are more common at gas seeps.

Another type of seep, the salt seep, was reported in 1990 at 650 m (2000 ft) on the floor of the Gulf of Mexico. The escape of gas through surface sediments has formed depressions that have filled with salt brine more than 3.5 times the usual salinity of seawater; the brine also contains methane gas. This extremely dense water forms brine lakes on the seafloor surrounded by large communities of chemosynthetic mussels.

Finding the remains of clam shells at some inactive vent sites, and knowing that it takes the shells approximately fifteen years to dissolve, researchers have estimated that active vents last about twenty years. This has been confirmed by radiometric dating. Therefore, if the colonies are to survive, they must colonize new vents as they develop. The clams have been shown to have high growth rates, becoming mature in four to six years. Their large body size is thought to allow them to produce many larvae as well as harbor large quantities of bacteria. It is estimated that larvae drifting for periods of several weeks to several months could be transported over hundreds of kilometers by the abyssal currents.

These self-contained communities are among the most profuse and productive in the world. Ninety-five percent of the nearly three hundred species found at these vents and seeps are new to science, and many are only distantly related to other earth creatures, appearing more closely related to ancient species. The presence of a plentiful food supply allows their rapid growth, despite the surrounding low temperatures and their remoteness from the photosynthetic layer at the sea's surface. The deep seafloor will never again be considered sparsely populated or inhospitable to life.

17.7   Sampling the Benthos

At low tide beaches all over the world are sampled by researchers armed with buckets and shovels. To understand the relationships between the organisms and their environment, one needs to know where the plants and animals are found on the beach with respect to the tide levels, or the vertical zonation pattern. In order to establish this pattern, it is necessary to determine the slope of the beach and to mark locations relative to mean sea level; this determination is done by surveying a **beach transect line** directly up the beach from low-tide level to a point above high tide. Trenches paralleling the transect line are dug on sandy beaches to reveal the infauna populations below a measured

Figure 17.16
A biological dredge used to collect epibenthic organisms from rocky substrate.

surface area, and on a rocky shore, surface counts of individuals within an area of specific size are made along the transect line. Once the transect line has been established, it is possible to return to the same place, season after season or year after year, to study seasonal changes in the populations or to determine the number of juveniles added yearly. Another type of study compares a natural area to an area stripped of all benthic organisms. The rate of repopulation and the sequence in which the plants and animals return give information on the relationships that exist among species as well as the recovery rates in the event of a catastrophe due to human error or natural occurrences.

To develop management techniques for the harvesting of benthic organisms such as shellfish, comparative studies are made between natural areas and harvested areas. Surveys determine the average size and age of the shellfish and the density of the populations under both conditions. From such data, researchers determine whether the harvesting is reducing the total population too quickly, whether it is removing too many individuals of spawning age or too many juveniles, and for how long an area should be open to repeated harvests.

The classic method of obtaining information on subtidal species is by using a bottom dredge towed by a slowly moving ship. A dredge is a metal frame to which a heavy net bag is attached; it is dragged over the seafloor, scraping up the organisms in its path (see fig. 17.16). This is a strictly qualitative sampling method, as the number and kinds of

Mussel Watch

Since 1986 the NOAA Mussel Watch Project has been collecting samples from mussels and oysters for chemical analyses, using them to monitor the concentrations of certain chemicals as indicators of human activity. More than two hundred U.S. coast and estuary sites are sampled each year; almost half of the sites are located in waters near urban areas within 20 km (12 mi) of population centers in excess of 100,000 people. Three sites are located in Hawaii and two in Alaska.

Analyses check for ten trace metals: arsenic, cadmium, chromium, copper, lead, mercury, nickel, selenium, silver, zinc, and five organic compounds: DDT (dichlorodiphenyltrichlorethane), chlordane, PCBs (polychlorinated biphenyls), PAHs (polycyclic aromatic hydrocarbons), and tributyl tin. The metals are all discharged into the environment as a result of industrialization, but the organic compounds are more difficult to categorize. Although DDT was banned in the United States in 1972 and U.S. chlordane use ended in 1983, DDT remains in the environment because of its resistance to degradation, and chlordane remains in the ground from its use in termite control. PCBs were banned in 1976 but devices containing these compounds are still in use. PAHs are found in fossil fuels such as oil, gasoline, and coal. Tributyl tins have been used as an antifouling agent in paint used on ships and underwater marine facilities. All of these metals and organic compounds can be toxic to marine life under some conditions.

Box figure 17.1
Mussel cluster in a rock crevice.

Because mussels and oysters are unable to move, they accumulate these compounds from the food they filter from their surrounding water and from the water itself at specific sites. Higher levels of contaminants are associated with sites closer to population centers, but high concentrations away from population centers are not necessarily the result of human actions. High concentrations of cadmium in mussels collected along the northern California coast are attributed to the upwelling of deep-ocean water that is naturally high in this metal.

Mussel Watch data show the contamination of coastal and estuarine waters to be decreasing in many cases; more decreases than increases in chemical concentrations were found between 1986 and 1990. Although five years of data is not sufficient to establish long-term trends, this decrease appears to support the limits being set on discharges of chemical contaminants and the laws supporting these limits.

---

organisms cannot be accurately related to the area sampled. It is possible to attach a measuring wheel to the dredge frame, so that when the dredge is on the bottom, the wheel measures the distance over which it is dragged; then knowing this distance and the width of the dredge, it is possible to compute the approximate size of the area sampled, assuming the dredge did not bounce or skip over a rough bottom.

Soft bottoms can be sampled with a bottom grab or a box corer (refer back to chapter 2). The grab can be designed to penetrate to a specific depth and to take a bite with a specific surface area. The sediment collected is washed through a series of mesh screens of varying sizes, and the organisms are collected and counted. Many samples need to be taken in order to determine the community structure of any single area of the seafloor.

Divers wearing scuba gear can sample and photograph bottom populations directly in depths to approximately 35 m (105 ft). A diver can place a frame of a specific area on the bottom, identify the species within it, and count individuals of each type. If the bottom is not disturbed, it is possible to return to the same plot and check it again. Other methods of observing the bottom without disturbing it include under-water photography and television cameras that are operated remotely from a ship or from a submersible.

Sampling the benthos of the bathyal, abyssal, and hadal zones is difficult and expensive, for it requires large research vessels, with or without submersibles, and extended periods of time at sea. The areas to be sampled are large and remote, and our knowledge of them is still incomplete. The recent discoveries of the vent communities have produced

Figure 17.17
Alaska king crab being unloaded from a crab pot.

totally unexpected results and have forced a reconsideration of the abundance and composition of deep-sea life based on sampling only a very few areas.

17.8 Practical Considerations: Harvesting the Benthos

The Animals

Benthic animals are a valuable part of the seafood harvest; they include crustaceans (crabs, shrimp, prawns, and lobsters) and the shellfish or mollusks (mainly the bivalved clams, mussels, and oysters). World harvests for 1990 were catches of 7.7 million metric tons of shellfish and 4.2 million tons of crustaceans. Because of the demand for shellfish and crustaceans, the catches are much more important in dollar value than the weight of the catches would suggest. In the United States, oysters are harvested in the Gulf of Mexico, southern New England, Chesapeake Bay, and Puget Sound; lobsters are fished in New England; crabs and clams are caught along all our coasts; and shrimp are caught in the Gulf area and off other coasts as well. In many cases, these fisheries make important contributions to the local economy.

Many of the problems of the finfish fisheries are repeated in the benthic fisheries. For example, between 1975 and 1980 more and more boats entered the king crab fishery of the Bering Sea; see figure 17.17. The huge catches of

1979 (70,000 metric tons) and 1980 (84,000 metric tons) were followed by a 1982 harvest of 17,000 metric tons, rapidly decreasing to 12,000 metric tons in 1983, and dropping to 7,000 metric tons in 1985. The rapid decrease in catch is apparently due to the classic ills of overfishing and to an insufficient knowledge of the king crab's natural history. King crab migrate across the floor of the Bering Sea, but not understanding how many crabs are migrating in any direction, and at what stage in their life they migrate, has made it difficult to implement an effective fishery conservation program. It is unlikely that the catches of 1979 and 1980 will ever be repeated, but a regulated fishery based on sustained yield may ensure long years of future fishing. The catch has slowly increased since 1985, reaching 15,000 metric tons in 1990.

Attempts to increase the harvest of crustaceans and mollusks have focused on aquaculture, or mariculture. Oysters, mussels, and clams are raised on aquaculture farms around the world. In Asia and Europe, raft culture of mollusks is popular. The larvae attach to ropes trailing below the rafts. This attachment keeps the shellfish in the water column, with abundant food and fewer predators. In Spain, aquaculture produced nearly 173,000 metric tons of mussels in 1990. Raft culture in Japan was responsible for the successful harvest of 249,000 metric tons of oysters in 1990. The Japanese culture scallops in hanging net cages, as well as on the bottom, resulting in a scallop industry of 250,000 metric tons a year. Raft culture methods are

Figure 17.18
Commercial raft culture of bay mussels in Puget Sound. Mussels adhere to ropes suspended from floating rafts.

beginning to be used in the United States (see fig. 17.18), but mussel and oyster farmers suffer from some of the same difficulties encountered by fish farmers, discussed in chapter 16.

The world harvest of shrimp produced in aquaculture ponds was 500,000 metric tons in 1990. In Japan, commercial shrimp culture, 3000 metric tons in 1990, has proved very successful, although the production costs are high. To produce 1 kg (2.2 lb) of shrimp during the spring–fall growing season requires 10 to 12 kg (22–26.41 lb) of diatoms, which must be raised to feed the young shrimp. However, the demand for the less than fully grown shrimp exceeds the supply. In Ecuador, the shrimp-farming harvest increased from 35,000 tons in 1985 to 76,000 metric tons in 1990, making it Ecuador's second-largest industry (next to oil) and generating more than 150,000 jobs. Shrimp are also being grown in Latin America by U.S. corporations taking advantage of the lower costs in those countries.

On a model aquaculture farm at Woods Hole Oceanographic Institution, biological oceanographer John Ryther developed a pilot plant in which the sewage effluent from a town of 50,000 contributed its nutrients to produce the algae needed to feed oysters, so as to harvest 800 metric tons of oyster meat. Because the oysters in the model farm produced large quantities of solid waste, marine worms

were introduced to feed on the oyster's waste. The worms were later harvested and sold for bait. Another waste product, industrial heat, may also have a use in aquaculture. In Long Island Sound, 30°C water from industrial cooling systems was used experimentally to increase oyster production. The success of these experimental projects has led to the idea of locating an aquaculture project in the 30°N–30°S latitude belt. Here, there is the advantage of warm water and year-round sunlight, which can be used with nutrients produced either from sewage or from an artificial upwelling created by pumping nutrient-rich deep water into the aquaculture pens. Such a project could be conducted in conjunction with ocean thermal energy conversion (OTEC), which is discussed in chapter 6.

Aquaculture projects for benthic species in the United States are faced with the same difficulties as those facing fish farms. High costs, strict licensing policies, the need for technology to replace hand labor, and the need for research to improve diet and disease control require considerable attention if we are to increase our seafood harvest in this way. The public must first want or need the products of aquaculture if these obstacles are to be overcome.

The Algae

Seaweeds are gathered from the wild in northern Europe, Japan, China, and Southeast Asia; they are also an important part of the Japanese aquaculture industry. The 1990 world estimate for total seaweed harvest was 4 million metric tons—15,000 metric tons of green algae, 1 million metric tons of red algae, and 2.7 million metric tons of brown algae.

Algae are good sources of vitamins and minerals but not of food calories, as most of the cellular material is indigestible. Certain species of green algae, known as sea lettuces, are used in seaweed soups, in salads, and as a flavoring in other dishes. A species of kelp, *Laminaria japonica,* is the **kombu** of Asia. Its blade is used in soups and stews, and it is used fresh, dried, pickled, and salted. It is also sweetened and shredded for use in candy and cakes. Another species of *Laminaria* is used in Europe in the same way. Historically along coastal areas, kelp was used as winter fodder for sheep and cattle and to mulch and fertilize the fields. Along the northwest coast of the United States and Canada, the stipes of bull kelp are made into pickles. The red alga *Porphyra* is the **nori** of Japan and the **laver** of the British Isles. It has been cultivated in Japan since 1700; today 400,000 metric tons are harvested yearly. Nori is used in soups and stews and is rolled around portions of rice and fish for flavor. Laver is fried or used in salads.

There are also important industrial uses for some algal products. **Algin** is extracted from brown algae, and **agar** and **carrageenan** are obtained from red algae. In California more than 160,000 tons of kelp are harvested every year for the production of algin. Algin derivatives are

Genetic Manipulation of Fish and Shellfish

Fisheries scientists in North America, Japan, and Northern Europe are using new gene-transfer and chromosome-manipulation techniques to manage and enhance the potential of ocean and coastal fisheries and to keep coastal waters and fish farms stocked with species of rapid development and high growth rate. Chromosome-manipulation techniques from the research laboratory are being used to produce sterile fish, fish of selected gender, and fish known as **triploids** that carry an extra third set of **chromosomes.**

Triploid salmon and trout are produced by exposing the eggs to temperature, pressure, or chemical shock shortly after fertilization. The shock interferes with the division of the egg nucleus; the egg retains an extra or third set of chromosomes. The third set is retained throughout the fishes' development and inhibits its sexual maturation. Salmon with a normal number of chromosomes grow, mature sexually, spawn, and die, while triploid fish do not mature sexually and reach a larger size since they are not spending energy on egg or sperm production. The improved growth and survival

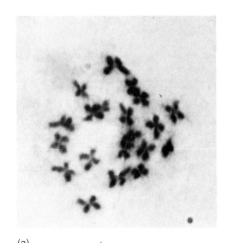
(a)

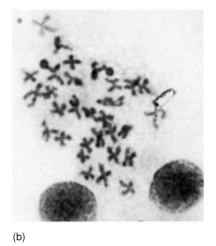
(b)

Box figure 17.2

(a) Chromosomes of the Pacific oyster (*Crassostrea gigas*) seen through a compound microscope. The diploid oyster has two sets of ten chromosomes. (b) The triploid oyster has three sets of ten chromosomes.

of these fish has led to widespread production of farmed triploid trout in the United Kingdom.

This same technique allows hybridization of salmon and trout. While normal chromosome-number hybrids between different species of salmon and trout do not survive, triploid hybrids do. In Idaho, a hybrid between rainbow trout and coho salmon has been shown to resist a common viral disease that causes serious economic loss to trout producers.

The gender of fish can be determined by chromosome manipulation and hormone treatment. Using a technique called gynogenesis, researchers expose fish sperm to ultraviolet light, denaturing its chromosomes. Eggs fertilized with the

---

used as stabilizers in dairy products, paints, inks, and cosmetics; and to strengthen ceramics, improve the consistency of plaster, and thicken jams as well as put a longer-lasting head on beer. Agar-producing algae are harvested in Japan, Africa, Mexico, and South America. Agar is used as a medium for bacterial culture in laboratories and hospitals, in addition to serving as an ingredient in desserts and in pharmaceutical products. Algae rich in carrageenan are gathered in the wild in New England and northeastern Canada; it is a stabilizer and emulsifier used to prevent separation in ice creams, salad dressings, soups, puddings, cosmetics, and medicines. About 1 million pounds of agar and 10 million pounds of carrageenan are used in the United States each year. As fishing declines, the culture of algae is being considered to provide jobs for marine communities; in 1992 the Canadian province of Quebec held its first government-sponsored symposium to explore algae culture.

Kelp Bioconversion

Because kelp grows at a great rate and is found over large areas of the world's oceans, some people have proposed that it be cultivated, harvested, and placed in tanks for

decomposition, which will produce methane gas for use as fuel and other chemical by-products. The use of a plant crop to harness the sun's energy for conversion to a fuel or energy source is called **bioconversion** (in this case, kelp bioconversion). The energy produced in such a system would be used first to power the harvesting and fuel production process. The remaining fuel either would be made available for other uses at the extraction site or it could be piped to other areas. Suggestions have been made to grow the kelp harvest on offshore raft systems, but no pilot plant of any size has been built, and the dollar and energy costs required are not known.

Biomedical Products

Many benthic organisms contain biologically active compounds that have potential practical use. Extracts of certain sponges yield anti-inflammatory and antibiotic substances; an anticoagulant has been isolated from red algae; and the antibiotic substance acrylic acid has been extracted from other seaweeds. Some corals produce antimicrobial compounds, and the sea anemone, *Anthopleura,* provides a cardiac stimulant. Certain polychaete worms produce a inactivated-chromosome sperm are briefly chilled, causing the eggs to develop with only the female's chromosomes. The result is all female offspring. Many female food fish grow bigger and live longer than males. Female flounder grow to twice the size of males, female coho salmon have firmer and more flavorful flesh, and an all-female brood of sturgeon would bring higher profits to caviar producers. Gynogenesis also magnifies genetic effects, making it possible to produce a superior strain in a much shorter period of time than is required for standard hybridization techniques. An alternative to gynogenesis is androgenesis, in which the egg chromosomes are inactivated and the sperm chromosomes are doubled.

It is possible to reverse the sex of female fish produced by gynogenesis by treating them with hormones. This procedure yields fish that are genetically female but functionally male, producing sperm with only female chromosomes. Matings of females with these males will result in all-female fish.

Current research programs extract genes for a specific characteristic from one fish and build it into another. Aquaculturists are transferring genes that boost the production of natural growth hormone into fish eggs. Although the foreign genes do not always function, when they do, the gene is passed on to the next generation. These genetically altered organisms, referred to as **transgenic** fish, are 20% to 46% faster-growing than the wild stock. Another program transfers genes that will increase the immunity of the fish and so boost their survival rate. Many farm-raised halibut and Atlantic salmon die in winter when their body fluids freeze. A gene from the Arctic winter flounder has been transferred into Atlantic salmon, and preliminary results indicate that these fish survive very cold temperatures better than normal salmon.

The commercial production of triploid, or "four-seasons," oysters has begun in the United States. Summer mortality and fluctuations in the marketability of oysters have kept the oyster industry seasonal. Normal oysters enter a summer reproductive phase during which their meat is poor in quality and unmarketable. Triploid oysters do not produce sperm or eggs but continue to grow steadily, becoming significantly larger than normal oysters and ready for the summer market. The blue mussel is the next mollusk considered ready to be changed to the triploid form.

There are no regulations in either the U.S. or Canada concerning transgenic fish, and researchers go to great pains to prevent their experimental stock from escaping to the wild environment, rigging pens with fences, screens, and alarm systems. It has been suggested that young transgenic fish be sterilized before they are placed in fish-farm pens, to avoid mixing wild and experimental fish. As the era of commercial transgenic fish approaches, many questions need to be answered. If mixing occurred between experimental fish and normally spawning fish, could it endanger the spawning population? Will the rapid inbreeding possible through gynogenesis produce populations more susceptible to disease or environmental problems? How will the higher growth rates of sterile fish affect a mixed population? Although these problems and questions must be addressed, chromosome and gene manipulation as a useful tool in aquaculture and fish management has arrived and is in use.

substance that kills some kinds of insects, and extracts of abalone and oyster act as antibacterial agents. A muscle relaxant has been isolated from the snail *Murex*. Even the organic adhesives used by organisms to attach themselves to hard surfaces are under investigation, for example, the cement secreted by mussels, which quickly hardens underwater, has possible dental uses to secure fillings and crowns, and good results have been obtained using this substance to repair the corneas and retinas of the eye. The strength and properties of these organic substances make them possibilities for closing wounds without stitches as well as for repairing the hulls of ships at sea. Corals are being investigated to discover why, when alive, they are not encrusted by larvae, algae, and other organisms. Perhaps corals produce substances that will be useful as antifouling compounds for boats, docks, pipes, and underwater equipment.

Collecting, extracting, identifying, testing, evaluating, and ultimately synthesizing active substances from the world's benthos is a time-consuming task. From discovery to pharmacy shelf takes ten to fourteen years and costs up to $100 million. However, scientists and pharmaceutical firms are continuing to look at the highly active compounds produced by marine organisms in the hopes that they may serve as models for the eventual development of new drugs.

The benthos is a remarkably diverse grouping of plants and animals; see figure 17.19. It is easily accessible in the shallow littoral zone, where it has been studied by scientists and students for hundreds of years. Yet we know very little about the benthos over most of the deeper-ocean floor. As we are able to see, sample, and understand more about the deep-sea benthos, we may expect to make other discoveries as surprising and as unexpected as the vent communities in the rift areas.

Many of the benthos have additional value to us as food resources, and some may even serve as energy resources. As we consider the uses we make of the oceans at present and the uses for which we need the oceans in the future, we must remember their impact on the benthic environment and its inhabitants. We gain little if we use one resource at the expense of another. The balance within the marine environment is fragile, easy to degrade, and difficult to reconstruct. The health of the ocean's plants and animals is important to us as a sign of the health of our planet. We are all linked together, and each of us affects the others.

(a)

(b)

(c)

(d)

Figure 17.19

The benthos are a large, varied group of animals living on or in the seafloor, especially the organisms of the tide pools and the intertidal areas of the rocky coasts. (a) The anemone, *Tealia crassicornis,* is a sessile carnivore. The graceful and beautiful nudibranchs or sea slugs include (b) the white *Dirona albolineata,* (c) the orange-flecked *Triopha carpenteri,* and (d) *Hermissenda crassicornis* with its orange-and-white striped tips. Starfish come in a remarkable diversity of shapes and sizes: (e) the bright orange bloodstar (*Herrica leviuscula*), the slender-armed *Evasterias troschelli,* the young, multiarmed *Solaster dawsoni, Mediaster aequalis* with its wide disk and broad arms, the leather star (*Dermasterias imbricata*), and the purple, rough-skinned *Pisaster ochraceus.* All starfish are carnivores and use their tube feet to hold and open the shellfish on which they feed (f). (g) The purple sea urchin (*Strongylocentrotus purpuratus*) and the green urchin (*S. droebachiensis*) are closely related to the sea stars but they are herbivores, clipping off the algae with their especially constructed mouth parts (h). (i) The pink sea scallop (*Chlamys hastata hericia*) lies open, as it filters organic particles from the seawater.

(e)

(f)

(g)

(h)

(i)

Summary

Benthic algae are anchored to firm substrates. These algae have a holdfast, a stipe, and photosynthetic blades but no roots, stems, or leaves. Algal growth along a rocky beach ranges from green algae at the surface through brown algae at moderate depths to red algae, which are found primarily below the low-tide level. Each group's pigments trap the available sunlight at these depths. Algae are generally classified by their principal pigment. The brown algae include the large kelps. Seaweeds provide food, shelter, and substrate for other organisms in the area. There are also benthic diatoms and a few seed plants, including eelgrass and mangroves.

Benthic animals are subdivided into the epifauna, which live on or attached to the bottom, and the infauna, which live buried in the substrate. Animals that inhabit the rocky littoral region are sorted by the stresses of the area into a series of zones. Organisms that live in the supralittoral (or splash) zone spend long periods of time out of water. The animals of the midlittoral zone experience nearly equal periods of exposure and submergence. These animals have tight shells or live close together to prevent drying out. The area is crowded, and competition for space is great. The lower littoral zone is a less stressful environment. It is home to a wide variety of animals. The organisms of the littoral zone are herbivores and carnivores, and each has its specialized life-style and adaptations for survival.

Symbiotic relationships are intimate, cooperative relationships between two dissimilar organisms. Mutualism, commensalism, and parasitism are types of symbiosis found in the marine environment.

The zonation of the organisms in the benthic region varies with local conditions. Tide pools provide homes for lower littoral zone organisms; they can also become extremely specialized habitats.

Mud, sand, and gravel areas are less stable than rocky areas. The size of the spaces between the substrate particles determines the porosity of the sediments. Some beaches have a higher organic content than others. Few algae can attach to soft sediments, and thus few grazers are found here. Eelgrass and surf grass provide food and shelter for specialized communities. Most organisms that live on soft sediments are detritus feeders or deposit feeders. Zonation patterns are not conspicuous along soft bottoms. Bacteria play an important role in the decomposition of plant material and its reduction to detritus. The bacteria themselves represent a large food resource.

The environment of the deep seafloor is very uniform. The diversity of species increases with depth, but the population density decreases. The microscopic members of the deep-sea infauna are the meiobenthos. Larger burrowers like sea cucumbers continually rework the sediments. The organisms of the epifauna are found at all depths.

Some organisms specialize in attaching to surfaces and others bore into them. Wood-borers are very destructive.

Tropical coral reefs are specialized, self-contained systems. The coral animals require warm, clear, clean, shallow water and a firm substrate. Photosynthetic dinoflagellates, called zooxanthellae, live in the cells of the corals and the giant clams. The reef exists in a complex but delicate biologic balance, which can be easily upset. Reefs have a typical zonation and structure associated with depth and wave exposure. Coral reefs are under great stress; human activities are their greatest threats. High-energy coastal benthic environments are two to ten times more productive than rain forest vegetation. Self-contained, deep-ocean benthic communities, made up of large, fast-growing animals, depend on chemosynthetic bacteria for the first step in their food chains. Communities are associated with hot-water vents and cold seeps.

Benthic organisms are sampled by hand in the intertidal zone; deeper samples are obtained with dredges, grabs, and corers. Sampling is both quantitative and qualitative. Divers, cameras, and television can be used to identify and count organisms without disturbing them.

Shellfish are valuable world food resources. Aquaculture can be used to increase harvests of shellfish and crustaceans. Experimental aquaculture projects have used nutrients from sewage and heat from industrial resources.

Algae are gathered in many countries and are cultivated in Japan. Some are used directly as food; others are used as stabilizers and emulsifiers in foods and other products. It has been suggested that kelp be cultivated, harvested, and used to produce fuel and other substances. Biologically active substances with potentially practical uses have been isolated from benthic organisms.

Key Terms

alga/algae	nemerteans
holdfast	brachiopod
stipe	nudibranch
kelp	mutualism
blade	symbiosis
epifauna	commensalism
infauna	parasitism
sessile	detritus
vertical zonation	meiobenthos
intertidal zonation	poganophora
mollusk	bioturbation
polyp	algin
zooxanthellae	agar
reef flat	carrageenan
reef crest	bioconversion
chemosynthesis	chromosome
beach transect line	triploid
kombu	transgenic
nori/laver	

Study Questions

1. Explain the relationship between hydrogen sulfide gas, bacteria, tube worms, and clams in a hydrothermal vent environment.
2. Distinguish between mutualism, commensalism, and parasitism in marine communities. Give an example of each.
3. In what ways are the benthic algae (seaweeds) adapted for life in the littoral and sublittoral zones? Consider their structure, pigments, and life requirements.
4. In what ways are the benthic algae important in the ocean environment?
5. Discuss the food-gathering strategies of motile and sessile organisms in the littoral and sublittoral zones.
6. Discuss the factors that are responsible for the littoral zonation of marine organisms along a rocky shore.
7. Design an original organism to inhabit the supralittoral, the littoral, or the sublittoral zone. Consider its requirements for food, shelter, and protection from predators, its adaptations to its environment, and its life history.
8. Why are some subtidal forms found in a tide pool high on a rocky beach, while other subtidal forms are not?
9. Why are there few benthic organisms on a beach made up of noncohesive sediments in a wave and surf area?
10. Discuss the importance of bacteria to benthic organisms.
11. Compare a square-meter sample of deep-sea benthos to a similarly sized sample from the rocky intertidal zone.
12. How are coral reefs able to support a rich and varied population, when the water surrounding the reef is clear and devoid of planktonic primary producers?
13. Compare the organisms found growing around deep-ocean hot-water vents and the organisms found around cold gas and oil seeps.
14. Discuss the genetic manipulation of fish and shellfish. Do the advantages of such techniques outweigh the possible disadvantages?
15. Compare photosynthesis and chemosynthesis; how are they similar and how are they different?

Suggested Readings

General

Lerman, M. 1986. *Marine Biology, Environment, Diversity and Ecology.* Benjamin-Cummings, Menlo Park, Calif. 535 pp.
Life in the Sea, readings from *Scientific American,* 1982. Freeman, San Francisco. 248 pp.
Niesen, T. M. 1982. *The Marine Biology Coloring Book.* Harper & Row, New York. 218 pp.
Sumich, J. L. 1992. *An Introduction to the Biology of Marine Life,* 5th ed. Wm. C. Brown, Dubuque, Ia. 449 pp.

The Plants and Animals

Alper, J. 1990. The Methane Eaters. *Sea Frontiers* 36 (6): 22–29. (Life at petroleum seeps.)
Ballard, R. D., and J. F. Grassle. 1979. Return to the Oases of the Deep. *National Geographic* 156 (5): 689–705. (Photographs of vent areas and life-forms.)
Birkeland, C. 1989. The Faustian Traits of the Crown of Thorns Starfish, *American Scientist* 77 (2): 154–63.
Childress, J., H. Felbeck, and G. Somero. 1987. Symbiosis in the Deep Sea. *Scientific American* 256 (5): 114–20.

Donavel, D. F. 1989. Needles and Pins. *Sea Frontiers* 35 (1): 18–25. (Sea anemones.)
Golden, F. 1991. Reef Raiders. *Sea Frontiers* 37 (1): 22–27.
Grall, G. 1992. Life on a Wharf Piling. *National Geographic* 182 (1): 95–115.
Grassle, J. F. 1987/88. A Plethora of Unexpected Life. *Oceanus* 31 (4): 41–46.
Hardy, A. 1959. *The Open Sea: Its Natural History,* Part 2: Fish and Fisheries. Houghton Mifflin, Boston. 322 pp.
Idyll, C. P. 1976. *Abyss—The Deep Sea and the Creatures That Live in It,* rev. ed. Crowell, New York. 428 pp.
Lapointe, B. E. 1989. Caribbean Coral Reefs: Are They Becoming Algal Reefs? *Sea Frontiers* 35 (2): 83–91.
Leal, J. H. 1991. Australia's Vast Offshore Maze. *Sea Frontiers* 37 (4): 48–51. (The Great Barrier Reef.)
Lobban, C. S., and M. J. Winne, eds. 1982. *The Biology of Seaweeds.* Univ. of California, Berkeley, Calif. 786 pp.
Lutz, R. A. 1991/92. The Biology of Deep-Sea Vents. *Oceanus* 34 (4): 75–83.
Peterson, C. H. 1991. Intertidal Zonation of Marine Invertebrates in Sand and Mud. *American Scientist* 79 (3): 236–49.
Ricketts, E. F., J. Calvin, and J. Hedgpeth. 1985 ed. *Between Pacific Tides.* Stanford Univ., Stanford, Calif. 678 pp.
Ryan, P. R., ed. Fall, 1984. *Oceanus* 27 (3). (Issue devoted to hot springs and cold seeps.)
Ryan, P. R., ed. Summer, 1986. *Oceanus* 29 (2). (Issue devoted to coral reefs.)
Swafford, D. 1991. Rx for a Sick Reef. *Sea Frontiers* 37 (6): 38–43.
Tennesen, M. 1992. Kelp: Keeping a Forest Afloat. *National Wildlife* 30 (4): 4–11.
Torrance, D. C. 1991. Deep Ecology: Rescuing Florida's Reefs. *Nature Conservancy* 41 (4): 8–17.
Tunnicliffe, V. 1992. Hydrothermal-Vent Communities of the Deep Sea. *American Scientist* 80 (4): 336–49.
Van Dover, C. L. 1987/88. Do "Eyeless" Shrimp See the Light of Glowing Deep-Sea Vents? *Oceanus* 31 (4): 47–52.
Ward, F. 1990. Florida's Coral Reefs Are Imperiled. *National Geographic* 178 (1): 115–32.

Harvesting the Benthos

Borgese, E. M. 1980. *The Story of Aquaculture.* Abrams, New York. 236 pp.
Chapman, V. J., and D. C. Chapman. 1980. *Seaweeds and Their Uses,* 3d ed. Chapman and Hall, New York. 334 pp.
Faulkner, D. J. 1992. Biomedical Uses for Natural Marine Chemicals. *Oceanus* 35 (1): 29–35.
Fischetti, M. 1991. A Feast of Gene-Splicing Down on the Fish Farm. *Science* 153 (5019): 512–13.
Iversen, E. S., and D. E. Jory. 1986. Shrimp Culture in Ecuador: Farms without Seed. *Sea Frontiers* 32 (6): 442–53.
Michael, R. G., ed. 1987. *Managed Aquatic Ecosystems.* Elsevier, New York. 166 pp.
Miura, A. 1980. Seaweed Cultivation: Present Practices and Potentials. In *Ocean Yearbook 2,* Borgese, E., and N. Ginsburg, eds. Univ. of Chicago, Chicago. pp. 57–68.
Ryther, J. H., and J. C. Goldman, C. E. Gifford, J. E. Huguenin, A. S. Wing, J. P. Clarner, L. D. Williams, and B. E. La Pointe. 1975. Physical Models of Integrated Waste Recycling—Marine Polyculture Systems. *Aquaculture* 5: 163–77.
Standish, K. A., Jr. 1988. Triploid Oysters Ensure Year-Round Supply. *Oceanus* 31 (3): 58–63.
Thorgaard, G. H., and S. K. Allen Jr. 1987. Chromosome Manipulation and Markers in Fishery Management. In *Population Genetics and Fishery Management,* Ryman, N., and F. Utter, eds. Washington Sea Grant Program, Univ. of Washington, Seattle. pp. 319–31.

Appendix

Scientific [or Exponential] Notation

The writing of very large and very small numbers is simplified by using exponents, or powers of 10, to indicate the number of zeros required to the left or to the right of the decimal point. The numbers that are equal to some of the powers of 10 are as follows:

1,000,000,000.	$= 10^9$	= one billion
1,000,000.	$= 10^6$	= one million
1000.	$= 10^3$	= one thousand
100.	$= 10^2$	= one hundred
10.	$= 10^1$	= ten
1.	$= 10^0$	= one
0.1	$= 10^{-1}$	= one tenth
0.01	$= 10^{-2}$	= one hundredth
0.001	$= 10^{-3}$	= one thousandth
0.000001	$= 10^{-6}$	= one millionth
0.000000001	$= 10^{-9}$	= one billionth

149,000,000 is rewritten by moving the decimal point eight places to the left and multiplying by the exponential number 10^8, to form 1.49×10^8. In the same way, 605,000 becomes 6.05×10^5.

A very small number such as 0.000032 becomes 3.2×10^{-5} by moving the decimal point five places to the right and multiplying by the exponential number 10^{-5}. Similarly, 0.00000372 becomes 3.72×10^{-6}.

To add or subtract numbers written in exponential notation, convert the numbers to the same power of 10. For example,

$$
\begin{array}{c}
1.49 \times 10^3 \\
+6.05 \times 10^2 \\
\hline
\end{array}
=
\begin{array}{c}
1.490 \times 10^3 \\
0.605 \times 10^3 \\
\hline
2.095 \times 10^3
\end{array}
=
\begin{array}{c}
14.90 \times 10^2 \\
6.05 \times 10^2 \\
\hline
20.95 \times 10^2
\end{array}
$$

$$
\begin{array}{c}
2.36 \times 10^3 \\
-1.05 \times 10^2 \\
\hline
\end{array}
=
\begin{array}{c}
23.60 \times 10^2 \\
-1.05 \times 10^2 \\
\hline
22.55 \times 10^2
\end{array}
=
\begin{array}{c}
2.360 \times 10^3 \\
-0.105 \times 10^3 \\
\hline
2.255 \times 10^3
\end{array}
$$

To multiply, the exponents are added and the numbers are multiplied:

$$
\begin{array}{c}
4.6 \times 10^3 \\
\times 2.2 \times 10^2 \\
\hline
\end{array}
$$

$$10.12 \times 10^5 = 1.012 \times 10^6$$

To divide, subtract the exponents and divide the numbers:

$$\frac{6.0 \times 10^8}{2.5 \times 10^3} = 2.4 \times 10^5$$

The following prefixes correspond to the powers of 10 and are used in combination with metric units:

Exponential value	Prefix	Symbol
10^{12}	tera	T
10^9	giga	G
10^6	mega	M
10^3	kilo	k
10^2	hecto	h
10^1	deka	da
10^{-1}	deci	d
10^{-2}	centi	c
10^{-3}	milli	m
10^{-6}	micro	μ
10^{-9}	nano	n
10^{-12}	pico	p
10^{-15}	femto	f
10^{-18}	atto	a

SI Units

The *Système international d'unités,* or International System of Units, is a simplified system of metric units (known as SI units) adopted by international convention for scientific use.

Basic SI Units

Quantity	Unit	Symbol
length	meter	m
mass	kilogram	kg
time	second	s
temperature	Kelvin	°K

Derived SI Units

Unit	Symbol	Quantity in SI base units	Expression
area	meter squared	m^2	m^2
volume	meter cubed	m^3	m^3
density	kilogram per cubic meter	kg/m^3	kg/m^3
speed	meter per second	m/s	m/s
acceleration	meter per second per second	m/s^2	m/s^2
force	newton	N	$(kg)(m)/s^2$
pressure	pascal	Pa	N/m^2
energy	joule	J	(N)(m)
power	watt	W	J/s ; (N)(m)/s

Length: The Basic SI Unit Is the Meter

Unit	Metric equivalents	English equivalents	Other
meter (m)	100 centimeters 1000 millimeters	39.37 inches 3.281 feet	0.546 fathom
kilometer (km)	1000 meters	0.621 land mile	0.540 nautical mile
centimeter (cm)	10 millimeters 0.01 meter	0.394 inch	
millimeter (mm)	0.1 centimeter 0.001 meter	0.0394 inch	
land mile (mi)	1609 meters	5280 feet	0.869 nautical mile
nautical mile (nm)	1852 meters	1.151 land miles 6076 feet	
fathom (fm)	1.8288 meters	6 feet	

Area: Derived from Length

Unit	Metric equivalents	English equivalents	Other
square meter (m^2)	10,000 square centimeters	10.76 square feet	
square kilometer (km^2)	1,000,000 square meters	0.386 square land mile	0.292 square nautical mile
square centimeter (cm^2)	100 square millimeters	0.151 square inch	

Volume: Derived from Length

Unit	Metric equivalents	English equivalents	Other
cubic meter (m^3)	1,000,000 cubic centimeters 1000 liters	35.32 cubic feet 264 U.S. gallons	
cubic kilometer (km^3)	1,000,000,000 cubic meters	0.239 cubic land mile	0.157 cubic nautical mile
liter (L)	1000 cubic centimeters	1.06 quarts, 0.264 U.S. gallon	
milliliter (mL)	1.0 cubic centimeter		

Mass: The Basic SI Unit Is the Kilogram

Unit	Metric equivalents	English equivalents
kilogram (kg)	1000 grams	2.205 pounds
gram (g)		0.053 ounce
metric ton, or tonne (t)	1000 kilograms 1,000,000 grams	2205 pounds
U.S. ton	907 kilograms	2000 pounds

Time: The Basic SI Unit Is the Second

Unit	Metric and English equivalents
minute	60 seconds
hour	60 minutes; 3600 seconds
day	86,400 seconds / 24 hours } mean solar day
year	31,556,880 seconds / 8765.8 hours / 365.25 solar days } mean solar year

Temperature: The Basic SI Unit Is the Kelvin

Reference points	Kelvin (°K)	Celsius (°C)	Fahrenheit (°F)
absolute zero	0	−273.2	−459.7
seawater freezes	271.2	−2.0	28.4
fresh water freezes	273.2	0.0	32.0
human body	310.2	37.0	98.6
fresh water boils	373.2	100.0	212.0
conversions	$°K = °C + 273.2$	$°C = \dfrac{(°F - 32)}{1.8}$	$°F = (1.8 \times °C) + 32$

Speed (Velocity): The Derived SI Unit Is the Meter per Second

Unit	Metric equivalents	English equivalents	Other
meter per second (m/s)	100 centimeters per second 3.60 kilometers per hour	3.281 feet per second 2.237 land miles per hour	1.944 knots
kilometer per hour (km/hr)	0.277 meter per second	0.909 foot per second	0.55 knot
knot (kt)	0.51 meter per second	1.151 land miles per hour	1 nautical mile per hour

Acceleration: The Derived SI Unit Is the Meter per Second per Second

Unit	Metric equivalents	English equivalents
meter per second per second per second (m/s^2)	12.960 kilometers per hour per hour 100 centimeters per second per second	3.281 feet per second per second 8053 miles per hour per hour

Force: The Derived SI Unit Is the Newton

Unit	Metric equivalents	English equivalents
newton (N)	100,000 dynes	0.2248 pound
dyne (dyn)	0.00001 newton	0.000002248 pound

Pressure: The Derived SI Unit Is the Pascal

Unit	Metric equivalents	English equivalents	Other
pascal (Pa)	1 newton per square meter 10 dynes per square centimeter		
bar	100,000 pascals 1000 millibars	14.5 pounds per square inch	0.927 atmosphere 29.54 inches of mercury
standard atmosphere (atm)	1.013 bars 101,300 pascals	14.7 pounds per square inch	29.92 inches of mercury

Energy: The Derived SI Unit Is the Joule

Unit	Metric equivalents	English equivalents
joule (J)	1 newton-meter 0.2389 calorie	0.0009481 British thermal unit
calorie (cal)	4.186 joules	0.003968 British thermal unit

Power: The Derived SI Unit Is the Watt

Unit	Metric equivalents	English equivalents
watt (W)	1 joule per second 0.2389 calorie per second 0.001 kilowatt	0.0569 British thermal unit per minute 0.001341 horsepower

Density: The Derived SI Unit Is the Kilogram per Cubic Meter

Unit	Metric equivalents	English equivalents
kilogram per cubic meter	0.001 grams per cubic centimeter (g/cm^3)	0.0624 pounds per cubic foot

Glossary

A

absorption taking in of a substance by chemical or molecular means; change of sound or light energy into some other form, usually heat, in passing through a medium or striking a surface.

abyssal pertaining to the great depths of the ocean below approximately 4000 m.

abyssal hill low, rounded submarine hill less than 1000 m high.

abyssal plain flat ocean-basin floor extending seaward from the base of the continental slope and continental rise.

abyssopelagic oceanic zone from 4000 m to the deepest depths.

accretion natural or artificial deposition of sediment along a beach, resulting in the buildup of new land.

adsorption attraction of ions to a solid surface.

advection horizontal or vertical transport of seawater, as by a current.

agar substance produced by red algae; the gelatinlike product of these algae.

algae marine and freshwater plants (including most seaweeds) that are single-celled, colonial, or multicelled, with chlorophyll but no true roots, stems, or leaves and with no flowers or seeds.

algin complex organic substance found in or obtained from brown algae.

alluvial plain flat deposit of terrestrial sediment eroded by water from higher elevations.

amphidromic point point from which cotidal lines radiate on a chart; the nodal, or low-amplitude, point for a rotary tide.

amplitude for a wave, the vertical distance from sea level to crest or from sea level to trough, or one-half the wave height.

anaerobic living or functioning in the absence of oxygen.

andesite type of volcanic rock that is intermediate in composition between basalt and granite; associated with subduction zones.

anion negatively charged ion.

anoxic deficient in oxygen.

Antarctic Circle *see* Arctic and Antarctic Circles.

antinode portion of a standing wave with maximum vertical motion.

aphotic zone that part of the ocean in which light is insufficient to carry on photosynthesis.

aquaculture cultivation of aquatic organisms under controlled conditions. *See also* Mariculture.

Arctic and Antarctic Circles latitudes 66 1/2°N and 66 1/2°S, respectively, marking the boundaries of light and darkness during the summer and winter solstices.

asthenosphere upper, deformable portion of the earth's mantle, the layer below the lithosphere; probably partially molten; may be site of convection cells.

atmospheric pressure pressure exerted by the atmosphere as a consequence of gravitational force exerted upon the column of air lying directly above any point on earth.

atoll ring-shaped coral reef that encloses a lagoon in which there is no exposed preexisting land and which is surrounded by the open sea.

attenuation decrease in the energy of a wave or beam of particles occurring as the distance from the source increases; caused by absorption, scattering, and divergence from a point source.

autotrophic pertaining to organisms able to manufacture their own food from inorganic substances. *See also* Chemosynthesis; Photosynthesis.

autumnal equinox *See* Equinoxes.

auxospore naked cell of a diatom, which grows to full size and forms a new siliceous covering.

B

backshore beach zone lying between the foreshore and the coast, acted upon by waves only during severe storms and exceptionally high water.

baleen whalebone; horny material growing down from the upper jaw of plankton-feeding whales; forms a strainer, or filtering organ, consisting of numerous plates with fringed edges.

bar offshore ridge or mound of sand, gravel, or other loose material, which is submerged, at least at high tide; located especially at the mouth of a river or estuary, or lying a short distance from and parallel to the beach.

barrier island deposit of sand, parallel to shore and raised above sea level; may support vegetation and animal life.

barrier reef coral reef that parallels land but is some distance offshore, with water between reef and land.

basalt fine-grained, dark igneous rock, rich in iron, magnesium, and calcium; characteristic of oceanic crust.

basin large depression of the seafloor having about equal dimensions of length and width.

bathyal pertaining to ocean depths between approximately 1000 m and 4000 m.

bathymetry study and mapping of seafloor elevations and the variations of water depth; the topography of the seafloor.

bathypelagic oceanic zone from 1000 m to 4000 m.

bathythermograph (BT) instrument used to determine water temperature as a function of depth (pressure).

beach zone of unconsolidated material between the mean low-water line and the line of permanent vegetation, which is also the effective limit of storm waves; sometimes includes the material moving in offshore, onshore, and longshore transport.

beach face section of the foreshore normally exposed to the action of waves.

Beaufort scale scale of wind forces by range of velocity; scale of sea state created by winds of these velocities.

benthic of the seafloor, or pertaining to organisms living on or in the seafloor.

benthos organisms living on or in the ocean bottom.

berm nearly horizontal portion of a beach (backshore) with an abrupt face; formed from the deposition of material by wave action at high tide.

berm crest ridge marking the seaward limit of a berm.

bilaterally symmetric having right and left halves that are approximate mirror images of each other.

bioconversion conversion of a plant crop storing the sun's energy to a fuel or energy source.

biogenous sediment sediment having more than 30% material derived from organisms.

bioluminescence production of light by living organisms as a result of a chemical reaction either within certain cells or organs or outside the cells in some form of excretion.

biomass amount of living matter, expressed in weight units, per unit of water surface or volume.

bioturbation reworking of sediments by organisms that burrow and ingest them.

blade flat, photosynthetic, "leafy" portion of an alga or seaweed.

bloom high concentration of phytoplankton in an area, caused by increased reproduction; often produces discoloration of the water. *See also* Red tide.

bore *see* Tidal bore.

breaker sea-surface water wave that has become too steep to be stable and collapses.

breakwater structure protecting a shore area, harbor, anchorage, or basin from waves; a type of jetty.

buffer substance able to neutralize acids and bases, therefore able to maintain a stable pH.

bulkhead structure separating land and water areas; primarily designed to resist earth sliding and slumping or reduce wave erosion at the base of a cliff.

buoy floating object anchored to the bottom or attached to another object; used as a navigational aid or surface marker.

buoyancy ability of an object to float due to the support of the fluid the body is in or on.

C

calcareous containing or composed of calcium carbonate.

calorie amount of heat required to raise the temperature of 1 gram of water 1°C.

calving breaking away of a mass of ice from its parent glacier, iceberg, or sea-ice formation.

capillary waves waves with wavelengths less than 1.5 cm in which the primary restoring force is surface tension.

carnivore flesh-eating organism.

carrageenan substance produced by certain algae that acts as a thickening agent.

cation positively charged ion.

cat's-paw patch of ripples on the water's surface, related to a discrete gust of wind.

centrifugal force outward-directed force acting on a body moving along a curved path or rotating about an axis.

chaetognaths free-swimming, carnivorous, pelagic, wormlike, planktonic animals; arrowworms.

chemosynthesis formation of organic compounds with energy derived from inorganic substances such as ammonia, sulfur, and hydrogen.

chloride atom of chlorine in solution, forming an ion with a negative charge.

chlorinity (Cl‰) measure of the chloride content of seawater in grams per kilogram.

chlorophyll group of green pigments that are active in photosynthesis.

chromosome one of the bodies in a cell that carry the genes in a linear order.

chronometer portable clock of great accuracy used in determining longitude at sea.

ciguatera toxin found in fish of tropical regions.

cilia microscopic, hairlike processes of living cells, which beat in coordinated fashion and produce movement.

coast strip of land of indefinite width that extends from the shore inland to the first major change in terrain that is unaffected by marine processes.

coastal circulation cell (drift sector, littoral cell) longshore transport cell pattern of sediment moving from a source to a place of deposition.

coccolithophore microscopic, planktonic alga surrounded by a cell wall with embedded calcareous plates (coccoliths).

cohesion molecular force between particles within a substance that acts to unite them.

colonial organism organism consisting of semi-independent parts that do not exist as separate units; groups of organisms with specialized functions that form a coordinated unit.

commensalism an intimate association between different organisms in which one is benefited and the other is neither harmed nor benefited.

compensation depth depth at which there is a balance between the oxygen produced by algae through photosynthesis and that consumed by them through respiration; net oxygen production is zero.

condensation process by which a vapor becomes a liquid or a solid.

conduction transfer of heat energy through matter by internal molecular motion; also heat transfer by turbulence in fluids.

conservative constituent component or property of seawater whose value changes only as a result of mixing, diffusion, and advection and not as a result of biological chemical processes; for example, salinity.

consumer animal that feeds on plants (primary consumer) or on other animals (secondary consumer).

continental crust crust forming the continental land blocks; mainly granite and its derivatives.

continental margin zone separating the continents from the deep-sea bottom, usually subdivided into shelf, slope, and rise.

continental rise gentle slope formed by the deposition of sediments at the base of a continental slope.

continental shelf zone bordering a continent, extending from the line of permanent immersion to the depth at which there is a marked or rather steep descent to the great depths.

continental shelf break zone along which there is a marked increase of slope at the outer margin of a continental shelf.

continental slope relatively steep downward slope from the continental shelf break to depth.

contour line on a chart or graph connecting the points of equal value for elevation, temperature, salinity, and so on.

convection transmission of heat by the movement of a heated gas or liquid; vertical circulation resulting from changes in density of a fluid.

convergence situation in which fluids of different origins come together, usually resulting in the sinking, or downwelling, of surface water and the rising of air.

copepod small, shrimplike members of the zooplankton.

coral colonial animal that secretes a hard outer calcareous skeleton; the skeletons of coral animals form in part the framework for warm-water reefs.

corange lines in a rotary tide, lines of equal tidal range about the amphidromic point.

core vertical, cylindrical sample of bottom sediments, from which the nature of the bottom can be determined; also the central zone of the earth, thought to be liquid or molten on the outside and solid on the inside.

corer device that plunges a hollow tube into bottom sediments to extract a vertical sample.

Coriolis effect apparant force acting on a body in motion, due to the rotation of the earth, causing deflection to the right in the Northern Hemisphere and to the left in the Southern Hemisphere; the force is proportional to the speed and latitude of the moving body.

cosmogenous sediment sediment particles with an origin in outer space; for example, meteor fragments and cosmic dust.

cotidal lines lines on a chart marking the location of the tide crest at stated time intervals.

covalent bond chemical bond formed by the sharing of one or more pairs of electrons.

cratons large pieces of the earth's crust that form the centers of continents.

crest *see* Berm crest, Reef crest, Wave crest.

crust outer shell of the solid earth; the lower limit is usually considered to be the Mohorovicic discontinuity.

crustacean member of a class of primarily aquatic organisms with paired jointed appendages and a hard outer skeleton; includes lobsters, crabs, shrimps, and copepods.

ctenophores transparent, planktonic animal, spherical or cylindrical in shape with rows of cilia; comb jellies.

current horizontal movement of water.

current meter instrument for measuring the speed and direction of a current.

D

deadweight ton (DWT) capacity of a vessel in tons of cargo, fuel, stores, and so on; determined by the weight of the water displaced.

declinational tide *see* Diurnal tide.

decomposer heterotrophic; microorganisms (usually bacteria and fungi) that break down nonliving organic matter and release nutrients, which are then available for reuse by autotrophs.

deep exceptionally deep area of the ocean floor, usually below 6000 m.

deep scattering layer (DSL) layer of organisms that move away from the surface during the day and toward the surface at night; the layer scatters or returns vertically directed sound pulses.

deep-sea reversing thermometer (DSRT) mercury-in-glass thermometer that records seawater temperature upon being inverted and retains its reading until returned to its upright position.

deep-water wave wave in water, whose depth is greater than one-half its wavelength.

delta area of unconsolidated sediment deposit, usually triangular in outline, formed at the mouth of a river.

demersal fish fish living near and on the bottom.

density property of a substance defined as mass per unit volume and usually expressed in grams per cubic centimeter or kilograms per cubic meter.

depth recorder *see* Echo sounder.

desalination process of obtaining fresh water from seawater.

detritus any loose material, especially decomposed, broken, and dead organic materials.

dew point temperature to which air must be cooled for its water vapor to condense.

diatom microscopic unicellular alga with an external skeleton of silica.

diatomaceous ooze sediment made up of more than 30% skeletal remains of diatoms.

diffraction process that transmits energy laterally along a wave crest.

diffusion movement of a substance from a region of higher concentration to a region of lower concentration (movement along a concentration gradient); may be due to molecular motion or turbulence.

dinoflagellate one of a class of planktonic organisms possessing characteristics of both plants and animals.

dispersion (sorting) sorting of waves as they move out from a storm center; occurs because long waves travel faster in deep water than short waves.

diurnal inequality difference in height between the two high waters or two low waters of each tidal day; the difference in speed between the two flood currents or two ebb currents of each tidal day.

diurnal tide (declinational tide) tide with one high water and one low water each tidal day.

divergence horizontal flow of fluids away from a common center, associated with upwelling in water and descending motions in air.

doldrums nautical term for the belt of light, variable winds near the equator.

downwelling sinking of water toward the bottom, usually the result of a surface convergence or an increase in density of water at the sea surface.

dredge cylindrical or boxlike sampling device made of metal, net, or both, which is dragged across the bottom to obtain biological or geological samples.

drift bottle bottle released into the sea for use in studying currents; contains a card identifying date and place of release and requesting the finder to return it with date and place of recovery.

drift sector *see* Coastal circulation cell.

dugong *see* Sea cow.

dune wind-formed hill or ridge of sand.

dynamic equilibrium state in which the sum of all changes is balanced and there is no net change.

E

earth sphere depth uniform depth of the earth below the present mean sea level, if the solid earth surface were smoothed off evenly (2440 m).

ebb current movement of a tidal current away from shore or down a tidal stream as the tide level decreases.

ebb tide falling tide; the period of the tide between high water and the next low water.

echolocation use of sound waves by some marine animals to locate and identify underwater objects.

echo sounder (depth recorder) instrument used to measure the depth of water by measuring the time interval between the release of a sound pulse and the return of its echo from the bottom. *See also* Precision depth recorder (PDR).

ecosystem the organisms in a community and the nonliving environment with which they interact.

eddy circular movement of water.

Ekman spiral in a theoretical ocean of infinite depth, unlimited extent, and uniform viscosity, with a steady wind blowing over the surface, the surface water moves 45° to the right of the wind in the Northern Hemisphere. At greater depths the water moves farther to the right with decreased speed, until at some depth (approximately 100 m) the water moves opposite to the wind direction. Net water transport is 90° to the right of the wind in the Northern Hemisphere. Movement is to the left in the Southern Hemisphere.

electrodialysis separation process in which electrodes of opposite charge are placed on each side of a membrane to accelerate the diffusion of particles across the membrane.

electromagnetic radiation waves of energy formed by simultaneous electric and magnetic oscillations; the electromagnetic spectrum is the continuum of all electromagnetic radiation from low-energy radiowaves to high-energy gamma rays, including visible light.

El Niño wind-driven reversal of the Pacific equatorial currents resulting in the movement of warm water toward the coasts of the Americas, so called because it generally develops just after Christmas.

entrainment mixing of salt water into fresh water overlying salt water, as in an estuary.

epifauna animals living attached to the sea bottom or moving freely over it.

epipelagic upper portion of the oceanic pelagic zone, extending from the surface to about 200 m.

episodic wave abnormally high wave unrelated to local storm conditions.

equator 0° latitude, determined by a plane that is perpendicular to the earth's axis and is everywhere equidistant from the North and South poles.

equilibrium tide theoretical tide formed by the tide-producing forces of the moon and sun on a nonrotating, water-covered earth.

equinoxes times of the year when the sun stands directly above the equator, so that day and night are of equal length around the world. The vernal equinox occurs about March 21, and the autumnal equinox occurs about September 22.

escarpment nearly continuous line of cliffs or steep slopes caused by erosion or faulting.

estuary semi-isolated portion of the ocean, which is diluted by freshwater drainage from land.

euphausiid planktonic, shrimplike crustacean. *See also* Krill.

evaporation process by which a liquid becomes a vapor.

evaporite deposit formed from minerals left behind by evaporating water, especially salt.

F

fast ice sea ice that is anchored to shore or the seafloor in shallow water.

fathom a unit of length equal to 1.8 m or 6 ft; used to measure water depth.

fault break or fracture in the earth's crust, in which one side has been displaced relative to the other.

fetch continuous area of water over which the wind blows in essentially a constant direction.

filament chain of living cells.

fjord narrow, deep, steep-walled inlet of the ocean formed by the submergence of a mountainous coast or by the entrance of the ocean into a deeply excavated glacial trough after the melting of the glacier.

flagellum long, whiplike extension from a living cell's surface that by its motion moves the cell.

floe discrete patch of sea ice moved by the currents or by the wind.

flood current movement of a tidal current toward the shore or up a tidal stream as the tide level increases.

flood tide rising tide; the period of the tide between low water and the next high water.

flushing time length of time required for an estuary to exchange its water with the open ocean.

fog visible assemblage of tiny droplets of water formed by condensation of water vapor in the air; a cloud with its base at the surface of the earth.

food chain sequence of organisms in which each is food for the next member in the sequence. *See also* Food web.

food web complex of interacting food chains; all the feeding relations of a community taken together; includes production, consumption, decomposition, and the flow of energy.

foraminifera minute, one-celled animals that usually secrete calcareous shells.

foraminiferal ooze sediment made up of 30% or more skeletal remains of foraminifera.

forced wave wave generated by a continuously acting force and caused to move at a speed faster than it freely travels.

foreshore portion of the shore that includes the low-tide terrace and the beach face.

fouling attachment or growth of marine organisms on underwater objects, usually objects that are made or introduced by humans.

fracture zone large, linear zone of irregular bathymetry of the seafloor, characterized by asymmetrical ridges and troughs; commonly associated with fault zones.

free wave wave that continues to move at its natural speed after its generation by a force.

friction resistance of a surface to the motion of a body moving (e.g., sliding or rolling) along that surface.

fringing reef reef attached directly to the shore of an island or a continent and not separated from it by a lagoon.

frustule siliceous external shell of a diatom.

G

generating force disturbing force that creates a wave, such as wind or a landslide.

geostrophic flow horizontal flow of water occurring when there is a balance between gravitational forces and the Coriolis effect.

glacier mass of land ice, formed by the recrystallization of compacted old snow, flowing slowly from an accumulation area to an area of ice loss by melting, sublimation, or calving.

graben a portion of the earth's crust that has moved downward and is bounded by steep faults; a rift.

grab sampler instrument used to remove a piece of the ocean floor for study.

granite crystalline, coarse-grained, igneous rock composed mainly of quartz and feldspar.

gravitational force mutual force of attraction between particles of matter (bodies).

gravity earth's gravity; acceleration due to the earth's mass is 981 cm/sec²; g is used in equations.

gravity wave water wave form in which gravity acts as the restoring force; waves with wavelengths greater than 2 cm.

groin protective structure for the shore, usually built perpendicular to the shoreline; used to trap littoral drift or retard erosion of the shore; a type of jetty.

group speed speed at which a group of waves travels (in deep water, group speed equals one-half the speed of an individual wave); the speed at which the wave energy is propagated.

guyot submerged, flat-topped seamount.

gyre circular movement of water, larger than an eddy; usually applied to a larger system.

H

habitat place where a plant or animal species naturally lives and grows.

hadal pertaining to the greatest depths of the ocean.

half-life time required for half of an initial quantity of a radioactive isotope to decay.

halocline water layer with a large change in salinity with depth.

harmonic analysis process of separating astronomical tide-causing effects from the tide record, in order to predict the tides at any location.

heat budget accounting for the total amount of the sun's heat received on earth during one year as being exactly equal to the total amount lost from the earth due to radiation and reflection.

heat capacity quantity of heat needed to raise the temperature of a unit mass (1 g) of a substance by 1°C.

height, of wave *see* Wave height.

herbivore animal that feeds only on plants.

heterotrophic pertaining to organisms requiring preformed organic compounds for food; unable to manufacture food from inorganic compounds.

higher high water higher of the two high waters of any tidal day in a region of mixed tides.

higher low water higher of the two low waters of any tidal day in a region of mixed tides.

high water maximum height reached by a rising tide.

holdfast organ of a benthic alga that attaches the alga to the seafloor.

holoplankton organisms living their entire life cycle in the floating (planktonic) state.

hook spit turned landward at its outer end.

horse latitudes regions of calms and variable winds that coincide with latitudes at approximately 30° to 35° N and S.

hot spot surface expression of a persistent rising jet of molten mantle material.

hurricane severe, cyclonic, tropic storm at sea, with winds of 120 km/hr (73 mph) or more; generally applied to Atlantic Ocean storms. *See also* Typhoon.

hydrocast process of obtaining water samples at predetermined depths.

hydrogen bond in water, the weak attraction between the positively polar hydrogen of one water molecule and the negatively polar oxygen of another water molecule.

hydrogenous sediment sediment formed from substances dissolved in seawater.

hydrologic cycle movement of water among the land, oceans, and atmosphere due to changes of state, vertical and horizontal transport, evaporation, and precipitation.

hydrothermal vent seafloor outlet for high-temperature groundwater and associated minerals; a hot spring.

hydrowire measured cable to which oceanographic instruments are attached to be lowered into the sea.

hypsographic curve graph of land elevation and ocean depth versus area.

I

iceberg mass of land ice that has broken away from a glacier and floats in the sea.

igneous rock rock formed by congealing rapidly or slowly from molten magma.

infauna animals that live buried in the sediment.

internal wave wave created below the sea surface at the boundary between two density layers.

intertidal *see* Littoral.

intertidal volume in an embayment, the volume of water gained or lost due to the rise and fall of the tide.

ion positively or negatively charged atom or group of atoms.

island arc chain of volcanic islands formed when plates converge at a subduction zone.

isobar line of constant pressure.

isobath contour of constant depth.

isohaline having a uniform salt content.

isopycnal having a uniform density.

isostasy mechanism by which areas of the earth's crust rise or subside until their masses are in balance, "floating" on the mantle. It is theorized that the continents and mountains are supported by low-density crustal "roots."

isothermal having a uniform temperature.

isotope atoms of the same element having different numbers of neutrons.

J

jellyfish semitransparent, bell-shaped pelagic organism, often with long tentacles bearing stinging cells.

jet stream a stream of air, between 30° and 50° N and about 12 km above the earth, coming from the west at an average speed of 100 km/h.

jetty structure located to influence currents or protect the entrance to a harbor or river from waves (U.S. terminology). *See also* Breakwater; Groin.

K

kelp any of several large, brown algae, including the largest known algae.

kinetic energy energy produced by the motion of an object.

knot a unit of speed equal to 0.51 m/sec or 1 nautical mile per hour.

krill term used by whalers for the small, shrimplike crustaceans found in huge masses in polar waters and eaten by baleen whales.

L

lag deposits large particles left on a beach after the small particles are washed away.

lagoon shallow body of water, which usually has a shallow, restricted outlet to the sea.

La Niña condition of colder-than-normal surface water in the eastern tropical Pacific.

larva immature juvenile form of an animal.

latent heat of fusion amount of heat required to change the state of 1 g of water from ice to liquid.

latent heat of vaporization amount of heat required to change the state of 1 g of water from liquid to gas.

lava magma, or molten rock, that has reached the earth's surface; the same material solidified after cooling.

lee shelter; the part or side sheltered from wind or waves.

lithogenous sediment sediment composed of rock particles eroded mainly from the continents by water, wind, and waves.

lithosphere outer, rigid portion of the earth; includes the continental and oceanic crust and the upper part of the mantle.

littoral (intertidal) area of the shore between mean high water and mean low water; the intertidal zone.

littoral cell *see* Coastal circulation cell.

littoral drift *see* Longshore transport.

longshore current current produced in the surf zone by waves breaking at an angle with the shore; runs roughly parallel to the shoreline.

longshore transport (littoral drift) movement of sediment by the longshore current.

loran navigational system in which position is determined by measuring the difference in the time of reception of synchronized radio signals; derived from the phrase "long-range navigation."

lower high water lower of the two high waters of any tidal day in a region of mixed tides.

lower low water lower of the two low waters of any tidal day in a region of mixed tides.

low-tide terrace flat section of the foreshore seaward of the sloping beach face.

low water minimum height reached by a falling tide.

lunar month time required for the moon to pass from one new moon to another new moon (approximately twenty-nine days).

M

magma molten rock material that forms igneous rocks upon cooling; magma that reaches the earth's surface is referred to as lava.

magnetic pole either of the two points on the earth's surface where the magnetic field is vertical.

manatee *see* Sea cow.

manganese nodules rounded, layered lumps found on the deep-ocean floor that contain as much as 20% manganese and smaller amounts of iron, nickel, and copper; a hydrogenous sediment.

mantle main bulk of the earth between the crust and the core; increasing pressure and temperature with depth divide the mantle into concentric layers.

mariculture cultivation of marine organisms under controlled conditions. *See also* Aquaculture.

maximum sustained yield maximum number or amount of a species that can be harvested each year without steady depletion of the stock; the remaining stock is able to replace the harvested members by natural reproduction.

meander turn or winding curve of a current that may detach from the main stream as an eddy.

mean sea level average height of the sea surface, based on observations of all stages of the tide over a nineteen-year period in the United States.

mean solar day *see* Solar day.

mechanical bathythermograph (BT) *see* Bathythermograph (BT).

meiobenthos very small animals living buried in the sediments of the seafloor.

meridian circle of longitude passing through the poles and any given point on the earth's surface.

meroplankton floating developmental stages (eggs and larvae) of organisms that as adults belong to the nekton and benthos.

mesopelagic oceanic zone from 200 m to 1000 m.

mesosphere the layer of atmosphere above the stratosphere; extends from about 50 to 90 km.

messenger weight, usually hinged and with a latch so it can be fastened around a wire, used to activate oceanographic instruments after they have been lowered to the desired depth.

microplankton net plankton, composed of individuals below 1 mm in size but large enough to be retained by a small mesh net.

mitigation coastal management concept requiring developers to replace developed areas with equivalent natural areas or to reengineer other areas to resemble areas prior to development.

mixed tide type of tide in which large inequalities between the two high waters and the two low waters occur in a tidal day.

Mohorovicic discontinuity (Moho) boundary between crust and mantle, marked by a rapid increase in seismic wave speed.

mollusks marine animals, usually with shells; includes mussels, oysters, clams, snails, and slugs.

monoculture cultivation of only one species of organism in an aquaculture system.

monsoon name for seasonal winds; first applied to the winds over the Arabian Sea, which blow for six months from the northeast and for six months from the southwest; now extended to similar winds in other parts of the world; in India, the term is popularly applied to the southwest monsoon and also to the rains that it brings.

moon tide portion of the tide generated solely by the moon's tide-raising force, as distinguished from that of the sun.

moraine glacial deposit of rock, gravel, and other sediment left at the margin of an ice sheet.

mutualism an intimate association between different organisms in which both organisms benefit.

N

nannoplankton plankton smaller than 10 μ, which will pass through an ordinary plankton net but can be removed from the water by centrifuging water samples.

nautical mile unit of length equal to 1852 m or 1.15 land miles.

neap tides tides occurring near the times of the first and last quarters of the moon, when the range of the tide is least.

nekton pelagic animals that are active swimmers; for example, adult squid, fish, and marine mammals.

nephelometer device that measures the scattering of light by particulate material in the sea.

neritic shallow-water marine environment extending from low water to the edge of the continental shelf. *See also* Pelagic.

net plankton *see* Microplankton.

node point of least or zero vertical motion in a standing wave.

nonconservative constituent component or property of seawater whose value changes as a result of biological or chemical processes as well as by mixing, advection, and diffusion; for example, nutrients and oxygen in seawater.

nudibranchs soft-bodies, gastropod mollusks; sea slugs.

nutrient in the ocean, any one of a number of inorganic or organic compounds or ions used primarily in the nutrition of primary producers; nitrogen and phosphorus compounds are examples.

O

oceanic pertaining to the ocean water seaward of the continental shelf; the "open ocean." *See also* Pelagic.

oceanic crust crust below the deep-ocean sediments; mainly basalt.

ocean ranching raising of salmon to a juvenile stage, release of the juveniles to sea, and harvest of adults on their return.

offshore direction seaward of the shore.

offshore current any current flowing away from the shore.

offshore transport movement of sediment or water away from the shore.

onshore direction toward the shore.

onshore current any current flowing toward the shore.

onshore transport movement of sediment or water toward the shore.

ooze fine-grained deep-ocean sediment composed of at least 30% sand or silt-sized calcareous or siliceous remains of small marine organisms, the remainder being clay-sized material.

orbit in water waves, the path followed by the water particles affected by the wave motion; also, the path of a body subjected to the gravitational force of another body, such as the earth's orbit around the sun.

orographic effect precipitation patterns caused by the flow of air over and around mountains.

osmosis tendency of water to diffuse through a semipermeable membrane to make the concentration of water on one side of the membrane equal to that on the other side.

osmotic pressure pressure that builds up in a confined fluid because of osmosis.

overturn sinking of more-dense water and its replacement by less-dense water from below.

oxygen minimum zone in which respiration and decay reduce dissolved oxygen to a minimum, usually between 800 and 1000 m.

ozone a form of oxygen that absorbs ultraviolet radiation from the sun.

P

parallel circle on the surface of the earth parallel to the plane of the equator and connecting all points of equal latitude; a line of latitude.

parasitism an intimate association between different organisms in which one is benefited and the other is harmed.

pelagic primary division of the sea, which includes the whole mass of water subdivided into neritic and oceanic zones; also pertaining to the open sea.

period *see* Tidal period; Wave period.

pH measure of the concentration of hydrogen ions in a solution; the concentration of hydrogen ions determines the acidity of the solution; $pH = -\log_{10}(H^+)$, where H^+ is the concentration of hydrogen ions in gram atoms per liter.

photic zone layer of a body of water that receives ample sunlight for photosynthesis; usually less than 100 m.

photo cell a device that converts light energy to electrical energy; used to determine solar radiation below the sea surface.

photophore luminous organ found on fish.

photosynthesis manufacture by plants of organic substances and release of oxygen from carbon dioxide and water in the presence of sunlight and the green pigment chlorophyll.

physiographic portrayal of the earth's features by perspective drawing.

phytoplankton microscopic plant forms of plankton.

pinniped member of the marine mammal group, characterized by four swimming flippers; for example, seals and sea lions.

plankton passively drifting or weakly swimming organisms.

plate tectonics theory and study of the earth's lithospheric plates, their formation, movement, interaction, and destruction; the attempts to explain the earth's crustal changes in terms of plate movements.

polar easterlies winds blowing from the poles toward approximately 60°N and S; winds are northeasterly in the Northern Hemisphere and southeasterly in the Southern Hemisphere.

polychaetes marine segmented worms, some in tubes, some free-swimming.

polyculture cultivation of more than one species of organism in an aquaculture system.

polyp sessile stage in the life history of certain members of the phylum coelenterata (Cnidaria); sea anemones and corals.

potential energy energy that an object has because of its position or condition.

precipitation falling products of condensation in the atmosphere, such as rain, snow, or hail; also the falling out of a substance from solution.

precision depth recorder (PDR) instrument used to obtain a continuous pictorial record of the ocean bottom by timing the returning echoes of sound pulses. *See also* Echo sounder.

primary coast coastline shaped primarily by land forces rather than sea forces.

primary production (primary productivity) amount of organic material synthesized by organisms from inorganic substances in unit time in a unit volume of water or in a column of water of unit area cross section and extending from the surface to the bottom.

prime meridian meridian of 0° longitude, used as the origin for measurements of longitude; internationally accepted as the meridian of the Royal Naval Observatory, Greenwich, England.

productivity amount of organic material synthesized by organisms in unit time in a unit volume of water. *See also* Primary production (primary productivity).

progressive tide tide wave that moves, or progresses, in a nearly constant direction.

progressive wave wave that moves, or progresses, in a certain direction.

protozoa minute, mostly one-celled animals.

pseudopodium flowing, temporary extension of the protoplasm of a cell, used in locomotion or feeding.

pteropod pelagic snail whose foot is modified for swimming.

pteropod ooze sediment made up of more than 30% shells of pteropods.

P-waves primary, or faster, waves moving away from a seismic event; can penetrate solid rock; consist of energy transmitted by alternate compressions and dilations of the material. *See also* S-waves.

pycnocline water layer with a large change in density with depth.

R

radar system of determining and displaying the distance of an object by measuring the time interval between transmission of a radio signal and reception of the echo return; derived from the phrase "Radio detecting and ranging."

radially symmetric having similar parts regularly arranged around a central axis.

radiation energy transmitted as rays or waves without the need of a substance to conduct the energy.

radiolarian ooze sediment made up of more than 30% skeletal remains of radiolarians.

radiolarians single-celled protozoans with siliceous skeletons.

radiometric dating determining ages of geological samples by measuring the relative abundance of radioactive isotopes and comparing isotope systems.

rafting transport of sediment, rocks, silt, and other land matter out to sea by ice, logs, and the like, with the deposition of the rafted material when the carrying agent disintegrates.

range *see* Tidal range.

red clay red to brown fine-grained lithogenous deposit, of predominantly clay size, which is derived from land, transported by winds and currents, and deposited far from land and at great depth; also known as brown mud and brown clay.

red tide red coloration, usually of coastal waters, caused by large quantities of microscopic organisms (generally dinoflagellates); some red tides result in mass fish kills, others contaminate shellfish, and still others produce no toxic effects.

reef offshore hazard to navigation made up of consolidated rock, with a depth of 20 m or less.

reef crest highest portion of a coral reef on the exposed seaward edge of the reef.

reef flat portion of a coral reef landward of the reef crest and seaward of the lagoon.

reflection rebounding of light, heat, sound, waves, and so on, after striking a surface.

refraction change in direction or bending of a wave.

relict sediments sediments deposited by processes no longer active.

residence time mean time that a substance remains in a given area before replacement, calculated by dividing the amount of a substance by its rate of addition or subtraction.

respiration metabolic process by which food or food-storage molecules yield the energy on which all living cells depend.

restoring force force that returns a disturbed water surface to the equilibrium level, such as surface tension and gravity.

ridge long, narrow elevation of the seafloor, with steep sides and irregular topography.

rift valley trough formed by faulting along a zone in which plates move apart and new crust is created, such as along the crest of a ridge system.

rip current strong surface current flowing seaward from shore; the return movement of water piled up on the shore by incoming waves and wind.

rise long, broad elevation that rises gently and generally smoothly from the seafloor.

rotary current tidal current that continually changes its direction of flow through all points of the compass during a tidal period.

rotary tide tide that is the result of a standing wave moving around the central node of a basin.

S

salinity measure of the quantity of dissolved salts in seawater. It is formally defined as the total amount of dissolved solids in seawater in parts per thousand (‰) by weight when all the carbonate has been converted to oxide, all the bromide and iodide have been converted to chloride, and all organic matter is completely oxidized.

salinity bridge (salinometer) instrument for determining the salinity of water by measuring the electrical conductivity of a water sample of a known temperature.

salt budget balance between the rates of salt supply and removal from a body of water.

salt wedge intrusion of salt water along the bottom; in an estuary, the wedge moves upstream on high tide and seaward on low tide.

sand spit *see* Spit.

satellite body that revolves around a planet; a moon; a device launched from earth into orbit around a planet of the sun.

scarp elongated and comparatively steep slope separating flat or gently sloping areas on the seafloor or on a beach.

scattering random redirection of light or sound energy by reflection from an uneven sea bottom or sea surface, from water molecules, or from particles suspended in the water.

sea same as the ocean; subdivision of the ocean; surface waves generated or sustained by the wind within their fetch, as opposed to swell.

sea cow (dugong, manatee) large herbivorous marine mammal of tropic waters; includes the manatee and dugong.

seafloor spreading movement of crustal plates away from the midocean ridges; process that creates new crustal material at the midocean ridges.

sea level height of the sea surface above or below some reference level. *See also* Mean sea level.

seamount isolated volcanic peak that rises at least 1000 m from the seafloor.

sea smoke type of fog caused by dry, cold air moving over warm water.

sea stack isolated mass of rock rising from the sea near a headland from which it has been separated by erosion.

sea state numerical or written description of the roughness of the ocean surface relative to wave height.

Secchi disk white disk used to measure the transparency of the water by observing the depth at which the disk disappears from view.

secondary coast coastline shaped primarily by marine forces or marine organisms.

sediment particulate organic and inorganic matter that accumulates in loose, unconsolidated form.

seiche standing wave oscillation of an enclosed or semienclosed body of water that continues, pendulum fashion, after the generating force ceases.

seismic pertaining to or caused by earthquakes or earth movements.

seismic sea wave *see* Tsunami.

seismic tomography using seismic data to produce computerized, detailed, three-dimensional maps of the boundaries between the earth's layers.

semidiurnal tide tide with two high waters and two low waters each tidal day.

semipermeable membrane membrane that allows some substances to pass through it but restricts or prevents the passage of other substances.

sessile permanently fixed or sedentary; not free-moving.

set direction in which the current flows.

shallow-water wave wave in water whose depth is less than one-twentieth the average wavelength.

shingles flat, water-warm pebbles, or cobbles found in beds along a beach.

shoal elevation of the sea bottom comprising any material except rock or coral (in which case it is a reef) and which may endanger surface navigation.

shore strip of ground bordering any body of water, which is alternately exposed and covered by tides and waves.

sidereal day time period determined by one rotation of the earth relative to a far distant star, about four minutes shorter than the mean solar day.

sigma-t abbreviated value of the density of seawater neglecting pressure and at a given temperature and salinity;

$\sigma_t = (density - 1) \times 1000$.

siliceous containing silica.

sill shallow area that separates two basins from one another or a coastal bay from the adjacent ocean.

slack water state of a tidal current when its velocity is near zero; occurs when the tidal current changes direction.

slick area of smooth surface water.

sofar channel natural sound channel in the oceans, in which sound can be transmitted for very long distances; the depth of minimum sound velocity; derived from the phrase "sound fixing and ranging."

solar constant rate at which solar radiation is received on a unit surface that is perpendicular to the direction of incident radiation just outside the earth's atmosphere at the earth's mean distance from the sun; equal to 2 cal/cm^2/min.

solar day time period determined by one rotation of the earth relative to the sun; the mean solar day is twenty-four hours.

solstices times of the year when the sun stands directly above 23 1/2°N or S latitude. The winter solstice occurs about December 22, and the summer solstice occurs about June 22.

sonar method or equipment for determining, by underwater sound, the presence, location, or nature of objects in the sea; derived from the phrase "sound navigation and ranging."

sorting *see* Dispersion.

sounding measurement of the depth of water beneath a vessel.

sound shadow area of the ocean into which sound does not penetrate because the density structure of the water refracts the sound waves.

specific gravity ratio of the density of a substance to the density of 4°C water.

specific heat ratio of the heat capacity of a substance to the heat capacity of water.

sphere depth thickness of a material speed uniformly over a smooth sphere having the same area as the earth.

spit (sand spit) low tongue of land, or a relatively long, narrow shoal extending from the shore.

spoil dredged material.

spore minute, unicellular, asexual reproductive structure of an alga.

spreading center region along which new crustal material is produced.

spring tides tides occurring near the times of the new and full moon, when the range of the tide is greatest.

standing crop biomass of a population present at any given time.

standing wave type of wave in which the surface of the water oscillates vertically between fixed points called nodes, without progression; the points of maximum vertical rise and fall are called antinodes.

steepness, of wave *see* Wave steepness.

stipe portion of an alga between the holdfast and the blade.

storm center area of origin for surface waves generated by the wind; an intense atmospheric low-pressure system.

storm tide (storm surge) along a coast, the exceptionally high water accompanying a storm, owing to wind stress and low atmospheric pressure, made even higher when associated with a high tide and shallow depths.

stratosphere the layer of the atmosphere above the troposphere where temperature is constant or increases with altitude.

subduction zone plane descending away from a trench and defined by its seismic activity, interpreted as the convergence zone between a sinking plate and an overriding plate.

sublimation transition of a substance from its solid state to its gaseous state without becoming a liquid.

sublittoral (subtidal) benthic zone from the low-tide line to the seaward edge of the continental shelf.

submarine canyon relatively narrow, V-shaped, deep depression with steep slopes, the bottom of which grades continuously downward across the continental slope.

submersible a research submarine, designed for manned or remote operation at great depths.

subsidence sinking of a broad area of the crust without appreciable deformation.

substrate material making up the base on which an organism lives or to which it is attached.

subtidal *see* Sublittoral.

summer solstice *see* Solstices.

sun tide portion of the tide generated solely by the sun's tide-raising force, as distinguished from that of the moon.

supralittoral benthic zone above the high-tide level that is moistened by waves, spray, and extremely high tides.

surf wave activity in the area between the shoreline and the outermost limit of the breakers.

surface tension tendency of a liquid surface to contract owing to bonding forces between molecules.

surimi refined fish protein used to form artificial crab, shrimp, and scallop meat.

swash zone beach area where water from a breaking wave rushes.

S-waves secondary, or slower, transverse seismic waves; cannot penetrate a liquid; consist of elastic vibrations perpendicular to the direction of travel. *See also* P-waves.

swell long and relatively uniform wind-generated ocean waves that have traveled out of their generating area.

symbiosis living together in intimate association of two dissimilar organisms.

T

tectonic pertaining to processes that cause large-scale deformation and movement of the earth's crust.

tektites particles with a characteristic round shape, derived from cosmic material.

terranes fragments of the earth's crust bounded by faults, each fragment with a history distinct from each other fragment.

terrigenous of the land; sediments composed predominantly of material derived from the land.

thermocline water layer with a large change in temperature with depth.

thermohaline circulation vertical circulation caused by changes in density; driven by variations in temperature and salinity.

thermosphere the layer of the atmosphere above the mesosphere; extends from 90 km to outer space.

tidal bore high-tide crest that advances rapidly up an estuary or river as a breaking wave.

tidal current alternating horizontal movement of water associated with the rise and fall of the tide.

tidal datum reference level from which ocean depths and tide heights are measured; the zero tide level.

tidal day time interval between two successive passes of the moon over a meridian, approximately 24 hours and 50 minutes.

tidal period elapsed time between successive high waters or successive low waters.

tidal range difference in height between consecutive high and low waters.

tide periodic rising and falling of the sea surface that results from the gravitational attractions of the moon and sun acting on the rotating earth.

tide wave long-period gravity wave that has its origin in the tide-producing force and is observed as the rise and fall of the tide.

tombolo deposit of unconsolidated material that connects an island to another island or to the mainland.

topography general elevation pattern of the land surface (or the ocean bottom). *See also* Bathymetry.

toxicant substance dissolved in water that produces a harmful effect on organisms, either by an immediate large dose or by small doses over a period of time.

trade winds wind systems occupying most of the tropics, which blow from approximately 30° N and S toward the equator; winds are northeasterly in the Northern Hemisphere and southeasterly in the Southern Hemisphere.

transform fault fault with horizontal displacement connecting the ends of an offset in a midocean ridge. Some plates slide past each other along a transform fault.

transgenic describing an organism that contains hereditary material from another organism incorporated into its genetic material.

transverse ridge ridge running at nearly right angles to the main or principal ridge.

trench long, deep, and narrow depression of the seafloor with relatively steep sides, associated with a subduction zone.

triploid condition in which cells have three sets of chromosomes.

trophic relating to nutrition; a trophic level is the position of an organism in a food chain or food (trophic) pyramid.

Tropics of Cancer and Capricorn latitudes 23 1/2° N and 23 1/2° S, respectively, marking the maximum angular distance of the sun from the equator during the summer and winter solstices.

troposphere the lowest layer of the atmosphere where the temperature decreases with altitude.

trough long depression of the seafloor, having relatively gentle sides; normally wider and shallower than a trench. *See also* Wave trough.

T-S diagram graph of temperature versus salinity, on which seawater samples taken at various depths are used to describe a water mass.

tsunami (seismic sea wave) long-period sea wave produced by a submarine earthquake or volcanic eruption. It may travel across the ocean for thousands of miles unnoticed from its point of origin and build up to great heights over shallow water at the shore.

tube worm any worm or wormlike organism that builds a tube or sheath attached to a submerged substrate.

turbidite sediment deposited by a turbidity current, showing a pattern of coarse particles at the bottom, grading gradually upward to fine silt.

turbidity loss of water clarity of transparency owing to the presence of suspended material.

turbidity current dense, sediment-laden current flowing downward along an underwater slope.

typhoon severe, cyclonic, tropic storm originating in the western Pacific Ocean, particularly in the vicinity of the South China Sea. *See also* Hurricane.

U

ultraplankton plankton that are smaller than nannoplankton; they are difficult to separate from water.

upwelling rising of water rich in nutrients toward the surface, usually the result of diverging surface currents.

V

vernal equinox *see* Equinoxes.

virus a noncellular infectious agent that reproduces only in living cells.

viscosity property of a fluid to resist flow; internal friction of a fluid.

W

water bottle device used to obtain a water sample at depth.

water budget balance between the rates of water added and lost in an area.

water mass body of water identified by similar patterns of temperature and salinity from surface to depth.

water type body of water identified by a specific range of temperature and salinity from a common source.

wave periodic disturbance that moves through or over the surface of a medium with a speed determined by the properties of the medium.

wave crest highest part of a wave.

wave height vertical distance between a wave crest and the adjacent trough.

wavelength horizontal distance between two successive wave crests or two successive wave troughs.

wave period time required for two successive wave crests or troughs to pass a fixed point.

wave ray line indicating the direction waves travel; drawn at right angles to the wave crests.

wave steepness ratio of wave height to wavelength.

wave train series of similar waves from the same direction.

wave trough lowest part of a wave.

westerlies wind systems blowing from the west between latitudes of approximately 30° and 60° N and 30° and 60° S; they are southwesterly in the Northern Hemisphere and northwesterly in the Southern Hemisphere.

wind wave wave created by the action of the wind on the sea surface.

winter solstice *see* Solstices.

Z

zenith point in the sky that is immediately overhead.

zonation parallel bands of distinctive plant and animal associations found within the littoral zones and distributed to take advantage of optimal conditions for survival.

zooplankton animal forms of plankton.

zooxanthellae symbiotic microscopic organisms (dinoflagellates) found in corals and other marine organisms.

Credits

Photographs

Chapter 12

Table of Contents and Opener: © David Muench; **12.9 a–c:** Courtesy Joe Lucas/Marine Entanglement Research Program National Marine Fisheries, NOAA; **12.9 d:** Courtesy R. Herron/Marine Entanglement Research Program/ National Marine Fisheries, NOAA; **12.10 a–f:** Courtesy Hazardous Materials Response Branch/Jerry Galt/NOAA; **12.11:** Courtesy Clean Sound Cooperative/Seattle; **12.12:** Port of Seattle; **12.13:** © Alex S. Maclean/Landslides; **Box 12.1:** U.S. Dept. of Interior Earth Observation Satellite Company Geological Survey; School of Oceanography/Univ. of Washington; **12.14:** Courtesy John Conomos, U.S. Geological Survey; **12.15:** NASA

Chapter 13

Table of Contents and Opener: © Dave B. Fleetham/Tom Stack & Associates; **13.6:** Courtesy James Sumich; **13.7:** Field Museum of Natural History, Chicago; **Box 13.1:** Courtesy P. Flanagan; **Box 13.2:** U.S. Army Corps of Engineers

Chapter 14

Table of Contents and Opener: © Eda Rogers; **Box 14.1a–d:** Courtesy David English, School of Oceanography, University of Washington. These images were formed from data of the NASA Global Ocean Color Data Set that was implemented by Gene Feldman of NASA's Goddard Space Flight Center. The CZCS data was processed using the University of Miami DSP system developed by Otis Brown and Robert Evans with maintenance support of NASA.

Chapter 15

Table of Contents and Opener: © William Amos; **15.1:** From A. Hardy, "The Open Sea," Houghton-Mifflin, Boston, 1956; William Collins, England; **15.4:** Courtesy Enge's Equipment Company, Morton Grove, IL; **15.7:** From A. Hardy, "The Open Sea," Houghton-Mifflin, Boston, 1956; William Collins, England; **15.9:** Courtesy Kendra Daly, School of Oceanography, University of Washington; **15.14 b:** School of Oceanography, University of Washington; **15.14 d:** Courtesy P. Brunner, Seattle Aquarium; **5.16 c:** Courtesy D. Perry, University of Washington; **15.17:** Courtesy by Mark Holmes, USGS, School of Oceanography, University of Washington; **15.18 a,b:** Photos Courtesy Rita Horner, School of Oceanography, University of Washington; **15.19:** Courtesy Rose Ann Cattolico, Department of Botany, University of Washington

Chapter 16

Table of Contents and Opener: © Molokini/Tom Stack & Associates; **16.1 a:** Courtesy Dr. Albert Erickson/Fisheries Research Institute/ University of Washington; **16.1 b:** Courtesy David Withrow/National Marine Fisheries Service, NOAA; **16.5 a:** Courtesy Leo J. Shaw, Seattle Aquarium; **16.5 b:** Courtesy Kemper, Seattle Aquarium; **16.5 c:** Courtesy Leo J. Shaw, Seattle Aquarium; **16.5 d:** Courtesy Kemper, Seattle Aquarium; **16.10 a–c:** Courtesy Leo J. Shaw, Seattle Aquarium; **16.10 d:** Courtesy Paulette Brunner, Seattle Aquarium; **16.10 e:** Courtesy Leo J. Shaw, Seattle Aquarium; **16.10 f:** Courtesy Gail Scott, Seattle Aquarium;

16.14 a,b: Courtesy Wayne Palsson, Washington State Dept. of Fisheries; **16.14 c:** Courtesy UNI Seafoods, Inc., Redmond, WA; **16.16:** Bruce W. Buls, Seattle, WA; **16.17:** Courtesy Sea Farm Washington, Inc.; **16.18:** Courtesy Anadromos, Inc., Corvallis, OR

Chapter 17

Table of Contents and Opener: © C. Garoutte/Tom Stack & Associates; **17.10 a–c,7.14:** Courtesy John Delaney, School of Oceanography, University of Washington; **17.15 a–c:** Courtesy Dr. James M. Brooks/Geochemical and Environmental Research Group, College Station, TX; **17.16:** Courtesy Kathy Newell, School of Oceanography, University of Washington; **17.17:** Courtesy Brad Matsen, National Fisherman Magazine; Seattle, WA; **17.18:** Courtesy Ken Chew, School of Fisheries, University of Washington; **17.19 a:** Courtesy Seattle Aquarium; **17.19 b:** Courtesy Leo J. Shaw, Seattle Aquarium; **17.19 c:** Courtesy Paulette Brunner, Seattle Aquarium; **17.19 d,e:** Courtesy Seattle Aquarium; **17.19 f,g:** Courtesy Leo J. Shaw, Seattle Aquarium; **17.19 h:** Courtesy Tina Link, Seattle Aquarium; **17.19 i:** Courtesy Leo J. Shaw, Seattle Aquarium; **Box 17.2 a and b:** Courtesy Washington Sea Grant, University of Washington

Illustrations

New cartography and line art by Cartographics, Eau Claire, WI.

Chapter 2

2.23: From M. N. Hill, ed., *The Sea*, Vol. 3. Copyright © 1963 by John Wiley & Sons, New York. Redrawn by permission of the publisher.

Chapter 3

3.10: From Deep Sea Drilling Project A-031, Scripps Institution of Oceanography.

3.14: From D. H. Tarling and J. C. Mitchell, "The Earth's Magnetic Polarity as a Function of Millions of Years Before Present," in *Geology*, 4:3(1976). The Geological Society of America. Reprinted by permission.

3.24 a–e: From R. S. Dietz and J. C. Holden, "Reconstruction of Pangea: Breakup and Dispersion of Continents, Permian to Present," in *Journal of Geophysical Research*, 75:26 (September 1970). American Geophysical Union, Washington, DC.

3.25 a–g: From Bambach, Scotese, and Ziegler, "Before Pangaea: The Geographics of the Paleozoic World" in *American Scientist*, Journal of Sigma Xi, Vol. 68, No. 1, Jan–Feb 1980. Copyright © 1980 Sigma Xi. Reprinted by permission.

Chapter 4

4.14: Adapted from Kinsler and Frey, *Fundamentals of Acoustics*, 2d ed., 1962, John Wiley & Sons, Inc., New York, by permission of the author.

Chapter 5

5.1: Adapted from A. N. Strahler, *Physical Geography*. Copyright © 1960 John Wiley & Sons, Inc., New York. Reprinted by permission of the author.

Chapter 7

7.4: From C. D. Keeling, R. B. Baastow, A. F. Carter, S. C. Piper, T. P. Whorf, M. Heimann, W. G. Mook and H. Roeloffzen, "A Three Dimensional Model of Atmospheric CO_2 Transport Based on Observed Winds: Observational Data and Preliminary Analysis." Appendix A, in *Aspects of Climate Variability in the Pacific and the Western Americas*. Geophysical Monograph, American Geophysical Union, Vol. 55, 1989 (Nov).

Chapter 11

11.23: From Inman and Frautschy, "Littoral Processes and the Development of Shorelines, 1965." Coastal Engineering Santa Barbara Specialty Conference. Reprinted by permission of the authors.

Chapter 14

14.10: From R. and M. Buchsbaum, *Basic Ecology*. Copyright © Boxwood Press, Pacific Grove, California. Adapted by permission of the author.

Chapter Opening Excerpts

Chapter 3: Excerpt from "Little Gidding" in *Four Quartets*, copyright 1943 by T. S. Eliot and Renewed 1971 by Esme Valerie Eliot, reprinted by permission of Harcourt Brace Jovanovich, Inc. and Faber & Faber, Ltd., London.

Chapter 4: From *The Selected Poetry of Robinson Jeffers* by Robinson Jeffers. Copyright 1925 and renewed 1953 by Robinson Jeffers. Reprinted by permission of Random House, Inc.

Chapter 6: From the book: *Kon-Tiki* by Thor Heyerdahl. © 1950, 1960, 1984. Used by permission of the publisher, Prentice-Hall/A division of Simon & Schuster, Englewood Cliffs, N.J.

Chapter 10: Excerpt from *Spring Tides* by Samuel Eliot Morison. Reprinted by permission of Curtis Brown, Ltd. Copyright © 1965 by Samuel Eliot Morison.

Chapter 12: From *Under the Sea Wind* by Rachel L. Carson. Copyright 1941 by Rachel L. Carson, Copyright renewed © 1969 by Roger Christie. A Truman Talley Book. Used by permission of Dutton Signet, an imprint of New American Library, a division of Penguin Books USA, Inc.

Chapter 13: Excerpt from "The Dry Salvages" in *Four Quartets*, copyright 1943 by T. S. Eliot and Renewed 1971 by Esme Valerie Eliot, reprinted by permission of Harcourt Brace Jovanovich, Inc. and Faber & Faber, Ltd., London.

Chapter 14: From "Sensuous Symbionts of the Sea" by Lewis Thomas, in *Natural History*, Vol. 80, No. 7. Courtesy American Museum of Natural History, New York, NY.

Chapter 16: Excerpt from Henry Beston, *The Outermost House*. Copyright © 1949 Henry Holt & Co., Inc., New York. Reprinted with permission.

Chapter 17: From *The Log from the Sea of Cortez* by John Steinbeck. Copyright 1941 by John Steinbeck and Edward F. Ricketts. Copyright renewed © 1969 by John Steinbeck and Edward F. Ricketts, Jr. Used by permission of Viking Penguin, a division of Penguin Books USA, Inc.

Index

A

Abyssal hills and seamounts, 57, 58
Abyssal plain, 57
Abyssopelagic zone, 340
Accreting, beaches, 300
Acoustically Navigated Underwater Survey
    System (ANGUS), 111–12
Adsorption, 145
Advection, and estuaries, 310
Agassiz, Alexander, 18
Age of earth, 30
Agulhas current, 239
Alaska current, 214
Alexander the Great, 5
Algae, 422
Alluvial plain, 288
Amphidromic point, tides, 268
Amplitude, wave, 235
Amundsen, Roald, 15, 17
Anaerobic bacteria, 149, 337
Anatomy of a beach, 295–98
Anatomy of Atlantic Ocean, 171
Anatomy of a wave, 235
Anchovies, 411–12
Andesite volcanoes, 99
Animals
    deep seafloor, 432–34
    fouling and boring organisms, 434–35
    of rocky shores, 425–30
    of soft substrates, 431–32
    symbiotic relationships, 430
    tide pools, 430–31
Animals, benthic marine, 425
Annual cycles, solar radiation, 163–64
Annual range, midocean sea surface
    temperatures, 165
Anoxic water, 149
Antarctic bottom water, 171
Antarctic Circle, 32
Antarctic convergence, 219
Antarctic intermediate water, 171
Antinodes, and standing waves, 249
Aphotic zone, 338
Aquaculture, 415
Arctic Circle, 32
Arctic convergence, 219
Aristotle, 5, 8
Armored beach, 298, 299
Artificial processes, beach changes, 303–4
Asthenosphere, 89

Atlantic Ocean currents, 215–16
Atlantic Ocean structure, 171–72
Atlantis II, 21
Atmosphere, structure of, 188
Atmosphere in motion, 192, 194
Atmospheric pressure, 188–89
Attenuation, of light, 127–28
Autumnal equinox, 32
Auxospore, centric diatom, 371
Average daily solar radiation value, different
    latitudes, 165

B

Backshore, 296
Bacon, Francis, 89
Bacteria, in plankton, 382–83
Baffin, William, 8
Balboa, Vasco Nuñez de, 8
Baleen whale, 376
Barrier islands, 291
Barrier reef, 58
Barriers and boundaries of life in ocean, 340
Bars, beach, 291, 298
Basalt-type rock, lithosphere, 88
Bathymetric chart, 37
Bathymetrics, 66–67
Bathymetric survey, 67
Bathymetry of the seafloor, 52–60
Bathypelagic zone, 340
Bathythermograph (expendable BT or XBT), 180
Beach, defined, 287
Beach dynamics, 299–304
Beach face, 297
Beach types, 298–99
Benguela current, 215
Benthic algae, general characteristics of, 422–23
Benthic zone, 340
Benthos classification summary, 435–36
Berms, 296–97
Biogenous ooze, 65–69
Biogenous sediments, seafloor, 64
Bioluminescence, 338
Biomass, 350
Biomedical products, benthic organisms, 446–47
Bioturbation, 434
Blades, seaweed, 422
Bloom, of diatoms, 371
Bores, tidal, 271
Bottom types, 340
Brachopods, 430

Brazil current, 215
Breakers, 245–46
Breakwaters, 303
Brown clay, 70
Buffon, George, 89
Buoyancy and flotation, and sea creatures,
    334–35

C

Calcareous ooze, 68
California current, 214
Calories, 120
Capillary waves, 234
Carbonate sediments, seafloor, 67
Carbon dioxide, changing levels of, 189–91
Carbon dioxide as buffer, 149, 150
Carbon dioxide cycle, 151–52
Castle bergs, 134
Centric diatoms, 368, 371, 372
Challenger expedition, 13, 14, 147
Changes of state, water, 120–22
Chart projections, 35–37
Charts and maps
    bathymetric charts, 37
    Franklin-Folger chart of Gulf Stream, 10
    mercator projection, 36
    New American Practical Navigator
        (Bowditch), 10
    northern Europe, 7
    Physical Geography of the Sea (Maury), 11
    physiographic chart, world's oceans, 62–63
    physiographic maps, 37
    topographic maps, 37
Chemical resources of sea, 153, 155
Chemical tracers, in seawater, 170
Chesapeake Bay, 328–29
Chlorinity, 147
Chlorophyll, 360
Ciguatera poisoning, 387
Circulation, ocean water, 339–40
Circulation patterns, in estuaries, 311
Classification, organisms, 341
Clouds and climate, 194
Cnidaria, 376
Coal, oil, and gas, as seabed resources, 77–78
Coastal circulation, 301–3
Coastal circulation cell, 301
Coasts, shores, and beaches
    anatomy of a beach, 295–98
    artificial processes, changing beaches, 303–4

beach dynamics, 299–304
beach types, 298–99
coastal circulation, 301–3
major zones, 286–87
natural processes, beach dynamics, 300
primary coasts, 288–91
secondary coasts, 291–95
types of coasts, 287–95
Coccolithophores, 69, 368
Coccoliths, 373
Coelenterata, 376
Cohesion, water, 123
Colonial forms, of jellyfish, 377
Color, sea creatures and, 338–39
Columbus, Christopher, 5, 6, 7
Commensalism, marine animals, 430
Commercial fisheries, 410–17
Commercial species, bony fish, 408
Compensation depth, 148
Composition of air, 188
Compressibility, water, 123
Conduction, 126
Conservative constituents, 143
Constant proportions, salt, 147
Continental drift theory, history of, 89–91
Continental margin, 52–57, 99, 101
Continental movement, Pangaea to present,
    105–6
Continental rise, 57
Continental shelf, 52–54
Continental shelf break, 54
Continuity of flow, and downwelling and
    upwelling, 170
Contours of elevation, 37
Controls, primary oceanic production, 350–53
Convection, 126
Convection cells, 91
Convergence, and ocean currents, 170, 219
Convergent plate boundaries, 96
Cook, James, 9–10, 38
Coordinated universal time, 39
Copepods, 374
Corange lines, tides, 268
Corer, and sediment sampling, 74, 76
Coriolis effect, 197
Cosmogenous sediments, seafloor, 67
Cotidal lines, 267
Covalent bonds, water molecules and, 120
Cratered coasts, 290
Cratons, 109
Crest, wave, 235
Cromwell current, 215
Croplankton, 368
Crustaceans, 374
Currents, tidal, 262
Current speed, 216–17
Cyanobacteria, 382

D

Darwin, Charles, 11, 58
DDT and other pesticides, 315, 318
Declinational tides, 264–66
Deep scattering layer, 129
Deep-sea chemosynthetic communities, 440–42
Deep Sea Drilling Program, 20
Deep seafloor animals, 432–34
Deep-sea reversing thermometer, 177

Deep-sea species, bony fish, 408–10
Deep-water waves, 236–38
Degradation, coastal environments, 314–22
Deltas, 288
Density
    earth, 86
    of seawater, 167
    sigma-t, 169
    water, 123–25
Density-driven ocean circulation, 168–69
Density structure and vertical circulation, oceans,
    167–69
Deposit patterns, seafloor sediments, 70–73
Depth recorder, 61, 129
Desalination, 155–57
Detritus feeders, 431
Dias, Bartholomeu, 6
Diatomaceous earth, 372
Diatomaceous ooze, 69
Diatoms, 69, 368–69
Diffraction, waves, 244–45
Dimethyl sulfide, role of, 192
Dinoflagellates, 368, 372
Discoverer, 20
Dispersion, waves, 237
Dissolved salts, 142–43
Dissolved salts in river water, 145
Dissolving ability, water, 125–26
Dittmar, William, 13, 147
Diurnal tide, 260
Divergence, ocean currents, 170, 219
Divergent plate boundaries, 96
Diving depths and durations, some marine
    mammals, 337
Dolphins and porpoises, 398
Downwelling zones, 170
Drake, Francis, 8
Dredges, and sediment sampling, 74, 75
Drift sector, beach, 300
Drilling plans, 110–15
Driving force, tectonic plate motion, 101
Drowned river valley, 288
Dry air, composition of, 188
Dugongs, 401–2
Dune coast, 290
Dynamic equilibrium, beach, 299
Dynamic tidal analysis, 262, 266–71

E

Earth and solar system
    age of earth, 30
    chart projections, 35–37
    early planet earth, 28, 30
    earth's shape, 33–34
    geologic time, 30–31
    hydrologic cycle, 41–42
    hypsographic curve, 44–45
    land and water distribution, 42
    latitude and longitude, 34–35
    latitude measurement, 37
    longitude and time, 37–39
    natural time periods, 31–33
    oceans, 42–44
    planets of solar system, 28–29
    reservoirs and residence time, 42
    solar system origin, 28
    water on earth's surface, 39–41

Earth layers, 86, 87
Earth movement
    continental margins, 99–101
    continental movement, history of theory,
        89–90
    continental movement, Pangaea to present,
        105–6
    drilling plans for further exploration of,
        110–11
    driving force, plate motion, 102
    evidence for crustal motion, 91–96
    evidence for new theory, 91
    history of the continents, 104–9
    hot spots, plates, 102–3
    hydrothermal vents, 111–13
    internal layers, 86
    internal structure, evidence of, 86–88
    isostasy, 89
    landmass movement, pre-Pangaea, 107–9
    lithosphere, layers, 88–89
    Pangaea breakup, 104
    plate boundaries, 96–97
    plate tectonics, 96–101
    polar wandering curves, 103–4
    pre-Pangaea, 104, 107–8, 109
    Project FAMOUS and, 111
    rates of plate motion, 102
    rift zones, 97–99
    subduction, 99, 100
    terranes, 109
Earth previous to Pangaea, 104, 107–8, 109
Earthquakes, and continental drift theory, 91
Earth's rotation, effect of, 195–97
Earth's seasons, 33
Earth's shape, 33–34
Earth's water supply, 41
East Australian current, 215
East Greenland current, 215
Ebb tide, 262
Echolocation and communication, 397
Echo sounder, 61
Echo sounding, 128–30
Eddies, 217–19
Ediz Hook, 305–6
Ehrenberg, Christian, 11
Ekman spiral, 216
Ekman transport, 216, 221
Electrodialysis desalination, 155–56
Electromagnetic radiation, 126
Electromagnetic spectrum, 126
El-Mas'údé, 5
El Niño, 204–6
Embayments with high evaporation rates, 313
Energy of waves, 240
Energy transmission, water, 126–30
Entrainment, estuaries, 310
Environmental Protection Agency (EPA), 315
Environmental zones, 340–41
Epifauna, 425
Epipelagic zone, 340
Episodic waves, 239–40
Equatorial Countercurrents, 215
Equilibrium surface, wave, 235
Equilibrium tidal theory, 262–66
Eratosthenes, 5
Escarpments, 97

Estuaries
  Chesapeake Bay, 328–29
  circulation patterns in, 311–12
  degradation, coastal environment, 314–22
  embayments with high evaporation rates, 313
  flushing time, 313–14
  marine wetlands, 324–25, 327–28
  Mediterranean pollution, 317
  oil spills, 320–22
  plastic trash in, 319–20
  San Francisco Bay development, 325–28
  in temperate zone, 312
  types of, 310–11
  water and sediment quality, 314–18
  western European water pollution, 315–16
Euphausiid, 374
Evaporation and precipitation patterns, oceans, 166–67
Evaporites, 145
Expendable BT (XBT), 180
Exxon Valdez, 320

F

Fast ice, 131
Fathoms, 61
Fault coasts, 291
Faults, 96
Fetch, and wave height, 238
Filaments, plankton, 368
Fish, 404–10
Fish farming, 415–17
Fjord or fjord type estuaries, 310
Fleming, Richard H., xiii–xvi, 18
Flood tide, 262
Florida current, 215
Flushing time, estuaries, 313–14
Fog, 135
Food train concentration, DDT, 318
Food web, 359, 360
Foraminiferans, 379
Forbes, Edward, 11
Forced wave, tide, 266
Forchhammer, Georg, 147
Foreshore, 296
Formaminifers, 69
Free waves, 237
Freshwater lid, and ocean density, 169
Fringing reef, coral, 58
Frisius, Gemma, 9, 38
Frobisher, Martin, 8
Frustule, diatom, 368
Fucoxanthin, 368

G

Galilei, Galileo, 9
Gama, Vasco da, 6
Gases
  in air and seawater, 149
  carbon dioxide as buffer, 149, 151
  carbon dioxide cycle, 151–52
  distribution with depth, 148–49
  measurement of, 152
  needed by sea creatures, 337
  oxygen balance, 152
Generating force, and waves, 234
Genetic manipulation, fish and shellfish, 446–47
Geochemical Ocean Section (GEOSECS) program, 170
Geologic time, 30–31
Geotrophic flow, ocean currents, 216

Global circulation changes, ocean currents, 220–23
Global Ocean Ecosystem Dynamic (GLOBEC) program, 22
Global primary oceanic food production, 353–54
Glomar Challenger, 21, 71, 93
Gondwanaland, 90
Graben, 97
Grab samplers, sediment, 74
Grand Banks of Newfoundland, 56
Granite-type rock, lithosphere, 88
Gravity waves, 234
Great circle, 35
Great whale
  characteristics of, 394
  estimated population of, 396
Greenhouse effect, 190
Greenwich mean time, 39
Gross, M. Grant, 315
Gross and net primary oceanic food production, 350
Gross primary food production, land and ocean, 355
Gross primary oceanic food production, 350, 355
Group wave speed, 237
Gulf Stream, 215, 216
Guyots, 57, 58
Gyres and current flow, 216–17

H

Half-life, 30
Halley, Edmund, 9
Halogens, 147
Harrison, John, 38
Harvesting algae, 445–46
Harvesting benthos animals, 444–45
Heard Island experiment, 136–37
Heat budget, 162–63
Heat capacity
  land and oceans, 164–65
  water, 122–23
Heating and cooling the earth's surface, 162–65
Heat transmission, water, 126
Heavy metal iron concentrations in sediments, 318
Height of waves, 235, 238–41
Henry, Prince of Portugal (the Navigator), 6
Hensen, Victor, 11
Hess, H. H., 91
Heterotrophic abilities, dinoflagellates, 373
High-energy environments, 439
High-pressure zone, 189
High water tide, 261
History of the continents, 104–9
Holdfast, seaweed, 422, 423
Holoplankton, 377, 379
Hot spots, tectonic plates, 102–3
Hudson, Henry, 8
Human conditions, and oceanic food production, 362
Humboldt, Alexander von, 11, 89
Humboldt Current, 215
Hurricanes, 203
Hydrocast, 176
Hydrogen bonds, 120
Hydrogenous sediments, seafloor, 67
Hydrologic cycle, 41–42
Hydrothermal vents, 111–13
Hydrowire, 175
Hypsographic curve, 44–45

I

Ice and fog, 130–35
Icebergs, 133–35
Igneous rocks, as source of salt, 143
Incidental catch, 415
Indian Ocean Expedition, 20
Indian Ocean structure, 172
Infauna, 425
Interaction, waves, 238
Interior of earth, 88
  evidence for internal structure, 86–88
Internal waves, 248–49
International date line, 35
International Decade of Ocean Exploration (DOE), 20
International Geophysical Year (IGY) program (1957–58), 20
Intertidal volume, and estuaries, 314
Intertidal zonation, marine animal sorting, 425
Inverse estuaries, 313
Ion exchange, 145
Ions, 125, 152
Isobars, 189
Isotasy, 89
Isotopes, 30

J

Jellyfish, 378
Jet streams, 202–3
Jetties, 303
Johansen, F. H., 15
JOIDES Resolution, 93, 104, 110–11, 112, 220
Joint Oceanographic Deep Earth Sampling (JOIDES), 110–11

K

Kelp bioconversion, 446
Kelps, seaweed, 422, 423
Kepler, Johannes, 9
Keulen, Johannes van, 6, 7
Knudsen, Martin, 18
Krill, 374
Krill-based ecosystem, 382
Kuroshio Current, 214

L

Laborador Current, 215
Lag deposits, 299
  on beaches, 298
Lamont Doherty Geophysical Laboratory, 223
Land and water distribution, 42
Landmass movements before Pangaea, 107–9
La Niña, 206
Latent heat of fusion, 121
Latent heat of vaporization, 121
Laurasia, 90
Lava coast, 290
Laws and treaties regarding seabed resources, 80
Layered oceans, 168, 171–72
Leading margin (edge) toward subduction zone, 101
Light, and oceanic food production, 350–51
Light and dark bottle experiment, and net and gross food production, 356, 357
Light transmission, water, 126–28
Lithogenous sediments, seafloor, 64
Lithographic plates, major, 96
Lithosphere, 88–89

Location systems
    chart projections, 35–37
    latitude and longitude, 34–35
    latitude measurement, 37
    longitude and time, 37–39
Longshore current, 300
Longshore transport, 300
Low-pressure zone, 189
Low-tide terrace, 297
Low water tide, 261
Luciferin and luciferase, 338
Lunar month, 33

M

Magellan, Ferdinand, 8
Magma, 91
Major coastal zones, 286–87
Major constituents of seawater, 143, 144
Mammals, 392–403
Manatees, 401–2
Manganese nodules, as seabed resource, 67, 78–79
Mantle, earth, 86
Marcet, Alexander, 147
Mariculture, 415
Marine archaeology, 23
Marine Mammal Protection Act (1972), 402
Marine toxins, 385–87
Marine wetlands, 324–25, 327–28
Matthews, D.H., 95–96
Maury, Matthew Fontaine, 10, 11, 213
Mean earth sphere depth, 45
Measuring ocean currents, 223, 225–26
Measuring ocean depths, 61–64
Measuring seawater gases, 152
Mediterranean pollution problems, 317
Mediterranean Sea water, 172
Megaplumes, 114
Meiobenthos, 434
Mercator projection, 36
Meridians, 35
Meroplankton, 377, 380, 381
Mesopelagic zone, 340
Mesosphere, 188
Messages in polar ice, 151
Messenger, 175
Midocean temperature profiles, Atlantic, Pacific, and Indian Ocean, 173
Milankovitch, Milutin, 221
Miscellaneous oceanic plants, 423–24
Miscellaneous toxic blooms, 386–87
Modern navigational techniques, 45, 48
Modification and mitigation, marine environment, 340–42
Moho, 88
Mohorovičić discontinuity, 88
Mollusks, 427
Monoculture, 415
Monsoon effect, 199, 201
Moon tide, 263
Moraine, 288
Motion of waves, 235–36
Mount St. Helens, 98, 101
Müller, Johannes, 11
Murray, John, 13
Mussel watch, 443
Mutualism, algae, 430

N

Nannoplankton, 368
Nansen, Fridtjof, 13–16
Natural processes, beach changes, 300
Natural time periods, earth, 31–33
Navigation from wave direction, 245
Neap tides, 265
Nekton
    anchovies, 411–12
    classification summary, 410
    commercial fisheries, 410–17
    commercial species, bony fish, 408
    deep-sea species, bony fish, 408–10
    dolphins and porpoises, 398
    fish, 404–10
    fish farming, 415–17
    mammals, 392–403
    Marine Mammal Protection Act (1972), 402
    problems and policies, commercial fishing, 414–15
    redfish, 414
    reptiles, 402–4
    salmon, 412–13
    sea cows, 401–2
    seals, sea lions, and walruses, 398–401
    sea otters, 401
    sea snakes, 403
    sea turtles, 403–4
    sharks and rays, 406–7, 414
    squid, 404
    tuna, 412
    whales and whaling, 392–98
Nemerteans, 430
Neritic sediments, seafloor, 70
Neritic zone, 340
Net circulation, estuary, 312
Net plankton, 368
Newton, Sir Isaac, 9
New ways to measure oceans, 179
Nodes, and standing waves, 249
Nonconservative constituents, 143
North Atlantic cool pool, 207
North Atlantic Current, 215
North Atlantic deep water, 171
North Atlantic Drift, 215
North Equatorial Current, 214, 215
North Pacific Current, 214
North Pacific gyre, 214
Norwegian current, 215
Nudibranchs, 430
Nutrient cycles, and oceanic food production, 352–53
Nutrients
    and oceanic food production, 351–52
    and sea creatures, 337

O

Ocean basin floor, 57–58
Ocean basins of world, major, 60
Ocean currents as source of energy, 226
Ocean density changes with depth, 167–68
Ocean Drilling Program, 110–11
Oceanic food production, 362
Oceanic zone, 340
Oceanography
    defined, 4

history of, 4–22
Ocean ranching, 416
Oceans, 42–44
Ocean salinities, 142
Oceans and atmosphere
    atmosphere in motion, 192, 194
    atmospheric pressure, 188–89
    changing levels, carbon dioxide, 189–91
    clouds on a nonrotating earth, 194–95
    El Niño, 204–6
    hurricanes, 203
    jet streams, 202–3
    monsoon effect, 199
    ozone problem, 191–92
    rotation, effects of, 195–97
    seasonal changes, atmosphere pressure, 199
    storm tides and surges, 206–8
    structure of atmosphere, 188–89
    sulfur compounds' role, 192
    topographic effects on winds, 199, 202–3
    wind band modification, 198–203
    wind bands, 197–98
Oceans as energy sources, 180–83
Oceans as environment for life
    barriers and boundaries to life, 340
    bioluminescence, 338
    bottom types, 340
    bouyancy and flotation, 334–35
    classification, organisms, 341
    color, 338–39
    environmental modification and mitigation, 341–42
    environmental zones, 340–41
    gases needed, 337
    nutrients, 337
    osmotic process, and sea creatures, 335–36
    pressure, and sea creatures, 336–37
    sunlight, 337–38
    temperature, and sea creatures, 336
    water circulation, 339–40
Ocean sediments, age and thickness of major, 93
Ocean structure
    annual cycles, solar radiation, 163–64
    Atlantic Ocean structure, 171–72
    changes with depth, 167–68
    chemical tracers in seawater, 170
    density-driven circulation, 168–69
    as energy source, 180–83
    evaporation and precipitation patterns, 166–67
    heat budget, 162–63
    heat capacity, 164–65
    heating and cooling earth's surface, 162–65
    Indian Ocean structure, 172
    layered oceans, 171–72
    Mediterranean Sea and Red Sea water, 172
    oceans compared, 172
    ocean thermal energy conversion, 181–83
    osmotic pressure as energy source, 180–81
    Pacific Ocean structure, 172
    surface processes, 167
    temperature measurements, 177–78, 160
    T-S diagrams and curves, 172, 174–75
    upwelling and downwelling, 169–70
    water properties determination, 175–76
Ocean surface processes, less dense water, 167
Ocean thermal energy conservation, 181–83
Ocean waste management, 115

Offshore, 296
Oil spills, 320–22
Onshore current, 30
Onshore transport, 300
Ooze, 68
Orographic effect, 202
Osmosis desalination, 156–57
Osmotic pressure, as ocean energy source, 180–81
Osmotic process, and sea creatures, 335–36
Outer core, earth, 86
Oxygen balance, seawater and, 152
Oxygen minimum, seawater, 149
Oyashio current, 214
Ozone problem, 191–92

P

Pacific Ocean currents, 214–15
Pacific Ocean structure, 172
Pacific Ocean wind distribution, 198
Pangaea, 90
    breakup of, 104
Parallels, 34
Parasitism, marine creatures, 430
Partially mixed estuaries, 310
Patterns of tides, 260–61
Pelagic sediments, seafloor, 70
Pelagic zone, 340
Pennate diatoms, 368, 371
Period, wave, 235
Permanent zones, convergence and divergence, currents, 219
Peru Current, 214
Phosphorite, as seabed resource, 77
Phosphorite sediment, seafloor, 67
Photic zone, 337
Photophores, 338
Photosynthesis, 350
Physiographic chart, world's oceans, 62–63
Physiographic map, 37
Phytoplankton, 350, 368–73
Plankton
    bacteria in plankton, 382–83
    ciguatera poisoning, 387
    classification summary, 383
    kinds of, 368–82
    marine toxins, 385–87
    miscellaneous toxic blooms, 386–87
    phytoplankton, 368, 371–73
    red tides, 385–86
    sampling plankton, 383–85
    zooplankton, 373–82
Plants, 422–24
    general characteristics, benthic algae, 422–23
    kinds of seaweeds, 423
    miscellaneous oceanic plants, 423–24
Plastic trash in the ocean, 319–20
Plate tectonics, 96–101
Plungers, breakers, 245
Poganophora, 434
Polar molecule, 120
Polar wandering curves, 103–4
Polyculture, 415
Ponce de León, Juan, 8
Posidonius, 5
Potential energy, waves, 240
Precision depth recorder, 129
Predicting tides and tidal currents, 271–74
Pressure

and sea creatures, 336–37
    and water density, 125
Primary coasts, 288–91
Primary oceanic food production measurement, 354–56
Prime meridian, 35
Problems and policies, commercial fishing, 414–15
Production and life
    controls on primary production, 350–53
    food chains and food webs, 356, 359, 361–62
    human concerns, oceanic production, 362
    measurement, primary production, 354–56
    nutrient cycles, 352–53
    nutrients and production, 351–52
    primary production, gross and net, 350
    standing crop, 350
    total production, 356–59
    trophic pyramids, 359–61
Progressive tides, 267
Progressive wind waves, 237
Project FAMOUS (French-American Mid-Ocean Undersea Study), 111
Pteropods, 69
Ptolemy, 5
P-waves, 87, 88
Pycnocline, 167
Pytheas, 5

R

Radiation, 126
Radiative fog, 135
Radiolarian, 69
Radiolarian ooze, 69
Radiolarians, 379
Radiometric dating, 30
Rafting, as sediment transport, 72
Rain shadow, 202
Rates of tectonic plate motion, 102
Red clay, 68–70
Redfish, 414
Red Sea water, 172
Red tides, 385–86
Reef coasts, 292
Reflection, waves, 243–44
Refraction, waves, 242
Relative size, baleen and toothed whales, 395
Release of energy, waves, 247
Relict sediments, 74
Reptiles, 402–4
Residence time, 42, 146–47
Respiration, and photosynthesis, 350
Restoring force, and waves, 234
Reverse osmosis desalination, 157
Ria coast, 288
Ridge and rises, seafloor, 58–59
Ridge and rise systems, ocean basin floor, 58
Rift (spreading) zones, 97–99
Rift valley ocean basin floor, 58
Rip currents, 246
Ripples, 234
Rising sea level, 294–96
Riviera Submersible Experiment (RISE), 112
Ross, James Clark, 11
Ross, John, 11
Rotary standing tide wave, 268

S

Salinity determination, 147–48

Salinometer, 148
Salmon, 412–13
Salps, 376
Salt, and water temperature, 124–25
Salt balance regulation, 145–46
Salt budget, estuary, 312
Salts
    constant proportions principle, 147
    dissolved salts, 142–43
    ocean salinities, 142
    residence time, 146–47
    salinity determination, 147–48
    salt balance regulation, 145–46
    salt sources, 143, 145
Salt wedge estuary, 310
Sampling benthos, 442–44
Sampling plankton, 383–85
San Andreas fault, 97, 98
Sand and gravel, as seabed resources, 75, 76
Sand spits and hooks, 292
San Francisco Bay development, 325–28
Santa Barbara, 304–5
Satellite oceanography, 46–48
Saturation value, 148
Scientific measurement, global primary food production, 358
Sea cows, 401–2
Seafloor
    bathymetry of, 52–61
    biogenous oozes, 68–69
    coal, oil, and gas resources of, 77–78
    continental margin, 52–57
    deep sea measurements, 61–64
    deposit patterns, 70–73
    deposit rate, 74
    laws and treaties regarding, 80
    manganese nodules resources of, 67, 78–79
    ocean basin floor, 57–58
    phosphorite resources of, 77
    red clay, 69–70
    ridges and rises, 58–59
    sand and gravel resources of, 75, 76
    sediment particle size, 73
    sediments, 64–75
    sediment sampling methods, 74–75
    sediment sources, 64, 66, 67
    sulfide mineral deposit resources of, 79
    sulfur resources of, 77
    trenches, 59, 61
Seafloor spreading, 91
Sea ice, 131–33
Seals, sea lions, and walruses, 398–401
Sea otters, 401
Sea ranching, 416
Sea smoke, 135
Sea snakes, 403
Seasonal changes, atmospheric pressure, 199
Seasonal zones, downwelling and upwelling, 219–20
Sea stacks, 291, 292
Sea surface topography, 65
Sea turtles, 403–4
Seawater chemical resources, 153, 155
Seawater nutrients, 152–53
Seawater organics, 153
Sea waves, 287
Seaweed kinds, 423
Secci disk, 128
Secondary coasts, 291–95

Sediment classification by particle size, 73
Sediment particle size and rates of deposit, 73–74
Sediments, on seafloor, 52, 64–75
Sediment sampling methods, 74–75
Sediment sources, seafloor, 64, 65, 67
Seismic wave, 87
Semidiurnal mixed tide, 261
Semidiurnal tide, 261
Sessile animals, 425
Shallow-water waves, 241–45
Sharks, as commercial catch, 414
Sharks and rays, 406–7
Shore, 287
Sigma-t, ocean density value, 169
Siliceous ooze, 68
Sill, 288
Sofar channel, 129
Solar system, 28–29
Sonar, 129
Sorting, waves, 237
Soundings, 61
Sound shadow zones, 129
Sound transmission, light, 128–30
Sources of salt, 143, 145
South Atlantic surface water, 172
South Equatorial Current, 214, 215
Spartina, 343–45
Specific heat, 122
Speed of waves, 236
Spillers, breakers, 245
Spit, 299
Spreading centers, above magma, 91
Spring tides, 265
Squid, 404
Standing crop, and biomass, 350
Standing waves, 249–50
Standing wave tides, 267–70
Steepness of waves, 241
Stipe, seaweed, 422, 423
Storm centers, and waves, 237
Storm tides and storm surges, 206–8
Strabo, 5
Stratosphere, 188
Subduction, 99, 100
Subduction zones, 91
Sublimation, 122
Submarine canyons, major, 55
Subtropical convergence, 219
Suess, Edward, 90
Sulfide mineral deposits, seabed, 79
Sulfur, as seabed resource, 77
Summer berm, 297
Summer sea surface temperatures, Northern
    Hemisphere, 164
Summer solstice, 32
Sunlight, and ocean plant life, 337–38
Sunlight, effect on seawater, 153
Sun tide, 264–65
Supersaturated water, 149
Surface currents
    Atlantic Ocean, 215–16
    convergence and divergence, 219–20
    current speed, 216–17
    eddies, 217–19
    Ekman spiral and Ekman transport, 216
    energy from currents, 226
    geostrophic flow, 216

global circulation changes, 220–23
gyre and current flow, 216–17
Indian Ocean, 216
measuring currents, 223, 225–26
Pacific Ocean, 214–15
permanent zones of convergence, 219
seasonal zone of upwelling and downwelling,
    219–20
western intensification of currents, 237
Surface tension
    water, 123
    waves and, 234
Surface weather map, 190
Surf zone, 245–48
Survival rates, salmonid spawners, 417
Swash, waves, 300
S-waves, 87, 88
Swell, 237
Symbiotic relationships, marine animals, 430

T

Taxonomic categories, some marine organisms,
    342
Taylor, Frank B., 90
Tektites, 67
Temperate zone estuaries, 312
Temperature, and sea creatures, 336
Temperature measurements, ocean, 177–78, 180
Terranes, 109
Terrigenous sediments, seafloor, 70
Thermocline, 167
Thermosphere, 188
Thomson, Wyville, 13
Tidal current tables, 273–74
Tidal datum, 261
Tidal day, 264
Tide pools, and benthic form zonation, 430–31
Tides
    declinational tides, 265–66
    dynamic tidal analysis, 266–71
    elliptical orbits and tides, 266
    equilibrium tidal theory, 262–66
    moon tide, 263
    predicting tides and tidal currents, 271–74
    progressive tides, 267
    spring tides and neap tides, 265
    standing wave tides, 267–70
    sun tide, 264–65
    tidal bores, 271
    tidal currents, 262
    tidal current tables, 273–74
    tidal day, 264
    tide levels, 261–62
    tide patterns, 260–61
    tide tables, 273–74
    tide wave, 264, 266–67
    tide waves in narrow basins, 270–71
Tides as source of energy, 274–78, 279
Tide tables, 273–74
Tide wave, 264, 266–67
Tide waves in narrow basins, 270–71
Time zone distribution, 40
Tombolo, 299
Topographic effect, on cloud vertical movement,
    199, 202–3
Trace elements, 143, 144
Transform fault, 97

ocean basin floor, 58
Transit Tracers in the Ocean (TTO), 170
Transport of water, surf zone, 246
Trenches, ocean basin floor, 58, 61
Tritium, 170
Trophic pyramids, various levels of production,
    359, 361
Tropical convergence, 219
Tropical coral reefs, 436–39
Tropical depression, 203
Tropical divergence, 219
Tropical Ocean and Global Atmosphere (TOGA),
    22
Tropic of Cancer, 32, 33
Tropic of Capricorn, 32, 33
Troposphere, 188
Trough, wave, 235
Troughs, beach, 298
T-S curves, 172, 174–75
Tsunamis, 247–48
Tuna, 412
Turbidites, 55
Turbidity currents, 55
Types of coasts, 287–95

U

Ultraplankton, 368
Undersea robotic technology, 278
U.S. Global Changes Research Program, 22
U.S. Joint Global Ocean Flux Study (JGOFS), 22
Universal sea state code, 242
Upwelling and downwelling, oceans, 169–70
Upwelling zones, 170

V

VENTS program, NOAA, 112
Vernal equinox, 32–33
Vertical zonation, marine animal sorting, 425
Vespucci, Amerigo, 6, 8
Vine, F. J., 95–96
Viruses in the oceans, 384
Viscosity, water, 123

W

Water and sediment quality, shore waters,
    314–18
Water budget, estuary, 312
Water masses, and T-S curves, 172
Water molecule, 120
Water on earth's surface. *See* Earth and solar
    system
Water properties
    changes in water states, 120–22
    cohesion, surface tension, and viscosity, 123
    compressibility, 123
    density, 123–26
    dissolving ability, 125–26
    energy transmission, 126–30
    fog, 135
    heat capacities, 122–23
    heat transmission, 126
    ice and fog, 130–35
    icebergs, 133–35
    light transmission, 126–28
    pressure and density, 125
    salt and density, 124–25

sea ice, 131–33
  sound transmission, 128–30
  temperature and density, 124
  water molecule, 120
Water properties determination, 175–76
Wavelength, 235
Waves
  anatomy of a wave, 235
  breakers, 245–46
  deep-water waves, 236–38
  diffraction, 244–45
  dispersion and group speed, 237
  energy of, 240
  energy release, 247
  episodic, 239–40
  height of, 238–41
  how a wave begins, 234
  internal waves, 248–49
  navigation from wave direction, 245

reflection, 243–44
refraction, 242–43
shallow-water waves, 241–45
standing waves, 249–50
steepness of, 241
and storm centers, 237
surf zone, 245–48
swell, 237
tsunamis, 247–49
water transport, 246
wave interaction, 238
wave motion, 235–36
wave speed, 236
Waves as sources of energy, 250–52
Wave trains, 237
Wegener, Alfred L., 90
Well-mixed estuaries, 310
Western European water pollution, 315–16
Western intensification, currents, 217

West Wind Drift, 214
Whale distribution, 393
Whales and whaling, 392–98
Wind bands, 197–98
  modifying, 198–203
Winds, nonrotating earth, 194–95
Wind speed and wave height, relationship
    between, 230
Winter berm, 297
World Ocean Circulation Experiment (WOCE),
    22
World ocean primary food production, 355
Worldwide search for carbon, 224

Z

Zenith, 38
Zooplankton, 373–82
Zooxanthellae algae, 436
Zulu time, 39